PERSPECTIVES IN PHOTOSYNTHESIS

THE JERUSALEM SYMPOSIA ON
QUANTUM CHEMISTRY AND BIOCHEMISTRY

Published by the Israel Academy of Sciences and Humanities,
distributed by Academic Press (N.Y.)

1. *The Physicochemical Aspects of Carcinogenesis* (October 1968)
2. *Quantum Aspects of Heterocyclic Compounds in Chemistry and Biochemistry* (April 1969)
3. *Aromaticity, Pseudo-Aromaticity, Antiaromaticity* (April 1970)
4. *The Purines: Theory and Experiment* (April 1971)
5. *The Conformation of Biological Molecules and Polymers* (April 1972)

Published by the Israel Academy of Sciences and Humanities,
distributed by D. Reidel Publishing Company (Dordrecht, Boston, Lancaster, and Tokyo)

6. *Chemical and Biochemical Reactivity* (April 1973)

Published and distributed by D. Reidel Publishing Company
(Dordrecht, Boston, Lancaster, and Tokyo)

7. *Molecular and Quantum Pharmacology* (March/April 1974)
8. *Environmental Effects on Molecular Structure and Properties* (April 1975)
9. *Metal-Ligand Interactions in Organic Chemistry and Biochemistry* (April 1976)
10. *Excited States in Organic Chemistry and Biochemistry* (March 1977)
11. *Nuclear Magnetic Resonance Spectroscopy in Molecular Biology* (April 1978)
12. *Catalysis in Chemistry and Biochemistry Theory and Experiment* (April 1979)
13. *Carcinogenesis: Fundamental Mechanisms and Environmental Effects* (April/May 1980)
14. *Intermolecular Forces* (April 1981)
15. *Intermolecular Dynamics* (Maart/April 1982)
16. *Nucleic Acids: The Vectors of Life* (May 1983)
17. *Dynamics on Surfaces* (April/May 1984)
18. *Interrelationship Among Aging, Cancer and Differentiation* (April/May 1985)
19. *Tunneling* (May 1986)
20. *Large Finite Systems* (May 1987)

Published and distributed by Kluwer Academic Publishers
(Dordrecht, Boston, London)

21. *Transport through Membranes: Carriers, Channels and Pumps* (May 1988)

VOLUME 22

PERSPECTIVES
IN PHOTOSYNTHESIS

PROCEEDINGS OF THE TWENTY-SECOND JERUSALEM SYMPOSIUM ON
QUANTUM CHEMISTRY AND BIOCHEMISTRY HELD IN
JERUSALEM, ISRAEL, MAY 15–18, 1989

Edited by

J. JORTNER

Department of Chemistry,
University of Tel-Aviv, Israel

and

B. PULLMAN

Institut de Biologie Physico-Chimique
(Fondation Edmond de Rothschild), Paris, France

KLUWER ACADEMIC PUBLISHERS

DORDRECHT / BOSTON / LONDON

Library of Congress Cataloging in Publication Data

Jerusalem Symposium on Quantum Chemistry and Biochemistry (22nd :
  1989)
    Perspectives in photosynthesis : proceedings of the Twenty second
  Jerusalem Symposium on Quantum Chemistry and Biochemistry held in
  Jerusalem, Israel, May 15-18, 1989 / edited by Joshua Jortner and
  Bernard Pullman.
        p.   cm. -- (The Jerusalem symposia on quantum chemistry and
  biochemistry ; v. 22)

    1. Photosynthesis--Congresses.   I. Jortner, Joshua.   II. Pullman,
  Bernard, 1919-   . III. Title.   IV. Series.
  QK882.J47   1989
  581.1'3342--dc20   89-27783

    ISBN-13: 978-94-010-6706-5   e-ISBN-13: 978-94-009-0489-7
    DOI: 10.1007/978-94-009-0489-7

Published by Kluwer Academic Publishers,
P.O. Box 17, 3300 AA Dordrecht, The Netherlands.

Kluwer Academic Publishers incorporates
the publishing programmes of
D. Reidel, Martinus Nijhoff, Dr W. Junk and MTP Press.

Sold and distributed in the U.S.A. and Canada
by Kluwer Academic Publishers,
101 Philip Drive, Norwell, MA 02061, U.S.A.

In all other countries, sold and distributed
by Kluwer Academic Publishers Group,
P.O. Box 322, 3300 AH Dordrecht, The Netherlands.

Printed on acid-free paper

TABLE OF CONTENTS

PREFACE

The Twenty-Second Jerusalem Symposium reflected the high standards of these distinguished scientific meetings, which convene once a year at the Israel Academy of Sciences and Humanities in Jerusalem to discuss a specific topic in the broad area of quantum chemistry and biochemistry. The topic at this year's Jerusalem Symposium was Perspectives in Photosynthesis, which constitutes a truly interdisciplinary subject of central interest to biophysicists, chemists and biologists.

The main theme of the Symposium was built around a conceptual framework for the acquisition, storage and useful disposal of energy in photosynthetic reaction centres. Emphasis was placed on the elucidation of primary charge separation processes in photosynthesis and their exploration within the framework of the electron transfer theory, on the interrelationship between structural data, inter-actions and electron transfer kinetics, and on the role of protein dynamics in primary processes in photosynthesis. The interdisciplinary nature of these research areas was deliberated by intensive and extensive interactions between scientists from different disciplines and between theory and experiment. This volume provides a record of the invited lectures at the Symposium.

Held under the auspices of the Israel Academy of Sciences and Humanities and the Hebrew University of Jerusalem, the Twenty-Second Jerusalem Symposium was sponsored by the Institut de Biologie Physico-Chimique (Fondation Edmond de Rothschild) of Paris. We wish to express our deep thanks to Baron Edmond de Rothschild for his continuous and generous support, which makes him a true partner in this important endeavour. We would also like to express our gratitude to the Administrative Staff of the Israel Academy and, in particular, to Mrs. Avigail Hyam for the efficiency and excellency of the local arrangements.

Joshua Jortner
Bernard Pullman

THE PHOTOSYNTHETIC REACTION CENTER FROM THE PURPLE BACTERIUM RHODOPSEUDOMONAS VIRIDIS: RELATION OF STRUCTURE AND FUNCTION

H. Michel* and J. Deisenhofer**
*Max-Planck-Institut fuer Biophysik
Abteilung Molekulare Membranbiologie
Heinrich-Hoffmann-Str. 7
D-6000 Frankfurt/M 71
West Germany
**Howard Hughes Medical Institute
and Department of Biochemistry
University of Texas
Southwestern Medical Center
5323 Harry Hines Boulevard
Dallas, Texas 75235
U.S.A.

ABSTRACT. The structure of the photosynthetic reaction centre from the purple bacterium Rhodopseudomonas viridis is described as determined by X-ray crystallographic analysis at 2.3 Å resolution. The most prominent feature is the basically symmetric architecture of the structure, which contains two nearly equivalent branches for the electron transfer across the membrane. Possible reasons for the use of only one pigment branch in the electron transfer are discussed.

1. INTRODUCTION

The primary charge separation of the biological light energy conversion takes place in the so-called photosynthetic reaction centres. These photosynthetic reaction centres consist of integral membrane proteins, which form the binding sites for the photosynthetic pigments. The pigments are needed for the light absorption (or energy transfer from the light-harvesting antennae) leading to the primary charge separation, and the subsequent electron transfer steps.

The best known reaction centres are those from the purple photosynthetic bacteria (for review see Feher and Okamura, 1978; Hoff et al., 1982; Parson, 1982). They are easy to isolate and relatively stable. In addition, they are the only membrane proteins (or membrane protein complexes) which could be obtained in a crystalline form in a way that a

1

J. Jortner and B. Pullman (eds.), Perspectives in Photosynthesis, 1–9.
© 1990 *Kluwer Academic Publishers.*

high-resolution X-ray crystallographic analysis could be performed successfully. Most of the reaction centres from purple bacteria contain three protein subunits which are called H (heavy), M (medium) and L (light) subunits according to their apparent molecular weights are determined by sodium dodecylsulphate polyacrylamide gel electrophoresis. Many reaction centres, e. g. that from Rhodopseudomonas (Rps.) viridis, contain a tightly bound cytochrome subunit. The cytochrome subunit is involved in the re-reduction of the photooxidized primary electron donor. The cytochrome subunit from the Rps. viridis reaction center contains four heme groups. The photosynthetic pigments and cofactors in the Rps. viridis reaction centre are four bacteriochlorophyll b molecules, two bacteriopheophytin-b molecules, one menaquinone as primary acceptor ("Q_A"), one non-heme-iron and one ubiquinone as secondary acceptor ("Q_B"). The reaction centers from most of the other purple bacteria contain bacteriochlorophyll-a instead of bacteriochlorophyll-b, bacteriopheophytin-a instead of bacteriopheophytin-b, and the menaquinone is replaced by another ubiquinone. The successful crystallization of the reaction center from Rps. viridis (Michel, 1982), yielding well ordered crystals in which the reaction centers retained their photochemical activity (Zinth et al., 1983) was an important step towards the elucidation of its three-dimensional structure. X-ray analysis of these crystals allowed the calculation of an electron density map at 3 Å resolution from which the arrangement of the chromophores could be determined (Deisenhofer et al., 1984). Subsequently the structure of the protein subunits (Deisenhofer et al., 1985) and details of the pigment-chromophore interactions (Michel et al., 1986) were presented. Recently, reaction centers from Rb. sphaeroides could also be crystallized (Allen and Feher, 1984, Chang et al., 1986). Their crystal structure could be determined with the help of the known structure of the reaction center from Rps. viridis (Chang et al., 1986; Allen et al., 1986, 1987). In the following we will describe and discuss pigment arrangement, protein structure and functional aspects of the reaction centres from *Rps.viridis* with the emphasis on the functional aspects.

2. RESULTS AND DISCUSSION

2.1 Pigment arrangement

The discovery of the quasisymmetric structure of the core of the reaction centre, and the linear arrangement of the four heme groups were the two major surprises of the structure analysis. The arrangement of the pigments and cofactors is shown in figure 1 (taken from Deisenhofer et al., 1984). The four heme groups (at the top of fig. 1) are related by a twofold local rotation axis which runs nearly perpendicular to the picture plane. The function of these hemes is to re-reduce the photooxidized primary electron donor which is a "dimer" of two non/covalently linked bacteriochlorophyll molecules ("special pair", just below the hemes in fig. 1). The existence of such a dimer had been postulated on the basis of EPR experiments (Norris et al., 1971), its detailed structure is now firmly established by the X-ray structure analysis. The ring systems of the two bacteriochlorophylls constituting the dimer are nearly parallel and they overlap with their pyrrole rings I. The plane to plane distance of these two bacteriochlorophylls is about 3.1 Å. The ring systems of all the chlorine pigments are also related by a local twofold rotation axis. This diad (vertically in fig. 1) runs through the special pair on the periplasmic side of the membrane and through the non-heme iron on the cytoplasmic side of the membrane. Since the two "accessory" bacteriochlorophylls and the two bacteriopheophytins are also related by the local diad. Two structurally equivalent branches are formed which could be used for electron transfer across the membrane. The symmetry is broken by the different arrangement of the phytyl chains in both branches. Additionally, in the original electron density map electron density for the secondary quinone, the ubiquinone Q_B, could not be found , since most of it is lost during the isolation and crystallization of the reaction centre, whereas excellent electron density was present for the menaquinone (Q_A). The two bacteriopheophytins are spectroscopically inequivalent absorbing light of different wavelengths. It is known that only the one absorbing light at longer wavelengths is involved in electron transfer. Comparison of absorbance spectra of crystals, taken with plane-polarized light (Zinth et al., 1983) shows that the bacteriopheophytin absorbine at the longer wavelengths is the one closer to the Q_A. These experments establish clearly that only one way (the right hand one in fig. 1) is used for light driven electron transfer across the photosynthetic membrane.

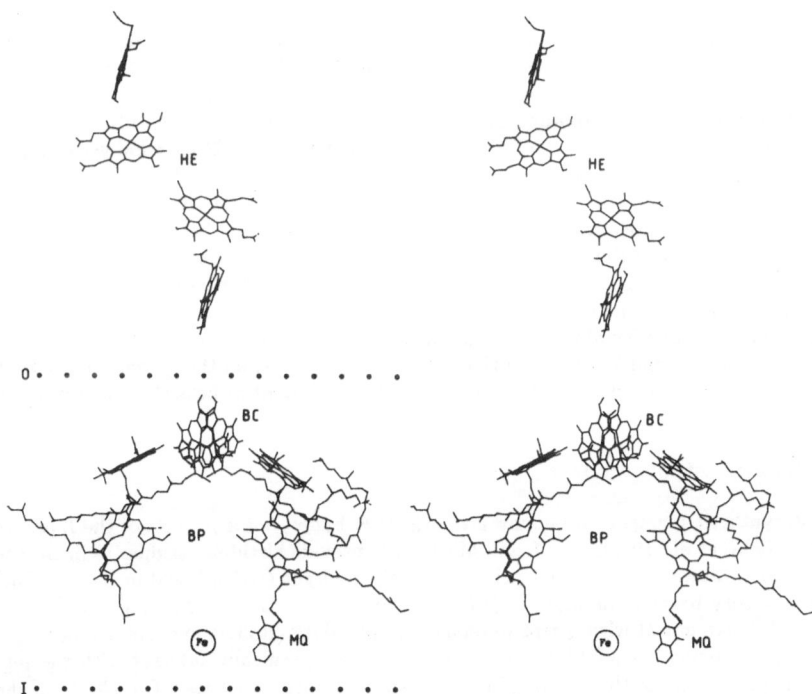

Figure 1 (stereo pair): Arrangement of the pigments in the photosynthetic reaction center from Rps. viridis, showing from top to bottom four heme groups (HE), four bacteriochlorophylls (BC), one non-heme iron (Fe), and one menaquinone (MQ). The approximate twofold symmetry axis relating the photosynthetic pigments runs vertically in the plane of the picture. The approximate position of the periplasmic and cytoplasmic face of the photosynthetic membrane is indicated by dotted lines.

In the refined electron density map weak, but significant electron density is present for the secondary quinone Q_B. Its binding site is symmetry related by the local diad to the binding site of Q_A. The electron has to be tranferred from Q_A to Q_B parallel to the surface of the membrane. There is no evidence for the participation of the non-heme iron in this electron transfer, since it can be removed without drastic changes in the kinetics of this electron transfer (Debus et al., 1985).

2.2 General Architecture and Protein Structure

Figure 2 shows a drawing of the polypeptide chains together with the chromophore model. The photosynthetic pigments are associated with the L and M subunits as had already been shown by biochemical and spectroscopical experiments (Feher and Okamura, 1978). They form the central part of the reaction center. The L and M subunits possess five long membrane spanning helices which are related by the same twofold axis as the pigments. The structural differences between the L and M subunits are mainly at the amino -terminus on the cytoplasmic side (M has a longer amino-terminus), in the connection of the first and second transmembrane helix (M shows an insertion of 7 amino acids, which gives rise to a short helix parallel to the membrane), in the connection of the fourth and fifth membrane spanning helix (M possesses an additional loop providing glu M232 as a bidentate ligand to the non-heme-ferrous iron atom) and at the carboxy-terminus (the M subunit from Rps. viridis contains subunit). The five transmembrane helices of the subunits L and M possess a remarkably open structure forming approximately half-cylinders. On both sides of the helical regions of the L and M subunits the polypeptide segments connecting the transmembrane helices and the terminal segments form flat surfaces perpendicular to the local diad. The cytochrome is bound to the surface close to the special pair on the periplasmic side. The H subunit possesses one membrane spanning helix close to its amino-terminus. Its carboxy-terminal domain is bound to the flat surface of the L-M complex on the cytoplasmic side. It is the only part of the reaction center with a significant amount of beta-sheet as secondary structure.

2.3 Protein-Pigment Interactions.

The photosynthetic pigments are bound into primarily hydrophobic pockets of the L and M subunits (Michel et al., 1986). The L and M subunits provide histidine residues as ligands to the magnesium atoms of the special pair bacteriochlorophylls (L173, M200 in Rps. viridis) and the accessory bacteriochlorophylls (L153, M180 in Rps. viridis). The Mg ions are five coordinated in agreement with recent resonance Raman data (Robert and Lutz, 1986). The protein, besides being a scaffold for the pigments, must specifically interact with the pigments to suppress one of the two possible electron transport pathways. The choice of the pathway could be influenced already at the special pair. Charged amino acids are not found in the vicinity of the special pair. However, in Rps. viridis we find hydroxyl groups of three amino acid side chains (tyr M195, thr L248, tyr M208) close to the special pair. They are located towards the branch which is used for electron transfer. Two of them form hydrogen bonds with carbonyloxygen atoms of the bacteriochlorophylls. On the side of the inactive branch we find only one polar side chain, his L168, in a position symmetry-related to tyr M195; his L168 also forms a hydrogen bond with the special pair. Sequence comparisons show that the hydrogen-bonding between special pair bacteriochlorophylls and protein must be different between Rps. viridis, Rb. sphaeroides and Rb. capsulatus.

The accessory bacteriochlorophylls are not hydrogen-bonded to the protein. However, the ring V carbonyl groups of both accessory bacteriochlorophylls seem to form hydrogen bonds with a firmly bound water molecule. These water molecules form additonal hydrogen bonds to the ε-nitrogens of histidines L173 or M200 respectively.

Both bacteriopheophytins are hydrogen-bonded to tryptophan residues (L100, M127) via their ring V ester carbonyls. That bacteriopheophytin which is an intermediate electron acceptor forms an additional hydrogen-bond with a, most likely protonated, glutamyl residue (L104). This glutamic acid seems to be of crucial importance for the light driven electron transfer. It is conserved between all three bacterial species and the D1 protein from photosystem II reaction centers.

The head group of Q_A is bound to a highly hydrophobic pocket. Trp M250 forms part of the binding site. Both ring systems are parallel, their distance is only 3.1 Å. Trp M250 also touches that bacteriopheophytin which is involved in light-driven electron transfer thereby "bridging" the pheophytin and quinone. In the symmetry related position phe L216 is found whose side chain is too small to bridge the bacteriopheophytin and the secondary quinone. The ferrous non-heme-iron atom is found half-way in between the primary and secondary quinone binding sites. It is bound to four histidine residues (L190, L230, M217, M264) and one glutamic acid (M232 in Rps. viridis). The primary quinone is hydrogen bonded to histidine M217 and the peptide nitrogen of alanine M258.

For Q_A the crystallographic data suggest that it forms hydrogen bonds to the protein with its two carbonyl oxygens. one to the S nitrogen of the iron ligand histidine L190, and a

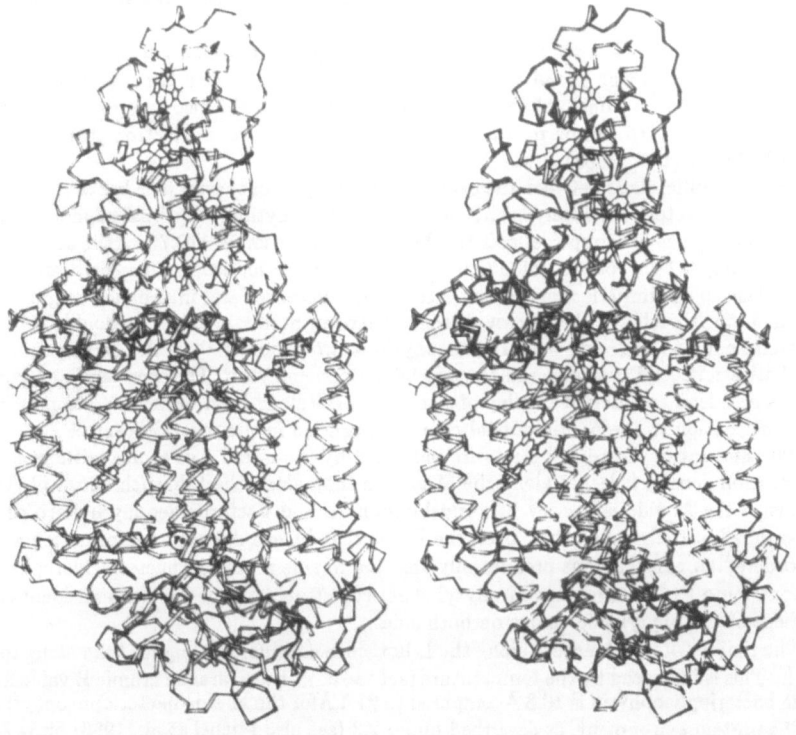

Figure 2 (stereo pair): Ribbon drawing of the polypeptide chains of the reaction center from Rps. viridis, together with the chromophore model (thin lines), showing the cytochrome (top), and the subunits L (middle left), M (middle right), and H (bottom, with the N-terminal helix extending from the cytochrome).

bifurcated hydrogen bond the OH- group of serine L223, and the peptide nitrogen of glycine L225. Additional major differences between the binding sites of Q_A and Q_B are the more polar nature of the Q_B site, and the presence of pathways through the protein which may be used for the protomotion of the doubly reduced Q_B. Especially glu L212 may be of crucial importance, its side chain may form a "rotating arm" for proton transfer.

The recently postulated difference in binding of the bacteriochlorophylls and Q_A between Rps. viridis and Rb. sphaeroides (Yeates et al., 1988, Allen et al., 1988) most likely are not real, but due to the considerably lower resolution and the lower accuracy of the Rb. sphaeroides reaction center data.

2.4 The use of only one pigment branch for electron transfer

It is evident that the reaction centres of present day's purple bacteria are derived from an ancestral , entirely symmetric reaction centre which was a homodimer with two absolutely equivalent pathways for electron transfer. One of these pathways could be switched off by a gene duplication and subsequent mutations which lead to the formation of a heterodimeric reaction centre. The possible advantages of having only one branch functioning in electron transfer will be discussed below. First, however, we will describe the structural differences of both pigment branches and their protein environment based on the X-ray structure at 2.3 Åresolution:

(i) The two bacteriochlorophyll ring systems of the special pair show a different degree of non-planarity. The one on the M side is considerably more deformed than the one on the L side. This can cause an unequal charge distribution between the two components of the special pair which in turn can be part of the reason for unidirectional electron transfer (see also Michel-Beyerle et al., 1988).

(ii) There are differences between the symmetry operations of special pair bacteriochlorophylls, accessory bacteriochlorophylls and the bacteriopheophytines. Optimal superposition of the tetrapyrrol rings of the special pair is achieved by a rotation of 179.7°, of the accessory bacteriochlorophylls by a rotation of -175,8° and for the bacteriopheophytines by a rotation of -173.2°. Due to the imperfect symmetry interatomic distances and interplanar angles are different in both branches. The closest distance between atoms involved in double bonds in the special and in the L-side bacteriopheophytin is 0.7 Åshorter than the corresponding distance between special pair and the M- side bacteriopheophytin. This and similar structural differences lead to different overlap of electronic orbitals in both branches, and may be another contribution to the unidirectional charge separation in the reaction centre.

(iii) The amount of disordered structure indicated by the number of atoms without significant electron density is larger along the M branch than along the L branch. Both phytyl side chains of the M side accessory bacteriochlorophylls and bacteriopheophytins are partially disordered at their ends, whereas on the L side they have a different conformation and are well ordered. A carotenoid is present only near the M side accessory bacteriochlorophyll. It may contribute to the difference in phytyl chain structure, since it prevents an identical arrangement of the phytyl side chains on both sides.

(iv) The rigidity of the structure along the L-branch is considerably higher than along the M-branch. This is indicated by the temperature factors, e. g. the averaged atomic B value for the L-side bacteriopheophytin is 10.3 Åcompared to 21.1 Åfor the M-side bacteriopheophytin.

(v) The protein evironment, as described under 2.3 (see also Michel et al., 1986) plays an important role not only as a scaffold but also by influencing electronic properties: Around the special pair there are more electron attracting polar residues and hydrogen bonds on the L side than on the M side. There are also more aromatic residues along the L side. This fact could lead to the higher rigidity of the structure on the L-side, since the aromatic residues are bulkier than the aliphatic ones, but also the presence of π-electrons might help in electron transfer. A prominent residue is of course tryptophan M250 bridging the L-side

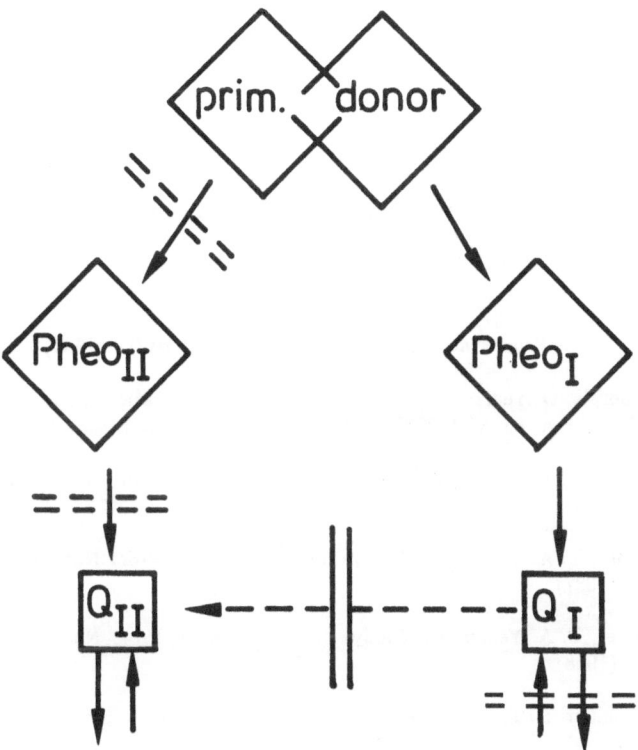

Figure 3: Schematic drawing of the situation in an ancient symmetric reaction centre, and the steps needed to convert it into the modern reaction centre. These steps are shown in broken lines. It is most important to open a way for electron transfer from Q_I to Q_{II}, which then become Q_A and Q_B, and to switch off protonation and release of Q_I.

bacteriopheophytin and Q_A.

We will now try to explain why one pathway for the light- driven electron transfer was switched off:

(i) A rather trivial explanation would be that the asymmetry in both pigment branches and the asymmetry of the protein environment cause an asymmetry in the distribution of electron in the special pair, especially in the excited state. Such a asymmetry can lead to a preferred release of an electron in one directional result in a faster find electron transfer step, thereby minimizing competing reactions. A higher quantum yield for the electron transfer can be achieved.

(ii) It is a clear advantage in the present day's reaction centres that the two quinones act in series, since only the released secondary quinone, Q_B, is a two-electron carrier. In a completely symmetric reaction centre the first excitation an electron would be transferred to the quinone at the end of one pigment branch. The resulting semiquinone is not stable and its electron is lost in the time range of seconds. Only if it receives a second electron it can be protonated and energy is stored in the form of the quinol. With two identical parallel electron transfer chains the propability for the second electron, to be funneled into the same chain to the same quinone as the first electron is only. A possible electrostatic repulsion by the negatively charged semiquinone might even decrease this probability. In a frequent situation the absorption of two photons then leads to the formation of two semiquinone in the same reaction centre and energy is not stored in a stable way. The way out of this dilemma is illustrated in figure 3: The two quinones must be switched in series. Protonation and release is allowed only to the final quinone. Thus it becomes a true Q_B in the electron transfer chain, as it is seen in the present day's reaction centres. A considerable increase in the efficiency of light energy conversion, especially under low light conditions, must result.

3. REFERENCES

1. J. P. Allen and G. Feher, Proc. Natl. Acad. Sci. U.S.A. 81, 4795-4799 (1984)

2. J. P. Allen, G. Feher, T. O. Yeates, D. C. Rees, J. Deisenhofer, H. Michel, and R. Huber, Proc. Natl. Acad. Sci. U.S.A. 83, 8589-8593 (1986)

3. J. P. Allen, G. Feher, T. A. Yeates, H. Komiya, and D. C. Rees, Proc. Natl. Acad. Sci. U.S.A. 84, 5730-5734 (1987)

4. J. P. Allen, G. Feher, T. A. Yeates, H. Komiya, and D. C. Rees, Proc. Natl. Acad. Sci. U.S.A. 84, 6162-6166 (1987).

5. J. P. Allen, G. Feher, T. A. Yeates, H. Komiya, and D. C. Rees, Proc. Natl. Acad. Sci. U.S.A. 85, 8487-8491 (1988).

6. C. H. Chang, M. Schiffer, D. Tiede, U. Smith, and J. R. Norris, J. Mol. Biol. 186, 201-203 (1985)

7. C. H. Chang, D. Tiede, J. Tang, K. Smith, J. Norris and M. Schiffer FEBS/Lett. 205, 82/86 (1986)

8. R. J. Debus, M. Y. Okamura, and G. Feher, Biophys. J., 47, 3a (1985)

9. J. Deisenhofer, O. Epp, K. Miki, R. Huber, and H. Michel, J. Mol. Biol. 180, 385-398 (1984)

10. J. Deisenhofer, O. Epp, K. Miki, R. Huber, and H. Michel, Nature 318, 618-624 (1985)

11. A. J. Hoff, In F. K. Fong, (ed.) Molecular Biology, Biochemistry, and Biophysics, Vol. 35, pp. 80-151, 31822-326, Springer, Berlin (1982)

12. H. Michel, J. Mol. Biol. 158, 567-572 (1982)

13. H. Michel, O. Epp and J. Deisenhofer, EMBO J. 5, 2445-2451 (1986)

14. M. E. Michel-Beyerle, M. Plato, J. Deisenhofer, H. Michel, M. Bixon, J. Jortner, Biochim. Biophys. Acta 932, 52-70 (1988)

15. W. W. Parson, Ann. Rev. Biophys. Bioeng. 11, 57-80 (1982)

16. J. R. Norris, R. A. Uphaus, H. L. Crespi, and J. J. Katz, Proc. Natl. Acad. Sci. U.S.A. 68, 625-628 (1971)

17. B. Robert and M. Lutz, Biochemistry 25, 2304-2309 (1986)

18. T. O. Yeates, H. Kemiya, A. Chirino, D. C. Rees, J. P. Allen and G. Feher, Proc. Natl. Acad. Sci. U.S.A. 85, 7993-7997 (1988)

19. W. Zinth, W. Kaiser and H. Michel, Biochim. Biophys. Acta 723, 128-131 (1983)

A STRUCTURAL BASIS FOR ELECTRON TRANSFER IN BACTERIAL PHOTOSYNTHESIS

James R. Norris[‡,§,*], Theodore J. DiMagno[§], Alexander Angerhofer[‡,*],
C.-H. Chang[†], Ossama El-Kabbani[‡] and M. Schiffer[†]
[‡]Chemistry Division, Argonne National Laboratory, Argonne, Illinois, 60439
[§]Department of Chemistry, University of Chicago, Chicago, Illinois, 60615
[*]The University of Stuttgart, Stuttgart, West Germany
[†]Biological, Environmental, and Medical Research Division, Argonne National Laboratory, Argonne, Illinois, 60439

ABSTRACT. Triplet data for the primary donor in single crystals of bacterial reaction centers of *Rhodobacter sphaeroides* and *Rhodopseudomonas viridis* are interpreted in terms of the corresponding x-ray structures. The analysis of electron paramagnetic resonance data from single crystals (triplet zero field splitting and cation and triplet linewidth of the primary special pair donor of bacterial reaction centers) is extended to systems of a non-crystalline nature. A unified interpretation based on frontier molecular orbitals concludes that the special pair behaves like a supermolecule in all wild-type bacteria investigated here. However, in heterodimers of *Rhodobacter capsulatus* (His^{M200} changed to Leu or Phe with the result that the M-half of the special pair is converted to bacteriopheophytin) the special pair possesses the EPR properties more appropriately described in terms of a monomer. In all cases the triplet state and cation EPR properties appear to be dominated by the highest occupied molecular orbitals. These conclusions derived from EPR experiments are supplemented by data from Stark spectroscopy of reaction centers from *Rb. capsulatus*. The most red-shifted Stark band in the *Rb. capsulatus* heterodimer is relatively intense and is interpreted as a "pure" charge transfer band within the special pair donor. This explanation locates the energy of $|Bchl^-_M>|Bchl^+_L>$ or $|Bchl^+_M>|Bchl^-_L>$, the internal charge transfer state, anywhere from ~.1 ev to ~.3 ev above $^{1*}[|Bchl_M>|Bchl_L>]$, the first excited singlet state of the primary donor. This internal charge transfer state has been invoked previously in the initial mechanism of electron transfer in photosynthesis.

1. INTRODUCTION[a]

The x-ray crystal structures of two photosynthetic bacterial reaction centers (RCs), *Rhodopseudomonas viridis* and *Rhodobacter sphaeroides*, are now known[1,2,3]. These

---

[a] Reaction center, (RC); bacteriochlorophyll, Bchl; bacteriopheophytin, Bphe, charge transfer, CT; electron paramagnetic resonance, (EPR); zero field splitting, (ZFS).

11

J. Jortner and B. Pullman (eds.), Perspectives in Photosynthesis, 11–21.
Kluwer Academic Publishers.

structures have confirmed that the primary donor in these RCs consists of a dimeric
bacteriochlorophyll special pair[4], [|Bchl$_M$>|Bchl$_L$>], with approximate C2 symmetry.
The subscripts M- and L- refer to the M-protein subunit and the L-protein subunit
respectively. The charge separation process is initiated by the absorption of light by the
special pair to form the excited singlet state of the special pair, designated
^{1*}[|Bchl$_M$>|Bchl$_L$>]. The primary electron transfer occurs to form an initial radical pair
species involving the special pair and a distant bacteriopheophytin in the L-protein
subunit in less than 3.5 ps.[5,6,7,8,9,10,11] When the special pair was proposed to explain
the magnetic resonance data over fifteen years ago,[4] the initial charge separation state
was invoked as Bchl$^+$Bchl$^-$ [12,13] where each Bchl is a member of the special pair.
Similar CT states have again been invoked in some recent, more sophisticated models of
the initial mechanism of electron transfer. Current detailed explanations of the initial
electron transfer reaction rate of bacterial RCs are based on electron transfer
theory.[14,15] Even though numerous experimental [5,6,7,16,17,18,19,20,21,22] and
theoretical [23,24,25,26] studies have addressed the initial charge separation in RCs, the
nature of the excited state of the primary donor still needs further illumination in order
to obtain general agreement on the theoretical model for electron transfer.

CT states can be probed by applying a large, electric field external to the RCs and
monitoring changes in spectroscopy and (or) the photosynthetic process. Kleuser and
Bücher[27] studied electric field effects on the change in the optical absorption of
chlorophyll a and chlorophyll b monolayers. Emrich *et al.*[28] used these results as well
as their own electrochromic measurements of carotenoids to explain the light-induced
absorption changes of spinach chloroplasts. Jackson and Crofts[29] studied the
electrochromic effects of the carotenoid by generated external electric fields in the dark
across membranes of *Rb. sphaeroides* chromatophores by ionic gradients operating
through ionophorous antibiotics. More recently, electrochromism has been used to study
the singlet excited state of the Bchl special pair. A large Stark effect was reported in the
Q_y absorption of the photosynthetic bacteria *Rb. sphaeroides* by deLeeuv *et al.*[16] The
value of the Stark effect in the special pair band was first quantitated by Lockhart *et*
al.[17] and later confirmed by Lösche *et al.*[18] Both groups suggested that the large
Stark effect may be due to a mixing of the charge-transfer state |$^+$Bchl$_M$>|$^-$Bchl$_L$> and (or)
|$^-$Bchl$_M$>|$^+$Bchl$_L$> with the excited singlet state ^{1*}[|Bchl$_M$>|Bchl$_L$>] of the
bacteriochlorophyll special pair.

2. EXPERIMENTAL APPROACH

In our opinion, the ultimate understanding of the initial steps in photosynthesis and
electron transfer will require detailed knowledge of numerous molecular parameters.
Of all such parameters the Hamiltonian matrix elements believed responsible for electron
transfer are generally regarded as the most desirable. The standard experimental
approach appropriate to the determination of the Hamiltonian matrix elements relevant
to electron transfer is the direct measurement by spectroscopy of the kinetics of electron
transfer. However, the direct and accurate connection between spectroscopic parameters
and structure remains rather limited. This paucity reflects the fact that few detailed
spectroscopic studies have been performed in single crystals of RCs. For that reason we

have examined the triplet state in single crystals of reaction centers where a very accurate connection between structure and EPR spectroscopy is readily achieved. Extension of EPR to non-crystalline environments is based on rigorous and well established magnetic resonance spectroscopy in single crystals.

To examine our earlier hypothesis of a CT within the special pair we have investigated RCs in which the special pair of $Rb.$ $capsulatus$ has been genetically modified through site-directed mutagenesis.[30] We have probed two genetically modified RCs in which histidine M200, the axial ligand to the M-side Bchl of the special pair, was replaced with either leucine or phenylalanine. This "simple" amino acid substitution results in the replacement of the M-side Bchl with Bphe, forming a Bchl-Bphe heterodimer.[8,30] The heterodimer RCs also differ from wild-type in quantum yield and rate of electron transfer. The quantum yield of the electron transfer from $D \rightarrow D^+ Bphe_L^-$ is ~50% in $His^{M200} \rightarrow$ Leu RCs, much lower than the ~100% yield for wild-type. The time constant for this electron transfer reaction is ~30 ps, about one order of magnitude slower than the 3.5 ps lifetime for wild-type.[8] These reasonably large differences in the electron transfer process make the heterodimer an excellent choice to test the credibility of various theoretical models of photosynthesis.

In particular, we have measured the Stark spectra and the triplet and cation EPR spectra of wild-type versus heterodimer containing RCs from $Rb.$ $capsulatus$. The differences and similarities in the Stark spectrum, in the triplet EPR data and in the cation EPR linewidths of these three RCs of $Rb.$ $capsulatus$ as well as those of $Rp.$ $viridis$ and $Rb.$ $sphaeroides$ are discussed. Ultimately, these experimental results are intended to clarify the nature of the primary donor and its role in the initial act of photosynthesis. Thus the next step is the interpretation of the spectroscopic measurements in terms of structure and electron transfer. In this regard, Closs et $al.$[31] have shown a connection between the matrix elements of electron transfer and triplet energy transfer using a frontier molecular orbital approach. Consequently, another method of examining the matrix elements of electron transfer is by indirect inference from the rate constants for triplet energy transfer. Ultimately both approaches will require incorporation of the precise structural details of the molecular complex in which electron or triplet transfer occurs. In this paper no rate constants are measured; instead, the spectroscopic properties of excited singlet states, excited triplet states and cation states are described that can be interpreted in terms of frontier molecular orbitals using an approach similar to that of Closs et $al.$[31] for transfer rates.

The EPR experimental conditions are similar to those reported previously.[4,12] Wild-type, $His^{M200} \rightarrow$ Leu, and $His^{M200} \rightarrow$ Phe reaction centers of $Rb.$ $capsulatus$ were prepared by standard [32] procedures.[8,30] The Stark spectroscopy and normal spectroscopy were performed in liquid nitrogen by previously developed methods [17,18,33]. Our experimental system was verified with $Rb.$ $sphaeroides$ where we obtained results identical to previous work.[17,18,19]

3. RESULTS

In $Rb.$ $sphaeroides$ the triplet state of the special pair exhibits approximate C2 symmetry since the triplet state is located almost equally on both halves of the special pair[34]. In

Rp. viridis the triplet state is located mainly on the L-side of the special pair,[34] the flatter half of the special pair according to the x-ray structure.[1] Table I summarizes the EPR parameters used in this discussion.[34]

The absorption, second derivative, and Stark spectra at 77 K for the Q_y absorption bands of wild-type, $His^{M200} \rightarrow$ Leu, and $His^{M200} \rightarrow$ Phe reaction centers from *Rb. capsulatus* have been determined by DiMagno *et al.*[35] The Q_y Stark spectra for wild-type *Rb. capsulatus* exhibits similar features as *Rb. sphaeroides R-26*[17,18,19,20,36] and *Rp. viridis*.[17,18,19] The absorption, second derivative, and Stark spectra for $His^{M200} \rightarrow$ Leu and $His^{M200} \rightarrow$Phe RCs are nearly identical to each other. The Stark spectrum of the heterodimers are similar to the Stark spectrum of wild-type *Rb. capsulatus* in the accessary Bchl and the Bphe spectral regions, but differ dramatically in the special pair region for both the absorption and the Stark spectrum. These features suggest that the $His^{M200} \rightarrow$ Leu and $His^{M200} \rightarrow$ Phe substitutions did not result in large structural changes.

Table II lists the angle, δ, between the transition moment vector, μ_{trans}, and the change in permanent dipole moment vector, $\Delta\mu_{app}$, as measured using the red most Stark band for the two *Rb. capsulatus* RCs where $\Delta\mu_{app}$ and the angle δ were determined as usual.[17,18] The results for wild-type *Rb. capsulatus* 11,530 cm^{-1} band at 77 K (δ=38.4°, $\Delta\mu_{app}$=6.7D) are the same as both *Rb. sphaeroides* and *Rp. viridis*[17,20] RCs within experimental error. Both $His^{M200} \rightarrow$Leu and $His^{M200} \rightarrow$ Phe RCs give the same value at ~11,000 cm^{-1} for δ and $\Delta\mu_{app}$ at 77 K (His200 → Leu: δ=20.3°, $\Delta\mu_{app} \geq$15.1D; $His^{M200} \rightarrow$ Phe: δ=22.6°, $\Delta\mu_{app} \geq$16.5D), but the angle δ and the value $\Delta\mu_{app}$ differ substantially from wild-type *Rb. capsulatus* RCs. This 11,000 cm^{-1} band is an additional band not present in wild type and can be interpreted as a transition to a "pure" charge transfer state. We point out that no absorption peak appears in either heterodimer at 11,000 cm^{-1}. We also note that the Stark bands of the heterodimer in the spectral region (~11,000 to ~11,500 cm^{-1}) normally diagnostic of the special pair are strong, unlike the weak absorption tails in this same spectral region.

4. DISCUSSION

To understand in elementary chemical terms what might be involved in describing the special pair and its role in electron transfer, a linear combination of simple molecular orbitals is usually employed. The excited states in the wild-type (subscript W) *Rp. viridis*, *Rb. sphaeroides* and *Rb. capsulatus* are expected to be linear combinations of the following frontier molecular orbitals;

$$|^*I>_W = |^*Bchl_M>|Bchl_L>,$$
$$|^*II>_W = |Bchl_M>|^*Bchl_L>,$$
$$|^*III>_W = |Bchl_M^+>|Bchl_L^->,$$
$$|^*IV>_W = |Bchl_M^->|Bchl_L^+>,$$

where * indicates either excited singlet or excited triplet. Possible excited state contributions to the heterodimer special pair mutant species of *Rb. capsulatus* (subscript H) are the following;

Table I
EPR Properties of Bacteriochlorophylls a and b and the Primary Donors in *Rp. viridis*, *Rb. sphaeroides* and *Rb. capsulatus* Reaction Centers

	D	E	TRIPLET ΔH_{pp}	CATION ΔH_{pp}
		(gauss)		
Bchl b in vitro	227	59	--	13.9
Bphe b in vitro	226	53	--	--
$Bchl_M Bchl_L$ (*viridis* - ^{1}H)	165	40	11.3	11.5*
$Bchl_M Bchl_L$ (*viridis* - ^{2}H)	165	40	6.8*	4.6*
Bchl a in vitro	240	57	--	13.0
Bphe a in vitro	274	48	--	13.5
$Bchl_M Bchl_L$ (*sphaeroides* - ^{1}H)	199	33	10.3*	9.6*
$Bchl_M Bchl_L$ (*capsulatus* - ^{1}H)	210	37	--	10.6
$Bphe_M Bchl_L$ (*capsulatus* - ^{1}H) (His → Phe)	225	64	--	12.0
$Bphe_M Bchl_L$ (*capsulatus* - ^{1}H) (His →Leu)	225	64	--	12.2

*, single crystal data.

Table II
A comparison of the direction and magnitude of the apparent dipole moments, $\Delta\mu_{app}$, in photosynthetic bacteria.[a]

Photosynthetic Bacteria	$\delta,°$	$\Delta\mu_{app}$, D	ν, cm^{-1}
Rb. capsulatus wild-type	38.4 ± 2.3	6.7 ± 1.0	11,530
Rb. capsulatus His^{M200} → Leu	20.3 ± 3.3	≥ 15.1	11,020
Rb. capsulatus His^{M200} → Phe	22.6 ± 2.4	≥ 16.5	11,147

Table III
Approximate Wavefunctions of the Triplet and Cation of the Primary Donor

	TRIPLET				CATION	
	I	II	III[a,b]	IV[a,b]	I[a,b]	II[a,b]
Rb. sphaeroides	32	55	13	0	38.5	61.5
Rp. viridis	0	77	23	0	11.5	88.5
Rb. capsulatus (wild-type)	c	c	c	c	c	c
Rb. capsulatus (heterodimer)	0	100	0	0	0	100

a, charge asymmetry is assumed and is not determined by EPR.
b, asymmetry can be reversed from that given here.
c, is very similar to *Rb. sphaeroides*.

$$|^*I>_H = |^*Bphe_M>|Bchl_L> \quad \text{(higher energy)},$$
$$|^*II>_H = |Bphe_M>|^*Bchl_L>,$$
$$|^*III>_H = |Bphe_M^+>|Bchl_L^-> \quad \text{(higher energy)},$$
$$|^*IV>_H = |Bphe_M^->|Bchl_L^+>,$$

where $*$ indicates either excited singlet or excited triplet. The heterodimer special pair states $|^*I>$ and $|^*III>$ are not considered likely participants in the eigenstates of the lowest first excited states because they are significantly higher energy due to the differences in redox potentials between bacteriopheophytin and bacteriochlorophyll.[39] Similarly, possible molecular orbitals that contribute to the wave function of the cation in wild type *Rp. viridis*, *Rb. sphaeroides*, and *Rb. capsulatus* are the following;

$$|^+I>_W = |^+Bchl_M>|Bchl_L>,$$
$$|^+II>_W = |Bchl_M>|^+Bchl_L>.$$

Finally, the possible MOs for the cation of heterodimer are the following:

$$|^+I>_H = |^+Bphe_M>|Bchl_L> \quad \text{(higher energy)}$$
$$|^+II>_H = |Bphe_M>|^+Bchl_L>$$

Here state $|^+I>$ is again not considered a likely participant in the working description of the heterodimer and thus we expect the system to behave like a monomeric Bchl species. Table III lists the **approximate** percentages of each wavefunction consistent with the magnetic resonance observations. Values in Table III for the triplet state are based on experiments in single crystals.[34]

The unusually small ZFS for the triplet state in *Rp. viridis* can be explained by ~77% of $|Bchl_M>|^{3*}Bchl_L>$ mixed with ~23% of $|Bchl_M^+>|Bchl_L^->$, i.e., a special pair mechanism.[34] Such a supermolecule mechanism is confirmed by phosphorescence measurements of Takiff and Boxer.[37] Consequently, alternative monomer (i.e., non-supermolecule) explanations involving different molecular orbitals induced by a special protein environment are ruled out. Because of the vector nature of the ZFS tensor, the 23% charge transfer state reduces the ZFS in magnitude but does not alter its direction significantly away from the L-side macrocycle. On the other hand the heterodimer special pair has ZFS completely characteristic of a monomer of Bchl. This indicates 100% $|Bphe_M>|^{3*}Bchl_L>$ and no participation of charge transfer. In both cases the C2 symmetry appears to be completely broken in the triplet state; however, from the point of view of the ZFS the special pair in *Rp. viridis* behaves like a supermolecule dimer whereas the heterodimer special pair behaves like a simple monomer. In both cases a generalized special pair model can explain the triplet (single crystals only) and cation (general) EPR linewidth equally well. The second moment of the EPR lineshape is given by

$$M_{sp} = f^2 M_m + (1-f)^2 M_m,$$

where M_{sp} and M_m are the second moments of the special pair and the monomer respectively. In these systems the observed linewidth is proportional to the square root of the second moment. In such a general model the fraction of occupancy of any one type of state (charge transfer, cation, triplet, or singlet) on the L-side is f and on the M-side is 1-f. The f can range from 0 to 1 where f = 0.5 describes the special pair with C2 symmetry. Thus, the triplet or cation delocalization can range from approximate C2 special pair as in *Rb. sphaeroides* to complete localization on the L-side of the special pair as in the heterodimer mutants of *Rp. capsulatus*. Using values of f determined from Table III the above equation offers a reasonable explanation of the experimental linewidths of Table I. In other words, the observed range of delocalization can be rationalized in terms of a simple frontier molecular orbital description.

The nature of the first excited singlet excited state can be examined by Stark spectroscopy. Based on such Stark spectroscopy here we present recent results for RCs of *Rb. capsulatus*. The magnitude of $\Delta\mu_{app}$ for RCs of *Rb. capsulatus* ($His^{M200} \rightarrow$ Leu) and *Rb. capsulatus* ($His^{M200} \rightarrow$ Phe) is larger than wild-type RCs for the red most band. For the Leu heterodimer this red most Stark band centered at ~11,020 cm^{-1} strongly suggests the formation of a charge-transfer state within the heterodimer that evolves within several hundred femtoseconds after the excitation flash.[8] The other Stark band at ~12,110 cm^{-1} is interpreted as the "normal" exciton band characteristic of the special pair. Previous calculations indicate that the relative energy of the charge-transfer state might be above the lowest excited singlet state of the special pair in wild-type RCs.[38] When the constituents of the charge-transfer state change from $Bchl_L^+Bchl_M^- \rightarrow Bchl_L^+Bphe_M^-$ in the heterodimer, the energy of the charge-transfer state will decrease by ~.1 to ~ .3 eV[39] and this should narrow the energy gap between the "normal" exciton and "pure" CT states. This decrease in energy of the charge-transfer state is expected to mix the corresponding oscillators in the heterodimer relative to wild-type RCs, thus giving some oscillator strength to the normally invisible charge transfer band. Such a picture is compatible with the extensive red tail in the normal optical spectrum in the heterodimer mutant. This conclusion is also consistent with the investigation by Kirmaier *et al.*,[8] who suggested the pure charge-transfer state ($^+Bchl_L^-Bphe_M$) in the heterodimer to explain the time resolved transient absorption difference spectra and the decreased rate of electron transfer (~30 ps) to the initial radical pair state.

The values for $\Delta\mu_{app}$ listed in Table II assume that all of the red most Stark lineshapes are pure second derivatives. The largest error in determining $\Delta\mu_{app}$ was in obtaining the weighted second derivative of the absorption spectrum, especially in the case of the broad featureless absorption spectra of the mutants. In the case of the two heterodimer RCs, no second derivative signal could be measured without significant noise. Thus, limits were established for the size of the second derivative as -2.4×10^{-7} for $His^{M200} \rightarrow$ Phe RC and -1.7×10^{-7} for $His^{M200} \rightarrow$ Leu RCs. These values correspond to minima for $\Delta\mu_{app}$ of 16.5 for $His^{M200} \rightarrow$ Phe and 15.1 for $His^{M200} \rightarrow$ Leu RCs. Approximately 40 Debye is an estimate of the value of $\Delta\mu_{app}$ for a pure charge separation within the special pair. Based on the estimated error in experimental and simulated data, we are only able to bracket the value for $\Delta\mu_{app}$ for the heterodimers between ~15 and ~40 Debye. We note that the value for $\Delta\mu_{app}$ is larger in the heterodimer than in wild-type, again consistent with an increase in the amount of charge-transfer character in the excited singlet of RCs of the heterodimers.

Although no crystal structure exists for *Rb. capsulatus*, linear dichroism shows that the orientation of pigments with respect to the membrane direction in wild-type, $His^{M200} \rightarrow$ Leu, and $His^{M200} \rightarrow$ Phe RCs are similar to each other and to RCs in *Rb. sphaeroides* R26.[40] Thus, the Stark data will be described using the crystal structure of *Rb. sphaeroides* RCs, which should act as a good model for *Rb. capsulatus* because *Rb.*

sphaeroides contains the same type of bacteriochlorophyll and bacteriopheophytin and a similar protein sequence in the vicinity of the special pair. From the crystal structure of *Rb. sphaeroides* R26 RCs, the angle between Q_Y transition moments of the members of the special pair is ~38°. The observed change in the angle δ between wild-type *Rb. capsulatus* and the *Rb. capsulatus* $His^{M200} \rightarrow$ Phe, $\delta_{wild-type} - \delta_{Phe}$, is 15.8°; and the change in the angle δ between wild-type *Rb. capsulatus* and the *Rb. capsulatus* $His_{M200} \rightarrow$ Leu, $\delta_{wild-type} - \delta_{Leu}$, is 18.1° in reasonable agreement with one half of 38° observed in the x-ray structure of *R26* RCs. The change of δ between the wild-type and heterodimer RCs can be attributed to the breaking of the apparent C_2 symmetry due to either the change in the Q_Y transition moment or the change in the direction of $\Delta\mu_{app}$ from a dimer to monomer-like special pair. Since a larger value for $\Delta\mu_{app}$ in heterodimer RCs, is observed, the change in δ most likely is due to the different direction of $\Delta\mu_{app}$ in dimeric-like versus monomeric-like donor. This observation supports the claim of charge-transfer n the initially excited singlet state in wild type bacterial reaction centers[12,13,34] and that this charge transfer state is above the first excited singlet as predicted by Parson and Warshel[38].

5. CONCULSIONS

Because of the difference in redox potentials between the bacteriochlorophyll and bacteriopheophytin, the sharing of spin density should be unequal in the oxidized heterodimer. This result is confirmed[41] in EPR experiments on the two heterodimers for both the triplet and the cation. The EPR linewidth from wild-type RC suggested a special pair dimer $^+[Bchl_L Bchl_M]$,[4] while the EPR linewidth from the heterodimer containing RCs indicated monomeric $Bchl^+_L$ electron donor.[41] The asymmetric charge distribution of the oxidized heterodimer confirm the notion that an internal CT state within the heterodimer special pair would be more favored due to the difference in redox potentials of the different heterodimer constituents. Both the Stark data and the transient absorption data suggest a rather asymmetric heterodimer special pair. In our measurements, δ appears to be more like monomeric pigments rather than the special pair of *Rb. sphaeroides, Rp. viridis*, and wild type *Rb. capsulatus*.[17,18,35]

The simple MO description that explains the size and directions of the ZFS also offers a simple explanation of both the triplet and cation epr linewidth. This frontier orbital approach suggests that triplet state is dominated by HOMO orbitals since certainly the cation is. The symmetry can be broken in two distinct ways: by small changes in the environment whereby the energies of the orbitals to be mixed between the two halves of the special pair are very similar, or by much larger changes where the two molecular orbitals are significantly different as in the heterodimer special pair where one half is bacteriopheophytin and one half is bacteriochlorophyll. In the former case the triplet behaves like a supermolecule, and in the latter case the triplet behaves as a simple monomer as clearly evidenced by the triplet zero field parameter. This also suggests that the triplet is dominated by the HOMO since the charge transfer in the supermolecule is less likely to involve the much higher triplet charge transfer state. The energy of the state $|Bchl^+_M \rangle |Bchl^-_L \rangle$ should be ~.1 to ~.3 ev above the lowest excited singlet state, $^{1*}[|Bchl_M \rangle |Bchl_L \rangle]$, in wild-type reaction centers if the red-most Stark band arises from a "pure" CT state, $|Bphe^-_M{>}|Bchl^+_L{>}$, in the heterodimer containing reaction centers as we have suggested here.

6. ACKNOWLEDGMENTS

The work performed in the Chemistry Division was supported by the U.S. Department of Energy, Office of Basic Energy Sciences, Division of Chemical Sciences under contract W-31-109-Eng-38. Work performed in the Biological, Environmental and Medical Research Division was supported by the Public Health Service Grant GM-36598 and by the Office of Health and Environmental Research under Contract W-31-109-Eng-38.

7. REFERENCES

1. Deisenhofer, J., Epp, O., Miki, K., Huber, R. and Michel, H.J. (1984) *J. Mol. Biol.*, 180, 385-398.
2. Chang, C.-H., Tiede, D., Tang, J., Smith, U., Norris, J. and Schiffer, M. (1986) *FEBS Lett.*, 205, 82-86.
3. Allen, J.P., Feher, G., Yeates, T.O., Komiya, H. and Rees, D.C. (1987) *Proc. Natl. Acad. Sci.USA*, 84, 5730-5734.
4. Norris, J.R., Uphaus, R.A., Crespi, H.L. and Katz, J.J. (1971) *Proc. Natl. Acad. Sci. USA*, 68, 625-628.
5. Woodbury, N.W., Becker, M., Middendorf, D. and Parson, W.W. (1985) *Biochemistry*, 24, 7516-7521.
6. Martin, J.-L., Breton, J., Hoff, A.J., Migus, A. and Antonetti, A. (1986) *Proc. Natl. Acad. Sci. USA*, 83, 957-961.
7. Fleming, G.R., Martin, J.L. and Breton, J. (1988) *Nature*, 333, 190-192.
8. Kirmaier, C., Holten, D., Bylina, E., and Youvan, D. (1988) *Proc. Natl. Acad. Sci. USA*, in press.
9. Kirmaier, C. and Holten, D. (1988) *FEBS Lett.*, 239, 211-218.
10. Rockley, M.G., Windsor, M.W., Cogdell, R.J. and Parson, W.W. (1975) *Proc. Natl. Acad. Sci. USA*, 72, 2251-2255.
11. Holten, D., Windsor, M.W., Parson, W.W. and Thornber, J.P. (1978) *Biochim. Biophys. Acta*, 501, 112-126.
12. Thurnauer, M. C., Katz, J. J., and Norris, J. R., (1975) *Proc. National Academy of Sciences USA* 72 (9), 3270-3274.
13. Norris, J. R., and Katz, J. J., *The Photosynthetic Bacteria*, Chapter 21, Clayton, R. K., and Sistrom, W. R., Eds., Plenum Press, 1978, pp. 397-418.
14. Marcus, R.A. and Sutin, N. (1985) *Biochim. Biophys. Acta*, 811, 265-322.
15. Jortner, J. (1980) *J. Am. Chem. Soc.*, 102, 6676-6686.
16. deLeeuv, D., Malley, M., Buttermann, G., Okamura, M. Y. and Feher, G. (1982) *Biophysical Journal*, 37, 111a.
17. Lockhart, D.J. and Boxer, S.G. (1987) *Biochemistry*, 26, 664-668.
18. Lösche, M., Feher, G. and Okamura, M.Y. (1987) *Proc. Natl. Acad. Sci. USA*, 84, 7537-7541.
19. Braun, H.P., Michel-Beyerle, M.E., Breton, J., Buchanan, S. and Michel, H. (1987) *FEBS Lett.*, 221, 221-225.
20. Lockhart, D.J. and Boxer, S.G. (1988) *Proc. Natl. Acad. Sci. USA*, 85, 107-111.
21. Boxer, S.G., Middendorf, T.R. and Lockhart, D.J. (1986) *FEBS Lett.*, 200, 237-241.
22. Meech, S.R., Hoff, A.J. and Wiersma, D.A. (1986) *Proc. Natl. Acad. Sci. USA*, 83, 9464-9468.

23. Creighton, S., Hwang, J.-K., Warshel, A., Parson, W.W. and Norris, J.R. (1988) *Biochemistry*, 27, 774-781.
24. Marcus, R.A. (1987) *Chem. Phys. Lett.*, 133, 471-477.
25. Scherer, P.O.J. and Fischer, S.F. (1987) *Chem. Phys. Lett.*, 141, 179-185.
26. Plato, M., Möbius, K., Michel-Beyerle, M.E., Bixon, M. and Jortner, J. (1988) *J. Am. Chem. Soc.*, 110, 7279-7285.
27. Kleuser, D. and Bücher, H. (1969) *Z. Naturforschg.*, 24b, 1371-1374.
28. Emrich, H.M., Junge, W. and Witt, H.T. (1969) *Z. Naturforschg.*, 24b, 1144-1146.
29. Jackson, J.B. and Crofts, A.R. (1969) *FEBS Lett.*, 4, 185-189.
30. Bylina, E. and Youvan, D. (1988) *Proc. Natl. Acad. Sci. USA*, 85, 7226-7230.
31. Closs, G.L., Piotrowiak, P., MacInnis, J. M., and Fleming, G. R., (1988) *J. Am. Chem. Soc.*, 110, 2652.
32. Prince, R.C. and Youvan, D.C. (1987) *Biochim. Biophys. Acta*, 890, 286-291.
33. Mathies, R. and Stryer, L. (1976) *Proc. Natl. Acad. Sci. USA*, 73, 2169-2173.
34. Norris, J. R., Budil, D. E., Gast, P., Chang, C-H., El-Kabbani, O., and Schiffer, M., (1989) *Proc. Nat. Acad. Sci.*, in press.
35. DiMagno, T. J., Bylina, E. J., Angerhofer, A., Youvan, D. C., and Norris, J. R., submitted to *Biochemistry*.
36. DiMagno, T.J. and Norris, J.R. (1988) unpublished data.
37. Takiff, L., and Boxer, S. G. (1988) *Biochim Biophys Acta*, 932, 325-334.
38. Parson, W.W. and Warshel, A. (1987) *J. Am. Chem. Soc.*, 109, 6152-6163.
39. Fajer, J., Brune, D.C., Davis, M.S., Forman, A. and Spaulding, L.D. (1975) *Proc. Natl. Acad. Sci. USA*, 72, 4956-4960.
40. Breton, J., Bylina, E. and Youvan, D. (1989) to be published.
41. Bylina, E., Kolaczkowski, S., Norris, J. and Youvan, D. (1989) to be published.

ORIENTATION OF THE PHEOPHYTIN PRIMARY ELECTRON ACCEPTOR AND OF THE CYTOCHROME B559 IN THE D1D2 PHOTOSYSTEM II REACTION CENTER

Jacques Breton
Service de Biophysique, Département de Biologie
CEN-Saclay, 91191 Gif-sur-Yvette Cedex, France.

INTRODUCTION

The primary photochemistry occurring in the photosystem II (PS II) of cyanobacteria, algae and plants is currently thought to involve a charge separation between the primary donor chlorophyll(s) of P680 and a pheophytin acceptor (Phe). This event is then followed by a stabilization step in which the electron proceeds from Phe to the first quinone acceptor. With respect to these primary reactions, PS II bears striking analogies with purple photosynthetic bacteria. In these organisms, the isolation and purification of a functional reaction center complex deprived of its associated antenna system have greatly facilitated our understanding of the structural and energetic factors governing the primary reactions. While the spatial organization of the bacterial reaction center cofactors (four bacteriochlorophylls, two bacterio-pheophytins, two quinones and one carotenoid) and protein has been inferred from various optical and EPR spectroscopic measurements (Breton and Verméglio, 1982; Breton and Nabedryk, 1987; Breton, 1988), the recent crystallographic determination of the structure of bacterial reaction centers at high resolution (Deisenhofer et al., 1985; Chang et al., 1986; Allen et al., 1987) has provided an invaluable picture of the detailed organization of the pigments and protein.

Until very recently, the purification of a minimal PS II reaction center comparable to that of purple bacteria had not been achieved and the photochemical activity was associated with a specific chlorophyll-protein complex bearing typically about 40 to 60 chlorophyll a per P680 and consisting of at least six polypeptide subunits with apparent molecular weights of about 47, 40, 30-34, 9 and 4 kDa (Velthuys, 1987). The two largest polypeptides carry antenna chlorophylls, whereas the two smallest ones are the

J. Jortner and B. Pullman (eds.), Perspectives in Photosynthesis, 23–38.

apoproteins of cytochrome (Cyt) b559. The two subunits in
the 30-34 kDa range are named D1 and D2. In view of the
analogies in the primary sequences of these polypeptides
with those of the L and M subunits, which constitute the
scaffold of transmembrane α-helices holding together in a
very precise geometry the cofactors within the reaction
centers from purple bacteria, it has been suggested that D1
and D2 play a similar role in forming the reaction center
of PS II (Trebst, 1986; Michel and Deisenhofer, 1986;
Michel and Deisenhofer, 1988). This hypothesis has been
proved correct by the recent isolation, using the detergent
Triton X-100, of a photoactive D1D2 PS II reaction center,
which also contains Cyt b559 (Nanba and Satoh, 1987).

Using the technique of low temperature linear
dichroism (LD), we have extensively investigated the
orientation of the pigments in a number of native PS II
fractions containing various complements of the six
polypeptides (Breton, 1986; Tapie et al., 1986; Breton and
Katoh, 1987), including studies of the D1D2 reaction center
(van Dorssen et al., 1987; Breton et al., 1988). In the
latter study, we have reported a large negative LD signal
of somewhat variable amplitude in the spectra of D1D2
particles prepared in different laboratories, an
observation which could indicate the presence in this
complex of a population of chlorophyll molecules which have
lost their native orientation during the isolation and
purification steps. This has prompted us to investigate the
LD of a newly developped D1D2 reaction center preparation
stabilized in the detergent lauryl maltoside and which
exhibits a high yield of primary photochemistry.

In the present report, the orientation of several of
the essential PS II cofactors in these stabilized D1D2
reaction centers is analyzed by comparing the changes in
the low temperature absorption (A) and LD spectra which
accompany the photoreduction of the pheophytin primary
acceptor and the transition of redox state of Cyt b559 to
equivalent changes measured in the biological membranes.

MATERIALS AND METHODS

The D1D2 reaction centers were prepared essentially as
described in (Nanba and Satoh, 1987) except that
chloroplasts from pea rather than spinach were used and
that the Triton X-100 in the NaCl gradient elution buffer
was replaced by 0.03% lauryl maltoside.

The D1D2 reaction centers were incorporated in
polyacrylamide gels, oriented by uniaxial squeezing of the
gels and the low temperature LD and A spectroscopies were
performed as previously described (Breton, 1985). For
photochemical reduction, the gels were preincubated with
100 mM Na dithionite in 60% glycerol for 24 hours at 4°C.

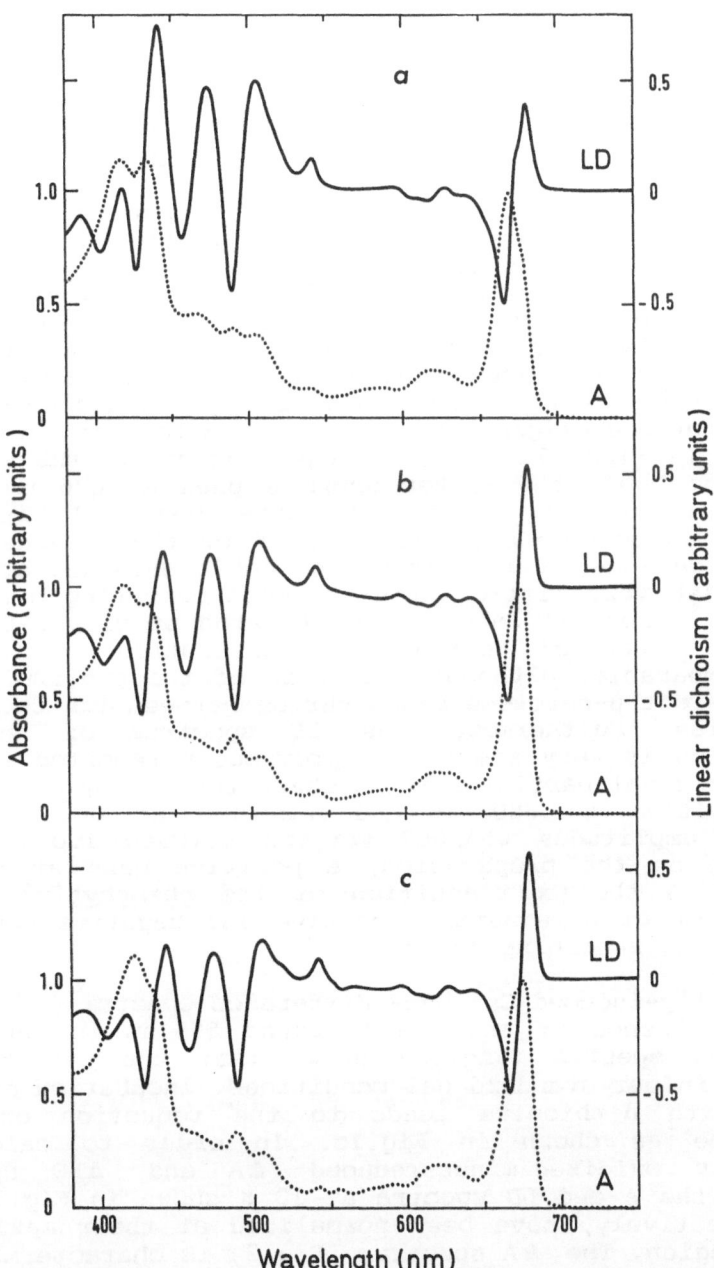

Fig.1: Absorption (·····) and linear dichroism (———)
spectra at 10 K for D1D2 reaction centers oriented in
polyacrylamide gels. (a) Classical preparation in Triton X-
100 without addition. New preparation stabilized with
lauryl maltoside (b) without addition and (c) after
incubation in Na dithionite.

RESULTS

1-Comparison with previously reported spectra

The original preparation of the D1D2 particle involves solubilization of PS II membranes in 4% Triton X-100 followed by chromatography and elution in a buffer containing 0.05% Triton X-100 (Nanba and Satoh, 1987). These particles are rather unstable and lose rapidly their photochemical activity when brought to room temperature. The report that the stability of D1D2 was greatly improved when Triton X-100 was exchanged for lauryl maltoside after the isolation of D1D2 (Seibert et al., 1988) suggested to us a slightly different strategy, i.e. to exchange the detergent on the purification column just before the elution step. A comparison of the absorption and LD spectra at 10 K for the original preparation and for the new one is depicted in Figs. 1a and 1b, respectively. In the main Qy absorption band, the latter shows a peak at 676 nm and a shoulder at 670 nm, whereas we have previously reported that the absorption spectrum at 10 K of the original D1D2 preparation shows a peak at 670 nm and a shoulder at about 678 nm with very little variation when comparing particles isolated in several laboratories (Breton et al., 1988). In the other spectral regions, the absorption spectra are quite comparable, although the ratio of the amplitudes for the different β-carotene bands varies between the two types of samples. Furthermore, the LD spectrum of the new preparation is very similar to previously reported spectra of the original particle. It is characterized by a S-shaped band positive at 680 nm and negative at 666 nm with relative amplitudes 666/680 varying between 1.0 and 1.5 depending on the preparation, a positive band at 542 nm assigned to the Qx transition of the pheophytins and a complex set of alternating positive and negative bands in the carotene absorbing region.

2-Chemically-induced Cyt b559 difference spectra

The absence of an α-band around 555 nm in the 10 K absorption spectra (Fig.1b) shows that the Cyt b559 is oxidized in our standard gel conditions. Incubation of such a gel with dithionite leads to the reduction of this cytochrome as shown in Fig.1c. In order to calculate chemically oxidized-minus-reduced ΔA and ΔLD spectra (Fig.2), the A and LD spectra at 10 K shown in Fig.1b and 1c, respectively, have been normalized at their maxima in the Qy region. The ΔA spectrum (Fig.2) is characterized by the bleaching of the symmetrical α-band at 555 nm, of the β-band at 526 nm and of the γ-band at about 425 nm (data not shown). The ΔLD spectrum exhibits a non symmetrical α-band peaking at 557 nm with a positive dichroism

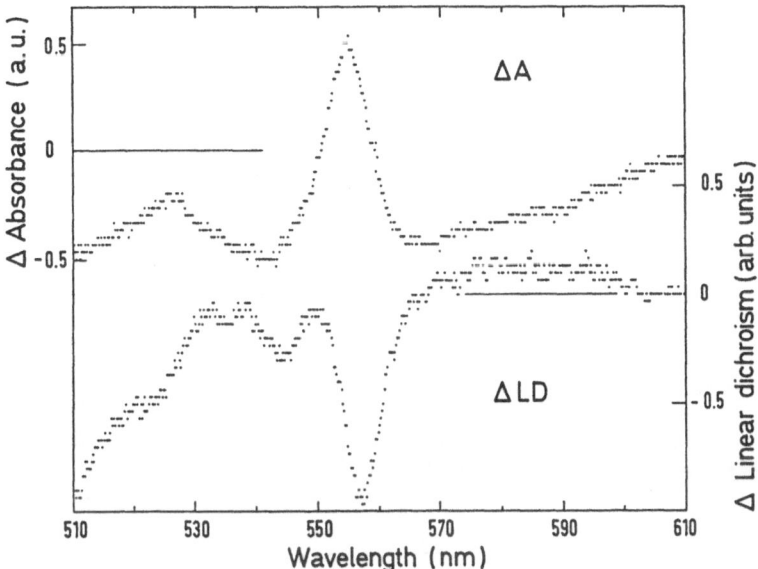

*Fig.2: Difference absorption (**Δ** A, top) and linear dichroism (**Δ** LD, bottom) calculated from the spectra shown in Figs.1b and c and presented in the region of absorption of the α- and β-bands of cytochrome b559.*

characteristic of a transition oriented preferentially perpendicular to the plane of the particle.

3-Photochemical reduction of the pheophytin acceptor

Three different sets of conditions for the reduction of Phe have been explored depending on the temperature of the sample at which the illumination was performed: at 250 K, at 10 K or during the cooling from 295 K to 10 K.

When a dithionite-treated gel containing the PS II reaction center is illuminated at 250 K, absorbance decreases at 680, 542, 512 and 420 nm and absorbance increases at 700, 650, 595 and 450 nm are detected (Fig.3a). These absorption changes are characteristic of the bleaching of Phe, together with the appearance of the pheophytin radical anion Phe$^-$ and the ensuing electrochromic shift of the nearby chromophores (Klimov et al., 1977; Ganago et al., 1982; Nanba and Satoh, 1987). The corresponding ΔLD spectrum (Fig.3a) shows positive(+) and negative(-) peaks at 686(-), 676.4(+), 650(-), 595(+), 542(-), 512(+), 440(-), 430(+) and 420(-) nm. In the Qy spectral region, inspection of the shape of the ΔA and ΔLD spectra clearly shows the composite nature of the

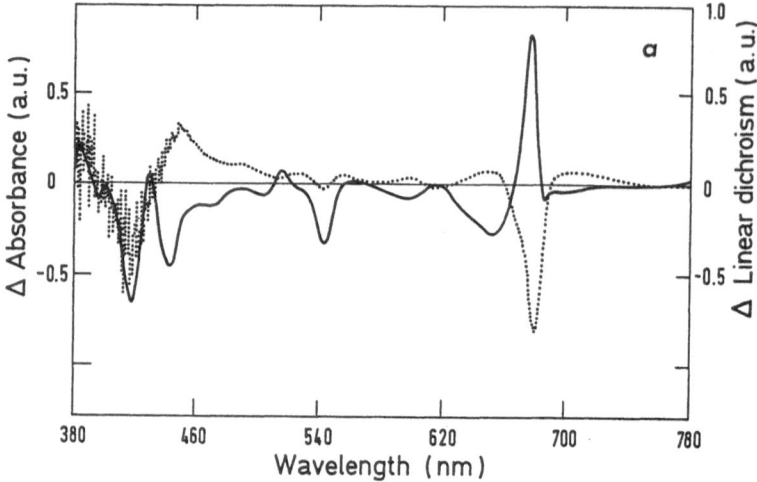

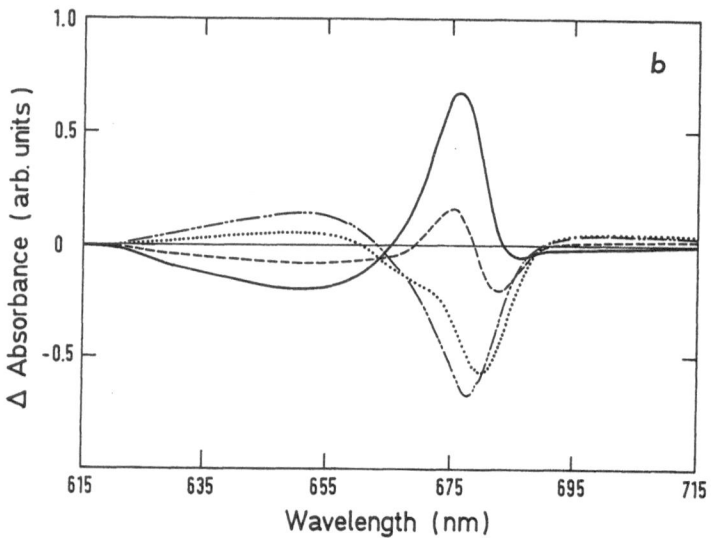

Fig.3: (a) Light-minus-dark ΔA *(·····) and* ΔLD
*(———) spectra at 250 K for the dithionite-reduced D1D2
reaction center. Illumination of the sample was performed
at 250 K. (b) In the Qy-absorbing region of the
chlorophylls, these* ΔA *and* ΔLD *spectra can be decomposed
in terms of a band shift (- - -) and a bleaching (-·-·-)*

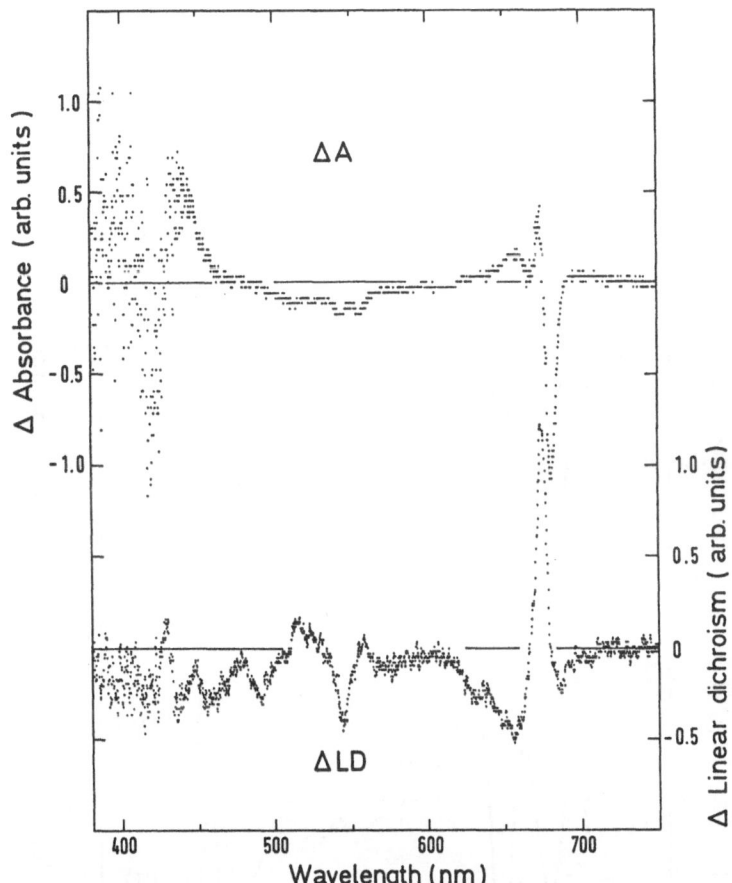

Fig.4: Light-minus-dark Δ*A (top) and* Δ*LD (bottom) spectra at 10 K for the dithionite-reduced D1D2 reaction center. Illumination of the sample was performed at 10 K.*

absorbance changes. More specifically, while the general shape of these spectra primarily reflects a large bleaching of a band located near 678 nm and polarized perpendicular to the main plane of the particle, the ~2.5-nm shift observed between the ΔA and ΔLD extrema as well as the negative ΔLD signal at 686 nm suggest the presence of a S-shaped signal polarized parallel to the particle plane. Assuming that only two components contribute to the absorbance changes in this spectral range, linear combinations of the ΔA and ΔLD spectra can be graphically compared until signals with optimally symmetrical shape are found. The result of such a decomposition, which is thought

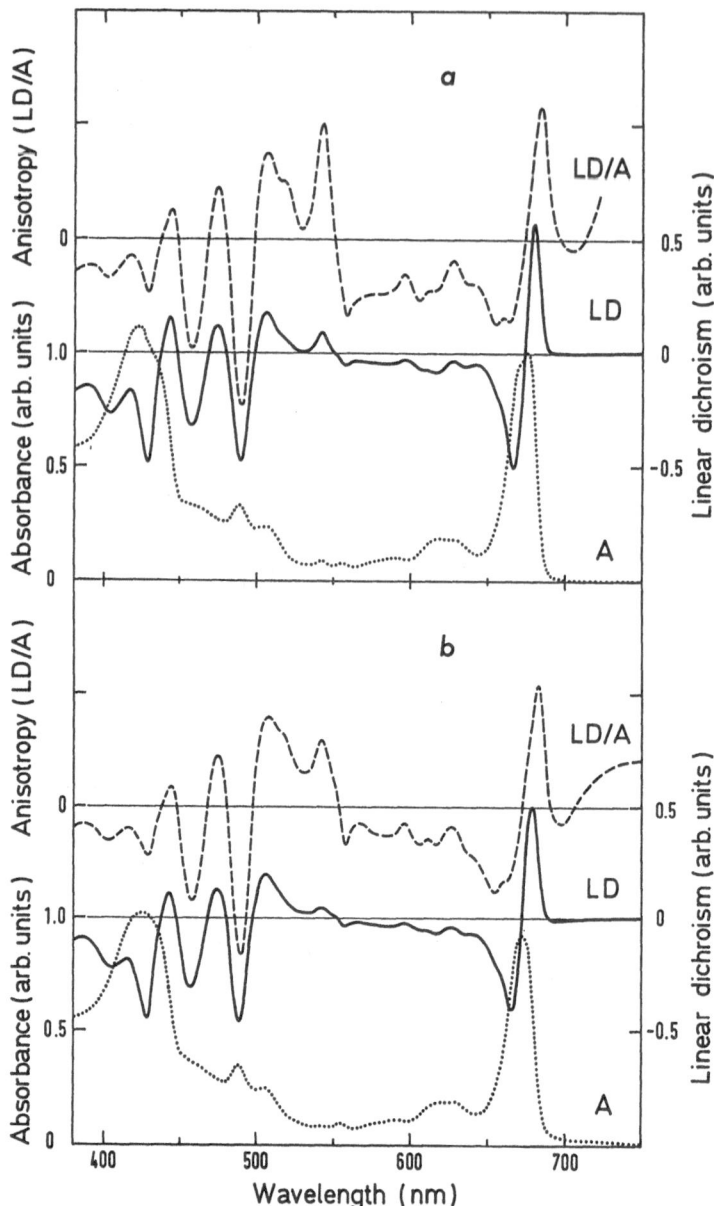

Fig.5: Same conditions as for the spectra shown in Fig.1c, with the sample cooled in the dark (a) or under illumination (b).

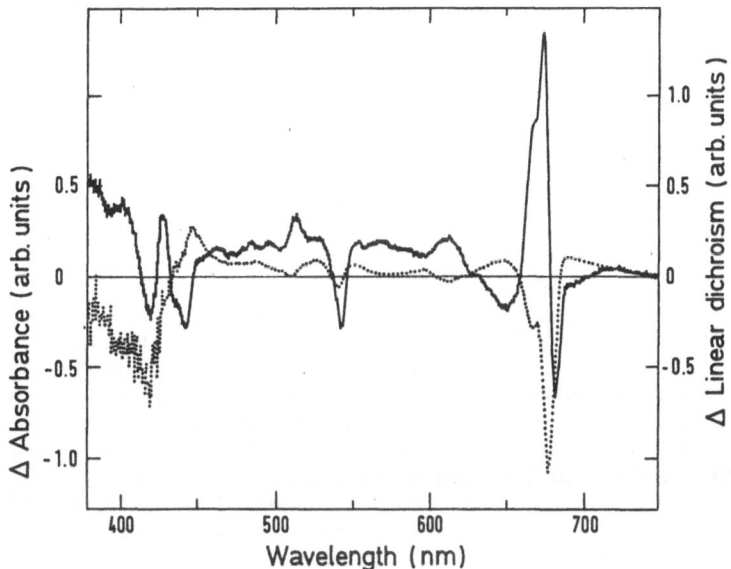

Fig.6: Light-minus-dark **Δ**A *(·····) and* **Δ**LD, *(_____)*
spectra calculated from the spectra shown in Figs. 5a and
5b.

to represent the bleaching and the band shift, is shown in
Fig.3b.

When the dithionite-treated sample is first cooled to
10 K and then illuminated, a small but clear bleaching of
Phe is observed at 542 nm. It is accompanied by a bleaching
of some Cyt b559 as can be seen in the corresponding **Δ**A
and **Δ**LD spectra (Fig. 4). In the Qy region, the disymmetry
between the **Δ**A and **Δ**LD spectra is much more pronounced
than in the 250 K spectra and the negative **Δ**LD component
at about 685 nm further confirms the presence of a band
shift.

When the sample is illuminated during the cooling,
which takes about 30 mn, the pronounced bleaching of Phe
which takes place can be detected directly on the spectra
(Figs. 5a and b). The corresponding **Δ**A and **Δ**LD spectra
are shown in Fig.6. It should be noticed that, in contrast
to the previous two cases in which only the photochemistry
is involved without physical disturbance of the sample
between the light and dark spectra, the latter set of
difference spectra (Fig.6) compares two successive cooling
cycles of the same sample.

DISCUSSION

1. Orientation of the pigments

The LD spectra of the original D1D2 preparation (Fig.1a) and of the new one (Fig.1b), in which the Triton X-100 detergent has been exchanged for lauryl maltoside, show a close similarity. This demonstrates both the same general organization of the pigments and an identical orientation of the anisotropy axis of these particles during the compression of the gel. From our previous work, it has been conluded that this anisotropy axis, which in the gel aligns along the compression direction, assumes an orientation *in vivo* along the normal to the membrane plane (van Dorssen et al., 1987; Breton et al., 1988). Together with the modification in the shape of the Qy absorption band at 10 K, the main difference between the two preparations is related to the structured spectral features in the region 460 to 520 nm where the dominant contribution is that of β-carotene, the only carotenoid present in these particles (Nanba and Satoh, 1987; Barber et al., 1987; Ghanotakis et al., 1989). More specifically, the new preparation contains more of the component at 489 nm, to which is associated a very large negative LD signal, than the original particle. The presence of a succession of closely spaced absorption bands associated with alternating positive and negative LD bands points to an unusual organization of the β-carotene, which is thought to be present at the ratio of only one molecule per reaction center (Nanba and Satoh, 1987). A probable explanation in terms of two populations of β-carotene with opposite orientation relative to the membrane plane and with a slightly variable ratio of the two populations has been previously proposed (van Dorssen et al., 1987; Breton et al., 1988). An alternative hypothesis, involving excitonic coupling between β-carotene molecules on two closely adjacent PS II reaction centers (R. van Grondelle, personal communication), is not supported by circular dichroism spectroscopy in the 450- to 500-nm region where only signals of very small magnitude are observed (data not shown).

In the Qy absorption region of the chlorophylls, the lauryl maltoside treatment has essentially no effect on the shape and position of the S-shaped LD signal (positive peak at 680 nm and negative peak at 666 nm; see Figs.1a and 1b). The former peak is assigned to the contribution from native chlorophyll forms, including P680 and Phe. The latter one is presently ascribed to chlorophyll molecules which have lost their native orientation, probably owing to the high concentration of Triton X-100 used for the extraction of the reaction center (Breton et al., 1988). This assignment is supported by the observation that the 666-nm negative LD band increases in amplitude upon aging of the gel at room temperature and that the corresponding difference ΔA and ΔLD spectra show a single 666-nm component with a negative dichroism in the Qy region (data not shown). A similar behaviour and an identical assignment have been recently reported for a chlorophyll form absorbing around 670 nm in a small PS I reaction center preparation containing approximately 10 chlorophylls per P700 (Breton and Ikegami, 1989).

2. Orientation of the cytochrome b559.

The first report on the measurement of the high potential Cyt b559 orientation *in vivo* came from light-induced difference absorption spectroscopy on magnetically oriented chloroplasts which were trapped at 77 K (Verméglio et al., 1980). It was observed that the shape of the α-band in the ΔLD spectrum is not symmetrical and the peak is at shorter wavelength than in the corresponding ΔA spectrum. It was thus concluded that the non-degenerate x and y axis within the heme plane are oriented at different angles with respect to the membrane plane and that the heme plane is oriented preferentially perpendicular to the photosynthetic membrane. This conclusion has been confirmed by EPR measurements (Bergstrom and Vanngard, 1981) and extended to the low potential form of Cyt b559 (Crowder et al., 1982; Bergstrom and Vanngard, 1982). As these two techniques probe different redox states of Cyt b559, these observations further show that the heme plane is not changing orientation during the redox transition.

The chemically-induced ΔA and ΔLD spectra in the α-band of Cyt b559 in D1D2, and notably the 2-nm shift between the wavelength of the peaks (Fig.2), are strikingly identical to those reported for the photooxidation of this cytochrome in intact chloroplasts. This demonstrates the validity of our previously proposed model for the orientation of D1D2 within the gel, in which the anisotropy axis of the isolated particle coincides with the normal to the membrane plane *in vivo*.

3. Orientation of the pheophytins.

The dichroism of the absorbance changes corresponding to the photoreduction of the primary electron acceptor pheophytin Phe in the D1D2 preparation (Fig.3a) is remarkably similar to that previously reported in PS II membranes (Ganago et al., 1982). In the Qy region, these spectra are clearly composed of a bleaching superimposed to the blue shift of a band. A simple decomposition, based on the same principles as described in (Ganago et al., 1982), leads to a symmetrical band shift centered at 679 nm and which exhibits a positive dichroism corresponding to a dipole oriented at a small angle from the plane assumed by the membrane *in vivo*. This spectral feature is assigned to the electrochromic effect of Phe⁻ on the transition of a nearby chlorophyll. The orientation close to the membrane plane and the location around 680-682 nm for the absorption maximum of the unperturbed Qy transition of this molecule, together with the known spectral characteristics of P680 (Doring et al., 1969; van Gorkom et al., 1975; Mathis et al., 1976) thus suggest that it is indeed P680 which shifts under the influence of Phe⁻. This hypothesis has been challenged by the report that the transient spectrum of P680⁺Phe⁻ still exhibits the same shift (Nuijs et al., 1986), although close inspection of the published ΔA spectrum clearly shows distortions, probably due to the presence of residual chlorophyll excited states, compared to the features seen in Fig.3 and in (Ganago et al., 1982). Thus, although our present data cannot rule out that the band shift affects the Qy transition of a chlorophyll molecule oriented like P680 and located close to P680, it still seems more likely that P680 is directly involved in this spectral feature.

The main Qy transition of Phe is found to absorb maximally at about 678 nm and the dichroism, although difficult to determine precisely owing to the overlapping band shift, allows a value of 45 to 65° to be estimated for the angle between this transition and the plane assumed by the membrane. This angle compares well with the values of 50 to 60° for the angle of the transition of the equivalent bacteriopheophytin in a number of reaction centers from purple bacteria (Breton, 1985; 1988; Breton et al., 1989). A small transition around 615 nm with the same dichroism (Figs.3a and 6) is assigned to a vibrational level. The Qx transition at 542 nm exhibits a large positive dichroism consistent with its orientation at a small angle from the membrane and indicating an identical geometry as in the bacterial reaction center (Breton, 1985; 1988; Breton et al., 1989). The negative dichroism of the small transition at 512 nm, which is resolved for the first time *in vivo*, demonstrates the y-polarization of this transition. In the Soret region, the large bleaching at 420 nm exhibits a positive dichroism. The appearing absorption bands at 700, 650, 595 and 450 nm, which all exhibit a negative LD, are assigned to the anion of Phe (Fujita et al., 1978).

The difference between the two sets of 10 K spectra shown in Figs.4 and 6 can arise from several factors such as the additional charge present on Cyt b559 when the photoreduction is performed at low temperature (Fig.4) and the possible mismatch at the normalization step or the degradation of some of the pigments when the photoreduction is performed during cooling (Fig.6). In this respect it is worth noting that the small band at 666 nm in the 10 K ΔA spectrum of Fig.6, to which corresponds a small shoulder in the 250 K ΔA spectrum of Fig.3a, is totally absent in the spectrum of Fig.4, corresponding to an illumination at 10 K when photodegradation of the pigments is minimum. The position and dichroism of these spectral features strongly suggest that they correspond to a bleaching of a few of the orientationally perturbed chlorophyll molecules absorbing at 666 nm.

Finally, the orientation of the second pheophytin present in the D1D2 reaction center and which is not photochemically active (Nanba and Satoh, 1987) can be determined by analyzing the dichroism of the photoreduced sample. As clearly seen in the corresponding LD/A spectrum (Fig.5b), the Qx transition of this molecule also absorbs at 542 nm and is oriented with the same geometry as that of the Phe molecule. Thus, the D1D2 PS II reaction center shows the orientation of both pheophytins to be the same as that observed in the bacterial reaction center (Breton, 1988). Furthermore, it is demonstrated here that both molecules absorb at 542 nm. Although this situation is very different from that found in the reaction center of purple bacteria, for which shifts of 12 to 15 nm between the absorption maxima of the two bacteriopheophytins (H_L and H_M) are currently observed, it can be rationalized in terms of the different amino acid sequences of the analogous polypeptides in the various reaction centers (Michel and Deisenhofer, 1988; Tiede et al., 1988). The genetically engineered replacement in *Rb. capsulatus* of the glutamic acid residue Glu L104, which in the wild-type is hydrogen bonded to the photoactive bacteriopheophytin H_L, by a glutamine or a leucine has been shown to shift the absorption maximum of H_L from 545 nm to 541 or 536 nm, respectively (Breton et al., 1989). In addition, the nature of the symmetrically located residue M131, which can interact with the inactive bacteriopheophytin H_M, has also an influence on its absorption maximum. In *Rb. capsulatus*, which contains a valine this maximum is at 530 nm while it appears at 533 nm in *Rb. sphaeroides* which has a threonine at the position M131 (Breton et al., 1989). Thus, it can be easily understood that the presence on D2 of a glutamine residue at the position equivalent to M131 shifts the maximum absorption of the nearby pheophytin to the same wavelength as that of the photoactive pheophytin which is hydrogen bonded to the glutamic acid on D1.

AKNOWLEDGEMENT

I sincerely thank Sandra Andrianambinintsoa for expert assistance in preparing the modified D1D2 reaction center.

REFERENCES

Allen J.P., Feher G., Yeates T.O., Komiya H. and Rees D.C. (1987)
Proc. Natl. Acad. Sci. USA 84, 5730-5734.

Barber J., Chapman D.J. and Telfer A. (1987)
FEBS Lett. 220, 67-73.

Bergström N.H.J. and Vänngärd T. (1981)
in: 'Photosynthesis II, Electron Transport and
Photophosphorylation',
(Akoyunoglou G., ed.), pp. 569-575,
Balaban International Science Services, Philadelphia.

Bergström N.H.J. and Vänngard T. (1982)
Biochim. Biophys. Acta 682, 452-456.

Breton J. (1985)
Biochim. Biophys. Acta 810, 235-245.

Breton J. (1986)
in: 'Encyclopedia of Plant Physiology', Photosynthesis III,
(Staehelin L.A. & Arntzen C.J., eds.), Vol. 19, pp. 319-326,
Springer Verlag, Berlin.

Breton J. (1988)
ISI Atlas of Science: Biochemistry 1, 323-328.

Breton J., Bylina E.J. and Youvan D.C. (1989)
Biochemistry, (in press).

Breton J., Duranton J. and Satoh K. (1988)
in: 'Photosynthetic Light-Harvesting Systems: Organisation and
Function'
(Scheer, H. and Schneider, S., eds.), pp. 375-386, W. de Gruyter,
Berlin.

Breton J. and Ikegami I. (1989)
Photosynth. Res. 21, 27-36.

Breton J. and Katoh S. (1987)
Biochim. Biophys. Acta 892, 99-107.

Breton J. and Nabedryk E. (1987)
in: 'Topics in Photosynthesis - The Light Reactions',
(Barber J., ed.), pp. 159-195, Elsevier, Amsterdam.

Breton J. and Verméglio A. (1982)
in: 'Photosynthesis : Energy Conversion by Plants and Bacteria',
(Govindjee, ed.), pp. 153-194, Academic Press, New York.

Chang C.-H., Tiede D., Tang J., Smith U., Norris J.
and Schiffer M. (1986)
FEBS Lett. 205, 82-86.

Crowder M.S., Prince R.C. and Bearden A. (1982)
FEBS Lett. 144, 204-208.

Deisenhofer J., Epp O., Miki K., Huber R. and Michel H. (1985)
Nature (Lond.) 318, 618-624.

Döring G., Renger G., Vater J. and Witt H.T. (1969)
Z Naturforsch. 24b, 1139-1143.

Fujita I., Davis M.S. and Fajer J. (1978)
J. Am. Chem. Soc. 100, 6280-6282.

Ganago I.B., Klimov V.V., Ganago A.O., Shuvalov V.A.
and Erokhin Y.E. (1982)
FEBS Lett. 140, 127-130.

Ghanotakis D.F., de Paula J.C., Demetriou D.M., Bowlby N.R.,
Petersen J., Babcock G.T. and Yocum C.F. (1989)
Biochim. Biophys. Acta 895, 44-53.

Klimov V.V., Klevanik A.V., Shuvalov V.A.
and Krasnovsky A.A. (1977)
FEBS Lett. 82, 183-186.

Mathis P., Breton J., Verméglio A. and Yates M. (1976)
FEBS Lett. 63, 171-173.

Michel H. and Deisenhofer J. (1986)
in: 'Encyclopedia of Plant Physiology', Photosynthesis III,
(Stachelin L.A. & Arntzen C.J., eds), Vol. 19, pp. 371-381,
Springer-Verlag, Berlin.

Michel H. and Deisenhofer J. (1988)
Biochemistry 27, 1-7.

Nanba O. and Satoh K. (1987)
Proc. Natl. Acad. Sci. USA 84, 109-112.

Nuijs A.M., van Gorkom H.J., Plijter J.J.
and Duysens L.N.M. (1986)
Biochim. Biophys. Acta 848, 167-175.

Seibert M., Picorel R., Rubin A.B. and Connolly J.S. (1988)
Plant Physiol. 87, 303-306.

Tapie P., Choquet Y., Wollman F.A., Diner B. and Breton J. (1986)
Biochim. Biophys. Acta 850, 156-161.

Tiede D.M., Budil D.E., Tang J., El-Kabbani O., Norris J.R.,
Chang C.H. and Schiffer M. (1988)
in: 'The Photosynthetic Bacterial Reaction Center: Structure and
Dynamics', (Breton J. & Verméglio A., eds.) Vol. 149, pp. 13-20.
NATO ASI Series, Plenum, New York.

Trebst A. (1986)
Z. Naturforsch. 41c, 240-245.

van Dorssen R.J., Breton J., Plijter J.J., Satoh K.
van Gorkom H.J. and Amesz J. (1987)
Biochim. Biophys. Acta 893, 267-274.

van Gorkom H.J., Pulles M.P.J. and Wessels J.S.C. (1975)
Biochim. Biophys. Acta 408, 331-339.

Velthuys B.R. (1987)
in: 'Topics in Photosynthesis , Light Reactions',
(Barber J., ed.), pp. 341-377, Elsevier, Amsterdam.

Verméglio A., Breton J., Barouch Y. and Clayton R.K (1980)
Biochim. Biophys. Acta 593, 299-311.

MECHANISM OF CHARGE SEPARATION IN PHOTOSYNTHETIC REACTION CENTERS: ELECTRIC FIELD EFFECTS ON THE INITIAL ELECTRON TRANSFER KINETICS

Steven G. Boxer,[1] David J. Lockhart,[1]
Christine Kirmaier[2] and Dewey Holten[2]

[1]Department of Chemistry
Stanford University
Stanford, California 94305
[2]Department of Chemistry
Washington University
St. Louis, Missouri 63130

ABSTRACT. The effect of an applied electric field on the kinetics of the initial electron transfer step in *Rb. sphaeroides* reaction centers has been measured in isotropic samples at 77 K. The rate of formation of H_L^- is slowed slightly upon application of an electric field of 10^6 V/cm. This result is consistent with earlier measurements of the electric field effect on the steady-state fluorescence [1,2]. No evidence was found for electron transfer down the M-side at the highest applied field.

1.   ELECTRIC FIELD EFFECTS ON THE INITIAL CHARGE SEPARATION KINETICS

1.1 *Electromodulated Kinetics in Isotropic Samples*

The intensity of fluorescence from an isotropic sample of photosynthetic reaction centers (RCs) is enhanced upon application of an electric field [1]. The change in fluorescence was found to be quadratic in the applied field, and the fluorescence in the field was found to become polarized [2]. These results were explained using the reaction scheme shown in Figure 1: for an isotropic sample there is a net decrease in the rate of the forward electron transfer reaction (rate

39

constant k_{et}) leading to a net increase in the competing fluorescence from ^1P. The issues we wish to address by a direct measurement of the field effect on the kinetics are the physical origin of the net decrease in electron transfer rate and the magnitude of the effect.

An electric field can affect the rate of an electron transfer reaction in a number of ways. The field will alter the free energy change for electron transfer, ΔG_{et}, because the energy of the dipolar product state $P^{\dagger}H_L^{\cdot-}$ [$\mu(P^{\dagger}H_L^{\cdot-})$ ~ 80 D] is sensitive to the field; the field may also affect the reorganization energy and the electronic coupling between ^1P and $P^{\dagger}H_L^{\cdot-}$ (P is the primary electron donor or special pair, H_L is the bacteriopheophytin on the L-side). In the original paper [1] we considered primarily effects on the free energy. Subsequently our group [2] and Bixon and Jortner [3] considered possible additional effects of the electric field on the electronic coupling if dipolar states such as $P^{\dagger}B_L^{\cdot-}$ mediate the interaction between ^1P and $P^{\dagger}H_L^{\cdot-}$. Bixon and Jortner have presented a detailed and stimulating analysis of the fluorescence results for both isotropic and oriented samples [3]. They explicitly considered the effect of the field on the energy of the $P^{\dagger}B_L^{\cdot-}$ state, leading to a change in the difference in energy between the mediating $P^{\dagger}B_L^{\cdot-}$ state and ^1P at the nuclear configuration of the crossing point of the potential energy surfaces of ^1P and the final product state $P^{\dagger}H_L^{\cdot-}$, as well as the field effect on the Franck-Condon term (the ΔG_{et} term). Their calculated value of $|\Delta F/F|$ is considerably larger than what was observed [1,2].

Two issues may complicate the analysis of the experimentally determined value of the change in fluorescence in the field, $|\Delta F/F|$. (1) The fluorescence in zero field is weak, and there could be some fluorescence at a wavelength appropriate for the special pair but from RCs with degraded function; consequently the zero-field value of the fluorescence could be too large and the experimentally determined $|\Delta F/F|$ would be lower than the true value. However, the values of $|\Delta F/F|$ obtained using RCs from many different preparations and from different laboratories differ by less than 20% of the originally reported value. (2) Although the externally applied electric field, F_{ext}, is known quite accurately, the internal field, F_{int}, is less well understood. The local field correction f accounts for this difference: $F_{int} = f \cdot F_{ext}$, where f is typically *greater* than unity (we also assume it is a scaler). Different models of the local field correction have been discussed [4,5], and it is reasonable to expect that $f \sim 1.2$-1.5. If the local field correction is on this order, then the descrepancy between the value predicted using the model of Bixon and Jortner [3] and the experimentally obtained value of $|\Delta F/F|$ [1,2] is more than one order of magnitude. As indicated above, the absolute magnitude of $|\Delta F/F|$ is subject to some uncertainty. However, the field dependence of $|\Delta F/F|$ is independent of most of these problems. At 77 K, for Q_A-containing *Rb. sphaeroides* RCs, $|\Delta F/F|$ is found to increase nearly quadratically with applied field between 1×10^5 V/cm and 1×10^6 V/cm [2]. The model of Bixon and Jortner [3] predicts a super-quadratic dependence on field up to about 1×10^6 V/cm, in Stark contrast to the experimental results [2].

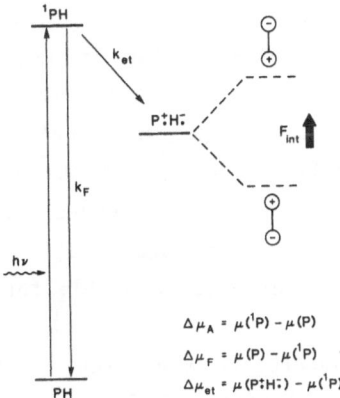

$$\Delta\mu_A = \mu(^1P) - \mu(P)$$
$$\Delta\mu_F = \mu(P) - \mu(^1P)$$
$$\Delta\mu_{et} = \mu(P^+_tH^-_t) - \mu(^1P)$$

Figure 1. Definition of the dipole moments of states, difference dipole moments, and rate constants. The solid lines are schematic energy levels in zero electric field. For an isotropic, immobilized sample in an applied electric field, the energy of the $P^+_tH^-_t$ state increases or decreases depending on the orientation of the $P^+_tH^-_t$ dipole relative to the field, F_{int}. The largest changes occur for dipoles oriented either parallel or antiparallel to the field and are illustrated with the dotted lines. The ground and excited states of P have nonzero permanent dipole moments; however, these are assumed to be much smaller than that of $P^+_tH^-_t$ and are not shown for simplicity.

In order to further our understanding of the mechanism of the initial electron transfer reaction we have undertaken preliminary measurements of the effects of an electric field on the *kinetics* of the initial charge separation step. Our basic interpretation of the fluorescence electric field effect illustrated in Figure 1 predicts a direct relationship between the steady-state fluorescence results and a time-resolved measurement, i.e., the relative change in the integral of the ^1P decay or $P^+_{\cdot}H^-_{\cdot}$ formation curve is predicted to be equal to $|\Delta F/F|$. Because we are working with an isotropic sample of RCs, the zero-field kinetics, which are approximately exponential to within the signal-to-noise [6-8], are expected to become non-exponential as the field is applied. It has been shown elsewhere both theoretically and experimentally (for the $P^+_{\cdot}Q^-_A$ charge recombination reaction) that the difference in the electron transfer kinetics in the field relative to those at zero field can be analyzed to recover the dependence of k_{et} on field [9-11]. Very good signal-to-noise is required for a detailed quantitative analysis of electromodulated kinetic data in an isotropic sample because the fractional changes are small. For example, in a quantitative analysis of electric field effects on the $P^+_{\cdot}Q^-_A$ decay in isotropic samples [11], signal-to-noise in excess of 2000:1 was required (note that the $P^+_{\cdot}Q^-_A$ dipole is about 130 D). Such a high signal-to-noise ratio is not currently possible for measurements on the fs→ps timescale.

Since the transient absorption experiment on the initial electron transfer kinetics in an electric field has not previously been described, it is helpful to illustrate the expected difficulties [see reference 11 for more details]. Figure 2A shows a calculated dependence of k_{et} on ΔG_{et} for the initial charge separation step which is consistent with the fluorescence electric field effect magnitude, angle dependence, field dependence, and lineshape [12]. Note that this dependence is considerably flatter than predicted by standard models. In Figure 2B, the calculated decay at zero field and in the field are compared, along with the difference decay, for a field of 10^6 V/cm. This difference decay was calculated assuming that the field only affects the energy of the product state dipole which is assumed to be 80 D. It is seen that if the observed value of $|\Delta F/F|$ is indicative of the change in k_{et}, then even at this very high applied field, the effect on the kinetics is expected to be quite small, just at the level of detectability. The calculated difference decay reaches a maximum between one and two 1/e times and decreases gradually out to several 1/e times. The relative change in the integral of the decay curve due to the presence of the field is expected to equal $|\Delta F/F|$ if the basic interpretation is correct.

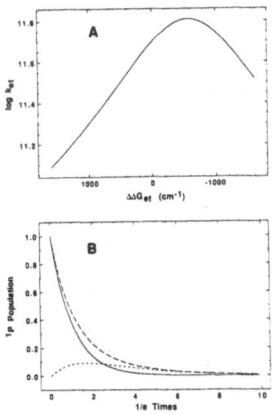

Figure 2. A) A plot of the dependence of $\log(k_{et})$ on ΔG_{et} which is consistent with the fluorescence electric field effect measurements assuming that the field only affects ΔG_{et} and that $\Delta\Delta G_{et}$ is equal to $-\Delta\mu_{et}\cdot F$ (for details of how this curve was obtained, see reference 12). B) Calculated decay curves for an isotropic sample using the dependence of k_{et} on ΔG_{et} given in (A) for zero field (solid line) and for a field of 10^6 V/cm (---). Also shown is the difference between the field on and field off decay curves (---) which shows that the maximum difference between the field on and off curves is expected to be small ($\sim < 10\%$ of the maximum amplitude) if the fluorescence electric field effect measurement of $|\Delta F/F|$ is indicative of the change in k_{et}.

By contrast, in the Bixon and Jortner model [3] the effects of a field on the decay kinetics are predicted to be much larger. This appears to be because in their model the field not only affects ΔG_{et} (treated using a standard vibronic coupling model and hence giving a substantially steeper dependence of k_{et} on ΔG_{et} and larger electric field effect than illustrated in Figure 2), but it also affects the electronic coupling between ^1P and $P^+_{\cdot}H_L^{\cdot-}$ by changing the energy of the mediating $P^+_{\cdot}B_L^{\cdot-}$ state. It is important to realize that with applied fields of 10^6 V/cm (10 meV/Å) the energy of a 50 D dipole such as $P^+_{\cdot}B_L^{\cdot-}$ can be increased or decreased by as much as about 800 cm^{-1} (100 meV) depending on orientation. Bixon and Jortner [3] estimate that the value of $|\Delta F/F|$ for an isotropic sample should be about 3.5 for an internal field of 10^6 V/cm at 77 K ($|\Delta F/F|$ is observed to be about 0.36 in an applied field of 10^6 V/cm at 77 K [2]), or equivalently, the area under the decay curve should increase by a factor of 4.5 upon application of the field. Such a large effect would be easily detected under the conditions used in the current experiment. Furthermore, as discussed above, the internal field is expected to be *greater* than the applied field, and we use *external* fields on the order of 10^6 V/cm. Consequently, the factor of 4.5 is a lower limit for an external field of 10^6 V/cm. Given the observed quadratic dependence of the fluorescence effect on field and assuming that f is between 1.2 and 1.5, the Bixon and Jortner model predicts an increase by a factor of 6 to 9 in the area under the ^1P decay or $P^+_{\cdot}H_L^{\cdot-}$ formation curve.

1.2 *Experimental Electromodulated Kinetics*

The electric field effect on the initial kinetics experiment was performed at 77 K on Q_A-containing *Rb. sphaeroides* RCs in PVA matrices; experimental details and analysis are presented elsewhere [13]. $|\Delta F/F|$ was measured on the sample (mounted in the dewar and cooled) just prior to the transient absorption measurements. Figure 3 shows one set of data that was obtained at an applied field of 10^6 V/cm (10 meV/Å) in the region of the Q_x electronic absorption bands of H_L (543 nm) and H_M (531 nm). At long times, the absorption change in the field and in the absence of the field are comparable, indicating that the field is not causing a significant ($\sim \leq 15\%$) reduction in the quantum yield of the initial charge separation step (note that for a reasonable ^1P radiative lifetime on the order of ns, the observed 36% increase in the fluorescence yield in a field of 10^6 V/cm corresponds to a change in the quantum yield of $P^+_{\cdot}H_L^{\cdot -}$ of less than 0.1% which would be undetectable by transient absorption; also, the Stark effect on the absorption of the excitation and probe beams is much too small to account for the differences in Fig. 3 [14]). The observation that the amplitude of the 543 nm bleach is consistently smaller in the presence of the field until several 1/e times (the zero field 1/e time is about 2 ps) indicates that the net rate of decay of ^1P and formation of $P^+_{\cdot}H_L^{\cdot -}$ is slightly slower with the field on than with the field off, consistent with the observation that the fluorescence intensity increases in the field.

The key result from the data in Figures 3 and 4 (see below) is that the electric field effect on the initial reaction is small, but not zero. Since it will be very difficult to obtain sufficient signal-to-noise for an isotropic sample to evaluate models for the field effect on the decay kinetics in great detail, it is useful to compare the integrated magnitude of the effect on the decay kinetics with $|\Delta F/F|$ measured on the same sample at the same field: the results agree to within a factor of two, indicating that the observed value of $|\Delta F/F|$ (equal to 0.36 in a field of 10^6 V/cm at 77 K) is not badly contaminated with fluorescence from damaged RCs. Furthermore, since comparable effects have been observed in steady-state fluorescence measurements and transient absorption measurements on the ps timescale, we can conclude that the observed fluorescence increase in the electric field at 77 K is not primarily from delayed fluorescence. We are therefore led to conclude that the disagreement between the experimental results and those calculated by Bixon and Jortner [3] reflects an oversimplification in the model used to derive the theoretical result, either in its basic formulation or in the choice of the values of various physical parameters, or in the local field correction. As mentioned earlier, if the local field correction is greater than unity as is likely (though not proven), then the disagreement is even worse, and we are forced to consider alternative models for the initial electron transfer step (see Section 1.3).

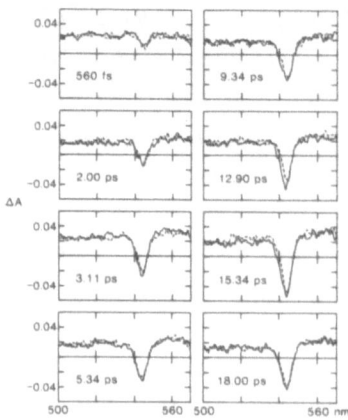

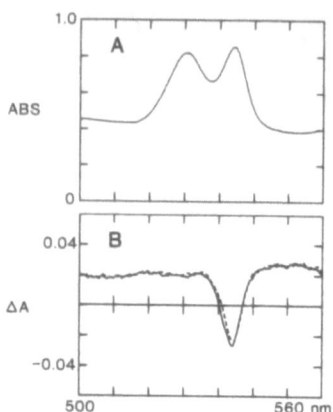

Figure 3.   Transient absorption spectra of *Rb. sphaeroides* RCs at 77 K in the region of the Q_x bands of H_L and H_M between zero and 18 ps (~ 0–9 1/e times) in zero field (solid lines) and in a field of 10^6 V/cm (---).

Figure 4.   (A) The absorption spectrum of *Rb. sphaeroides* RCs in PVA at 77 K.   B) The average of the field on (solid lines) and field off (---) transient absorption spectra obtained between zero and 18 ps (the spectra used in the average were not separated from each other by a constant time interval, i.e. the number of spectra per ps used in the average is not a constant).   The amplitude of the bleaching centered near 543 nm is smaller in the presence of the field, indicative of a net slowing of the $^1P \rightarrow P^+H_L^-$ electron transfer reaction which is consistent with the observed increase in the intensity of the competing fluorescence from ^1P.   There is no evidence for the formation of H_M^- as would be indicated by an additional bleaching at 531 nm.   The origin of the apparent slight shift and narrowing of the bleaching in the presence of the field in not yet known, but it is too large to be accounted for by the usual Stark effect on the absorption in this region [14].

1.3 *Can we make the electron go down the M-side in an electric field?*

We have also addressed the question of whether the electron will go down the wrong M-side in a very large electric field. Examination of the x-ray structure coordinates [15] shows that the direction of the $P^+B_L^-$ dipole moment is nearly antiparallel to the $P^+B_M^-$ dipole moment (the angle between them is estimated to be ~155°). This fortuitous situation can be exploited since for those RCs oriented in the field such that the $P^+B_L^-$ state energy is most increased by the field (resulting in the least favorable situation for L-side electron transfer in the P^+B^--mediated model) the $P^+B_M^-$ state energy is most decreased (approximately), and vice versa. It is important again to note the magnitude of the effect: assuming a 50 D dipole for both $P^+B_L^-$ and $P^+B_M^-$ based on the x-ray structure, for a field of 10^6 V/cm, the energy *spread* between the two hypothetical mediating states can be changed by more than 1600 cm^{-1} (200 meV), a significant amount on the scale of energy differences believed to be relevant to this problem [3]. Note that working with isotropic samples is the key to this experiment: the L- and M-sides are not differentially affected in samples oriented such that the local C_2 axis is roughly parallel to the applied electric field direction (e.g. lipid bilayers [16] or Langmuir-Blodgett films [17]). The origin of unidirectional electron transfer is not well understood, and likely involves a combination of structural, electronic and energetic factors. In this experiment we can differentially perturb the two potential electron transport chains by affecting the energies of the P^+B^- states which are of central importance in most superexchange models for the initial reaction.

As shown in Figures 3 and 4, we find no evidence within our signal-to-noise for electron transfer to H_M, as would be indicated by a bleaching at 531 nm, for an isotropic sample at an applied field of 10^6 V/cm at 77 K. There are several possible interpretations for this interesting negative result, among them the following. (i) The $P^+B_M^-$ state energy is so much higher than the $P^+B_L^-$ state at zero field that even for the most favorable orientations its energy is not low enough to be relevant. (ii) The distances along the L-side are so much closer than along the M-side that $P^+B_M^-$-mediated electron transfer down the

M-side will always have a small electronic coupling. (iii) The electronic asymmetry of ^{1}P state is so large that the energy of the $P^+_\cdot B^-_M\cdot$ state is essentially irrelevant [12]. (iv) States of the type $P^+_\cdot B^-_\cdot$ are not the key to the superexchange interaction between ^{1}P and $P^+_\cdot H^-_L\cdot$.

1.4 *Possible elaborations of the superexchange mechanism*

We have three admittedly fragmentary pieces of evidence suggesting that there is a problem with the superexchange mechanism as specifically formulated to interpret the electric field effect experiments [3] (several other groups have also discussed superexchange involving $P^+_\cdot B^-_\cdot$ [18–22], but have not considered electric field effects in detail). (i) the electric field dependence of $|\Delta F/F|$ predicted using a model in which *both* the electronic coupling and ΔG_{et} are significantly affected by the field [3] is very different from that observed experimentally; (ii) the observed value of $|\Delta F/F|$ is much less than predicted in the quantitative model presented by Bixon and Jortner [3], and this is confirmed, at least at a qualitative level, by the direct measurement of the effect of an electric field on the initial charge separation step presented in Section 1.1; (iii) there is no evidence that electron transfer can be forced down the M-side in an isotropic sample in a very large electric field.

We would like to consider several possible alternatives. (i) The local field correction is grossly different from what is expected. (ii) The treatment using the superexchange model in reference [3] considers the effect of the applied electric field on both the electronic coupling between ^{1}P and $P^+_\cdot H^-_L\cdot$ (via the energy of the $P^+_\cdot B^-_L\cdot$ mediating state, Eqn. III.4 of reference [3]) and on ΔG_{et} (via the energy of the $P^+_\cdot H^-_L\cdot$ product state). However, an additional effect of the field must be considered. A vertical displacement of the $P^+_\cdot H^-_L\cdot$ energy surface affects not only ΔG_{et} but also the electronic coupling in a superexchange model by changing the vertical difference in energy between the mediating $P^+_\cdot B^-_L\cdot$ state and ^{1}P by changing the nuclear configuration at which the reactant and product state energy surfaces cross (i.e. the nuclear configuration at which the relevant vertical energy differences are calculated is dependent on field). The quantitative consequences of this are discussed elsewhere [13], but this effect can significantly *reduce* the effect of the field on the electronic coupling. (iii) The state which mediates the interaction between ^1P and $P^+_{.}H_L^-$ in the superexchange model is not $P^+_{.}B_L^-$ but is rather a neutral excited state, for example, the singlet excited state of B_L. In this picture the initial electron transfer step is formulated as a virtual energy transfer process. The ^1B_L state appears to be quite non-polar: the absorption Stark effect on this transition in RCs [4,5] is quite small, being comparable with a monomeric bacteriochlorophyll in a simple polymer [14]. In this model the field affects ΔG_{et} as well as the electronic coupling due to the dependence of the energy of the $P^+_{.}H_L^-$ product state on field (as in (ii) above). In some cases, the absence of a field effect on the energy of the mediating state in such a model actually makes the effect on the electronic coupling larger than when the mediating state is highly dipolar, but in a way that tends to mitigate the change in k_{et} due to the effect on ΔG_{et} alone [13].

Recent experimental work comparing the rates of electron, hole and triplet energy transfer in a series of structurally related synthetic compounds suggests that the electronic coupling is comparable for electron and hole transfer [23] and the square root of this value for triplet energy transfer [24]. It is evident, however, that the energies, occupancies and to some extent shapes of the orbitals which mediate the interaction between the donor and acceptor on the intervening molecular bridge are not identical in these cases, and that the spatial extents of the wavefunctions in the anions, cations, and neutral triplet are not the same. This encourages us to suggest that the electronic coupling calculated for P^1B^*-mediated superexchange may not be too small to achieve rapid electron transfer (the actual value relative to that for $P^+_{.}B_L^-$-mediated superexchange will depend on the details of the electron densities in the states ^1P, $P^+_{.}B_L^-$, P^1B^* and $P^+_{.}H_L^-$). Whereas the absolute energies of states such as $P^+_{.}B^-$ are not certain, we know from the absorption spectrum that the state P^1B^* is not far in energy above ^1P. Consequently the formal aspects of the analyses presented earlier [3,18-22] may well survive in this new mechanism with modified values for the electronic coupling. However, the physical nature of the relevant states and consequently the sensitivity to external or internal electric field perturbations will be quite different.

One of the attractive features of the $P^+_\cdot B^-_\cdot$-mediated proposal (among many) is that unidirectional electron transfer could be partly explained by differential stabilization or destabilization of $P^+_\cdot B_L^-$ and $P^+_\cdot B_M^-$ by the local electrostatic field in the RC protein [19,25]. In such a model the internal electric field is playing a role analogous to the external field in the experiments described in Section 1.3. If the $P^1 B^*$-mediated model proposed here is correct, this electrostatic mechanism for producing unidirectionality is less important. We are left then with structural asymmetry (different distances among the reactive components on the M- and L-sides) and electronic asymmetry in the excited state of the special pair as the likely causes of the symmetry breaking. The latter is likely the result of environmental differences in the vicinity of the special pair on the M- and L-sides.

Acknowledgements: We acknowledge very useful discussions with Professors Bixon, Jortner and Ulstrup. This work was supported in part by grants from the National Science Foundation to both S.G.B. and D.H.. S.G.B. is the recipient of a Presidential Young Investigator Award.

2.   REFERENCES

1.   Lockhart, D.J. and Boxer, S.G., Chem. Phys. Lett. (1988) 144,243-50.

2.   Lockhart, D.J., Goldstein, R.F. and Boxer, S.G., J. Chem. Phys. (1988) 89, 1408-1415.

3.   Bixon, M. and Jortner, J., J. Phys. Chem. (1988) 92, 7148-7156.

4.   Lockhart, D.J. and Boxer, S.G., Biochem. (1987) 26, 664-668.

5.   Lösche, M., Feher, G. and Okamura, M.Y., Proc. Natl. Acad. Sci. (1987) 84, 7537-7541.

6.   Woodbury, N.W., Becker, M., Middendorf, D., Parson, W.W., Biochem. (1985) 24, 7516-521.

7.   Breton, J., Martin, J.L., Fleming G.R., Lambry, J.C., Biochem. (1988) 27, 8276-8284

8.   Kirmaier, C., and Holten, D., FEBS Lett. (1988) 239, 211-218.

9.   Boxer, S.G., Goldstein, R.A. and Franzen, S., Photoinduced Electron

Transfer; Fox, M. A., Chanon, M., Eds.; Elsevier Press., Vol. B., p. 163-215, 1988.

10. Boxer, S.G., Lockhart, D.J., Franzen, S., In Photochemical Energy Conversion (J.R. Norris, Jr., and D. Meisel, Eds) Elsevier Press, p. 196-210, 1989.

11. Franzen, S., Goldstein, R.F. and Boxer, S.G., J. Phys. Chem., submitted.

12. Lockhart, D.J. and Boxer, S.G. (1989) submitted.

13. Lockhart, D.J., Kirmaier, C., Holten, D. and Boxer, S.G., (1989) submitted.

14. Lockhart, D.J., Boxer, S.G., Proc. Natl. Acad.Sci.(1988) 85, 107-11.

15. Deisenhofer, J., Epp. O., Miki, K., Huber, R. and Michel, H., J. Mol. Biol. (1984) 180, 385-398.

16. Gopher, A., Blatt, Y. Schönfeld, M., Okamura, M.Y., Feher, G., and Montal, M., Biophys. J. (1985) 48, 311-320.

17. Popovic, Z.D., Kovas, G.J., Vincett, P.S., Alegria, G. and Dutton, P.L., Biochim. Biophys. Acta (1986) 851, 38-48.

18. Scherer, P.O.J., Fischer, S.F., Chem Phys. Lett. (1986) 131, 153-9.

19. Plato, M., Mobius, K., Michel-Beyerle, M.E., Bixon, M., Jortner, J., J. Am. Chem. Soc. (1988) 110, 7279-7285.

20. Warshel, A., Creighton, S., Parson, W.W., J. Phys. Chem. (1988) 92, 2696-2701.

21. Marcus, R.A., Chem. Phys. Lett. (1988) 146, 13-21.

22. Won, Y., Friesner, R.A., Biochim. Biophys. Acta (1988) 935, 9-18.

23. Johnson, M.D., Miller, J., Green, S., Closs, G.L., J. Phys. Chem. (1989) 93, 1173-1176.

24. Closs, G.L., Piotrowiak, P., MacInnis, J.M., Flemming, G.R., J. Am. Chem. Soc. (1988) 110, 2652-2653.

25. Creighton, S., Hwang, J.-K., Warshel, A., Parson, W.W., Norris, J., Biochemistry (1988) 27, 774-781.

ELECTROSTATIC ANALYSIS OF THE MIDPOINTS OF THE FOUR HEMES IN THE BOUND CYTOCHROME OF THE REACTION CENTER OF RP. VIRIDIS

M.R. GUNNER AND BARRY HONIG
Dept. of Biochemistry and Molecular Biophysics.
Columbia University, New York, N.Y. 10027 U.S.A.

ABSTRACT. The electrostatic potential throughout the Rp. viridis reaction center protein (RC) was obtained by solving the Poisson-Boltzmann equation with a finite difference algorithm. The energetic cost of oxidizing each of the 4 hemes of the bound cytochrome subunit was examined. The calculations provide quite good agreement with the experimentally determined relative heme midpoints, thus accounting for the pattern of two low and two high potential cytochromes. The success of the method allows the factors that generate the differences the in situ electrochemistry of the hemes to be identified. In order of importance these appear to be: (1) the presence of acidic and basic residues on the surface of the cytochrome subunit; (2) differences in the stabilization of the charge on the heme by solvent; (3) the nature of the axial ligands; (4) interheme interactions.

Introduction

Electron transfer proteins make use of a small number of redox cofactors to carry out a wide range of functions. Thus, the electrochemistry of molecules such as hemes, chlorophylls, flavins, and quinones are modified by their association with protein. This provides a flexible range of midpoints, allowing fine control of the reaction ΔGs in these proteins which are central to the cell's energy metabolism. In addition, the free energy of electron transfer affects the rates of reaction (Jortner (1980), Marcus & Sutin (1985), Miller et al. (1984), Gunner et al. (1988), Gunner & Dutton (1989). Thus, modulation of the reaction ΔG can control the yields of competing reactions as well as the final equilibria (Gunner & Dutton (1989). It is therefore of great interest to understand how proteins influence the chemistry of specific redox sites.

This is an especially interesting problem in the reaction center protein of photosynthetic bacteria (RC) which contains 2 bacteriochlorophylls (B_M and B_L), 2 bacteriopheophytins (H_M and H_L), and 2 quinones (Q_A and Q_B) with differences in electrochemistry and reactivity that clearly depend solely on the site in the protein to which they are bound. Another striking example of protein induced effects is provided by the four identical c type hemes in the cytochrome subunit, a feature of several reaction center proteins. In the Rp. viridis RC, interactions of the hemes with the protein yield a range of midpoints spanning >400 meV. Analysis of the in situ midpoints within this cytochrome provides an ideal test case for theoretical calculations of site electrochemistry. Success in reproducing the experimental data may justify the use of these methods for predicting the midpoints of L-branch bacteriochlorophyll monomer (B_L), the M-branch monomer (B_M) and bacterio-pheophytin (H_M) which cannot be measured directly. These values represent important

J. Jortner and B. Pullman (eds.), Perspectives in Photosynthesis, 53–60.

unknowns in the understanding of RC function. For example assumptions about the relative midpoint potential of B_M and B_L are implicit in most theoretical discussions of the observed asymmetry of electron transfer (Bixon & Jortner (1988), Moser et al. (1988))

There are several methods allowing theoretical analysis of cofactor midpoints within a protein (Warwiker & Watson, H.C. (1982), Gilson et al. (1985), Gilson & Honig (1988a,b), Warshel & Russell (1984), Churg & Warshel (1986)). These evaluate the cost of adding or removing electrons from the redox sites given the shape of the protein, particularly in relation to solvent, and the distribution of charges on the sites. These theoretical calculations can provide an analysis of the relative contributions of different interactions between the protein and redox sites as well as an estimate of cofactor behavior at specific sites.

Methods

Values of the electrostatic potential inside and around the RC were calculated with the DelPhi program which solves the Poisson-Boltzmann equation using a finite difference algorithm (Klapper et al. (1986), Gilson et al (1987), Nicholls & Honig (1989)). The starting point for the analysis reported here is the x-ray crystal structure of the RC from Rp. viridis (Brookhaven Data Bank Structure 1PRC) (Deisenhofer et al. (1985), Michel et al. (1986)). A complete set of protons were added to the structure with the xplor molecular modeling program (Brunger et al. (1987)). The positions of the heavy atoms in the crystal structure were unchanged. In the DelPhi program, each atom is assigned a radius (e.g. Brooks, et al. (1983)). These values are mapped onto a 65^3 grid so that each grid point is identified as being located inside, outside or at the surface of the protein. A standard probe radius of 1.4 Å defines the water accessible regions. A dieletric constant of 80 is assigned outside the protein, while the interior points are assigned a value of 4 . This choice of internal dieletric implicitly accounts for the energy of electronic polarization as well as local movements in the protein upon introduction of a charge (Gilson & Honig (1986)). The assigned partial atomic charge on each atom is also distributed to the neighboring grid points. Given this distribution of charge and dieletric, DelPhi calculates the potential throughout the grid.

The mathematical simplification provided by use of a grid allows rapid solution of the problem. Current versions of DelPhi provide well converged potential distributions for the RC in less than a minute of computer time on a Convex C1. Two minutes are needed to map the 20,000 atoms of the protein onto the grid. The numerical method need not cause significant loss of microscopic information about the protein, provided the grid spacing is comparable to atomic spacing. For a large protein this is accomplished by calculating a series of potential maps of increasingly fine scale that are focused on a point of interest and by averaging the results of computations that differ only in the position of the protein relative to the grid (Gilson et al. 1987)).

The work reported here represents the results of a series of calculations involving successive 'focussing' where in each calculation the cofactor of interest lies near the center of the grid. In the first of a series the entire protein lies well within the boundaries of the grid. In successive runs the protein is expanded so that it extends beyond the grid dimensions. The potentials at the new grid boundaries are fixed by the potential at the these locations calculated in a previous, coarser grid run where they were well inside. While no precise information can be gained about positions outside of the grid region, an estimate of the influence of the charges and protein shape in these regions is contained in the assigned boundary values. In the first calculation the RC fills 30% of the total grid, providing a spacing of 0.15 grids/Å. In successive runs of DelPhi the protein is expanded in relation to the grid until only a third lies inside. The final resolution is 1.5 grids/Å. The reported results are the average of 4 focussed runs each starting with the protein in a different position relative to the grid.

The partial charges on the individual atoms of the protein were taken from the Charmm data set (Brooks et al. (1983)). This provides charges on the protein backbone (amide N and H, carbonyl C and O and alpha C), as well as at side chain atoms with the exception of non-polar carbons and hydrogens. All histidines are assumed to have zero net charge. The axial ligands to the hemes are either histidine nitrogens or the sulfur of methionine. In the Charmm charge set the partial charge on an unprotonated histidine nitrogen is -0.4 while a methionine sulfur carries a charge of -0.12. It should be noted that 3 alanines and the C terminal lysine are missing from the coordinates of the cytochrome subunit. The apparent terminal carbon and oxygen on lysine c332, are charged as an intrachain carboxyl. As the missing segment contains a lysine, normally assigned a formal positive charge, and a negatively charged carboxy terminal, the omission does not effect the total subunit charge. In addition, the electron distribution on the cofactors will influence the potential felt by neighboring redox sites. No charges were added to neutral cofactors. The charge on an oxidized heme was given by +0.25 on each of the nitrogens.

The value for the potential throughout the protein, in the absence of charge on the hemes, provides the contribution of charges within the protein to the energy of oxidation/reduction using:

$$\Delta G = \frac{1}{2} \sum_{\text{all atoms}} \Phi(i) \, \Delta q(i) \tag{1}$$

Where Δq is the change in partial charge at each atom on oxidation/reduction and $\Phi(i)$ is the potential at that atom. In this preliminary analysis, the oxidation of the cytochrome hemes is assumed to cause a $\Delta q = +0.25$ on each of the heme nitrogens with no change elsewhere. This distribution was suggested by Warshel for the treatment of cytochrome solvation energy (Churg & Warshel (1986)).

The contribution of specific parts of the protein to the potential at a given site can be evaluated given $\Phi(i)$ calculated when only the sites of interest are charged. The influence of the acidic and basic residues was evaluated from $\Phi(i)$ at the heme nitrogens when a charge of -0.5 was placed on each carboxylate oxygen of aspartic and glutamic acids, +1 on the ε-amino of all lysines, and +0.5 on each arginine terminal nitrogen. The influence of the +1 charge on the low potential hemes on the midpoints of the neighboring high potential hemes was included by calculating $\Phi(i)$ at the nitrogens of sites 1 and 3 when there was a +0.25 charge on the nitrogens of hemes 2 and 4 (heme number given by position relative to P).

The ΔG of placing charges at particular locations in a protein is not solely determined by interactions with other charges in the protein. The response of the medium surrounding the protein can also make a significant contribution to the effective ΔG of electron transfer (Krassner (1977), Marcus and Sutin (1985), (Churg & Warshel (1986))). This will be termed the solvation energy. It is a result of the reorganization of the solvent in response to the change in redox state which lowers the total energy of the system. This polarization of solvent can occur in response to charges buried within the protein as well as those on the surface because of the relatively long effective range of electrostatic interactions. In these calculations, the solvation energy for each heme is defined as the difference in the energy of the system with that heme charged when the protein is surrounded by a dieletric of 80 and that found when the solvent as well as the protein has a dielectric of 4 (Gilson & Honig (1988b)).

Results

The goal of this work is to understand the differences in redox potential when a c type heme is bound at different sites in the reaction center protein of Rp. viridis. Figure 1 shows the measured redox potentials for each of the sites as recently determined optically (Drachavia et al. (1988), Alegria & Dutton (1988)) and by EPR (Nitschke & Rutherford (1989)). There is remarkable agreement on both the heme assignments and midpoints among the three groups.

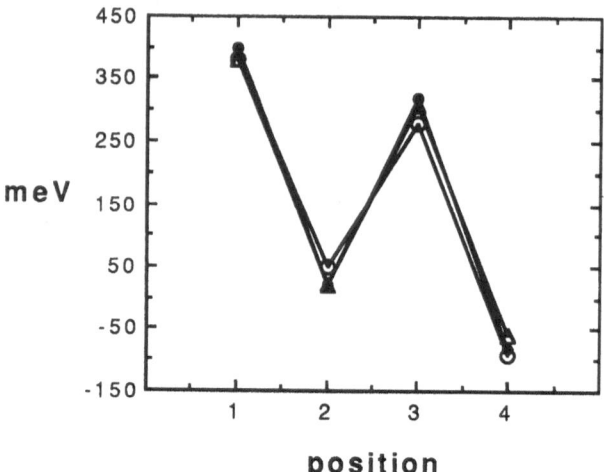

Figure 1. Midpoint potentials for the four hemes of the tightly bound cytochrome c subunit of the RC from Rp. viridis as determined by Alegria (1988) (o) , Rutherford (1989) (•), and Drachevia et al. (1988) (Δ).

Figure 2 compares the calculated midpoints with the average values measured for the 4 cytochromes. The general features of the measurements are well reproduced by the calculations. The relative midpoints are in the same order as in the experiments with the distinction between low and high potential hemes clearly seen. The relative numerical values also agree quite well with those measured. The calculated midpoint span is 100 meV greater than observed (565 meV vs. 465 meV), a difference of only 20%.

Figure 3 provides an analysis of several factors that contribute to the variations in the calculated heme midpoints:

1) The distribution of acidic, aspartic and glutamic acid, and basic, lysine and arginine, residues around the protein creates a range of ΔG_{ox} at the different sites of ≈300 meV. This appears to be an important determinate of the in situ midpoint of the lowest and highest potential hemes.

2) The variation in the solvation energy of the different hemes, representing the stabilization of the charge on the heme by the surrounding water, is also significant,

spanning ≈150 meV. Heme 4 (heme numbering is given by the position relative to P), seen to be the most exposed in the crystal structure has the largest solvation energy while heme 1, closet to P, the least.

3) The relative midpoint of heme 2 appears to be strongly influenced by its having two axial histidine ligands which lowers its midpoint due to the replacement of a methionyl sulfur axial ligand with a histidine nitrogen, bearing a greater negative charge.

4) Positive charges on adjacent hemes contribute to the midpoint of all but the first heme oxidized. Thus, the effect of oxidation of the low potential hemes, is to raise the midpoint of the high potential hemes by 80 to 100 meV. The midpoint of the non-adjacent hemes is increased by less than 15 meV.

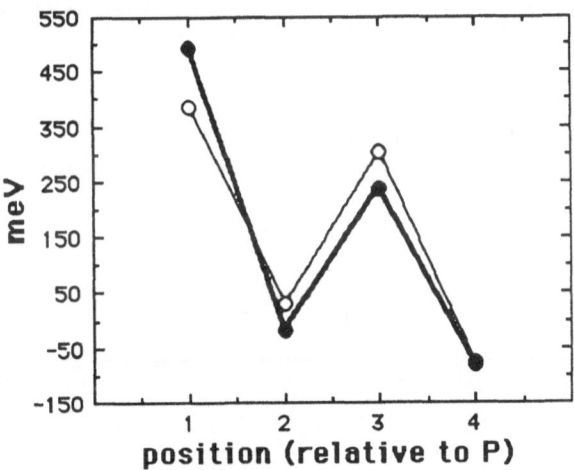

Figure 2. Comparison of the theoretical (•) and measured (o) midpoints of the 4 hemes. The values shown in figure 1 were averaged for the experimental results. The calculated numbers were obtained using equation 1, where $\Delta q=+0.25$ for each of the 4 heme nitrogens on heme oxidation and zero elsewhere. $\Phi(i)$ was obtained with the program DelPhi. The factors included in the analysis were: the charges on the protein, the energy of solvating the +1 charge on the heme, and the effect on the high potential hemes of a +1 charge on the low potential hemes. A constant of 343 meV was added to the calculated values. It is only possible to compare relative values for theory and experiment as the reference states are different for the two sets of numbers.

The energy of interaction between charges on the different hemes can be converted into an effective dieletric constant as defined by:

$$\varepsilon_{effective} = \frac{\Phi(i, \ \varepsilon=1)_{coulombic}}{\Phi(i, \varepsilon_{in}=4, \ \varepsilon_{out}=80)_{P\text{-}B}} \tag{2}$$

where

$$\Phi(i,\varepsilon=1)_{coulombic} = \frac{14\ q(i)}{r} \qquad \qquad (3)$$

where r is the distance in angstroms and $\Phi(i)$ is in meV/electron. The Poisson-Boltzmann calculations provide $\Phi(i)_{P-B}$ at a heme Fe in the presence of a +1 charge on another Fe given a uniform internal protein dieletric of 4 with an external value of 80. It was found that the effective dieletric was between 13 to 19 for the interaction of adjacent hemes (see table 1). However, $\varepsilon_{effective}$ increases as the distance between the sites increases. This seemingly paradoxical behavior does not require an inhomogenous interior dieletric for the protein. Rather, it is caused by the high dieletric solvent surrounding the protein. As the distance between charges in a low dieletric environment adjacent to a higher dieletric region gets longer the apparent value of ε between these charges approaches that of the external dieletric (Jackson (1975)).

Table 1. The effective dieletric constant for the interaction of the cytochrome hemes calculated with equation 3.

	heme 2	target site heme 3	heme 4
source	ε	ε	ε
heme 1	13 (14Å)	41 (28Å)	82 (41Å)
heme 2		19 (15Å)	58 (40Å)
heme 3			17 (14Å)

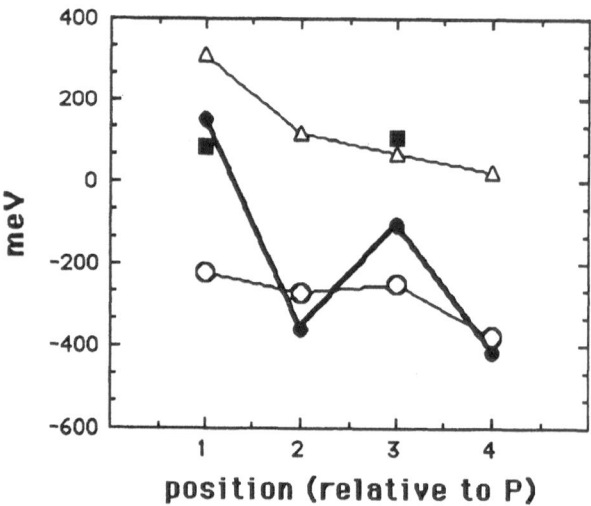

Figure 3. The influence on the theoretical midpoint of the hemes of acidic and basic residues (Δ), the energy for solvating the +1 charge on the oxidized heme (O), and the effect on the high potential hemes of the +1 charge on the low potential hemes (■). The total calculated value (•), as in figure 2 is shown for comparison.

Discussion

The ability of these calculations to qualitatively and quantitatively reproduce the experimental midpoints of the 4 hemes of the cytochrome subunit of the RC suggests that the assumptions made in the analysis are appropriate. One important contribution to the energetics of electron transfer that was treated in a very simplified way is the reorganization of the protein and cofactors on electron transfer. The use of a uniform dieletric of 4 implicitly accounts for an averaged electronic polarization and protein rearrangement upon reaction (Gilson & Honig (1986)). However, microscopic information about the location of these motions is lost. The success of the analysis suggests that there are no major rearrangements of the protein on reaction. This is not surprising, especially for the low potential hemes which continue to reduce P+ at cryogenic temperatures.

A second set of simplifying assumptions were made in the treatment of the solvent surrounding the RC. It appears that the essential features of water that contribute to both the solvation of charge as well as the energy of charge-charge interactions near the protein interface can be modeled effectively by its bulk dieletric constant (Jayaram et al. (1989), Gilson & Honig (1988a,b)). In addition, it is assumed that water surrounds the whole protein. This is likely to be reasonable for the exposed cytochrome subunit. However, this would be inappropriate for the rest of the reaction center which is surrounded either naturally by membrane or in functional, isolated preparations by detergent.

The work reported here reproduces the relative midpoints of the 4 hemes of the Rp. viridis cytochrome c subunit with an analysis that considers only the electrostatic potential of the system. It is found that the in situ electrochemistry of each heme arises from a sum of electrostatic contributions from several sources rather than from a single factor. A clear example can be seen in the relative midpoints of the two low potential cytochromes. The use of two histidines as axial ligands of heme 2 tends to lower its midpoint relative to the other hemes which are liganded by a methionine and a histidine. Despite this, it is heme 4 that has the lowest measured and calculated potential, apparently because the distribution of charged amino acids as well the relative solvation energy favors its oxidation . The distribution of charged acidic and basic residues as well as the exposure of the heme to solvent appear to control the total span of midpoints. Both factors significantly raise the midpoint of heme 1 and lower the midpoint of heme 4. The difference in potential at the two sites of intermediate midpoints (heme 2 and heme 3) is apparently controlled by the differences in their axial ligands.

Thus, this work demonstrates that experimentally established midpoints in the RC protein can be calculated by a complete electrostatic analysis of the protein with a limited number of simplifying assumptions. This finding should allow the prediction of the midpoints of cofactors that can not be measured to be attempted with some confidence.

Acknowledgements.

We would like to thank Guillermo Alegria for helpful discussion and Anthony Nicholls for timely modifications of DelPhi. This work was supported by N.S.F. DMB88-03489. M.R.G. is supported by a NIGMS Research Fellowship.

References

Alegria, G. & Dutton, P.L. (1988) Biophys. J. 53, 615a.

Bixon, M. and Jortner (1988) 92, J. Chem. Phys. 7148-7156.

Brooks, B.R., Bruccoleri, R.E., Olafson, B.D., States, D.J., Swaminathan, S., and Karplus, M. (1983) J.Comput. Chem 4, 187.

Brunger, A.T., Kuriyan, J. and Karplus, M. (1987) Science 235, 458-460.

Churg, A.K. and Warshel, A (1986) Biochemistry 25, 1675-1681.

Deisenhofer, J., Epp, O., Miki, K., Huber, R., and Michel, H. (1988) Nature 318, 618-624.

Dracheva, S. M., Drachev, L.A., Konstantinov, A.A., Semenov, A.Y. Skulachev, V.P., Arutjunjan, A.M. Shulavlov, V.A., and Zaberezhnaya, S.M. (1988) Eur. J. Biochem. 171, 253-264.

Gilson, M.K. and Honig, B. (1988a) Proteins 3, 32-52.

Gilson, M.K. and Honig, B. (1988b) Proteins 4, 7-18.

Gilson, M.K. and Honig, B. (1986) Biopolymers 25, 2097-2199.

Gilson, M.K., Rashin, A., Fine, R., and Honig, B. (1985) J. Mol. Biol. 183, 503-516.

Gilson,. M.K., Sharp, K.A., and Honig, B. H. (1987) J. Comp. Chem. 9, 327-335.

Gunner, M.R. and Dutton, P.L. (1989) J. Am. Chem. Soc. 111, 3400-3412.

Gunner, M.R., Robertson, D.E. and Dutton, P.L. (1986) J. Phys. Chem. 90, 3783-3795.

Jackson, John D. (1975) Classical Electrodynamics, John Wiley & Sons, New York.

Jayram, B, Fine, R., Sharp, K., and Honig B. J. Phys. Chem. in press.

Jortner, J. (1980), Biochim. Biophys. Acta 811, 265-322.

Kassner, R.J. (1977) J. Am. Chem. Soc. 99, 4351-4355.

Klapper, I., Hagstrom, R., Fine, R., Sharp, K., and Honig, B. (1986) Proteins 1, 47-59.

Marcus, R.A., and Sutin, N. (1985) Biochim. Biophys. Acta 811, 265-322.

Michel, H. Epp, O. and Deisenhofer, J. (1986) The EMBO J. 5, 2445-2451.

Miller, J.R., Beitz, J.V., and Huddleston, R.K. (1984) J.Amer. Chem. Soc. 106, 5057-5068.

Moser, C.A., Alegria, G., Gunner, M.R., and Dutton, P.L. (1989) in J.R. Norris, Jr., and D. Meisel (eds.), Photochemical Energy Conversion, Elsevier, New York, pp. 221-231.

Nicholls, A. and Honig, B. (1989) manuscript in preperation.

Nitschke, W. and Rutherford, A.W. (1989) Biochemistry 28, 3161-3168.

Warshel, A. and Russell, S.T. (1984) Q. Rev. Biophys. 17, 283-422.

Warwicker, J. and Watson, H.C. (1982) J. Mol. Biol. 157, 671-679.

PRIMARY PHOTOCHEMISTRY IN REACTION CENTERS FROM THE HIS^M200→LEU MUTANT OF *RHODOBACTER CAPSULATUS*

C. KIRMAIER*, L.M. McDOWELL*, E.J. BYLINA[†‡],
D.C. YOUVAN[†], and D. HOLTEN*

*Department of Chemistry, Washington University, St. Louis, MO 63130
†Department of Chemistry, Massachusetts Institute of Technology,
Cambridge, MA 02139
‡Current Address: Biotechnology Program, Pacific Biomedical Research Center,
University of Hawaii at Manoa, Honolulu, Hawaii 96822

ABSTRACT. The initial transient state observed immediately following excitation of reaction centers from the His^M200→Leu mutant of *Rb. capsulatus* with a 350-fs flash is the intradimer charge transfer state $[BChl_{LP}{}^+BPh_{MP}{}^-]$. This state has a lifetime of about 14 ps and decays with nearly equal probabilities via electron transfer to BPh_L and via rapid internal conversion (charge recombination) to the ground state. Essentially the same behavior is observed at room and liquid N_2 temperature. These results provide important insights into how the electronic properties of the primary electron donor dictate the initial stage of the charge separation process.

1. Introduction

The charge separation process in bacterial photosynthetic reaction centers (RCs) is initiated by excitation of the primary electron donor (P), a dimer of bacteriochlorophyll (BChl) molecules. The excited dimer (P*) transfers an electron to the bacteriopheophytin associated with the L polypeptide (BPh_L) with an ~3 ps time constant. $BPh_L{}^-$ in turn transfers an electron to the primary quinone (Q_A) with a time constant of ~200 ps. The overall quantum yield of this charge separation process is near unity [1]. The importance of interactions between the chromophores and specific amino acid residues in determining the electronic properties of the cofactors, and the kinetics and yield of charge separation, is a key but relatively unexplored area. With the three dimensional structures of two bacterial RCs now in hand [2-4], powerful site-directed mutagenic and spectroscopic techniques allow us to begin to address the role of the protein in the primary photochemistry. In this article we describe the results of time-resolved studies on an RC having a site-directed mutation in the vicinity of the primary electron donor.

The crystal structures of RCs from *Rhodopseudomonas viridis* and *Rhodobacter sphaeroides* have shown that the central Mg^{2+} ion of the $BChl_{MP}$ component of P is coordinated to a histidine residue from the M polypeptide, His M200 in *Rps. viridis* and His M202 in *Rb. sphaeroides* [2,3]. This interaction is depicted in Fig. 1 for the *Rps. viridis* RC. Changing the analogous His^M200 to leucine in *Rb. capsulatus* yields an RC in which $BChl_{MP}$ is replaced by its Mg-free BPh analog, and which thus contains a BChl/BPh heterodimer (D) [5]. We have also found that the initial stage of the charge separation

61

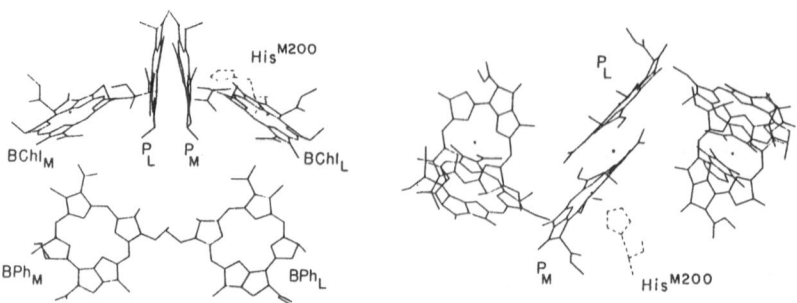

Figure 1. Two views of the co-factors in RCs from *Rps. viridis*, showing histidine residue M200 (dashed) which has been changed to leucine in RCs from *Rb. capsulatus*. The right-hand view looks from P along the C_2 axis toward the Fe atom (dot in center), with $BChl_L$, BPh_L, and Q_A on the right, and $BChl_M$ and BPh_M on the left. (The *Rps. viridis* crystal structure coordinates were provided by J. Deisenhofer.)

process is profoundly different in the $His^{M200} \rightarrow Leu$ mutant from that observed in wild-type RCs, and involves a new transient state [6,7]. Excitation of the heterodimer results in the rapid (<350 fs) formation of the intradimer charge transfer (CT) state [$BChl_{LP}^+$-BPh_{MP}^-], which is observed to have a lifetime of about 14 ps [7]. Decay of this intradimer CT state occurs via two routes: rapid internal conversion (charge recombination) to the ground state (~50% yield), and electron transfer to the normal BPh_L acceptor, forming state $D^+BPh_L^-$ also with an ~50% yield. State $D^+BPh_L^-$ subsequently transfers an electron to Q_A with the same high (~100%) yield as in wild type RCs, giving an overall yield of change separation (i.e., $D^+Q_A^-$ formation) about half that of the native system. In this article we review these observations, and also present the results of subpicosecond measurements on the initial electron transfer reaction in $His^{M200} \rightarrow Leu$ RCs at low temperature.

2. Materials and Methods

Preparation of wild-type RCs from U43(L228BamHI) and $His^{M200} \rightarrow Leu$ RCs followed published procedures [5]. Subpicosecond transient absorption and photodichroism studies were performed as previously described [6,8]. RCs in 10 mM potassium phosphate buffer pH 7.4/0.05% LDAO/0.2 mM sodium ascorbate were flowed through a 2 mm pathlength cell and maintained at ~10°C during experiments, except for low temperature measurements which employed RCs in PVA films or glycerol/water glasses. Details of low-temperature sample preparation and cryogenic techniques can be found elsewhere [8].

3. Results and Discussion

3.1 COMPOSITION AND GROUND STATE ABSORPTON SPECTRUM OF THE RC

The first indication that there is an altered pigment content in the $His^{M200} \rightarrow Leu$ RC is found in its ground state absorption spectrum, which is shown in comparison to the spectrum of wild-type RCs in Fig. 2. The spectra have been normalized at ~800 nm; note that this normalization results in virtually identical spectra in the Soret region as well. Compared to wild-type RCs (dashed spectrum), the 850-nm absorption in the mutant (solid spectrum) is significantly reduced, but tails to longer wavelengths, while the absorption between 750 and 800 nm has increased. In the Q_X region, the absorption near 600 nm, normally attributed to the accessory BChls and P in wild-type RCs, is weaker in the mutant, while the 530-550-nm absorption of the BPhs is stronger.

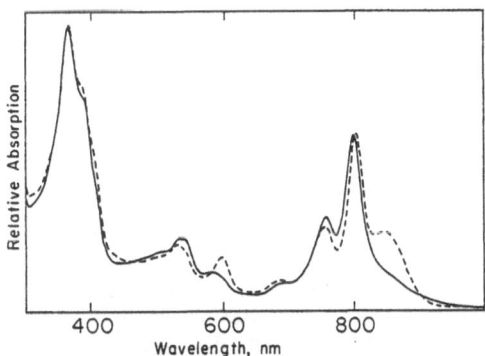

Figure 2. Room temperature absorption spectra of RCs from the His^{M200}→Leu mutant (——) and wild-type (----) *Rb. capsulatus.*

The composition of the mutant RC was assayed chemically by quantitatively extracting the chromophores into acetone:methanol (7:3) and analyzing the Q_Y absorption bands of the pigment extracts. This procedure affords an unambiguous determination of the BChl/BPh ratio, since it depends only on the extinction coefficients of the two chromophores in this solvent mixture, values which have been measured previously [9]. Extractions on several different preparations of His^{M200}→Leu RCs yield a BChl/BPh ratio of 1.0 ± 0.1 [5,6]. This can be compared with a BChl/BPh ratio of 1.9 ± 0.1 determined in side-by-side extractions on wild-type RCs, a value which is essentially the expected ratio of 2/1. Determination of the total number of pigments per RC requires the use of an extinction coefficient of some band in the RC spectrum. Since extinction coefficients have not been determined for *Rb. capsulatus*, we have used the *Rb. sphaeroides* extinction coefficient at 802 nm of 288 mM^{-1} cm^{-1} [10]. Using this value, a total pigment content of 5.8 ± 0.2 is found for several different preparations of both *Rb. capsulatus* wild-type and His^{M200}→Leu RCs [5,6]. Based on these pigment assays, the ground state absorption spectrum, and the site of the mutation, the logical deduction is that the His^{M200}→Leu RC contains a BChl-BPh heterodimer in place of the BChl-BChl dimer that serves as the primary electron donor in wild-type RCs [5].

As a further test of the composition of the His^{M200}→Leu RC, we have integrated the ground state absorption spectra of mutant and wild-type RCs between 700 and 1100 nm. Using several different preparations of wild-type and His^{M200}→Leu RCs with samples again adjusted in concentration so that their spectra compare as in Fig. 2, the integrations (in wavenumbers) reveal only a small (≤4%) reduction in the Q_Y oscillator strength in the His^{M200}→Leu RC compared to wild-type. Given that the Q_Y oscillator strength of BPh is lower than that of BChl by 20-30% [9,11], the replacement of one BChl by a BPh should reduce the total Q_Y oscillator strength of a six-pigment RC by only 4-5%. Thus, the results of the pigment extractions and the integrated Q_Y oscillator strength of the ground state spectrum of the mutant are in excellent agreement.

Together these results give a clear indication that the His^{M200}→Leu RC is homogeneous in its chemical composition. It is virtually impossible to obtain a BChl/BPh ratio of 1 and a total pigment content of 6 with some combination of wild-type and denatured RCs. If, for example, the sample contained 50% wild-type RCs and 50% RCs having no dimer, then such a sample would have a BChl/BPh ratio of 1.5 and an average of 5 pigments per RC. Furthermore, as we shall see, the large amplitudes of the absorbance changes and the uniform (single-exponential) kinetic behavior for each step of the primary photo-chemistry in the mutant indicate that the $His^{M200} \rightarrow Leu$ RCs are photochemically homogeneous as well.

3.2 THE CHARGE SEPARATION PROCESS

The familiar absorption changes that charac-terize the primary photochemistry in wild-type RCs, $P^* \rightarrow P^+BPh_L^- \rightarrow P^+Q_A^-$, can be seen in Figs. 3B-5B [6,7,12]. Like the $^1(\pi,\pi^*)$ excited states of porphyrins, chlorins and bacteriochlorins [13-15], the P^* spectrum consists of a flat featureless absorption inter-rupted by bleaching of the ground state absorptions (Fig. 3B). An additional indica-tion that P^* is a singlet state is the observa-tion of stimulated emission, which appears in the 600-fs spectrum as the extra absorption decrease on the red side of the bleaching in the 850-nm band of P (Figs. 3B and 5B). The stimulated emission decays with a time constant of 3.5 ± 0.6 ps as the electron moves from P^* to BPh_L (inset to Fig. 5B). The reduction of BPh_L is accompanied by bleaching of its Q_X (and Q_Y) absorption band at 542 nm (and 765 nm) and the appearance of its anion band centered at 665 nm (Fig. 4B and inset). Both of these features decay with a time constant of about 200 ps as an electron moves from BPh_L^- to Q_A. The high yield of charge separation is apparent from the fact that the bleaching of the 850-nm absorption of P does not recover as an elect-ron moves from P^* to BPh_L to Q_A. State $P^+Q_A^-$ subsequently decays on a much longer time scale by slow charge recombination ($\tau \sim$ 145 ms) [5,6].

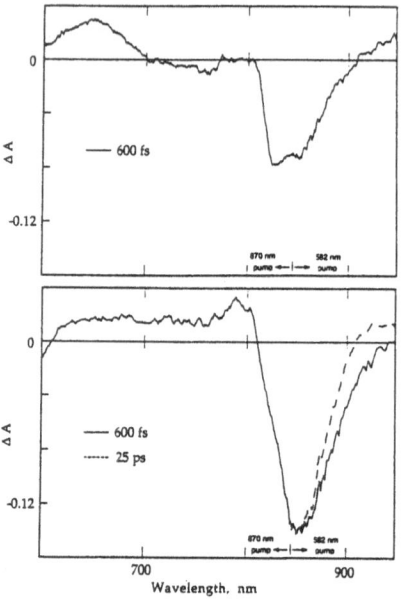

Figure 3. Spectra taken 600 fs after excitation of $His^{M200} \rightarrow Leu$ (A), and wild-type (B), RCs with 350-fs flashes. For both excitation wave-lengths employed, the flashes were polarized at 45° with respect to the probe pulses to minimize dichroism of the absorption changes.

The most profound difference between the primary photochemistry in $His^{M200} \rightarrow Leu$ and wild-type RCs occurs immediately after photon absorption [6,7]. This can be seen in the 600-fs spectrum of $His^{M200} \rightarrow Leu$ RCs (Fig. 3A), which is unlike the P^* spectrum of wild-type RCs in two significant regards. The pronounced absorption band near 650 nm (Fig. 3A) is not readily reconciled with a $^1(\pi,\pi^*)$ state of D, or, for that matter, a $^1(\pi,\pi^*)$ state of any of the six chromophores. Additionally, stimulated emission is not apparent in the very early time spectra of the $His^{M200} \rightarrow Leu$ RC (Fig. 5A). These observations suggest that the 600-fs spectrum of the mutant is due to a state other than D^*. The first clue to the identity of this early-time transient in the mutant is that, except for a slight blue shift, the 650-nm band of Fig. 3A is very similar to the 665-nm band of BPh_L^- seen in the spectra of $P^+BPh_L^-$ in wild-type RCs (Fig. 4B, 10-ps) and $D^+BPh_L^-$ in $His^{M200} \rightarrow Leu$ RCs (see Fig. 4A, 45-ps and discussion

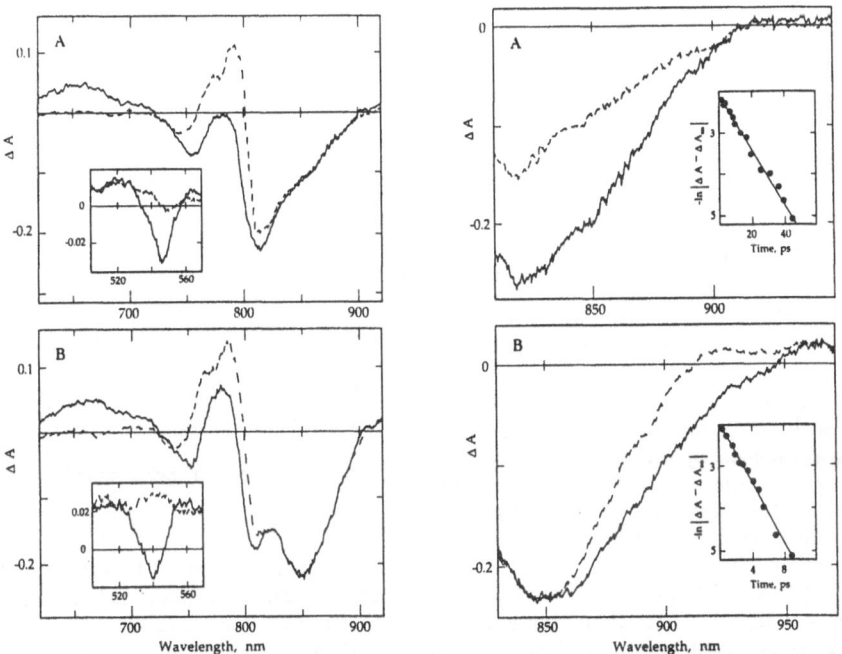

Figure 4. (left) Transient absorption difference spectra acquired using a 350-fs 582-nm excitation flash. (A) Spectra taken 45 ps (——) and 1.9 ns (----) after excitation of His^{M200}→ Leu RCs. (B) Spectra taken 10 ps (——) and 1.9 ns (----) after excitation of wild-type RCs. The mutant RCs were approx. a factor of 1.7 times more concentrated than the wild-type RCs; all other experimental conditions were identical.

Figure 5. (right) Transient absorption difference specta taken 670 fs (——) and 1.9 ns (----) after a 350-fs 582-nm excitation flash, for His^{M200}→Leu RCs (A), and wild-type RCs (B). (A) inset shows a logarithmic plot of the absorption changes averaged over the interval 840-850 nm; a 15.1 ±1.3 ps lifetime for [BChl$_{LP}$$^+BPh_{MP}$$^-$] is derived from the data shown. (B) inset shows a logarithmic plot of the data between 880 and 890 nm in wild-type RCs. From these data a 3.5 ± 0.6 ps P* lifetime is obtained.

below). A similar anion band is also found in this region of the P$^+$BPh$_L$$^-$ spectrum in RCs from other bacteria [1], and for BPh$^-$ and BChl$^-$ *in vitro* [16].

A simple analysis [7], taking into account that BPh is easier to reduce than BChl by 200-300 mV *in vitro* [16], leads to the assignment of the 650-nm band in the 600-fs spectrum of His^{M200}→Leu RCs to the reduced BPh component of the heterodimer CT state [BChl$_{LP}$$^+BPh_{MP}$$^-$]. The transient absorption in the 650-nm region decays with dual exponential kinetics (Fig. 6), with time constants of 12 ± 1.5 and 100 ± 20 ps. The shorter time constant is in good agreement with the value of 14 ± 3 ps measured for the observed 50% decay of the initial bleaching of the long-wavelength ground state absorption of the heterodimer (Fig. 5A). This 50% decay of bleaching indicates that the intradimer CT state [BChl$_{LP}$$^+BPh_{MP}$$^-$] decays by about half via internal conversion (charge recombination) to the ground state. The formation of $D^+BPh_L^-$ accounts for the other 50% of the decay of $[BChl_{LP}^+BPh_{MP}^-]$.

State $D^+BPh_L^-$ is hence responsible for the absorption changes observed immediately after $[BChl_{LP}^+BPh_{MP}^-]$ has decayed, and also for the slower (~100 ps) component of the anion-region decay kinetics (Fig. 6). Consistent with this assignment, the 45-ps spectrum of the $His^{M200} \rightarrow$ Leu RC shows the 665-nm anion band and the 542-nm Q_X band bleaching characteristic of state $D^+BPh_L^-$ (Fig. 4A and inset). Both of these features decay with a time constant of about 100 ps, as an electron moves to Q_A. Electron transfer from BPh_L^- to Q_A in the $His^{M200} \rightarrow$ Leu mutant occurs with the same high yield as in wild-type RCs, as evidenced by the lack of recovery of the 850-nm heterodimer bleaching during this process (i.e., on the timescale of 45 ps to several ns; Fig. 4A). In fact, electron transfer from BPh_L^- to Q_A

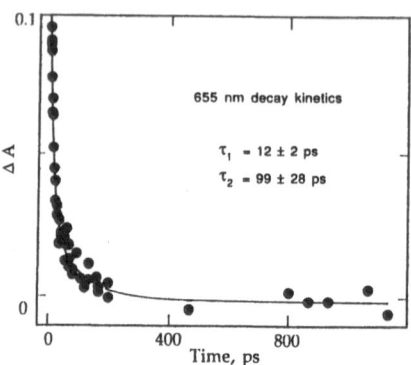

Figure 6. Decay kinetics of the 650-660-nm transient absorption in $His^{M200} \rightarrow$ Leu RCs, measured with 870-nm excitation flashes polarized perpendicular to the probe flashes.

occurs with a somewhat faster time constant in the mutant compared to wild-type RCs (τ ~115 ps vs ~200 ps [6]). Similarly, charge recombination between D^+ and Q_A^- (τ ~100 ms) is slightly faster than charge recombination between P^+ and Q_A^- (τ ~145 ms) [5,6]. It is not clear whether these differences reflect alterations in the electronic or Franck-Condon factors for these processes in the two RCs.

Picosecond photodichroism (photoselection) measurements further demonstrate that the primary photochemistry in the $His^{M200} \rightarrow$ Leu mutant is characterized by two anion-bearing transient states, namely $[BChl_{LP}^+BPh_{MP}^-]$ and $D^+BPh_L^-$. The photodichroism spectra of Fig. 7 show the absorption changes polarized parallel (solid) or perpendicular (dashed) to the heterodimer long-wavelength transition pumped with 350-fs 870-nm excitation flashes. In the analysis of the photodichroism results ([7] and below) we have assumed the following: a) the Q_Y direction of each chromophore lies along its N_1-N_3 axis; b) the

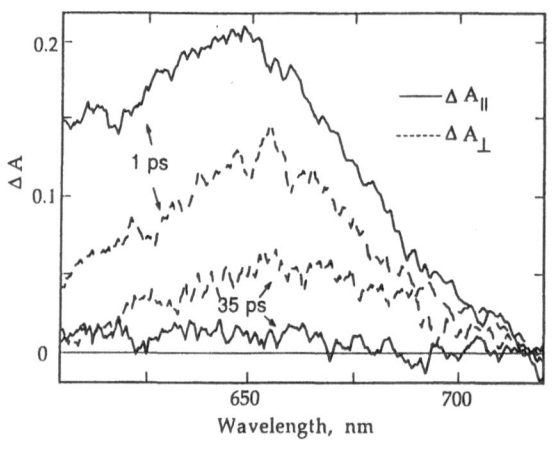

Figure 7. Photodichroism spectra in the anion-band region at time delays of 1 and 35 ps. These spectra give the transient absorption polarized parallel (solid) and perpendicular (dashed) to the transition dipole direction of the long-wavelength ground state absorption band pumped by 350-fs 870-nm excitation flashes. These photodichroism spectra were calculated from raw spectra, acquired with vertical/vertical (V/V) and vertical horizontal (V/H) relative polarizations of the pump and probe pulses, using the formulas $\Delta A_\parallel = 2(V/V) - (V/H)$ and $\Delta A_\perp = 3(V/H) - (V/V)$ [7,8,21]. The angle θ between the net transition dipole giving rise to the absorbance change at a particular wavelength and the 870-nm transition is given by $\tan^2\theta = \Delta A_\perp / \Delta A_\parallel$.

transition(s) giving rise to the broad absorption bands (620-690 nm) of BPh and BChl anions are polarized in the Q_Y direction of the macrocycles, as indicated by *ab initio* calculations [17]; c) the geometrical arrangement of the pigments in the mutant RC is basically the same as found in the crystal structures of the *Rps. viridis* and *Rb. sphaeroides* RCs; and d) the Q_Y direction of the heterodimer is given by the vector sum of its constituent Q_Y directions. The latter common assumption, which follows from simple exciton theory, is supported by linear dichroism studies on *Rps. viridis* single crystals [18] and oriented RCs of *Rps. viridis* and *Rb. sphaeroides* [19].

The photodichroism spectra of the anion band centered at 665-nm in the 35-ps spectrum, which we assign to state $D^+BPh_L^-$, indicate that the angle between this anion transition of BPh_L and the Q_Y transition of D is ~65°. This is in excellent agreement with the angle of 60-65° expected for the BPh_L anion based on the structures of the *Rps. viridis* and *Rb. sphaeroides* RCs. This result also is in accord with our previous measurement that the 665-nm band of BPh_L^- in the transient spectrum of $P^+BPh_L^-$ in *Rb. sphaeroides* is polarized more perpendicular than parallel with respect to the 870-nm band of P [8]. A similar result has been found in steady-state linear dichroism studies of the BPh_L anion band in oriented *Rps. viridis* RCs or whole cells trapped with BPh_L reduced [20,21]. A spatial arrangement of the pigments in His^{M200}→Leu RCs not too different from the structure of wild-type RCs thus seems reasonable. This is further supported by recent linear dichroism measurements which show that the bands in the ground state spectrum of His^{M200}→Leu RCs, both before and after photo-oxidation, are polarized essentially the same as in wild-type RCs [23].

The 650-nm anion band in the 1-ps spectrum is mainly parallel polarized; the data yield an angle of ~35° between this transition and the Q_Y transition of D. Again, based on the model for wild-type RCs, we expect the anion band of the BPh_{MP} component of the heterodimer to be polarized at an angle of 20-25°, which is in reasonable accord with our observation. These photodichroism results thus support our assignment of the 650-nm band present at early times in His^{M200}→Leu RCs to the reduced BPh component of the heterodimer CT state $[BChl_{LP}^+BPh_{MP}^-]$.

The state diagram of Fig. 8 summarizes the sequence of events we propose follows excitation of His^{M200}→Leu RCs. The intra-dimer CT state is probably not the state reached directly upon absorption of a photon (i.e., this state is not responsible for the main ground state absorption between 820 and 900 nm.) Rather, the initial excited state may be most simply described as the neutral low-energy exciton state of the heterodimer. We propose that this initial excited state, which we will call D*, rapidly (<350 fs) relaxes to the intradimer CT state $[BChl_{LP}^+BPh_{MP}^-]$. Our proposed ordering of the states in the mutant (Fig. 8) compared to wild-type RCs is in accord with simple energetic considerations. Since BPh *in vitro* is easier to reduce than BChl by 200-300 mV [16], it is reasonable to suppose that if the intradimer states

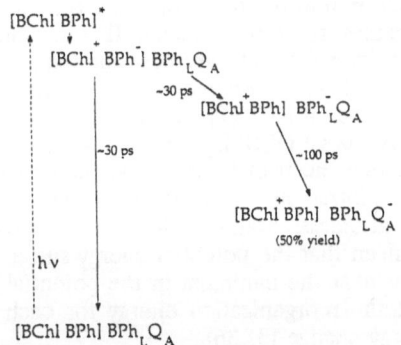

Figure 8. Relative ordering of the states which are proposed to participate in the photochemistry of His^{M200}→Leu RCs.

$[BChl^+BChl^-]$ lie slightly above P* in wild-type RCs, as has been suggested [24-28], then the intradimer CT state $[BChl_{LP}^+BPh_{MP}^-]$ may lie below D* in the His^{M200}→Leu mutant. The fact that we observe the same initial photochemistry in the mutant at low temperature (see below) supports this interpretation. Our results on the His^{M200}→Leu RC thus lend support to current discussion that the internal CT states of P in wild-type RCs lie reasonably close to the low energy exciton state of the dimer, and can therefore readily mix with the exciton state to give P* CT character [24-33].

One can imagine positive consequences of P* having some intradimer CT character, such as building the directionality of charge separation toward the pigments on the L polypeptide within the initial excited state [26,30]. We can see, however, that if the CT character within the excited dimer is too great, or if (as in the mutant) the "pure" intradimer CT state actually forms, this may lead to rapid radiationless decay competing effectively with electron transfer to BPh$_L$. Based on our measured 14-ps lifetime of [BChl$_{LP}$$^+BPh_{MP}$$^-$], during which time there is an ~50% yield of internal conversion, we can estimate that decay of [BChl$_{LP}$$^+BPh_{MP}$$^-$] to the ground state has a rate of about (30 ps)$^{-1}$. This is at least ten times faster than the inherent rate of decay of P* to the ground state. Internal conversion (charge recombination) within [BChl$_{LP}$$^+BPh_{MP}$$^-$], leading to ground state repopulation, could be very rapid due to a Franck-Condon factor which may be larger than that for decay of a state which has only partial or no charge transfer character associated with it. It is possible that the loss of the Mg•••His^{M200} interaction may result in a less rigid structure of the heterodimer compared to P, which may facilitate deactivation of the intradimer CT state. A less rigid dimer structure may also contribute to the broad long-wavelength absorption of the heterodimer (Fig. 2) and to the unusual shape of the bleaching in the difference spectrum of the intradimer CT state (Fig. 3A).

3.3 EFFECTS OF TEMPERATURE ON HIS^{M200}→LEU RC PHOTOCHEMISTRY

The temperature dependence of the lifetime of [BChl$_{LP}$$^+BPh_{MP}$$^-$] provides additional insights into the mechanism of the initial stage of charge separation in His^{M200}→Leu RCs. We find that the lifetime of the intradimer CT state in 60% glycerol is essentially the same at 285 K (18 ± 4 ps) and 78 K (16 ± 4 ps) [34]. We also find that the roughly 50/50 branching of the two decay pathways of [BChl$_{LP}$$^+BPh_{MP}$$^-$], ground state recovery and electron transfer to BPh$_L$ (Fig. 8), remains essentially unchanged at low temperature. Together, these results mean that the inherent rate constants for two decay routes of the intradimer CT state are basically temperature independent.

Based on the energetics in wild-type RCs, and our proposed ordering of the states in His^{M200}→Leu RCs (Fig. 8), [BChl$_{LP}$$^+BPh_{MP}$$^-$] in the mutant probably lies a few tenths of an eV above D$^+$BPh$_L$$^-$ and slightly more than 1 eV above the ground state. The observation that electron transfer and internal conversion (charge recombination) decay routes of the intradimer CT state are temperature independent indicates that both processes are activationless. Standard theoretical approaches more specifically would take activationless to mean that the potential energy surfaces for both D$^+$BPh$_L$$^-$ and the ground state cross at or near the minimum in the potential surface for [BChl$_{LP}$$^+BPh_{MP}$$^-$]. This also implies that the reorganization energy for each reaction should be roughly equal to the free-energy change [35,36].

Perhaps the simplest way of achieving an activationless curve-crossing arrangement would be for the potential surface for the intradimer CT state to be extremely shallow. A flexible structure of the heterodimer could, for example, yield a shallow potential surface for [BChl$_{LP}$$^+BPh_{MP}$$^-$], perhaps containing multiple minima with small barriers arising from a distribution of distances between the two macrocycles of D. (Such internal dimer motion has been suggested to be important in the spectral and photochemical properties of the primary donor in wild-type RCs [37-40].) As noted above, a flexible structure for the heterodimer may also need to be invoked to understand the long-wavelength absorption spectrum of D.

Alternatively, the observation of a temprature independent lifetime for the intradimer CT state may not necessarily imply that the potential surfaces for $D^+BPh_L^-$ and the ground state cross near the minimum of that for $[BChl_{LP}^+BPh_{MP}^-]$. Rather, the potential surfaces of the relevant states all could be nested, in which case the decay of the intradimer CT state could be considered internal conversion of an excited state of a strongly-coupled dimer to the ground electronic state. This is, in fact, how one often describes radiationless deactivation of CT excited states of large molecules, including metal <->ring and ring <->ing excited states of metalloporphyrin monomers and dimers. These states are thought to facilitate the rapid (in many cases <10 ps) radiationless decay of the CT states in many such complexes [41-46]. The lifetimes of these states also are generally independent of temperature. Such considerations not only have implications for the primary photochemistry in His^{M200}→Leu RCs, but may also be relevant to understanding the temperature dependence and mechanism of the electron transfer reactions in native RCs.

4. Acknowledgements

This work was supported by National Science Foundation Grants DMB 86-04693 (DH) and DMB 86-09614 (DY), and Department of Agriculture Grant 87-CR-CR-1-2328 (DY).

5. References

1. C. Kirmaier and D. Holten (1987) *Photosyn. Res.* 13, 225-260.
2. H. Michel, O. Epp and J. Deisenhofer (1986) *EMBO J.* 5, 2445-2451.
3. T.O. Yeates, H. Komiya, A. Chirino, D.C. Rees, J.P. Allen and G. Feher (1988) *Proc. Natl. Acad. Sci USA* 85, 7993-7997.
4. C.-H.Chang, D. Tiede, J. Tang, U. Smith, J. Norris and M. Schiffer (1986) *FEBS Lett.* 205, 82-86.
5. E.J. Bylina and D.C. Youvan (1988) *Proc. Natl. Acad. Sci. USA* 85, 7226-7230.
6. C. Kirmaier, D. Holten, E.J. Bylina and D.C. Youvan (1988) *Proc. Natl. Acad. Sci. USA* 85, 7562-7566.
7. C. Kirmaier, E.J. Bylina, D.C. Youvan and D. Holten (1989) *Chem. Phys. Lett.* (in press).
8. C. Kirmaier, D. Holten and W.W. Parson (1985) *Biochim. Biophys. Acta* 810, 49-61.
9. M. van der Rest and G. Gingras (1974) *J. Biol. Chem.* 249, 6446-6453.
10. S.C. Straley, W.W. Parson, D.C. Mauzerall, and R.K. Clayton (1973) *Biochim. Biophys. Acta* 305, 597-609.
11. J.S. Connolly, E.B. Samuel and F.E. Janzen (1982) *Photochem. Photobiol.* 36, 565-574.
12. C. Kirmaier and D. Holten (1988) *FEBS Lett.* 239, 211-218.
13. D. Holten, M. Gouterman, W.W. Parson, M.W. Windsor and M. Rockley (1976) *Photochem. Photobiol.* 23, 415-423.
14. D. Holten, C. Hoganson, M.W. Windsor, C.G. Schenck, W.W. Parson, A. Migus, R.L. Fork and C.V. Shank (1980) *Biochim. Biophys. Acta* 592, 461-477.
15. J. Rodriguez, C. Kirmaier and D. Holten (1989) *J. Am. Chem. Soc.* (in press).
16. J. Fajer, D.C. Brune, M.S. Davis, A. Forman and L.D. Spaulding (1975) *Proc. Natl. Acad. Sci. USA* 72, 4956-4960.
17. J.D. Petke, G.M. Maggiora, L.L. Shipman and R.E. Christofferson (1981) *Photochem. Photobiol.* 33, 663-671.

18. W. Zinth, M. Sander, J. Dobler and W. Kaiser (1985) in *Antennas and Reaction centers of Photosynthetic Bacteria*, M.E. Michel-Beyerle (ed.) Springer, Berlin, pp. 97-102.
19. J. Breton (1988) in *The Photosynthetic Bacterial Reaction Center - Structure and Function*, J. Breton and A. Vermeglio (eds.) Plenum, New York, pp. 59-69.
20. G. Paillotin, A. Vermeglio and J. Breton (1979) *Biochim. Biophys. Acta* **545**, 249-264.
21. D.M. Tiede, Y. Choquet and J. Breton (1985) *Biophys. J.* **47**, 443-447
22. A. Vermeglio, J. Breton, G. Paillotin and R. Cogdell (1978) *Biochim. Biophys. Acta* **501**, 514-530.
23. J. Breton, E.J. Bylina and D.C. Youvan (1989) *Biochemistry* (in press).
24. A. Warshel, S. Creighton and W.W. Parson (1988) *J. Phys. Chem.* **92**, 2696-2701.
25. P.O.J. Scherer and S.F. Fischer (1989) *Chem. Phys.* **131**, 115-127.
26. M. Plato, K. Mobius, M. Michel-Beyerle, M. Bixon, and J. Jortner (1988) *J. Am. Chem. Soc.* **110**, 7279-7285.
27. Y. Won and R.A. Friesner (1987) *J. Phys. Chem.* **92**, 2214-2219.
28. M.A. Thompson and M.C. Zerner (1988) *J. Am. Chem. Soc.* **110**, 606-607.
29. S.R. Meech, A.J. Hoff and D.A. Wiersma (1986) *Proc. Natl. Acad. Sci. USA* **83**, 9464-9468.
30. S.G. Boxer, T.R. Middendorf and D.J. Lockhart (1986) *FEBS Lett.* **200**, 237-241.
31. D. Tang, R. Jankowiak, G. J. Small and D.M. Tiede (1989) *Chem. Phys.* **131**, 99-113.
32. D.J. Lockhart and S.G. Boxer (1988) *Proc. Natl. Acad. Sci. USA* **85**, 107-111.
33. M. Loesche, G. Feher and M.Y. Okamura (1987) *Proc. Natl. Acad. Sci. USA* **84**, 7537-7541.
34. L.M. McDowell, C. Kirmaier and D. Holten (unpublished results).
35. R.A. Marcus and N. Sutin (1985) *Biochim. Biophys. Acta* **811**, 265-322.
36. M. Bixon and J. Jortner (1986) *J. Phys. Chem.* **90**, 3795-3800.
37. A. Warshel (1980) *Proc. Natl. Acad. Sci. USA* **77**, 3105-3109.
38. C. Kirmaier and D. Holten (1988) in *The Photosynthetic Bacterial Reaction Center - Structure and Function*, J. Breton and A. Vermeglio (eds.) Plenum, New York, pp. 219-228.
39. Y. Won ana R.A. Friesner (1988) in *The Photosynthetic Bacterial Reaction Center - Structure and Function*, J. Breton and A. Vermeglio (eds.) Plenum, New York, pp. 341-349.
40. Y. Won and R.A. Friesner (1989) *Israel J. Chem.* (in press).
41. I. Fujita, J. Fajer, C.K. Chang, C.-B. Wang, M.A. Bergkamp and T.L. Netzel (1982) *J. Phys. Chem.* **86**, 3754-3759.
42. D. Holten and M. Gouterman (1985) in *Optical Properties and Structures of Tetrapyrroles*, G. Blauer and H. Sund (eds.) deGruyter, Berlin, pp. 64-90.
43. X. Yan and D. Holten (1988) *J. Phys. Chem.* **92**, 409-414.
44. J. Rodriguez, L. McDowell and D. Holten (1988) *Chem. Phys. Lett.* **147**, 235-240.
45. O. Ohno, N. Ishikawa, H. Matsuzawa and H. Kobayashi (1989) *J. Phys. Chem.* **93**, 1713-1718.
46. O. Bilsel, J. Rodriguez and D. Holten (unpublished results).

A NOTE ON SINGLET-SINGLET AND SINGLET-TRIPLET EXCITON INTERACTIONS IN THE BACTERIAL PHOTOSYNTHETIC REACTION CENTER

A.J. HOFF, E.J.J. GROENEN and E.J. LOUS
Department of Biophysics and Centre for the Study of the
Excited State of Molecules, Huygens Laboratory of the State
University
P.O. Box 9504
2300 RA   Leiden
The Netherlands

ABSTRACT. The optical absorbance spectrum of photosynthetic reaction centers is made up of strongly-mixed absorptions of the chlorophyllous cofactors. This note discusses the questions whether the spectrum can be decomposed in contributions of individual cofactors, or combinations of cofactors. It is furthermore discussed whether certain features of the triplet-minus-singlet absorbance difference spectrum give a clue to the degree of localization of the triplet state of the primary electron donor.

1. Introduction

Following the elucidation of the structure of the reaction center (RC) of the photosynthetic bacteria *Rhodopseudomonas* (*Rps.*) *viridis* and *Rhodobacter* (*Rb.*) *sphaeroides* in atomic detail [1,2], several attempts have been made to explain their optical absorption, LD and CD spectra using the exciton formalism and the coordinates for the six chlorophyllous pigments of the RC [3-6]. We will label these pigments according to the system of Feher and coworkers [7]. Two bacteriochlorophylls (BChl), D_A and D_B from the primary electron donor D, two BChls, B_A and B_B, lie at different sides at van der Waals distance of D and two bacteriopheophytins (Bph), ϕ_A and ϕ_B, lie next to the B BChls (Fig. 1). The subscripts A and B refer to the two 'chains' that are related by an approximate C_2 symmetry. Chain A constitutes the active electron transport chain: upon photoexcitation of D, an electron is transferred to ϕ_A and subsequently to the secondary quinone acceptors, first ϕ_A, then ϕ_B.

71

J. Jortner and B. Pullman (eds.), Perspectives in Photosynthesis, 71–80.
© 1990 *Kluwer Academic Publishers.*

1. Schematic drawing of the reaction center of *Rps. viridis.* Symbols explained in the text.

Optical difference spectroscopy in several forms is an important tool to investigate the functioning of the RC. The various steps of the electron transport reaction can be followed by recording the time evolution of absorbance changes as a function of wavelength following excitation of D by a short laser pulse. Interpretation of the kinetic traces is, of course, only meaningful if the bands in the absorbance difference (ΔA) spectrum can be assigned with confidence to the electron transport components. For several bands, e.g. the lowest wavelength band, this is readily done. For the crowded region around 810 and 830 nm for BChl a and BChl b containing RC, respectively, this is much more difficult because of the strong excitonic interactions between the D and B pigments. For example, the band at 850 nm (*Rps. viridis*) has been assigned to the high-energy exciton band of D, D(+) [8] and alternatively to a mixture of D(+) and one of the accessory BChls [6]. For the study of the primary events it is clearly important to assess the validity of these and similar assignments both experimentally and theoretically. Experimentally the approach is to compare different forms of difference spectroscopy that are sensitive to different states of the RC pigment complex, such as oxidized-minus-reduced (D^+ - D) and triplet-minus-singlet (^{3}D - ^{1}D or T - S) ΔA spectra, with and without the use of polarized light. Theoretically, ΔA changes have to be calculated using some kind of exciton theory and structural data (either measured, or surmized) and compared to the observed changes. From this blend of theory and experiment one may hope to obtain a better understanding of the significance and meaning of the observed absorption changes.

In this note we will address two questions related to the interpretation of ΔA spectra: first, is it possible to make a statement on the relative contribution of the D(+) band to bands in the 810/850 nm region? Secondly, is it possible to draw a conclusion on the degree of localization of the ^{3}D state from the relative intensities of a band around 810 nm (840 nm in *Rps. viridis*) attributed to a singlet-singlet absorption of D and a band around 820 nm (870 nm in *Rps. viridis*) attributed to a triplet-triplet transition of ^{3}D?

2. The Relative Contribution of the D(+) Transition to Other Optical Transitions

Consider an initial state ψ_i and a final state ψ_f with transition moment $\mu = \langle\psi_f|\mu|\psi_i\rangle$. Let the final state be composed of two states ϕ_1 and ϕ_2: $\psi_f = \lambda_1\phi_1 + \lambda_2\phi_2$. Then

$$\langle\psi_f|\mu|\psi_i\rangle = \lambda_1\langle\phi_1|\hat{\mu}|\phi_i\rangle + \lambda_2\langle\phi_2|\hat{\mu}|\phi_i\rangle = \lambda_1\mu_{1i} + \lambda_2\mu_{2i} \tag{1}$$

and the oscillator strength is proportional to

$$|\mu_{fi}|^2 = \lambda_1^2|\mu_{1i}|^2 + \lambda_2^2|\mu_{2i}|^2 + \lambda_1\lambda_2\ \mu_{1i}.\mu_{2i} . \tag{2}$$

The product term $\lambda_1\lambda_2\ \mu_1.\mu_2$ does not allow the decomposition of $|\mu_{fi}|^2$ into contributions of ϕ_1 and ϕ_2 given by the squares of the coefficients λ_1 and λ_2. Thus, the statement that " λ_i^2 represents the contribution of $|\mu_i|^2$ to the oscillator strength of the observed transition, $|\mu_{fi}|^2$" [8] is in general not correct.

Can one nevertheless make a, perhaps approximate, statement on the relative contribution of a subset of transition moments to the observed transition? In other words, can one single out the D(+) contribution from a strongly-mixed transition? Let us consider the six excition states of the RC. If Φ_i is the product function $\phi_1\phi_2...\phi_i^*...\phi_6$, where the asterisk denotes the excited state of the i-th RC pigment, then the j-th excition state is given by

$$\Psi_j = \Sigma_i\ a_{ji}\Phi_i \qquad\qquad j = 1,2\ ...\ 6 . \tag{3}$$

Let now ϕ_1, ϕ_2, ... ϕ_6 represent the wavefunction of the pigments Φ_B, B_B, D_B, D_A, B_A and Φ_A in this order. Since the interaction between D_A and D_B is much stronger than that between D_A or D_B and the other pigments, it is useful to define

$$\Phi_+ = \frac{1}{\sqrt{2}}\ (\Phi_3 + \Phi_4)$$

$$\Phi_- = \frac{1}{\sqrt{2}}\ (\Phi_3 - \Phi_4) \tag{4}$$

where for simplicity we have assumed that the dimer is completely symmetric, also with respect to its environment. We may now substitute the inverse relations $\Phi_3 = \frac{1}{2}\sqrt{2}(\Phi_+ + \Phi_-)$ and $\Phi_4 = \frac{1}{2}\sqrt{2}(\Phi_+ - \Phi_-)$ into Eqn. 3:

$$\Psi_j = a_{j1}\Phi_1 + a_{j2}\Phi_2 + a_{j3}\tfrac{1}{2}\sqrt{2}(\Phi_+ + \Phi_-) + a_{j4}\tfrac{1}{2}\sqrt{2}(\Phi_+ - \Phi_-) + a_{j5}\Phi_5 + a_{j6}\Phi_6$$
$$= a_{j1}\Phi_1 + a_{j2}\Phi_2 + \tfrac{1}{2}\sqrt{2}(a_{j3} + a_{j4})\Phi_+ + \tfrac{1}{2}\sqrt{2}(a_{j3} - a_{j4})\Phi_- + a_{j5}\Phi_5 + a_{j6}\Phi_6.$$

The dipole strength of the $j \leftarrow 0$ transition (0, initial ground state; j, final state) is then

$$|\mu_{j0}|^2 = a_{j1}^2|\mu_{10}|^2 + a_{j2}^2|\mu_{20}|^2 + \tfrac{1}{2}(a_{j3} + a_{j4})^2|\mu_{+0}|^2 + \tfrac{1}{2}(a_{j3} - a_{j4})^2|\mu_{-0}|^2$$
$$+ a_{j5}^2|\mu_{50}|^2 + a_{j6}^2|\mu_{60}|^2 + 2a_{j1}a_{j2}\mu_{10}\cdot\mu_{20} + \sqrt{2}a_{j1}(a_{j3} + a_{j4})\mu_{10}\cdot\mu_{+0}$$
$$+ \sqrt{2}a_{j1}(a_{j3} - a_{j4})\mu_{10}\cdot\mu_{-0} + 2a_{j1}a_{j5}\mu_{10}\mu_{50} + 2a_{j1}a_{j6}\mu_{10}\cdot\mu_{60}$$
$$+ \text{other cross terms.} \tag{5}$$

Again, we are not allowed to neglect the cross terms. This may be illustrated by considering only the quadratic terms. Then the contribution of e.g. the $+ \leftarrow 0$ transition to the $j \leftarrow 0$ transition is given by

$$\text{contribution of D(+)} = \tfrac{1}{2}(a_{j3} + a_{j4})^2|\mu_{+0}|^2$$
$$= \tfrac{1}{2}(a_{j3} + a_{j4})^2\{\tfrac{1}{2}|\mu_{30}|^2 + \tfrac{1}{2}|\mu_{40}|^2 + \mu_{30}\cdot\mu_{40}\}. \tag{6}$$

Neglecting the cross product we have

$$\text{contribution of D(+)} = \tfrac{1}{2}(a_{j3} + a_{j4})^2\{\tfrac{1}{2}|\mu_{30}|^2 + \tfrac{1}{2}|\mu_{40}|^2\}$$
$$= \tfrac{1}{2}(a_{j3} + a_{j4})^2|\mu_{30}|^2 \tag{7}$$

for a symmetric dimer. We now compare Eqns. 6 and 7 using the results of the exciton calculation in [6]. Considering the 848.4 nm transition in Table 1 of [6] we find for a_{j3} and a_{j4} 0.567 and 0.592, respectively. With $|\mu_{30}|^2 = 49.1$ D^2 Eqn. 7 yields a D(+) contribution of 33 D^2, i.e. 33/38.2 x 100 = 86% of the total strength of the 848.4 nm band. If we apply Eqn. 6, however, then the cross term cancels to a large extent the first two terms and the contribution of D(+) is only 7.9 D^2, i.e. 21%. The reason for this large difference is, of course, that neglecting the cross product in (7) essentially means neglecting the exciton interaction between D_A and D_B. Since their transition moments are almost antiparallel, the cross term almost cancels the monomer dipole strengths: the $+ \leftarrow 0$ transition is quasi-forbidden.

The above makes clear that it is not allowed to neglect the cross terms in (5) a priori. Consequently, a meaningful statement about the 'contribution' of D(+) to the oscillator strength of the $j \leftarrow 0$ transition can only be made working out the full expression of Eqn. 5. If for a

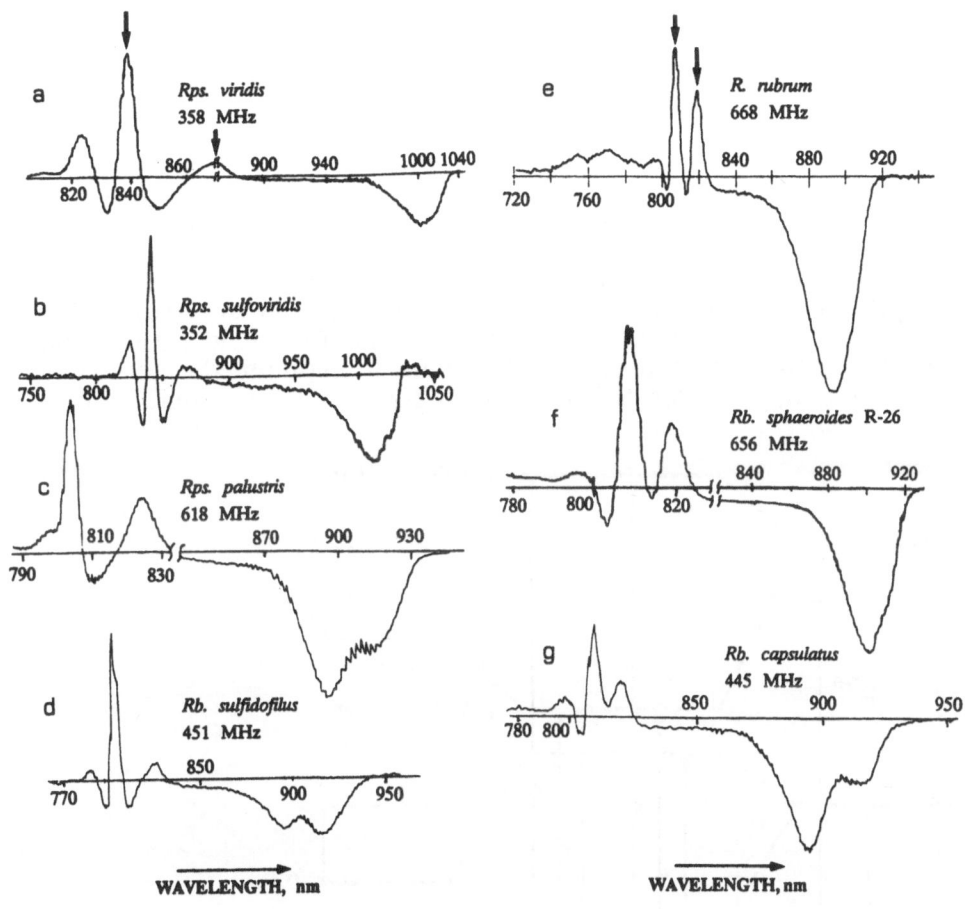

2. Triplet-minus-singlet absorbance difference spectra of Rhodo-spirillaceae. Arrows indicate the bands ascribed to a singlet-singlet (left) and triplet-triplet (right) transition. From [10].

given combination of amplitude and direction of the six transition moments μ_o the sum of the terms is much less than the sum of the quadratic terms, then and only then is it meaningful to entertain the notion of a 'D(+) contribution'.

3. Localization of ^{3}D

The triplet-minus-singlet spectrum of *Rps. viridis* shows a band at 870 nm that is attributed to a triplet-triplet (T-T) absorption of ^{3}D [6]. Such a band is present in the T-S spectra of most photosynthetic purple bacteria (Fig. 2). Remarkably, the intensity of this band relative to that of the strong positive band around 840 or 810 nm seems to be a function of the localization of the primary donor triplet. In *Rps. viridis* it was found that ^{3}D is localized on D_A [6], whereas in *Rb. sphaeroides* R-26 ^{3}D was reported to be largely delocalized over D_A and D_B [4,9]. The ratio of the amplitude of the T-T band to that of the 810/840 nm band is 0.38 for *Rb. sphaeroides* and 0.12 for *Rps. viridis*. Could one use this relative intensity of the T-T band to estimate the percentage of localization, as suggested in [10]? To answer this question we must

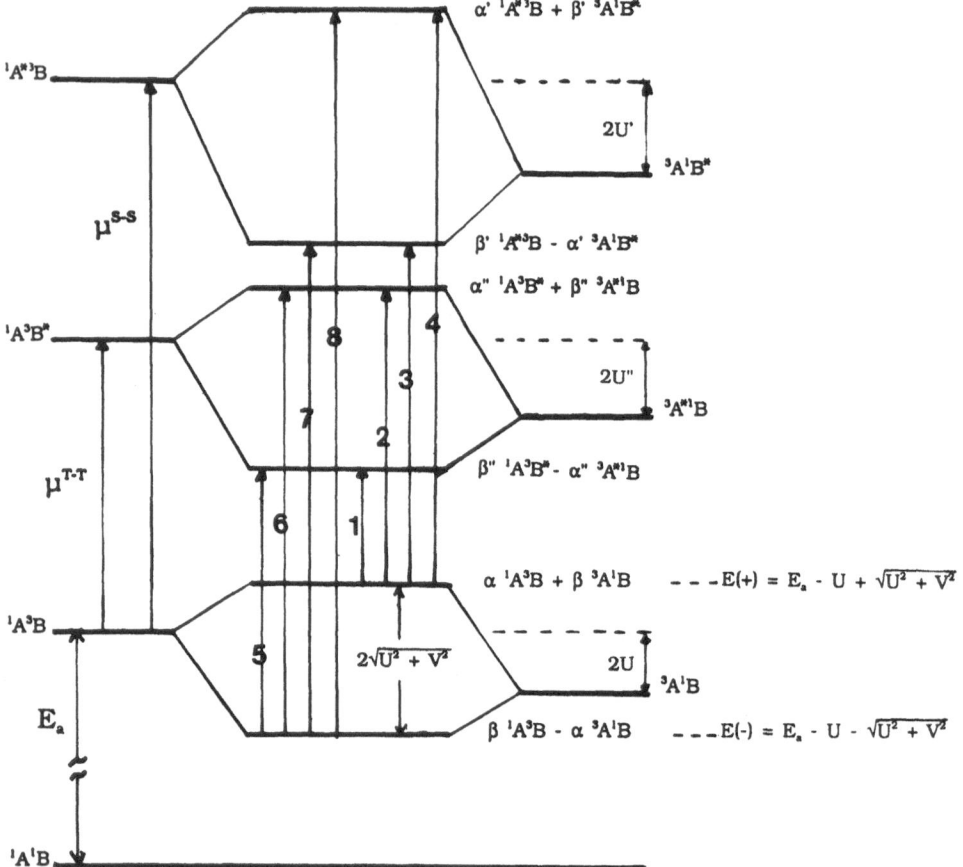

3. Energy level scheme of ^{3}D. For symbols see text.

examine the exciton coupling scheme for RC in which a triplet state is present on D.

Fig. 3 depicts the energy levels of the dimer. We consider transitions from the ^3D state to the doubly-excited $^3D^*$ state. Since D_A and D_B are excitonically coupled the triplet is either on D_A or on D_B and the coupled states are the symmetric and anti-symmetric linear combinations of these states. In Fig. 3 allowance is made for site-splitting of $^1D_A^3D_B$ and $^3D_A^1D_B$, i.e. the dimer is assumed to be asymmetric, either because of slightly different molecular structure of D_A and D_B, e.g. puckering of the macrocyle, or because of environmental effects. The coefficients α and β are determined by the amplitude of the splitting 2U between the uncoupled ^3D_B and ^3D_A states:

$$\alpha = \cos \phi$$
$$\beta = \sin \phi$$
$$\text{tg } \phi = \pm(-U + \sqrt{U^2 + V^2})/V, \tag{8}$$

where 2V is the exciton splitting for U = 0.

There are two species of doubly-excited states: either the singlet D_A or D_B is excited to the first excited singlet state, these are the states $^1D_A^{*3}D_B$ or $^3D_A^1D_B^*$, or the triplets ^3D_A or ^3D_B are excited to their first excited state, the $^1D_A^3D_B^*$ and $^3D_A^{*1}D_B$ states. These are allowed electronic transitions, so these states will also be excitonically coupled, with coefficients α', β', and α'', β'' that depend on the site-splitting U' and U'', respectively.

The coupled states exhibit a total of eight transitions, with transition moments given by $(A \equiv D_A, B \equiv D_B)$

$$\mu_1 = <\alpha\ ^1A^3B + \beta\ ^3A^1B|\hat{\mu}|\beta''\ ^1A^3B^* - \alpha''\ ^3A^{*1}B>$$
$$= \alpha\beta''\ \mu_B^{T\text{-}T} + (\beta\beta'' - \alpha\alpha'')DT - \alpha''\beta\ \mu_A^{T\text{-}T}$$

$$\mu_2 = <\alpha\ ^1A^3B + \beta\ ^3A^1B|\hat{\mu}|\alpha''\ ^1A^3B^* + \beta''\ ^3A^{*1}B>$$
$$= \alpha\alpha''\ \mu_B^{T\text{-}T} + (\alpha\beta'' + \alpha''\beta)DT + \beta\beta''\ \mu_A^{T\text{-}T}$$

$$\mu_3 = <\alpha\ ^1A^3B + \beta\ ^3A^1B|\hat{\mu}|\beta'\ ^1A^{*3}B - \alpha'\ ^3A^1B^*>$$
$$= \alpha\beta'\ \mu_A^{S\text{-}S} + (\beta\beta' - \alpha\alpha')DT - \alpha'\beta\ \mu_B^{S\text{-}S}$$

$$\mu_4 = <\alpha\ ^1A^3B + \beta\ ^3A^1B|\hat{\mu}|\alpha'\ ^1A^{*3}B + \beta'\ ^3A^1B^*>$$
$$= \alpha\alpha'\ \mu_A^{S\text{-}S} + (\alpha\beta' + \alpha'\beta)DT + \beta\beta'\ \mu_B^{S\text{-}S}$$

$$\mu_5 = <\beta\ ^1A^3B - \alpha\ ^3A^1B|\hat{\mu}|\beta''\ ^1A^3B^* - \alpha''\ ^3A^{*1}B>$$
$$= \beta\beta''\ \mu_B^{T\text{-}T} - (\alpha''\beta + \alpha\beta'')DT + \alpha\alpha''\ \mu_A^{T\text{-}T}$$

$$\mu_6 = <\beta\ ^1A^3B - \alpha\ ^3A^1B|\hat{\mu}|\alpha''\ ^1A^3B^* + \beta''\ ^3A^{*1}B>$$
$$= \alpha''\beta\ \mu_B^{T\text{-}T} + (\beta\beta'' - \alpha\alpha'')DT - \alpha\beta''\ \mu_A^{T\text{-}T}$$

$$\tag{9}$$

$$\mu_7 = \langle \beta \ ^1A^3B - \alpha \ ^3A^1B | \hat{\mu} | \beta' \ ^1A^3B - \alpha' \ ^3A^1B' \rangle$$

$$= \beta\beta' \ \mu_A^{S-S} - (\alpha'\beta + \alpha\beta')DT + \alpha\alpha' \ \mu_B^{S-S}$$

$$\mu_8 = \langle \beta \ ^1A^3B - \alpha \ ^3A^1B | \hat{\mu} | \alpha' \ ^1A^3B + \beta' \ ^3A^1B' \rangle$$

$$= \alpha'\beta \ \mu_A^{S-S} + (\beta\beta' - \alpha\alpha')DT - \alpha\beta' \ \mu_B^{S-S}.$$

Here, μ^{S-S} and μ^{T-T} are the singlet-singlet and triplet-triplet transition moments of the monomers A or B, respectively, and DT represents the double transitions $^1A^3B \leftarrow ^3A^1B$ or $^3A^1B' \leftarrow ^1A^3B$.

Eqns. 9 simplify if we assume that the double transitions have a very low probability and that the singlet-singlet and triplet-triplet transitions are equal and parallel: $\mu_A^{S-S} = \mu_B^{S-S} = \mu^{S-S}$ and $\mu_A^{T-T} = \mu_B^{T-T} = \mu^{T-T}$. Then

$$\begin{aligned}
\mu_1 &= (\alpha\beta'' - \alpha'\beta)\mu^{T-T}, & \mu_5 &= (\alpha\alpha'' + \beta\beta'')\mu^{T-T} \\
\mu_2 &= (\alpha\alpha'' + \beta\beta'')\mu^{T-T}, & \mu_6 &= (\alpha''\beta - \alpha\beta'')\mu^{T-T} \\
\mu_3 &= (\alpha\beta' - \alpha'\beta)\mu^{S-S}, & \mu_7 &= (\alpha\alpha' + \beta\beta')\mu^{S-S} \\
\mu_4 &= (\alpha\alpha' + \beta\beta')\mu^{S-S}, & \mu_8 &= (\alpha'\beta - \alpha\beta')\mu^{S-S}.
\end{aligned} \quad (10)$$

It is seen that $\mu_2 = \mu_5$ and $\mu_4 = \mu_7$. Furthermore, for a symmetric dimer with $U = U' = U'' = 0$ and consequently $\alpha = \alpha' = \alpha'' = \beta = \beta' = \beta'' = \frac{1}{2}\sqrt{2}$, i.e. complete delocalization, we have $\mu_1 = \mu_3 = \mu_6 = \mu_8 = 0$, $\mu_2 = \mu_5 = \mu^{T-T}$ and $\mu_4 = \mu_7 = \mu^{S-S}$. For a triplet completely localized in all states $\alpha = \alpha' = \alpha'' = 1$ and $\beta = \beta' = \beta'' = 0$, or vice versa, the same result obtains. Thus, a completely localized and a completely delocalized triplet both give just two transitions, viz. μ^{S-S} and μ^{T-T} when the exciton interactions in all states are the same, $V = V' = V''$ (Fig. 4a). When $V \neq V'$, V'' and for a completely delocalized triplet, the two T-T and S-S transitions each split into two equally intense lines separated by $V - V'$ and $V - V''$, respectively (Fig. 4b). The transitions for a completely localized triplet (site splitting large compared to V, V', V'') do not change.

Let us now assume that at very low temperatures (the condition of an ADMR-monitored T-S experiment) only the lowest excitonic state is populated. We allow the transition moments μ_A^{S-S}, μ_B^{S-S}, and μ_A^{T-T}, μ_B^{T-T} to make an angle θ_S and θ_T, respectively. We then have

$$\begin{aligned}
\mu_5 &= \alpha\alpha'' \ \mu_A^{T-T} + \beta\beta'' \ \mu_B^{T-T} \\
\mu_6 &= -\alpha\beta'' \ \mu_A^{T-T} + \alpha''\beta \ \mu_B^{T-T} \\
\mu_7 &= \beta\beta' \ \mu_A^{S-S} + \alpha\alpha' \ \mu_B^{S-S} \\
\mu_8 &= \alpha'\beta \ \mu_A^{S-S} - \alpha\beta' \ \mu_B^{S-S}.
\end{aligned} \quad (11)$$

For the primary donor, the S-S transitions are almost parallel, so that the minus-combination $|\mu_8|^2$ is weak. We may assume that the T-T transition lies in the macrocycle planes [6], which are almost parallel, so that also the T-T transition moments will be approximately parallel with $\theta_T \approx \theta_S = \theta$ and hence also $|\mu_6|^2$ is small. Taking

$|\mu_A{}^{S\text{-}S}| = |\mu_B{}^{S\text{-}S}| = |\mu^{S\text{-}S}|$ and $|\mu_A{}^{T\text{-}T}| = |\mu_B{}^{T\text{-}T}| = |\mu^{T\text{-}T}|$ we then obtain from Eqn. 11 for the two dominant bands

$$|\mu_5|^2 = |\mu^{T\text{-}T}|^2\{(\alpha\alpha'')^2 + (\beta\beta'')^2 + 2\alpha\alpha''\beta\beta'' \cos \theta\}$$
$$|\mu_7|^2 = |\mu^{S\text{-}S}|^2\{(\alpha\alpha')^2 + (\beta\beta')^2 + 2\alpha\alpha'\beta\beta' \cos \theta\}. \qquad (12)$$

The coefficients α, α' and α'' need not be equal, since the interaction of the excited molecule with its environment will be somewhat larger in the excited state than in the ground state (the former has a more extended wavefunction). Let us first assume that $\alpha \neq \alpha' = \alpha''$. The ratio of the intensities of the two bands, R, is then R = $|\mu_5|^2/|\mu_7|^2$ = $|\mu^{T\text{-}T}|^2/|\mu^{S\text{-}S}|^2$, independent of the degree of localization.

Since experimentally the ratio appears to depend effectively on the triplet localization, we must have $\alpha' \neq \alpha''$ (and thus $\beta' \neq \beta''$). Fig. 2 displays T-S spectra of a number of photosynthetic bacteria. If we take the triplet state of *Rps. viridis* localized and that of *Rb. sphaeroides* delocalized we see that $R_{deloc}/R_{loc} > 1$. Since for *Rps. viridis* $\alpha \sim 1$ and α', $\alpha'' < 1$ we then get that $\alpha' > \alpha''$, i.e. the doubly-excited triplet states $^1D_A{}^3D_B{}^\ast$ are somewhat more localized than the mixed excited state $^1D_A{}^\ast{}^3D_B$. This is not unreasonable, since delocalization of the triplet state depends on orbital overlap of the wavefunctions of D_A and D_B, which will be somewhat stronger for the more extended $^1D_A{}^\ast$ than for ^1D_A.

From the above it will be clear that the ratio R completely depends on subtle differences in the localization of the doubly-excited states and that it is hazardous to draw from R a conclusion on the degree of localization per se as attempted in [10].

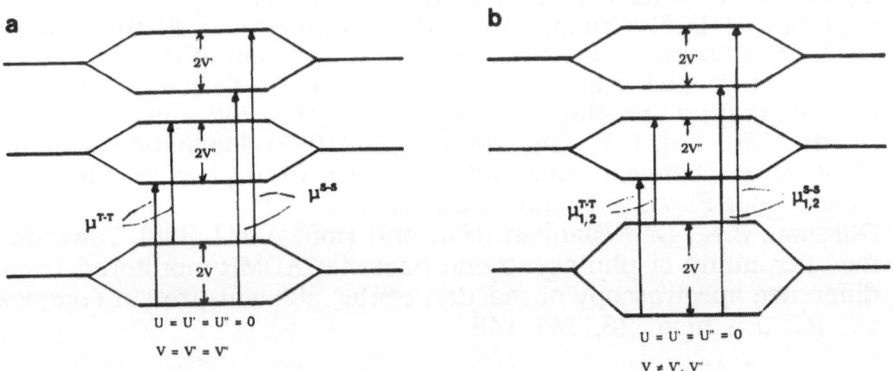

4. Energy level scheme for a symmetric primary donor dimer. a) All exciton couplings V, V', V" are equal, b) V $\neq$ V', V".

REFERENCES

1. Deisenhofer, J., Epp, O., Miki, H., Huber, R. and Michel, H. (1985) 'Structure of the protein subunits in the photosynthetic reaction centre of *Rhodopseudomonas viridis* at 3   resolution', Nature (London) 318, 618-624.
2. Allen, J.P., Feher, G., Yeates, T.O., Komiya, H. and Rees, D.C. (1987) 'Structure of the reaction center from *Rhodobacter sphaeroides* R-26: The cofactors', Proc. Natl. Acad. Sci. USA 84, 5730-5734.
3. Knapp, E.W., Scherer, P.O.J. and Fischer, S.F. (1986) 'Model studies of low-temperature optical transitions of photosynthetic reaction centers. A-, LD-, CD-, ADMR- and LD-ADMR-spectra for *Rhodopseudomonas viridis*', Biochim. Biophys. Acta 852, 295-305.
4. Scherer, P.O.J. and Fischer, S.F. (1987) 'Model studies to low-temperature optical transitions of photosynthetic reaction centers. II. *Rhodobacter sphaeroides* and *Chloroflexus aurantiacus*', Biochim. Biophys. Acta 891, 157-164.
5. Vasmel, H., Amesz, J. and Hoff, A.J. (1986) 'Analysis by exciton theory of the optical properties of the reaction center of *Chloroflexus aurantiacus*', Biochim. Biophys. Acta 852, 159-168.
6. Lous, E.J. and Hoff, A.J. (1987) 'Exciton interactions in reaction centers of the photosynthetic bacterium *Rhodopseudomonas viridis* probed by optical triplet-minus-singlet polarization spectroscopy at 1.2 K monitored through absorbance-detected magnetic resonance', Proc. Natl. Acad. Sci. USA 84, 6147-6151.
7. Allen, J.P. and Feher, G. (1988) 'Structure of the reaction center from *Rhodobacter sphaeroides* R-26 and 2.4.1' in J. Breton and A. Verméglio (eds.), The Photosynthetic Reaction Center. Structure and Dynamics, Plenum Press, New York, pp. 5-11.
8. Breton, J. (1985) 'Orientation of the chromophores in the reaction center of *Rhodopseudomonas viridis*. Comparison of the low-temperature linear dichroism spectra with a model derived form X-ray crystallography', Biochim. Biophys. Acta 810, 235-245.
9. Norris, J.R., Lin, C.P. and Budil, D.E. (1987) 'Magnetic resonance of ultrafast chemical reactions. Examples from photosynthesis', J. Chem. Soc. Faraday Trans. I. 83, 13-17.
10. Dijkman, J.A., Den Blanken, H.J. and Hoff, A.J. (1988) 'Towards a new taxonomy of photosynthetic bacteria: ADMR-monitored triplet difference spectroscopy of reaction center pigment-protein complexes', Isr. J. Chem. 28, 141-148.

DETERMINATION OF THE EXCHANGE INTERACTION IN THE PRIMARY RADICAL ION PAIR IN REACTION CENTERS

W. LERSCH, E. LANG, R. FEICK, W.J. COLEMAN*,
D.C. YOUVAN* and M.E. MICHEL-BEYERLE
*Institut für Physikalische und Theoretische Chemie, TU München,
Lichtenbergstr. 4, 8046 Garching, FRG*
*Department of Chemistry, Massachusetts Institute of Technology,
Cambridge, Massachusetts 02139, USA*

ABSTRACT

The recombination dynamics of the primary radical ion pair state P^F in bacterial reaction centers has been investigated with the particularly sensitive technique of Reaction Yield Detected Magnetic Resonance (RYDMR). Data are presented on the values of the exchange interaction J and the triplet recombination rate k_T in the state P^F of quinone-depleted reaction centers from *Rb. sphaeroides* (strain R26), *Chloroflexus aurantiacus* and *Rb. capsulatus* (T=277K). The results show surprisingly little variations of J and k_T among the species investigated. The recombination dynamics of P^F has also been studied in genetically modified reaction centers from *Rb. capsulatus* where the tryptophane residue (M250) neighbouring both the bacteriopheophytin (BPh_A) and the quinone (Q_A) acceptor had been replaced by either leucine, phenylalanine or glutamic acid via site-directed mutagenesis. A slight increase of J and k_T in the order Leu $\rightarrow$ Phe $\rightarrow$ Trp $\rightarrow$ Glu is observed although the overall effect amounts to less than a factor of 2.

1. Introduction

Investigate electron transfer (ET) mechanisms in the reaction center (RC) means not only measure transfer times with up to femtosecond resolution but also probe the energetic and dynamic properties of the radical pair (RP) stages mediating between subsequent ET processes. Consider the first spectroscopically resolved RP state in bacterial RCs consisting of the oxidized primary donor $BChl_2^+$ and the reduced primary acceptor BPh_A^- (figure 1). This state, called P^F, is populated from $^1BChl_2^*$ within several picoseconds and decays within about 200 ps to yield the secondary RP state $BChl_2^+Q_A^-$ [1]. As shown in figure 1, P^F is not just a short-lived intermediate but has its own dynamics, which is hidden under native conditions but which becomes apparent under conditions where the ET to the quinone is blocked. In that case the lifetime of P^F increases by almost two orders of magnitude. The singlet RP state initially formed either decays to the ground state (rate constant k_S) or undergoes intersystem crossing (via the hyperfine interaction (HFI)) to the RP triplet state which in turn decays to the local triplet state $^3BChl_2^*$ of the dimer (rate constant k_T). In general the RP singlet and triplet states are not degenerate but are separated by the exchange interaction J and the magnetic spin dipole interaction (zero field parameters D and E) between the unpaired electron spins.

J. Jortner and B. Pullman (eds.), Perspectives in Photosynthesis, 81–90.
© 1990 *Kluwer Academic Publishers.*

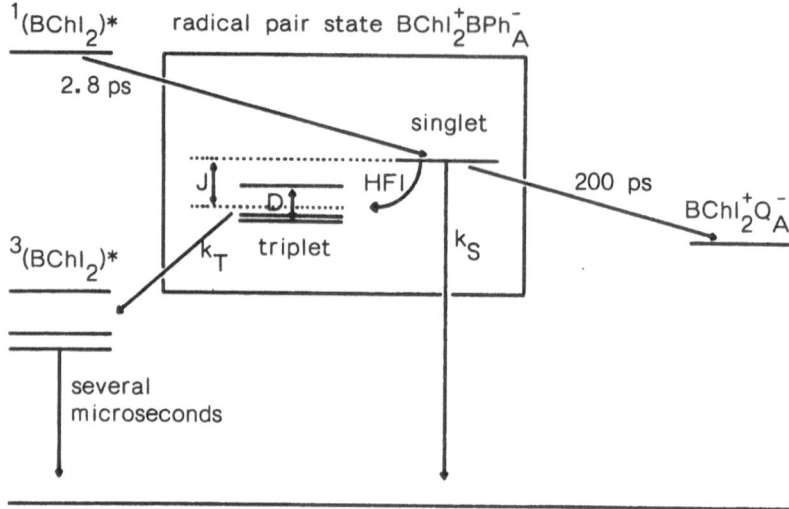

Figure 1. Electron transfer steps in bacterial reaction centers. The time constant of 2.8 ps for the first electron transfer step was taken from reference [5]. All time constants refer to room temperature. The parameters governing the spin dynamics in the radical pair state $BChl_2^+ BPh_A^-$ are defined as follows: HFI = hyperfine interaction; J (D) = exchange interaction (zero field parameter of the magnetic spin dipole interaction) between the unpaired electrons; k_S (k_T) = recombination rate constant for singlet (triplet) phased radical pairs. The energy level spacings in the figure are not true to scale.

The exchange interaction J and the triplet recombination rate k_T deserve special interest. Not only, because they characterize a magnetic interaction and a spin-dependent recombination pathway in the RP state P^F, respectively, but also because their temperature dependences contain information on free energy distances of P^F to other excited states and on factors that determine the mechanism of the primary ET step [2-4]. Although this interrelationship with the forward ET stimulates the interest in J and k_T, one has to be careful in interpreting parameters related to the recombination dynamics in quinone-depleted or quinone-reduced RCs in terms of the forward ET process. As was pointed out by several authors structural relaxation may play an unpredictable role in going over from a picosecond forward ET process to nanosecond recombination dynamics [4,6,7].

The aim of this article is to work out the theoretical and the experimental basis for the determination of J and k_T, to give limits on their accuracy and to provide data on the spread of their values over different bacterial species (*Rhodobacter sphaeroides, Chloroflexus aurantiacus* and *Rhodobacter capsulatus*). In particular we investigated the sensitivity of J and k_T to site-directed mutagenesis of the trytophane residue M 250 located between BPh_A and Q_A in RCs from *Rhodobacter capsulatus*.

2. Theory

In the following we consider only RC preparations where the forward ET to the quinone has been blocked. The extent to which triplet products are formed from the initially singlet phased RP state P^F depends among other things on the value of the singlet-triplet energy gap given by J and on the value of the triplet recombination rate constant k_T. By splitting the three triplet sublevels of P^F in an external magnetic field the triplet yield (Φ_T) changes in a manner characteristic of the values of J, k_T and the other parameters that enter the spin dependent recombination dynamics of P^F. Under favourable conditions J can be directly determined from the position of a resonance in the magnetic field dependence of the triplet recombination yield, which is due to level anticrossing between the singlet state and one of the triplet states at a field strength $B_0 \simeq J$. In general, the parameters characterizing the recombination dynamics of P^F can be extracted only from numerical simulations of the magnetic field dependence of the triplet yield according to the theory of magnetic field effects in RP reactions, reviewed e.g. in [8].

As was shown in [9], J and k_T can be determined even more sensitively by using a technique known as Reaction Yield Detected Magnetic Resonance (RYDMR) [10]. The RYDMR method allows detection of the EPR signal of reactive pairs of paramagnetic species, which escape observation by time-resolved EPR techniques due to their short lifetime of only a few nanoseconds. This is achieved by detecting the effect of the microwaves applied to the sample not as a change of the magnetization of the sample but instead as a change of the product yield of a spin-selective reaction of the pairs of paramagnetic species investigated. Applied to the state P^F in bacterial RCs this means that resonant irradiation of microwaves under the conditions of an EPR experiment results in a characteristic variation of e.g. the triplet recombination yield [11]. Like the above mentioned magnetic field effect on the triplet yield in low static magnetic fields, the RYDMR signal can be evaluated to yield the parameters governing the spin dynamics of P^F.

Shape and amplitude of the RYDMR signal strongly depend on the microwave power applied to the sample. At low microwave powers the triplet yield in the center of resonance typically increases as a function of the strength B_1 of the microwave field until a maximum is reached for $B_1 \simeq J$. For larger values of B_1 the yield drops rapidly below its value in the absence of microwaves. The shape of this so-called "B_1-spectrum" (B_0 is kept fixed) resembles the shape of the magnetic field dependence of the triplet yield at low static magnetic field strengths B_0. This is not surprising since the resonance in the B_1-spectrum around $B_1 \simeq J$ is due to level anticrossing between the singlet state and one of the triplet levels split by the Zeeman interaction with the microwave field in the rotating frame. The decrease of the "relative triplet yield" $\Phi_T(B_0,B_1)/\Phi_T(B_0,0)$ to values below 1 at high microwave powers ($B_1 >> J$) indicates the onset of spin locking in the microwave field [12]. Corresponding to the variation of the amplitude of the RYDMR signal the shape of the "RYDMR-spectrum" (=dependence of the relative triplet yield on B_0 for fixed B_1 and fixed microwave frequency) changes from an upright to an inverted structure as shown in figure 2.

The values of J and k_T can be roughly estimated from the structure of the B_1-spectrum. J is determined by the position of the maximum [13] whereas the amplitude of the maximum and the "width" of the spectrum are indicative of the value of k_T (the larger k_T, the lower is the maximum and the greater the width). To precisely evaluate J and k_T large-scale numerical simulations of the B_1-spectrum and/or of RYDMR-spectra taken at different microwave powers have to be carried out along the lines of the RYDMR-theory outlined elsewhere [12]. As for the simulations to be discussed below two details concerning the actual calculations may be worth mentioning:

1) In the absence of specific hyperfine data for *Rhodobacter capsulatus* and *Chloroflexus aurantiacus* we used the hyperfine parameters valid for *Rhodobacter sphaeroides* for all

species investigated.

2) Anisotropic parts of the spin Hamiltonian, e.g. the magnetic spin dipole interaction between the unpaired spins, were tested to play no decisive role in the interpretation of data from *Rb. sphaeroides* [12]. Lacking further information, they were disregarded in all calculations for the other species.

3. Materials and Methods

RYDMR- and B_1-spectra were recorded by detecting microwave induced variations of the triplet recombination yield with a delay of 100 ns to photoexcitation of the sample (excitation wavelength: 600 nm). The triplet yield was measured optically either as a bleaching of the dimer band around 860 nm (*Rb. sphaeroides* and *Chloroflexus aurantiacus*) or, after quenching of the dimer triplet excitation by a nearby carotenoid, as an absorption increase of a carotenoid triplet-triplet transition around 570 nm (*Rb. capsulatus*). For details of the experimental procedure see reference [12].

Quinone-depleted RCs of *Rb. sphaeroides, Chloroflexus aurantiacus* and *Rb. capsulatus* were prepared as described in the following references :

	RC isolation	quinone extraction	mutagenesis
Rb. sphaeroides	[14,15]	[14,15]	-
Chloroflexus aur.	[16]	[17]	-
Rb. capsulatus	[18,19]	[20]	[19]

Samples were suspended in 10-20 mM Tris buffer, pH 8.0 . During the measurements, the sample temperature was kept at 277±1 K.

4. Results and Discussion

In order to extract the parameters J and k_T from an experimental RYDMR-spectrum it turned out to be most useful to calculate a series of theoretical spectra using closely spaced parameter sets and to test which of them coincide with the measured spectrum to within the 2σ limits of the experimental errors. The spread of the parameter values yielding suitable theoretical spectra depends on the signal to noise ratio and on the microwave power range to which the RYDMR-spectrum belongs. For a given signal to noise ratio determination of J and k_T is most accurate for "intermediate" microwave powers around the maximum of the B_1-spectrum [9].

At high microwave powers the shape of the inverted signal is mainly determined by the value of the microwave field strength and - in spite of an excellent signal to noise ratio - does not permit sensitive evaluation of J or k_T. At very low microwave powers the sensitivity of the signal shape to slight variations of J or k_T is typically not as high as in the "intermediate" power region. Moreover the signal to noise ratio - especially in the inverted wings of the signal - tends to be poor (cf. spectrum (a) in figure 2).

The accuracy to which J and k_T can be determined also depends on how precise the value of the microwave field strength B_1 is known. In our set-up the uncertainty of B_1 amounts to almost 20 % since we are not equipped to measure B_1 directly via the spin-echo technique. Within its experimental limits B_1 is treated here as a fit parameter therefore.

The aspects mentioned so far pertain to the evaluation of J and k_T from a single RYDMR-spectrum. Surprisingly, the parameter values extracted from RYDMR-spectra taken at different microwave powers do not necessarily agree, not even within their respec-

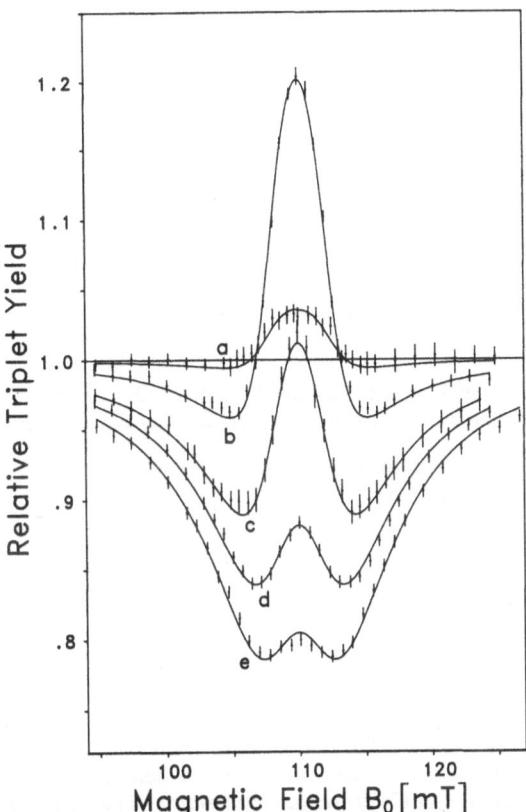

Figure 2. Experimental RYDMR spectra of quinone-depleted reaction centers from *Chloro-flexus aurantiacus* for different microwave field strengths B_1 at a sample temperature of 277 ± 1 K. Reaction centers suspended in Tris buffer (pH 8.0) were adjusted to an optical density of 1.0 at 865 nm. The spectra were taken with the RYDMR spectrometer described in [12] (microwave frequency : 3.07 GHz). Each data point constitutes an average of 1000 measurements (error bars = 95% confidence limits). The full curves are theoretical fits to the data. Each spectrum was adjusted separately by varying the fit parameters B_1, k_S, k_T and J. The following sets of parameters were finally used to simulate the spectra (for comparison, the experimentally determined values of B_1 are also listed):

a: $B_1=0.70$ mT ($B_1^{exp}=0.64\pm0.13$ mT), $k_S=5.0\cdot10^7$ s^{-1}, $k_T=7.0\cdot10^8$ s^{-1}, J=2.30 mT;
b: $B_1=1.75$ mT ($B_1^{exp}=2.08\pm0.40$ mT), $k_S=5.0\cdot10^7$ s^{-1}, $k_T=6.5\cdot10^8$ s^{-1}, J=2.25 mT;
c: $B_1=3.00$ mT ($B_1^{exp}=3.15\pm0.63$ mT), $k_S=5.0\cdot10^7$ s^{-1}, $k_T=8.5\cdot10^8$ s^{-1}, J=2.30 mT;
d: $B_1=3.48$ mT ($B_1^{exp}=3.57\pm0.71$ mT), $k_S=5.0\cdot10^7$ s^{-1}, $k_T=9.5\cdot10^8$ s^{-1}, J=2.20 mT;
e: $B_1=3.90$ mT ($B_1^{exp}=3.86\pm0.77$ mT), $k_S=6.0\cdot10^7$ s^{-1}, $k_T=9.5\cdot10^8$ s^{-1}, J=2.20 mT.

tive scatter. This can be due to a) experimental drifts, b) approximations in the theoretical description of the signals, e.g. the disregard of anisotropic interactions (see above) or c) the neglect of intrinsic inhomogeneities of RCs such as for instance a distribution of the value of J. In any case this inconsistency adds to the uncertainty of J and k_T and deserves further investigation.

To exemplify the evaluation of J and k_T consider the sequence of RYDMR-spectra of quinone-depleted RCs from the green bacterium *Chloroflexus aurantiacus* shown in figure 2. The spectra were taken at microwave powers between 34 W (a) and 1056 W (e). Even the highest power available was not sufficient to completely invert the spectrum. By evaluating only the maximum positive RYDMR-signal (spectrum (b)) J and k_T can be determined to be J=2.25±0.05 mT and k_T=6.0±0.05 · 10^8 s^{-1}, respectively. Taking into consideration the whole sequence of spectra the smallest ranges to which J and k_T can be confined are given by: 2.20 mT ≤ J ≤ 2.30 mT and 6.5 · 10^8 s^{-1} ≤ k_T ≤ 9.5 · 10^8 s^{-1}. These can be compared to the corresponding ranges determined for quinone-depleted RCs from *Rb. sphaeroides*, the only photosynthetic organism that has been investigated with the RYDMR-technique so far.

Table 1. Comparison of the forward ET rate k_1 and of the recombination parameters J and k_T in quinone-depleted RCs from *Chloroflexus aurantiacus* and *Rb. sphaeroides*, strain R26 (T=277±1 K).

	k_1 (10^{11} s^{-1})	J (mT)	k_T (10^8 s^{-1})
Chloroflexus aur.	1.25 [21]	2.25±0.05	8.0±1.5
Rb. sphaeroides	3.59±0.26 [5]	1.10±0.20 [12]	5.0±1.0 [12]

For comparison, values of the forward ET rate k_1 have been included in this table. Interestingly, k_1 seems to be correlated neither to J nor to k_T, whereas J and k_T both increase in going over from *Rb. sphaeroides* to *Chloroflexus aurantiacus*. The singlet recombination rate k_S has been evaluated from the absolute amount of triplet products and the lifetime of P^F to be k_S=5.0±1.0 · 10^7 s^{-1} for both species [22].

To find out the spread of the values of J and k_T over different bacterial species we also investigated RCs from the purple bacterium *Rb. capsulatus*, which phylogenetically is very closely related to *Rb. sphaeroides*. Figure 3 shows the B_1-spectrum of quinone-depleted RCs from *Rb. capsulatus*, wild type, in comparison with the B_1-spectra of quinone-depleted RCs from *Rb. sphaeroides*, strain R26, and *Chloroflexus aurantiacus*. From the positions of the maxima and from the respective widths of the spectra together with information from table 1 it can be inferred that both J and k_T slightly increase in the order *Rb. capsulatus* → *Rb. sphaeroides* → *Chloroflexus aurantiacus* (It should be emphasized, however, that this conclusion rests on the tacit assumption that none of the remaining recombination parameters, e.g. k_S, varies significantly among the three species.). The overall variation of J and k_T among the species investigated is surprisingly small, however.

Apart from being labels for intergeneric differences of the recombination dynamics J and k_T are also sensitive probes for variations of the recombination dynamics under the influence of external factors such as temperature, pH or electric fields applied to the sample. In particular, energetic or structural changes caused by site-directed mutagenesis of amino acid residues neighbouring the constituents of the RP can be detected from variations of J and/or k_T. As an example we investigated the recombination dynamics in quinone-depleted RCs from *Rb. capsulatus* where the Trp^{M250} residue located between BPh_A and Q_A had been replaced by leucine, phenylalanine and glutamic acid, respectively, in order to study its

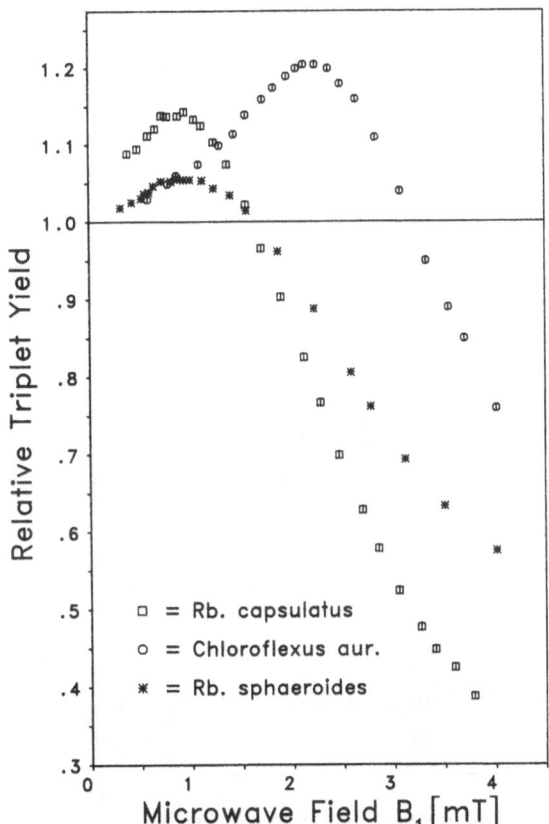

Figure 3. Experimental B_1-spectra of quinone-depleted reaction centers from *Chloroflexus aurantiacus*, *Rb. capsulatus* and *Rb. sphaeroides* , strain R26, at a sample temperature of 277 ± 1 K ($B_0=109.2\pm0.05$ mT). The positions of the maxima and the widths of the spectra allow a comparison of the values of J and k_T among the three species. For details see text.

role in the ET from BPh_A to Q_A. Analysis of the RYDMR data has not been finished yet but from the B_1-spectra shown in figure 4 it can be concluded that J and k_T slightly increase in the order Leu → Phe → Trp → Glu. The overall effect on the recombination dynamics, however, appears to be small.

In conclusion, we have applied the RYDMR technique to evaluate the exchange interaction J and the triplet recombination rate k_T in the RP state P^F in quinone-depleted RCs from the purple bacteria *Rb. sphaeroides* and *Rb. capsulatus* and the green bacterium *Chloroflexus aurantiacus*. We have also investigated effects on the recombination dynamics caused by site-directed mutagenesis of the Trp^{M250} residue in quinone-depleted RCs from *Rb. capsulatus*. All measurements were performed at a sample temperature of 277 ± 1 K. On condition that the singlet recombination rate k_S, the hyperfine and the magnetic spin dipole interaction are practically invariant among the species investigated we come to the following conclusions:

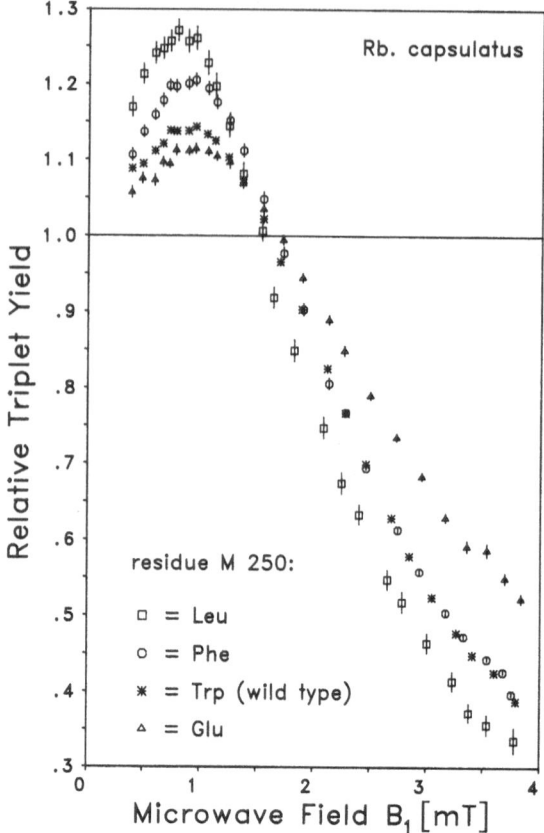

Figure 4. Experimental B_1-spectra of quinone-depleted RCs from *Rb. capsulatus* at a sample temperature of 277±1 K (B_0=109.2±0.05 mT). Comparison of wild type RCs with genetically modified RCs where the tryptophane residue M250 close to both BPh_A and Q_A has been replaced by leucine, phenylalanine or glutamic acid.

1) The overall variations of J and k_T are small (not more than a factor of two).
2) The forward ET rate k_1 (taken from the literature) seems to be correlated neither to J nor to k_T.
3) J and k_T tend to vary in parallel.

These results should be taken as preliminary since data analysis has not been fully completed yet. Temperature dependent measurements will complement the data and provide additional insight into ET mechanisms in the RC.

Acknowledgement

Financial support by the Deutsche Forschungsgemeinschaft (SFB 143) is gratefully acknowledged. We wish to thank Petra Tappermann for assistance in the RYDMR measurements.

References

[1] Parson, W.W. and Ke, B. (1982) "Primary Photochemical Reactions",in Govindjee (ed.), Photosynthesis, Volume 1: Energy Conversion by Plants and Bacteria, Acad. Press, New York, pp. 331-385.

[2] Ogrodnik, A., Remy-Richter, N., Michel-Beyerle, M.E. and Feick, R. (1987) "Observation of activationless recombination in reaction centers of *Rb. sphaeroides*. A new key to the primary electron-transfer mechanism", Chem. Phys. Lett. 135, 576-581.

[3] Bixon, M., Jortner, J., Michel-Beyerle, M.E., Ogrodnik, A. and Lersch, W. (1987) "The role of the accessory bacteriochlorophyll in reaction centers of photosynthetic bacteria: intermediate acceptor in the primary electron transfer ", Chem. Phys. Lett. 140, 626-630.

[4] Michel-Beyerle, M.E., Bixon, M. and Jortner, J. (1988) "Interrelationship between primary electron transfer dynamics and magnetic interactions in photosynthetic reaction centers", Chem. Phys. Lett. 151, 188-194.

[5] Martin, J.-L., Breton, J., Hoff, A.J., Migus, A., and Antonetti, A. (1986) "Femtosecond spectroscopy of electron transfer in the reaction center of the photosynthetic bacterium *Rhodopseudomonas sphaeroides* R26: Direct electron transfer from the dimeric bacteriochlorophyll primary donor to the bacteriopheophytin acceptor with a time constant of 2.8±0.2 psec", Proc. Natl. Acad. Sci. USA 83, 957-961.

[6] Marcus, R.A. (1988) "Early Steps in Bacterial Photosynthesis. Comparison of three mechanisms", in Breton, J. and Vermeglio, A. (eds.), The Photosynthetic Bacterial Reaction Center - Structure and Dynamics, NATO ASI Series, Series A: Life Sciences, Vol. 149, Plenum Press, New York, pp. 389-398.

[7] Woodbury, N.W.T. and Parson, W.W. (1984) "Nanosecond fluorescence from isolated photosynthetic reaction centers of *Rhodopseudomonas sphaeroides*", Biochim. Biophys. Acta 767, 345-361.

[8] Hoff, A.J. (1981) "Magnetic field effects on photosynthetic reactions", Quart. Rev. Bioph. 14, 599-665.

[9] Lersch, W. (1987) Thesis, TU München

[10] Frankevich, E.L. and Kubarev, S.I. (1982) "Spectroscopy of Reaction Yield Detected Magnetic Resonance", in Clarke, R.H. (ed.), Triplet State ODMR Spectroscopy, John Wiley, New York, pp. 137-183.

[11] Bowman, M.K., Budil, D.E., Closs, G.L., Kostka, A.G., Wraight, C.A. and Norris, J.R. (1981) "Magnetic resonance spectroscopy of the primary state, P^F, of bacterial photosynthesis", Proc. Natl. Acad. Sci. USA 78, 3305-3307.

[12] Lersch, W. and Michel-Beyerle, M.E. (1989) "RYDMR - Theory and Applications", in Hoff, A.J. (ed.), Advanced EPR in Biology and Biochemistry, Elsevier, Amsterdam, in press.

[13] Tang, J. and Norris, J.R. (1983) "Theoretical calculations of microwave effects on the triplet yield in photosynthetic reaction centers", Chem. Phys. Lett. 94, 77-80.

[14] Okamura, M.Y., Isaacson, R.A. and Feher, G. (1975) "Primary acceptor in bacterial photosynthesis: Obligatory role of ubiquinone in photoactive reaction centers of *Rhodopseudomonas sphaeroides*", Proc. Natl. Acad. Sci. USA 72, 3491-3495.

[15] Feick, R. (1980), Thesis, Universität Freiburg.

[16] Shiozawa, J.A., Lottspeich, F. and Feick, R. (1987) "The photochemical reaction center of *Chloroflexus aurantiacus* is composed of two structurally similar polypeptides", Eur. J. Biochem. 167, 595-600.

[17] Feick, R., in preparation.

[18] Prince, R.C. and Youvan, D.C. (1987) "Isolation and spectroscopic properties of photochemical reaction centers from *Rhodobacter capsulatus*", Biochim. Biophys. Acta 890,
286-291.
[19] Bylina, E.J. and Youvan, D.C. (1988), Proc. Natl. Acad. Sci. USA 85, 7226-7230.
[20] Coleman, W.J., Bylina, E.J. and Youvan, D.C., in preparation.
[21] Becker, M., Middendorf, D., Woodbury, N.W., Parson, W.W. and Blankenship, R.E.
(1986) "Picosecond Electron Transfer and Stimulated Emission in Reaction Centers of
Rhodobacter sphaeroides and *Chloroflexus aurantiacus*", in Fleming, G.R. and Siegman,
A.E. (eds.), Ultrafast Phenomena V, Springer, Berlin, pp. 374-378.
[22] Volk, M., Scheidel, G. and Ogrodnik, A., private communication.

PHOTOELECTRON TRANSFER REACTIONS: PORPHYRINS-QUINONE. FT- AND CW- TIME RESOLVED- EPR SPECTROSCOPIES.

H. Levanon
Department of Physical Chemistry
and the Fritz Haber Research Center for Molecular Dynamics
The Hebrew University
Jerusalem, 91904
Israel

ABSTRACT. Electron transfer reactions between porphyrins and chlorophylls in their photoexcited triplet state (electron donors) and duroquinone (electron acceptor) were studied by laser excitation-Fourier transform (FT)-EPR spectroscopy (~ 10 ns time resolution). This highly time-resolved EPR spectroscopy enables to differentiate between the triplet mechanism (TM) and radical-pair mechanism (RPM) which contribute to the overall CIDEP mechanism associated with the electron transfer process.

1. INTRODUCTION

Fast EPR has the advantage of unambiguously identifying the paramagnetic intermediates which directly (doublets) or indirectly (triplets) participate in the main photochemical pathway in photosynthesis. This is schematically demonstrated by Eqs. 1-2.

$$[AB]Q + h\nu \rightarrow [A^*B]Q \rightarrow [A^{\cdot+}B^{\cdot-}]Q \rightarrow [A^{\cdot+}B]Q^{\cdot-} \tag{1}$$

$$[AB]Q^{\cdot-} + h\nu \rightarrow [A^*B]Q^{\cdot-} \rightarrow [A^{\cdot+}B^{\cdot-}]^S Q^{\cdot-} \rightarrow [A^{\cdot+}B^{\cdot-}]^{T^o}Q^{\cdot-} \xrightarrow{\text{RP-ISC}} [A^T B]Q^{\cdot-} \tag{2}$$

$$\downarrow \text{SO-ISC}$$

$$[A^T B]Q^{\cdot-}$$

The former process describes the case where the final charge separation (electron transfer) is between the primary donor (A) and the quinone (Q). The second process describes the case where the quinone is reduced prior to photoexcitation. Such prereduction results in the production of a triplet state originating from triplet radical-pair (RP) which is different, in line shape and spin dynamics, form that formed via spin-orbit intersystem crossing (SO-ISC). Although, in principle, the triplet state can be formed in the process (1), its detection *in vivo* systems is remote. Inspection of Eqs. 1-2 shows that most of the intermediate states are paramagnetic, i.e., doublets and triplets with different lifetimes, spanning from ps to ms time range. It is clear therefore, that time-resolved EPR spectroscopy is most appropriate in monitoring some of the processes given by Eqs. 1-2. Of particular interest is the very recently application of Fourier transform (FT)-EPR spectroscopy combined with selective laser

J. Jortner and B. Pullman (eds.), Perspectives in Photosynthesis, 91–97.

excitation in the ~10 ns time regime [1-4]. The celebrated light driven process in primary photosynthesis is the charge separation and electron transfer to produce the anion radical of the quinone (doublet), where the precursor is the photoexcited singlet. On the other hand, it has been established that the photoexcited triplet state, although not in the main route of *in vivo* photosynthesis, may serve as a diagnostic probe for structure and spin dynamics [5]. Unfortunately, because of narrow bandwidth, FT-EPR is not capable yet to monitor triplets, thus leaving time-resolved CW-EPR detection as the most suitable EPR spectroscopy.

In this presentation we shall discuss an *in vitro* example where time resolved EPR spectroscopy is applied in two modes: i. FT-EPR spectroscopy in studying the spin dynamics of doublets in electron transfer reactions, in the liquid phase, between porphyrinoid chromophores and duroquinone (DQ); ii. Time resolved CW-EPR spectroscopy in studying the photoexcited triplet state of the porphyrinoid chromophores oriented in nematic liquid crystals. Combining the results gained in i. and ii. enables one to differentiate, in a direct way, the mechanisms of chemically induced dynamic spin polarization (CIDEP) associated with the electron spin dynamics [6]. Thus, the time evolution of transient EPR spectra of the anion radical, starting at sufficiently early time ($\tau_d \sim 10$ ns) after the laser pulse, show conspicuous changes in the intensity of the various hyperfine components, $I(m_I)$, due to non-Boltzmann spin populations. These intensity variations provide details of the CIDEP mechanisms under which the radical is produced [1-4,7]. A full analysis of the results provides with the following dynamic processes: i. Spin lattice relaxation rates of the triplet precursors in solution; ii. Donor-acceptor electron transfer reaction rates; iii. Spin lattice relaxation rates of the acceptor radical.

2. EXPERIMENTAL

2.1 FT-EPR Experiments

FT-EPR spectra were taken with a home-built (ANL) pulsed spectrometer [1,7] employing a rectangular TE_{102} cavity and quadrature detection. The FT experiments were combined with selective laser excitation at the Q-band optical transition of the particular chromophore. Light pulses of 1-2 mJ/pulse at a repetition rate of 100 Hz, were produced by an excimer laser pumping a dye laser (Lambda Physik). All FT-EPR experiments were carried out in the liquid phase (ethanol) by monitoring the anion radical of the duroquinone.

2.2 Time-Resolved CW EPR Experiments

Selective laser excitation-diode-detection CW EPR experiments (HUJ) were carried out for triplet detection by utilizing a Varian E-12 EPR spectrometer coupled to a Nd:Yag laser pumping a dye laser (Quanta Ray). Photoexcitation into the triplet state was achieved by selective excitation at the Q-band optical transition of the chromophore by light pulses of ~10 mJ/pulse at a repetition rate of 10 Hz. Chromophores were first oriented in the nematic phase of the liquid crystal (parallel orientation), and then quickly frozen to the desired low temperature. Rotating the sample by $\pi/2$ gives rise to the perpendicular orientation [8].

2.3 Material and Sample Preparation

Porphyrins were purchased from Midcentury Chemical Co. Pyrochlorophyll and Zn-pyrochlorophyll were prepared as described previously [4], and duroquinone was further purified by recrystallization from petroleum ether. Liquid crystal E-7 was purchased from BDH and all other solvents were of analytical grade and were used without further purification.

3. RESULTS AND DISCUSSION

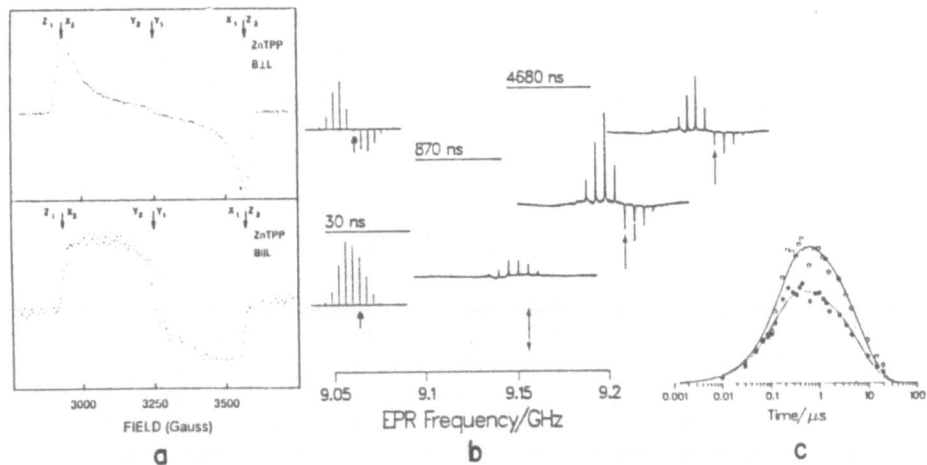

Figure 1. a. Diode-detection CW triplet EPR spectrum of ZnTPP (T ~116 K) ~10^{-3} M, in E-7 liquid crystal oriented perpendicular and parallel to the external magnetic field. The spectra were taken 400 ns after the laser pulse (564 nm, 20 mJ/pulse) and microwave power of 50 mW. The arrows indicate the canonical orientations. The low-field absorption and high-field emission correspond to out-of-plane spin polarizations $A_z > 0$, and a positive ZFS parameter, D. These results in the absorption character of the FT-EPR spectrum in early times. b. FT-EPR spectra of DQ $\cdot^{-}$ in ethanol at ~256 K, employing ZnTPP as electron donor, at different delay times (indicated) between the excitation and detection. The arrows indicate the central hyperfine line ($m_I = 0$). The arrows at the bottom spectra indicate the resonance frequency of the central line. Magnetic field: 3263.8 G. Microwave pulsewidth: 12 ns. Bandwidth of the cavity is ~30 MHz. It should be noticed that the highest frequency line corresponds to the lowest field in a magnetic field representation. The stick spectra (left) are the calculated intensities of 13 hyperfine lines utilizing Eq. 9. c. Time dependence of TM ($\cdot$) and RPM (O) contributions to the CIDEP effect of DQ $\cdot^{-}$ produced by ZnTPP^T. The TM and RPM polarizations were determined employing Eq. 9 (see text and stick spectrum in trace b.), and those data were analyzed in terms of Eq. 10 (smooth curves). Best fit parameters: $T_1 = 7.0\text{x}10^{-6}$ s, $k_{RPM} = 5.7\text{x}10^6$ s^{-1}, $k_{TM} = 7.9\text{x}10^6$ s^{-1};

In Figure 1b we show the time evolved FT-EPR spectra of DQ $\cdot^{-}$ generated by electron transfer from the photoexcited triplet of ZnTPP via the reaction:

$$ZnTPP + h\nu \rightarrow ZnTPP^* \tag{3}$$

$$ZnTPP^* \xrightarrow{\text{SO-ISC}} ZnTPP^T \tag{4}$$

$$ZnTPP^T + DQ \xrightarrow{k_{et}} ZnTPP \cdot^{+} + DQ \cdot^{-} \tag{5}$$

The results show that the polarization of the DQ $\cdot^{-}$ strongly depends on the selectivity of the intersystem crossing (ISC) channels which populate the triplet state sublevels of the porphyrin. This is reflected via the triplet mechanism (TM) in the transient EPR spectra detected at very early times. For example, the early spectrum at 30 ns shows that the polarization direction is in the absorption mode. At later times the spectral changes, i.e., the appearance of mixed absorption/emission lines in the spectrum, indicate that radical-pair mechanism (RPM) starts to dominate. These observations imply that the electron transfer can compete with the spin relaxation within the triplet spin states of the precursor [9], i.e., $k_{et} > 10^7$ s^{-1}. In Figure 1a we also show the triplet spectra of ZnTPP oriented in a nematic liquid crystal. Line shape analysis of the triplet spectrum [4,10] provides the magnetic ZFS parameters, D and E (including the absolute sign of D), as well as the population rates to the triplet sublevels, A_i (i = x, y, z). The contribution of TM to the polarization effect of the anion radical is given by Eq. 6 [11,12]:

$$P_{TM} \propto -[D (A_x + A_y - 2A_z) + 3E (A_y - A_x)] \tag{6}$$

If the early FT-EPR spectrum is governed by TM, then the polarization direction of the anion radical will be determined by the magnetic and triplet spin dynamic parameters of the precursor. In the present example of ZnTPP (Figure 1): from the triplet line shape analysis, D > 0 and $A_x:A_y:A_z \approx 0:0:1$, resulting in a positive polarization, i.e., net absorption of the early FT-EPR spectrum of $DQ^{\cdot-}$. Since these parameters vary among different chromophores, it is expected that the early FT-EPR spectra of the radical anion should reflect the magnetic and dynamic properties of the triplet chromophores. This certainly bears important implications on structural and dynamical properties of electron donors.

As indicated above, the time evolved FT-EPR spectra shown in Figure 1 originate from both TM and RPM. We analyze the temporal changes in the line intensities in terms of the TM and RPM models [13]. We fit the experimental results with stick spectra taking into consideration both RPM (including Δg between the donor and the acceptor radicals) and TM. The difference in g-values, at X-band EPR frequencies, corresponds to ~4 Gauss or ~2 proton hyperfine splittings of $DQ^{\cdot-}$ [14]. The expression for the line spectrum of $DQ^{\cdot-}$ due to RPM [15] is given by:

$$P_{RPM} (n, \Delta g) = I(n) * F(\Delta g - n)/(2 \sqrt{|F(\Delta g - n)|}) \tag{7}$$

where P_{RPM} (n,Δg) is the relative polarization of hyperfine line n of $DQ^{\cdot-}$ resulting from a single-line radical (donor) with g-factors differing by Δg, F(Δg,n) is the resonant frequency difference between hyperfine line n and the donor radical, while I(n) is the spectral intensity due to the binomial distribution of line intensities in unpolarized $DQ^{\cdot-}$:

$$I(n) = 12!/[(n + 6)! * (6-n)!] \tag{8}$$

The total relative polarization (including TM) for the stick spectrum is given by:

$$P_{Total} (n, \Delta g, tp) = (1 - |tp|) * P(n, \Delta g) + tp * I(n) \tag{9}$$

where tp is the relative contribution of TM to the total polarization.

Numerical fits were carried out on each set of experiments employing Eq. 9. The minimum-error fits give the TM and RPM, as a function of time, shown in Figure 1 (right trace) for the $DQ^{\cdot-}$

produced by the ZnTPP. The time dependence of the TM and RPM contributions were fit, when possible, by the function [1]:

$$p_j = A_j \left[\exp(-k_j t) - \exp(-t/T_1) \right] \qquad (10)$$

where j denotes the appropriate TM or RPM values, k_j^{-1} is the risetime of TM or RPM, while A_j is the relative amplitude, and T_1 is the DQ $\cdot^-$ spin-lattice relaxation time. Because these fits include all measurable line intensities, they give much better measures of TM and RPM than the earlier estimates based on intensities of only one or two lines per spectrum [3,16]. Thus, the rate k_{RPM} equals k_{et} while $k_{TM} = k_{et} + {}^3T_1^{-1}$. The ${}^3T_1$ derived from the k_{TM} are compatible with estimates of ${}^3T_1$ [12,17,18]

The experiments described above have been carried out with different porphyrin and chlorophyll precursors, and the spectral analysis of the triplets (liquid crystal) and doublets (ethanol solution) are summarized in Tables I and II. The results of FT-EPR spectroscopy combined with light excitation, show that the triplet spin memory transfer is very sensitive to: i. The branching ratio of the selective ISC populating rates, $A_x : A_y : A_z$, for $T_i \leftarrow S_1$. ii. The difference in g-values (Δg) between the cation radical of the donor (g = 2.0025 [19,20]) and the anion radical of the acceptor (g = 2.0049 [14]). These observations of spin memory effects in electron transfer reactions are of general importance since they allow differentiation and elucidation of the exact mechanism of the electron transfer process regarding both the donor, in its excited state, and the acceptor. It is clear that these experiments should be extended to model and *in vivo* systems where the triplet precursor can originate via RP-ISC. Such a mechanism of triplet production, which differs form SO-ISC, occurs in the reaction center and the resulting in triplet can be identified by EPR spectroscopy.

Table I. ZFS Parameters[a], Zero-Field Population and Depopulation Rates[a] of Triplet Precursors

Compound	$A_x : A_y : A_z$	$k_x : k_y : k_z$	$\lvert D \rvert$[b]	$\lvert E \rvert$[b]	Ref.
ZnTPP	~0 : ~0 : ~1		+ 0.0298	0.0098	3,10
MgTPP	>0 : >0 : ~0		0.0295	0.0075	3
ZnpChl*a*	~1 : ~0 : ~1		+ 0.0309	0.0038	4
ZnpChl*a*		.5 : .5 : 1			21
ZnChl*a*		.5 : .5 : 1	0.0306	0.0042	21
ZnChl*b*	.3 : .7 : 1		0.0328	0.0032	22
pChl*a*	.8 : 1 : .25		+ 0.0281	0.0028	4
pChl*a*		1 : .1 : .2			21
Chl*a*	.9 : 1 : .4		+ 0.0286	0.0038	22-24

[a] typical values, see also reference [24].
[b] in cm^{-1}.

Table II. Electron Transfer Rates, k_{et}, of the Reaction ^3P + DQ, and SLR Times of ^3P and DQ $^{..}$

Precursor (P)	k_{et} [a]	T_1 [b]	$^3T_1$ [b]
ZnTPP	11.4×10^8	7.00	.46
MgTPP	6.1×10^8	3.30	.23
ZnpChl$a$	1.87×10^8	7.25	
pChl$a$	8.84×10^8	8.80	

[a] $M^{-1}s^{-1}$. From best-fit analysis, Eq 10.
[b] μs. From best-fit analysis, Eq. 10.

4. ACKNOWLEDGEMENT

This presentation summerizs the work in refernces 3 and 4 which was carried out in collaboration with Professor J. R. Norris, Dr. M. K. Bowman, Dr. T. J. Michalski, Dr. A. Angerhofer (ANL), Dr. M. Toporowicz, A. Regev and A. Berman (HUJ). The work described above was supported by the U.S. Department of Energy, Office of Basic Energy Sciences, Division of Chemical Sciences under contract W-31-109-Eng-38. The research described herein was partially supported by a U.S.-Israel BSF grant (No. 86-00020, H.L.) and by a DFG grant (No. 132/8-1, H. L.); The DFG fellowship (AN-160/1, A.A.) and the support of the Israel National Council for Research and Development (M.T.) are highly acknowledged. The Fritz Haber Research Center is supported by the Minerva Gesellschaft für die Forschung, GmbH, München, FRG.

5. REFERENCES AND NOTES

1. Massoth, R. J. *Ph.D. Thesis,* University of Kansas, (1987).
2. T. Prisner, O. Dobbert, K. P. Dinse and H. van Willigen, *J. Am. Chem. Soc.* 110,   1622 (1988).
3. A. Angerhofer, M. Toporowicz, M. K. Bowman, J. R. Norris and H. Levanon,
    J. Phys. Chem. 92, 7164 (1988).
4. M. K. Bowman, M. Toporowicz, J. R. Norris, T. J. Michalski, A. Angerhofer and H. Levanon,
    Israel J. Chem. 28, 215 (1988).
5. J. R. Norris, this volume.
6. For a recent review on fast EPR detection in liquids, see e.g., A. D. Trifunac, R. G. Lawler,
    D. M. Bartels and M. C. Thurnauer, *Prog. Reaction Kinetics,* 14, 43 (1986).
7. A. Angerhofer, R. J. Massoth and M. K. Bowman, *Israel J. Chem.* in press.
8. H. Levanon, *Rev. Chem. Intermed.* 8, 287 (1987).
9. P. W. Atkins, A. J. Dobbs and K. A. McLauchlan, *Chem. Phys. Lett.* 29, 616 (1974).
10. O. Gonen and H. Levanon, *J. Phys. Chem.* 89, 1637 (1985).
11. S. K. Wong, D. A. Hutchinson and J. K. S. Wan, *J. Chem. Phys.* 58, 985 (1973).
12. P. W. Atkins and G. T. Evans, *Mol. Phys.* 27, 1633 (1974).
13. K. M. Salikhov, Yu. N. Molin, R. Z. Sagdeev and A. L. Buchachenko in:
    Spin Polarization and Magnetic Effects in Radical Reactions,
    (Yu.N. Molin, Ed.); Elsevier: Amsterdam, (1984).
14. Landolt-Börnstein, New Series, Group 12, Vol. 9, Springer-Verlag (Berlin, 1980).

15. F. J. Adrian, *J. Chem. Phys.* **54**, 3918 (1975).

16. M. Pluschau, A. Zaht, K. P. Dinse and H. van Willigen, *J. Chem. Phys.* in press.

17. S. I. Weissman, *J. Chem. Phys.* **29**, 1189 (1958).

18. D. E. Budil, *Ph.D. Thesis* University of Chicago, (1987).

19. J. Fajer, D. C. Borg, A. Forman, R. H. Felton, L. Vegh and D. Dolphin, *Anal. New York Acad. Sci.* **206**, 349 (1973).

20. A g-value of 2.0025 is also assumed for the cation radicals of pChl*a* and ZnpChl*a* (Dr. J. Fajer private communication).

21. R. H. Clarke, D. R. Hobart and W. R. Leenstra, *J. Am. Chem. Soc.* **101**, 2416 (1979).

22. M. C. Thurnauer and J. R. Norris, *Chem. Phys. Lett.* **47**, 100 (1977).

23. A. Regev, O. Gonen and H. Levanon, (unpublished results).

24. H. Levanon and J. R. Norris, *Molecular Biology and Biophysics* (Fong F. K. Ed.) Springer-Verlag: Berlin, **35**, 155 (1982).

STRUCTURE AND MARKER MODE OF THE PRIMARY ELECTRON DONOR STATE ABSORPTION OF PHOTOSYNTHETIC BACTERIA: HOLE BURNED SPECTRA

D. Tang, S. G. Johnson, R. Jankowiak, J. M. Hayes, G. J. Small
Ames Laboratory-USDOE and Dept. of Chemistry
Iowa State University, Ames, Iowa 50011 USA

and

D. M. Tiede
Chemistry Division, Argonne, National Laboratory, Argonne, IL 60439 USA

ABSTRACT. Structured photochemical hole burned spectra are presented for P870 and P960 of the reaction centers (RC) of *Rhodobacter sphaeroides* and *Rhodopseudomonas viridis* . A special pair marker mode (ω_{sp}) Franck-Condon progression is identified for both P870 and P960. Zero-phonon holes are reported which yield P870* and P960* decay times in good agreement with the time domain values. This agreement suggests that vibrational thermalization occurs prior to the primary charge separation process. The theory of Hayes and Small [1], embellished for the marker mode progression, is shown to account for the primary donor state absorption and burn-wavelength dependent hole spectra. Site excitation energy selection is used to establish correlation between a higher energy RC state and P* for both bacteria.

I. Introduction

The recently revealed structures of the reaction center (RC) of *Rps. viridis* [2-4] and *Rb. sphaeroides* [5-7] have led to even greater activity [8,9] directed towards understanding the primary charge separation process which is triggered by excitation of the primary electron donor states P960* and P870*, respectively. It is generally accepted that the lowest energy component of the special pair Q_y transition contributes significantly to the electronic structure of P*. This component is often referred to as P$_-$ since the Q_y transition dipoles of the monomers, which comprise the pair (P_L, P_M), would be anti-parallel in the simplest model. It should be noted, however, that agreement on the structures of the higher energy Q_y states of the six-pigment RC has not been arrived at on the basis of semi-empirical calculations [10,11].

In this paper we present new photochemical hole burned (PHB) spectra for P960 and P870 which reveal that their underlying structures are very similar, identify a low frequency marker mode for both special pairs, reveal the origin of the homogeneous broadening for the P960 and P870 absorption profiles and determine the decay times of P960* and P870* from their zero-point levels at 4.2 K. The decay times are used to address the question of whether or not ther-

99

J. Jortner and B. Pullman (eds.), Perspectives in Photosynthesis, 99–120.
© 1990 *Kluwer Academic Publishers*.

malization of vibrational modes occurs prior to charge separation. In addition, a site excitation energy correlation effect for *Rps. viridis* and *Rb. sphaeroides* is observed which identifies an excited RC state that is correlated with P_-. An assignment of the former to a state which is significantly contributed to by P_+ (upper special pair component) is suggested.

The λ_B (burn-wavelength)-dependent PHB spectra of Boxer and his group [12,13] first revealed that there is a significant homogeneous broadening contribution to the P870 and P960 absorption profiles. This result attracted considerable attention [14,17] since the broad (~ 400 cm^{-1}) unstructured holes observed could be a manifestation of ultra-fast electronic relaxation [15,16] and/or significant protein-pigment geometry changes [1,17] which accompany electronic excitation of P. We present further evidence [18,19] that proves that the latter is the correct model.

We recently reported structured transient PHB spectra (4.2K) for P960 for three glass-detergent host systems [18,19]. The structure included two relatively broad (~ 100 cm^{-1}) holes denoted as X and Y, with Y assigned as a ~ 130 cm^{-1} ($\equiv \omega_{sp}$ in what follows) vibronic hole which builds to higher energy on X. The ω_{sp} mode was also identified from the spectrum of PHB⁻Q⁻ (B ≡ BChl monomer, H ≡ BPheo, Q ≡ quinone) produced by narrow line laser irradiation into P960 of PBHQ⁻ [19]. We note that the ω_{sp} feature in the (PBH⁻Q⁻ - PBHQ⁻) difference spectrum had been observed much earlier by Vermeglio and Paillotin [20] under non-line narrowing conditions; however, it was not assigned as a vibration. In addition, in refs. [18,19] a weak but relatively sharp (~ 10 cm^{-1}) zero-phonon hole (ZPH) coincident with λ_B was observed when λ_B is located in the vicinity of the low energy shoulder (see below) of the P960 absorption profile for PBHQ. It was concluded that X correlates with this shoulder and, furthermore, that the 10 cm^{-1} hole is the ZPH of X. Hole X was assigned as the intramolecular zero-point level (ω_{sp}^0) of P960. The ZPH width of ~ 10 cm^{-1} yielded a P960* decay time of 1 ps at 4.2 K which is in reasonable agreement with the 0.7 ± .1 ps value measured in the time domain at 10 K [21,22]. The weak intensity of the ZPH relative to X (~ 1:100) was ascribed to moderately strong linear electron-phonon coupling (Huang-Rhys factor S ~ 2) involving low frequency protein phonons [1,17]. Arguments, based mainly on the experimental gating employed and the λ_B-dependence of the hole spectra, were presented for the PHB structure (incl. the ZPH) not being due to impurity, e.g. stable PBH⁻Q⁻ afforded by irradiation of PBHQ⁻ in the presence of cytochrome. Observation of the same structure for P870 of *Rb. sphaeroides* would provide an even stronger argument against impurity since the cytochrome is absent and especially if the ZPH holewidth were to yield a decay time for P870* in good agreement with the time domain value of 1.2 ± .1 ps at 10 K [21,22]. The results presented here provide definitive proof that the low energy shoulder of P870 and P960 absorption profiles as well as the structure reported in refs. [18,19] for the PHB spectra of P960 are intrinsic to the RC.

II. Experimental

SAMPLE PREPARATION

Fresh samples of reaction centers (RC) from *Rps. viridis* and *Rb. sphaeroides* were prepared by dissolving RC crystals in suitably buffered hosts. Details concerning crystallization procedure can be found in Ref. 5. *Rps. viridis* and *Rb. sphaeroides* were prepared either in glycerol:water glass (2:1, hereafter referred to as glycerol glass) with 0.1% LDAO detergent, 10 mM Tris, pH = 8.0 and/or in glycerol glass in the presence of 0.8% n-octyl-β-D glucopyranoside detergent, 10 mM Tris, 1 mM BDTA, pH = 8.0. RC samples in polyvinyl alcohol films (PVOH) (~ .1 mm thick) were also utilized. The optical density (OD) of the samples utilized in this paper was less than .5 at the peak of the primary donor state absorption.

MEASUREMENTS

The block diagram of the experimental system employed is shown and described in Figure 1. Burn irradiation (linewidth 0.2 cm^{-1}) was provided by the Raman shifted (H_2 gas) output of an excimer-pumped dye laser, Lambda Physik EMG 102 and FL-2002, respectively. A pulse repetition rate of 16 and 20 Hz was utilized for *Rb. sphaeroides* and *Rps. viridis*, respectively. Raman shifted pulse energies utilized were ≤ .4 mJ (focused to either a .3 cm diameter spot or a .2 cm x .8 cm spot). Samples were mounted and cooled to T = 4.2 K in a Janis model 8-DT super vari-temp liquid helium cryostat. All PHB spectra reported here for P870 and P960 are for delta transmission (ΔT) changes ≤ 20%, although by utilizing higher pulse energies bleaching of ~ 80% could be obtained. Measurements were made by employing a Stanford Research SR250 boxcar averager interfaced with a IBM-PC compatible computer. The gate delay (triggered by the 10 nsec laser pulse) for the boxcar was 2-3 ms and the gatewidth was 150 µsec. The ΔT and ΔA (absorbance) spectra were obtained by subtracting laser-on and laser-off spectra. Interference from the laser was eliminated by the gating and by using a mechanical shutter between the sample and the photomultiplier tube. In addition, any laser scatter would be detected as a broad range baseline deviation and specific interference with any sharp features would be negligible. By varying the gate delay it was determined that the lifetimes of the charge separated bottleneck state P^+BHQ^- for *Rps. viridis* and *Rb. sphaeroides* (4.2 K) are 8 ± 1 ms and 34 ± 3 ms, respectively. A large number of burn wavelength values were employed, only several examples of representative spectra are shown. Absorption spectra were obtained with a Bruker IFS 120 HR Fourier-transform infrared (visible) spectrometer operating at a resolution of 4 cm^{-1}.

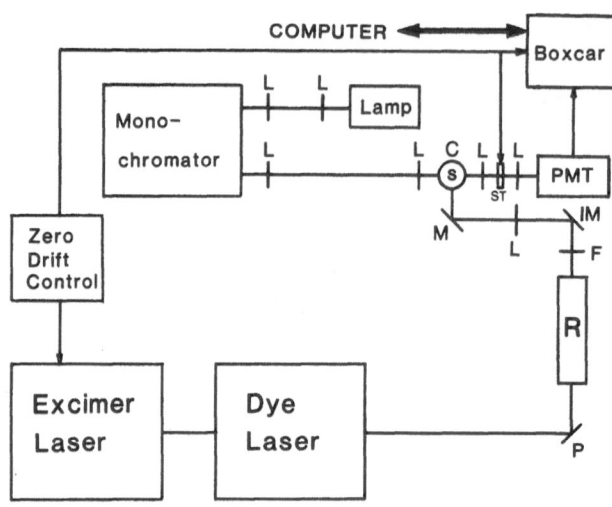

Fig. 1.   Block diagram of experimental apparatus. L= lens, M = mirror, IM = infrared
            mirror, ST = mechanical shutter, R = Raman shifter, S = sample, P = prism, C =
            liquid helium cryostat, F = filter, PMT= photomultiplier tube.

III. Results and Discussion

Transient PHB spectra for the Q_y-region of *Rps. viridis* and *Rb. sphaeroides* will be presented
for two host/detergent combinations; glycerol/LDAO and PVOH/LDAO for the former and
glycerol/NGP and PVOH/NGP for the latter. The 4.2 K absorption spectra are shown in
frames A and B of Fig. 2. The FWHM of P960 and P870 are given in the caption. These
spectra and those for *Rps. viridis* in glycerol/NGP and PVOH/NGP (not shown) show that
PVOH produces a significant blue shift and increase in the FWHM of the PED state
absorption profile relative to glycerol [17] host. In spectrum 1 of frame A the bands at 810
and 790 nm are labeled in the usual way as H_L and H_M while the band at 835 nm is labeled
as B_L and B_M (unresolved). Of course such assignments should be viewed as approximate
since electronic structure calculations suggest significant coupling between the Q_y-transitions
of the six-pigment aggregate [11]. The shoulder at ~ 852 nm of spectrum 1 has been assigned
by Vermeglio et al. [20] to P_+, the upper dimer component of the Q_y-transition of the special

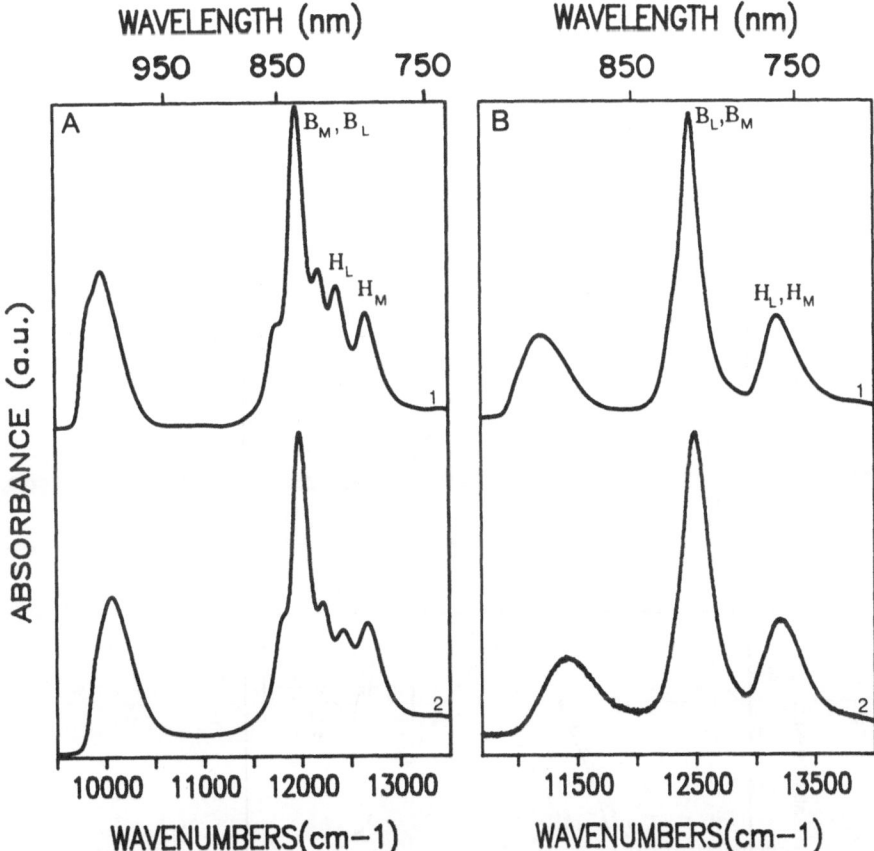

Fig. 2.   Absorption spectra of the Q_y region, T = 4.2 K: A) *Rhodopseudomonas viridis*:
1.) glycerol/LDAO (FWHM = 420 cm^{-1}), 2.) PVOH/LDAO (470 cm^{-1}). B)
Rhodobacter sphaeroides: 1.) glycerol/NGP (470 cm^{-1}), 2.) PVOH/NGP (550
cm^{-1}).

pair. We present data later that are consistent with this assignment. Comparison of the frame
A and B spectra shows that the H_L and H_M transitions are not resolved for *Rb. sphaeroides*
and, furthermore, that the "P_+" transition, which appears on the low E side of the (B_M,B_L)
band, is difficult to discern for *Rb. sphaeroides*. It is also evident that the reduction in
inhomogeneous broadening provided by glass hosts (relative to PVOH) leads to improved
spectral resolution. This is also the case for the transient PHB spectra, vide infra.

Absorption profiles of P960 and P870 are shown in Fig. 3 for glycerol/NGP and
glycerol/LDAO glass solvents, respectively. The low E shoulder of P960 is more evident than
for P870 (the second derivative spectrum of the P870 profile clearly reveals the shoulder, see

insert spectrum). Further evidence [18,19] for the shoulder being the origin (ω_{sp}^0) band of the PED state absorption profile is presented below.

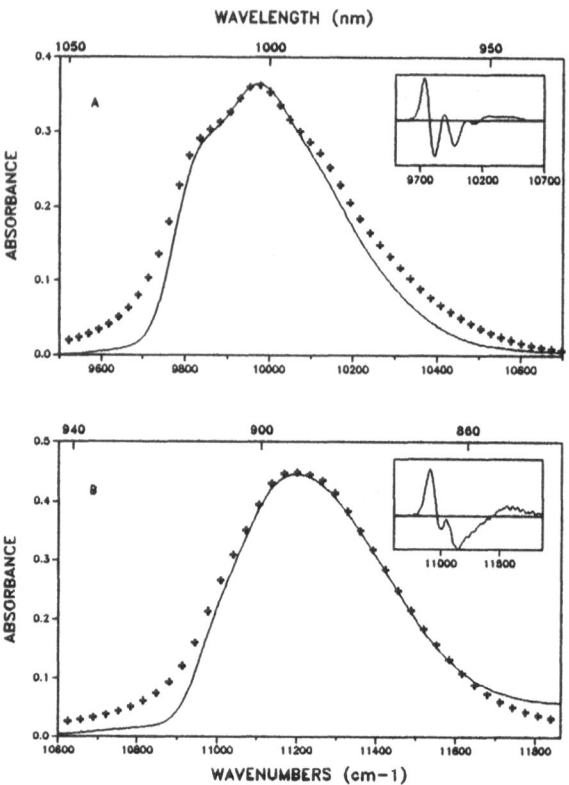

Fig. 3.   Calculated and experimental absorption spectra for: A) P960. Parameters for calculated spectrum (++++): Γ (one phonon profile width) = 50 cm^{-1}, ω_m (mean phonon frequency) = 40 cm^{-1}, S (Huang-Rhys factor) = 1.5, Γ_I (inhomogeneous linebroadening) = 120 cm^{-1}, ω_{sp} = 150 cm^{-1}, S_{sp} = 1.1. B) P870. Parameters for calculated spectrum: Γ = 50 cm^{-1}, ω_m = 35 cm^{-1}, S = 2.0, Γ_I = 130 cm^{-1}, ω_{sp} = 125 cm^{-1}, S_{sp} = 1.55. Four overtones of ω_{sp} were utilized in both cases.

A. Transient hole burned spectra of P870 and P960

Transient ΔA PHB spectra for P870 (glycerol/NGP) are shown in Fig. 4 for four λ_B-values. ΔT spectra are also presented for $\lambda_B = 907$ and 910 nm in order to illustrate the improved resolution provided by the ΔT-mode when the absorbance is sufficiently high. The ZPH coincident with λ_B becomes more pronounced as λ_B is decreased from the optimum value. Such behavior is consistent with the theory of hole burning in the presence of moderately strong linear electron-phonon coupling and inhomogeneous broadening [1,17], vide infra. The ZPH widths and their relationship to the depopulation decay time of P870[*] are discussed in III.B. In spectra 2a and 2b of Fig. 4 one observes a broader hole (indicated by the dashed arrow) which is displaced by ~ 125 cm^{-1} to higher energy of the ZPH (λ_B). It is hole Y (ω_{sp}^1) for P870.

The vibronic hole structure associated with ω_{sp} is more apparent for P960, as shown by Fig. 5 (glycerol/LDAO), primarily because ω_{sp} (P960) is significantly higher than for P870 (by about ~ 20%) and the P960 absorption profile is narrower than the P870 profile (420 vs. 475 cm^{-1} FWHM, glycerol/glass). In Fig. 5 three quanta of ω_{sp} ~ 150 cm^{-1} are discernible (see dashed arrows) and the ZPH coincident with λ_B is observable in spectra 2 and 3b. The ZPH for P960 is generally more difficult to detect than for P870 because it is considerably broader, cf. section III.B.

The spectra of Figs. 3-5 demonstrate that the underlying structure of the PED state absorption profile for photosynthetic bacteria is strongly contributed to by the ω_{sp}-progression with a Huang-Rhys factor, S_{sp}, in the vicinity of unity. However, the weakness of the ZPH belonging to the broader ω_{sp}^0 (origin) hole indicates that the coupling of the $P_- \leftarrow P$ transition to lower frequency phonons must be moderately strong. As discussed in section III.B., the suggestion that P_- undergoes ultra-fast electronic relaxation into CT states prior to formation of P^+BH^- [15] finds no support from our data. Thus, we proceed to investigate the degree to which the theory of Hayes et al. [17] can account for the absorption and PHB spectra of P870 and P960. This theory was developed to explain the previously reported [12,13] unstructured hole spectra of P870 and P960 and, to this end, the mean phonon frequency (ω_m) approximation was utilized. Because the Franck-Condon factors for the intramolecular modes of BChl are very small (≤ 0.04) [23,24], they were reasoned to be unimportant for understanding the principal features of the PHB spectra of P_- [17]. However, with the observation of the ω_{sp}-progression, the theory must be augmented to include the electron-vibration coupling due to ω_{sp}. The procedure for doing so is straightforward and is briefly outlined in what follows.

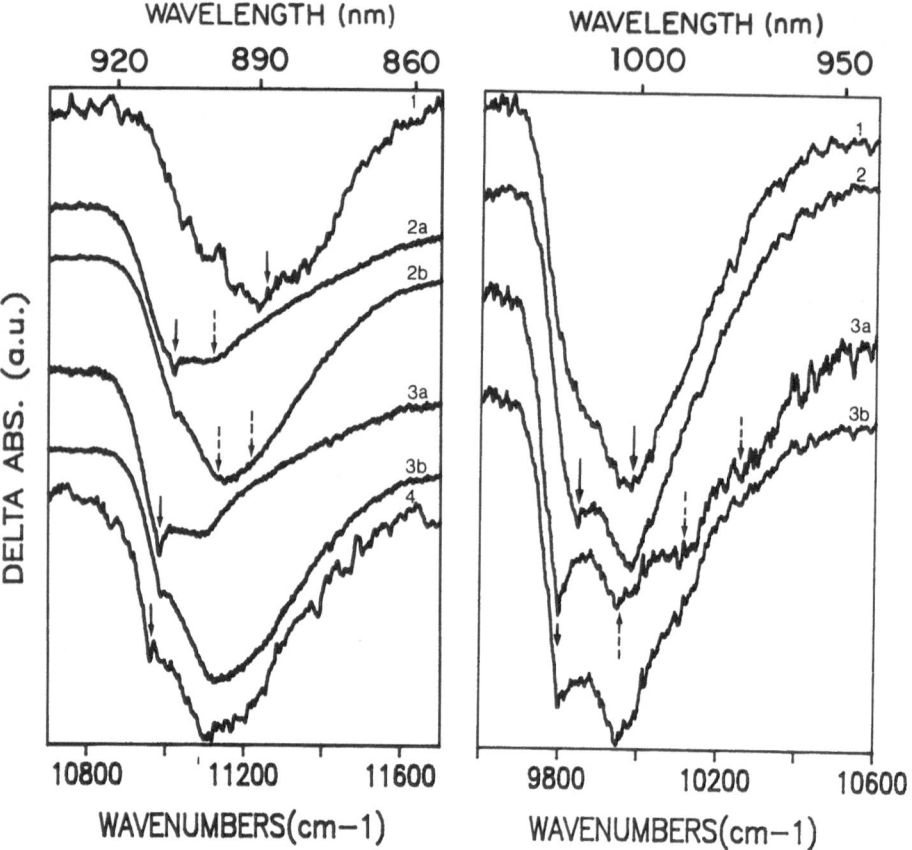

Fig. 4. Hole burned spectra for P870, T = 4.2 K. Solid arrows locate λ_B = 890, 907, 910, and 912 nm for spectra 1-4. All spectra are ΔA except for 2a and 3a which are ΔT spectra corresponding to 2b and 3b, respectively. Resolution ≤ 8 cm^{-1}. Dashed arrows in 2a and 2b indicate approximate positions of ω_{sp}^1 and ω_{sp}^2 satellites.

Fig. 5. Hole burned spectra for P960, T = 4.2 K. Solid arrows locate λ_B = 1000, 1015, 1020 nm for spectra 1-3. All spectra are ΔA except for 3a which is ΔT spectrum corresponding to 3b. Resolution = 8 cm^{-1}. Dashed arrows in 3a indicate 1st, 2nd, and 3rd quanta of ω_{sp} satellite holes.

Within the mean phonon frequency approximation Hayes et al. [17] express the low temperature absorption profile of a single site as

$$L(\Omega-\nu) = e^{-S}\ell_0(\Omega-\nu) + \sum_{r=1}^{\infty} \frac{S^r e^{-S}}{r!}\ell_r(\Omega-\nu-r\omega_m).\tag{1}$$

where ν is the zero-phonon transition frequency and ω_m is the mean frequency for phonons which couple to the electronic transition. The Huang-Rhys factor is S and the Franck-Condon factors for the $r = 0, 1, ...$ phonon transitions are governed by the Poisson distribution $\{S^r e^{-S}/r!\}_r$. Thus, the Franck-Condon factor for the zero-phonon transition is exp(-S); its profile is a Lorentzian (l_0) with a FWHM = γ, which is the homogeneous linewidth of the zero-phonon line. The lineshape for the one-phonon profile is l_1 and is centered at $\nu + \omega_m$ with a FWHM of Γ. It is well known that the one-phonon profiles for electronic transitions of molecules imbedded in amorphous solids carry a width of about 30 cm^{-1} and the profiles for antenna Chl a and b are no exception [23]. To a good approximation the profile can be taken to be a Gaussian. Equation 1 is valid for coupling to a pseudo-localized phonon or a distribution of host phonons governed by a suitable density of states. For the latter case and a one-phonon profile governed by a Gaussian, the width of the r-phonon profile (centered at $\nu +$ rω_m) is given by $\Gamma_r = r^{1/2}\Gamma$. In order to derive an analytic expression for the hole profile Lorentzians for l_r (r≥1) were used [1] with widths governed by the Gaussian values, i.e. $r^{1/2}\Gamma$. Since that work it has been shown that the differences in the hole spectra calculated with Lorentzians and Gaussians are small [25].

Now Eq. 1 is readily modified to include coupling to the ω_{sp}-mode:

$$L(\Omega-\nu) = \sum_{j=0}^{\infty} \frac{S_{sp}^j e^{-S_{sp}}}{j!}\left[e^{-S}\ell_0^j(\Omega-\nu-j\omega_{sp})\right.$$
$$\left. + \sum_{r=1}^{\infty} \frac{S^r e^{-S}}{r!}\ell_r^j(\Omega-\nu-r\omega_m-j\omega_{sp})\right].\tag{2}$$

where S_{sp} is the Huang-Rhys factor for ω_{sp}. In writing Eq. 2 the reasonable assumption that the electron-phonon coupling (S) is independent of the ω_{sp}-mode occupation number j is made. Similarly, the mean phonon frequency ω_m and Γ, the width of the one-phonon profile, are to be considered independent of j. However, the homogeneous linewidths γ_j of the zero-phonon l_0^j functions may differ due, for example, to rapid vibrational relaxation of the ω_{sp}^j (j ≥ 1) levels.

For disordered hosts a Gaussian distribution of zero-phonon transition frequencies of width Γ_I is the appropriate choice but in order to obtain an analytic expression for the hole profile a Lorentzian is utilized, $N_0(\nu-\nu_m)/N$ where N is the total number of absorbers and ν_m is the mean zero-phonon frequency. The absorption spectrum is calculated as the convolution of this distribution function with the single site absorption profile $L(\Omega-\nu)$. We define the absorption cross-section, laser intensity and photochemical quantum yield as σ, I and ϕ. Then following a burn for time τ

$$N_\tau(\nu - \nu_m) = N_0(\nu - \nu_m)e^{-\sigma I\phi\tau L(\omega_B - \nu)}. \tag{3}$$

where ω_B is the laser burn frequency and $L(\omega_B-\nu)$ is given by Eqn. 2. To obtain the absorption spectrum, A_τ, following the burn we must convolve Eqn. 3 with $L(\Omega-\nu)$ and integrate over ν. For notational simplicity, Eq. 1 rather than Eq. 2 is employed in what follows. The modifications of the resulting hole shape function necessary to take into account the ω_{sp}-progression will simply be stated. Thus,

$$A_\tau(\Omega) = \sum_{r=0}^{\infty} \frac{S^r e^{-S}}{r!} \int d\nu N_0(\nu - \nu_m)e^{-\sigma I\phi\tau L(\omega_B - \nu)} \ell_r(\Omega-\nu - r\omega_m). \tag{4}$$

For simplicity the short-burn-time limit is employed so that the exponential can be expanded as $1 - \sigma I\phi\tau L(\omega_B-\nu)$. This approximation need not be made, although the resulting expressions are very cumbersome if it is not. The hole spectrum in the short-burn-time limit is simply

$$A_0(\Omega) - A_\tau(\Omega) = \sigma I\phi\tau \sum_{r,r'=0}^{\infty} \left(\frac{S^r e^{-S}}{r!}\right)\left(\frac{S^{r'} e^{-S}}{r'!}\right)$$
$$\cdot \int d\nu N_0(\nu - \nu_m)\ell_r(\Omega-\nu - r\omega_m)\ell_{r'}(\omega_B - \nu - r'\omega_m). \tag{5}$$

Because we are interested in holes whose widths are comparable to Γ_I we cannot assume that $N_0(\nu-\nu_m)$ is constant in Eqn. 5. Integration of Eqn. 5 yields

$$[A_0 - A_\tau](\Omega) = \frac{\sigma\phi\tau}{3(2\pi)^2} \sum_{r,r'=0}^{\infty} \left[\left(\frac{S^r e^{-S}}{r!} \right) \left(\frac{S^{r'} e^{-S}}{r'!} \right) \right]$$

$$\cdot \left[\left\{ \frac{\Gamma_I + \Gamma_r}{(\Omega - \nu_m - r\omega_m)^2 + \left[\frac{\Gamma_I + \Gamma_r}{2} \right]^2} \right\} \left\{ \frac{\Gamma_r + \Gamma_{r'}}{(\Omega - \omega_B + \omega_m(r' - r))^2 + \left[\frac{\Gamma_r + \Gamma_{r'}}{2} \right]^2} \right\} \right.$$

$$+ \left\{ \frac{\Gamma_I + \Gamma_{r'}}{(\omega_B - \nu_m - r'\omega_m)^2 + \left[\frac{\Gamma_I + \Gamma_{r'}}{2} \right]^2} \right\} \left\{ \frac{\Gamma_r + \Gamma_{r'}}{(\Omega - \omega_B + \omega_m(r' - r))^2 + \left[\frac{\Gamma_r + \Gamma_{r'}}{2} \right]^2} \right\}$$

$$+ \left\{ \frac{\Gamma_I + \Gamma_{r'}}{(\omega_B - \nu_m - r'\omega_m)^2 + \left[\frac{\Gamma_I + \Gamma_{r'}}{2} \right]^2} \right\} \left\{ \frac{\Gamma_I + \Gamma_r}{(\Omega - \nu_m - r\omega_m)^2 + \left[\frac{\Gamma_I + \Gamma_r}{2} \right]^2} \right\} \right]. \quad (6)$$

The qualitative implications of Eqn. 6 are discussed by Hayes et al. [17]. Model calculations with realistic values for Γ, ω_m, γ and Γ_I are given in the same paper for various values of S ranging from 0.5 (weak coupling) to 8.0 (strong coupling). In Eqn. 6, $\Gamma_0 = \gamma$ and $\Gamma_r = r^{1/2}\Gamma$ ($r \geq 1$). For strong coupling ($S \geq 2$) and $\omega_B \sim \nu_m$, the intensity of the ZPH relative to the broad hole is given approximately by $\exp(-2S)$. For this value of ω_B, the ZPH is located near the center of the broad and more intense hole upon which it is superimposed. For $\Gamma_I \geq S\omega_m$ a burn with ω_B located on the low and high energy sides of the absorption profile produces broad hole profiles that are shifted to the blue and red, respectively, of ω_B.

The modifications of Eqn. 6 required to take into account the ω_{sp}-progression are as follows: first, an additional double summation $\sum_{j,j'=0}(\exp(-S_{sp})S_{sp}^j/j!)(\exp(-S_{sp})S_{sp}^{j'}/j'!)$ must be included and $\Gamma_r, \Gamma_{r'}$ replaced everywhere by Γ_{rj} and $\Gamma_{r'j'}$, respectively. The latter damping constants are defined as $\Gamma_{rj} = \Gamma_r$ for $r \geq 1$ and $= \gamma_j$ for $r = 0$. Thus, γ_j determines the relaxation frequency of the zero-phonon level associated with the jth member of the ω_{sp}-progression; second, the energy denominators are modified by the replacements $r\omega_m \rightarrow r\omega_m + j\omega_{sp}$ and $r'\omega_m \rightarrow r'\omega_m + j'\omega_{sp}$.

For the calculations it was found sufficient to terminate the j- and r-sums at 4 and 10, respectively. It should be emphasized that the PHB spectra provide good first approximations for the values of ω_{sp}, S_{sp}, and S (since the ratio of the intensity of the ZPH to the more intense ω_{sp}^0 phonon sideband hole (hole X) is $\sim \exp(-2S)$ for $\omega_B \sim \nu_m$ [17]) and a direct measurement of γ_0, cf. section III.B. Furthermore, the Stokes shift is given by $\sim 2S\omega_m$. Our absorption data (specifically the energy of the ω_{sp}^0 low E shoulder for P870 and P960) together with the low T fluorescence spectra for P870 [26] and P960 [27] provide approximate values of 140 and 150 cm^{-1} for the Stokes shift of P870 and P960. Thus, an estimate for ω_m is available. Furthermore, the low T absorption linewidth of the PED state is given roughly by $S\omega_m + S_{sp}\omega_{sp} + \Gamma_I$ so that an estimate for Γ_I can be made. The point is that in fitting the absorption and PHB spectra one cannot vary the values for ω_{sp}, S_{sp}, ω_m, S and Γ_I too far from the estimated values.

The calculated λ_B-dependent hole spectra for P870 (glycerol/NGP) and P960 (glycerol/LDAO) are given in Figs. 6 and 7, respectively. The parameter values utilized are given in the captions. We note in particular that ω_{sp} = 125 and 150 cm^{-1} for P870 and P960, respectively. Spectra 1, 4, 5, 6 of Fig. 6 can be directly compared with spectra 1, 2b, 3b, 4 of Fig. 4 while spectra 2, 3, 4 of Fig. 7 can be compared with spectra 1, 2, 3b of Fig. 5. The λ_B-dependence of the PHB spectra is reasonably well accounted for by the theory; in particular, the loss of line narrowing and elimination of the ZPH as λ_B is decreased from the value corresponding to ~ v_m. These features of the λ_B-dependence are principally the consequence of an increasing probability for multi-phonon excitation (non-line narrowing) as λ_B is decreased from the v_m-value. For a sufficiently low value of λ_B, essentially all the structure in the spectrum is lost and further reduction of λ_B produces no further change in the spectrum. In the calculations, allowance was made for sub-ps decay of the ω_{sp}^j (j ≥ 1) levels with the decay proportional to j^{-1} (Fermi-Golden rule prediction with cubic intermolecular anharmonicity). The decay times of ω_{sp}^1 for P870 and P960 were set equal to 260 and 350 fs but no significance should be attached to the difference in these values. In the absence of sub-ps decay the calculations predict that vibronic satellite ZPH [28] associated with the ω_{sp}-progression should be observed for λ_B-values that produce a ZPH in the ω_{sp}^0 band. Repeated attempts to observe such features met with no success for both P870 and P960. Additional arguments for sub-ps decay are given in the following sub-section.

The calculated absorption spectrum for P870 and P960, which correspond to the PHB spectra of Figs. 6 and 7, are given in Fig. 3 where they are compared with the experimental spectra. Agreement is reasonable except on the low E tail. The disagreement on the low E tail is primarily the result of utilizing a Lorentzian for the Γ_I-distribution function. Unfortunately, utilization of a Gaussian precludes derivation of an analytic expression for the hole profile and, as a consequence, would lead to a significant increase in computation time. A comparable disagreement on the low E side between the calculated and observed PHB spectra exists for the same reason. We hasten to add, however, that the important features of the absorption and λ_B-dependent hole spectra are accounted for by the theory and that calculations with a Gaussian for the Γ_I-distribution function are not expected to lead to significant changes in the values of the parameters given in the captions to Figs. 6 and 7. Thus, we conclude that approximately 70% and 30% of the P870 and P960 absorption widths

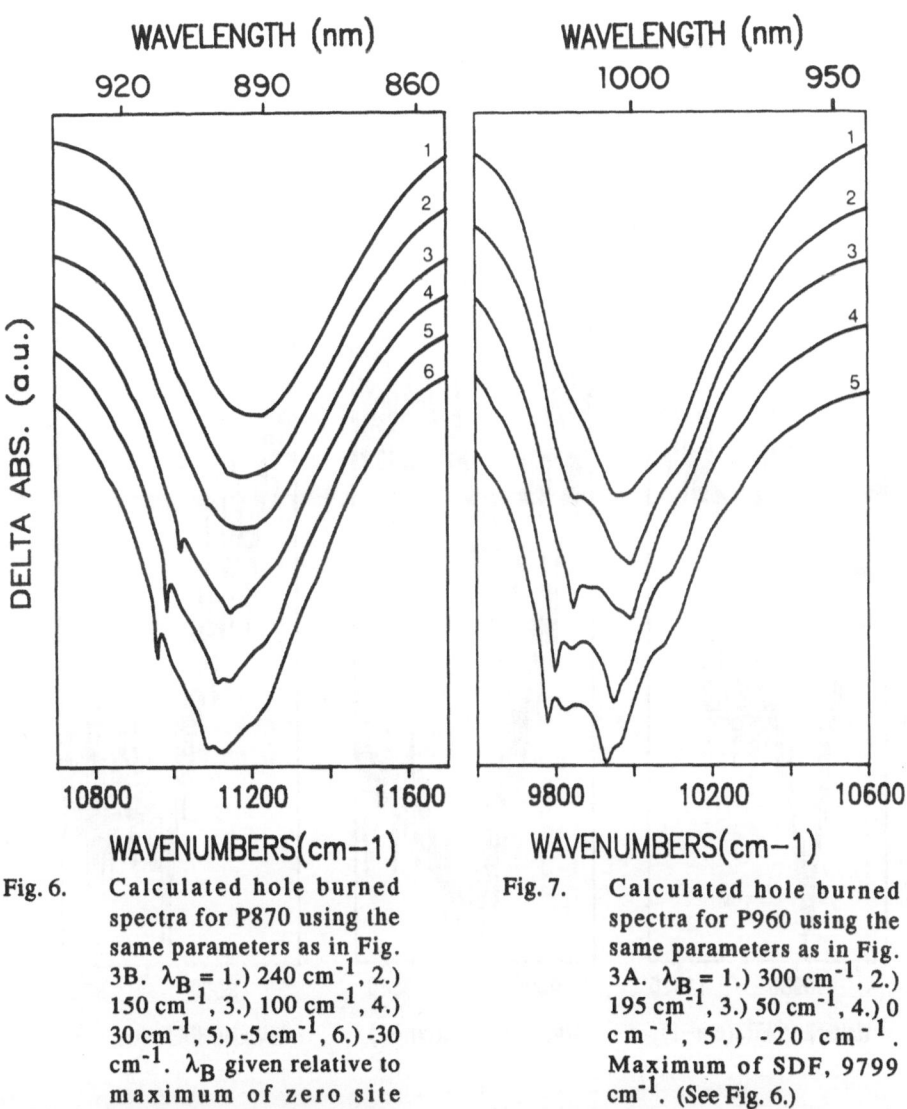

Fig. 6.   Calculated hole burned spectra for P870 using the same parameters as in Fig. 3B. λ_B = 1.) 240 cm^{-1}, 2.) 150 cm^{-1}, 3.) 100 cm^{-1}, 4.) 30 cm^{-1}, 5.) -5 cm^{-1}, 6.) -30 cm^{-1}. λ_B given relative to maximum of zero site distribution function (SDF). Maximum of SDF, 10995 cm^{-1}.

Fig. 7.   Calculated hole burned spectra for P960 using the same parameters as in Fig. 3A. λ_B = 1.) 300 cm^{-1}, 2.) 195 cm^{-1}, 3.) 50 cm^{-1}, 4.) 0 cm^{-1}, 5.) -20 cm^{-1}. Maximum of SDF, 9799 cm^{-1}. (See Fig. 6.)

are due to homogeneous broadening (from the linear electron-phonon and $-\omega_{sp}$ mode coupling) and inhomogeneous broadening, respectively.

B. Zero-phonon holewidths

For P870 (glycerol/NGP) an average of several scans (2 cm^{-1} read resolution) for λ_B in the range 907-912 nm, see Fig. 3, yielded a ZPH width of 8.5 ± 2.0 cm^{-1} (corrected for read

resolution). Typical profiles are shown in Fig. 8a. This width yields a P870* decay time of
1.3 ± 0.3 ps, which is in good agreement with the time domain value of 1.2 ± 0.1 ps at 10 K
[21,22]. A similar procedure for P960 (glycerol/LDAO) with λ_B in the range 1016-1021 nm,
see Fig. 8, yielded a P960* decay time of 0.8 ± 0.1 ps (ZPH width of 13.0 ± 1.5 cm^{-1}). A
typical ZPH profile is given in Fig. 8 along with profiles for glycerol/NGP and PVOH/LDAO.

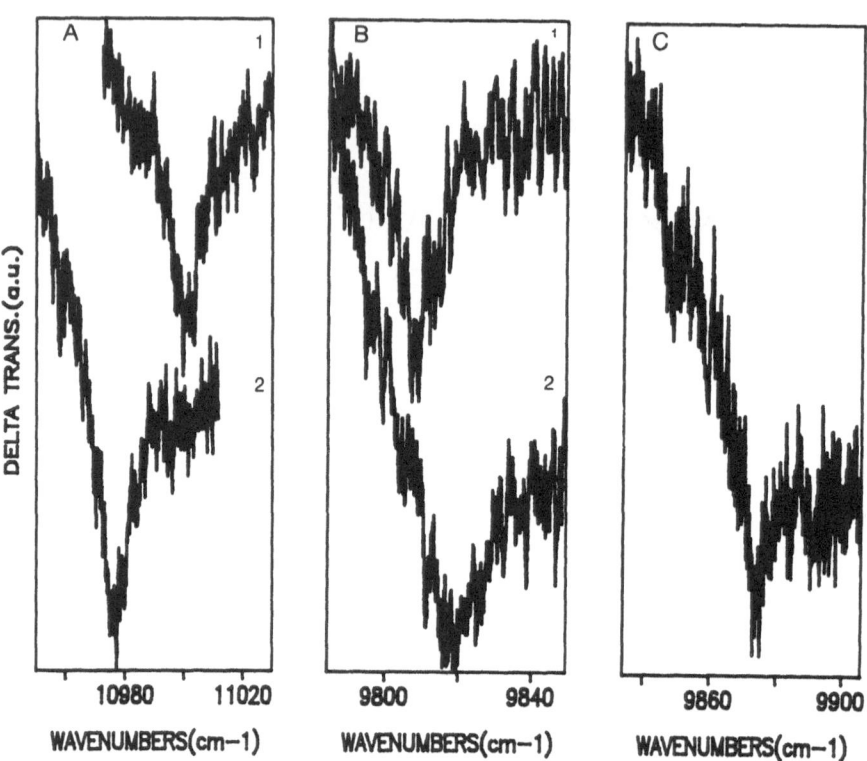

Fig. 8.   ZPH at 4.2 K for A) P870 1.) λ_B = 909 nm (11001 cm^{-1}), 2.) λ_B = 911 nm (10977
          cm^{-1}); B) P960 (glycerol/LDAO) 1.) λ_B = 1019.6 nm (9808 cm^{-1}), 2.) λ_B =
          1018.3 nm (9820 cm^{-1}); C) P960 (PVOH/LDAO) λ_B = 1012.6 nm (9876 cm^{-1})
          (PVOH). Resolution = 2 cm^{-1} for all spectra.

The close agreement between the PHB and time domain decay times is interesting since the ZPH width measures decay from <u>zero-point</u> while the time domain experiments initially prepare P^* vibrationally excited. If charge separation occurred to a significant extent prior to thermalization (at 10 K), one would not expect agreement [29,30]. The observed agreement suggests that thermalization occurs on a sub-ps time scale. This suggestion is consistent with the photon echo results of Meech et al. [31]. These results and the absence of vibronic satellite holes associated with the ω_{sp}-progression provide justification for inclusion of sub-ps decay of the ω_{sp}^j ($j \geq 1$) zero-phonon levels in the calculations of the P870 and P960 PHB spectra. It should be emphasized that it was carefully determined that scattered laser light (0.2 cm^{-1} linewidth) did not interfere with the observed ZPH (see experimental section).

C.   Site excitation energy correlation effects in the transient spectra

Recently nonphotochemical hole burning experiments on the antenna complex of the green algae *Prosthecochloris aestuarii* have proven that there is a high degree of site excitation energy correlation between different exciton components of a subunit characterized by strong excitonic interactions [32]. It is with this in mind that we consider the PHB spectra for *Rps. viridis* (PVOH/LDAO) and *Rb. sphaeroides* (PVOH/NGP) shown in Figs. 9 and 10, respectively. For all of the host/detergent systems utilized <u>only one feature of the Q_y-region of the transient spectra lying higher in energy than P960 and P870 exhibits an observable dependence on λ_B</u>. It is the hole (bleach [20]) that corresponds to the low E shoulder (LES) of the (B_M, B_L) "monomer" absorption band. The λ_B-dependence is most pronounced for PVOH host films because of the additional ~ 120 cm^{-1} of inhomogeneous broadening they provide relative to the glass hosts. In both Figs. 9 and 10 the right and left vertical lines are centered at the centroids of the LES and P_ holes for the highest λ_B-value used. The LES holes for *Rps. viridis* and *Rb. sphaeroides* are located at 848.0 nm (11790 cm^{-1}) and 811.3 nm (12330 cm^{-1}) for this λ_B-value. It is apparent that the LES hole tracks λ_B in a similar manner to the P_ hole for both bacterial RC. Therefore, there is a significant degree of positive correlation between the site excitation energies of the P_ and LES states. The LES state for *Rps. viridis* has been assigned by Vermeglio et al. [20] to P_+, the upper dimer component of the Q_y-transition of the special pair. For both bacterial RC our linearly polarized transient spectra (not shown) have confirmed that the LES state carries a polarization that is close to opposite of the polarization for P_ [33]. Because of the aforementioned studies on *P. aestuarii* and because we are not aware of any line narrowing studies on molecular systems that establish any correlation for excited states of different electronic parentage, we suggest that our results provide support for the LES state assignment of Vermeglio et al. [20]. The bands in Figs. 9 and 10 lying to higher energy of the LES hole are due to the electrochromic shifting of the "accessory" B and H pigment absorption bands produced by formation of P^+BHQ^-. It is interesting that these bands are independent of λ_B. At this time, however, one

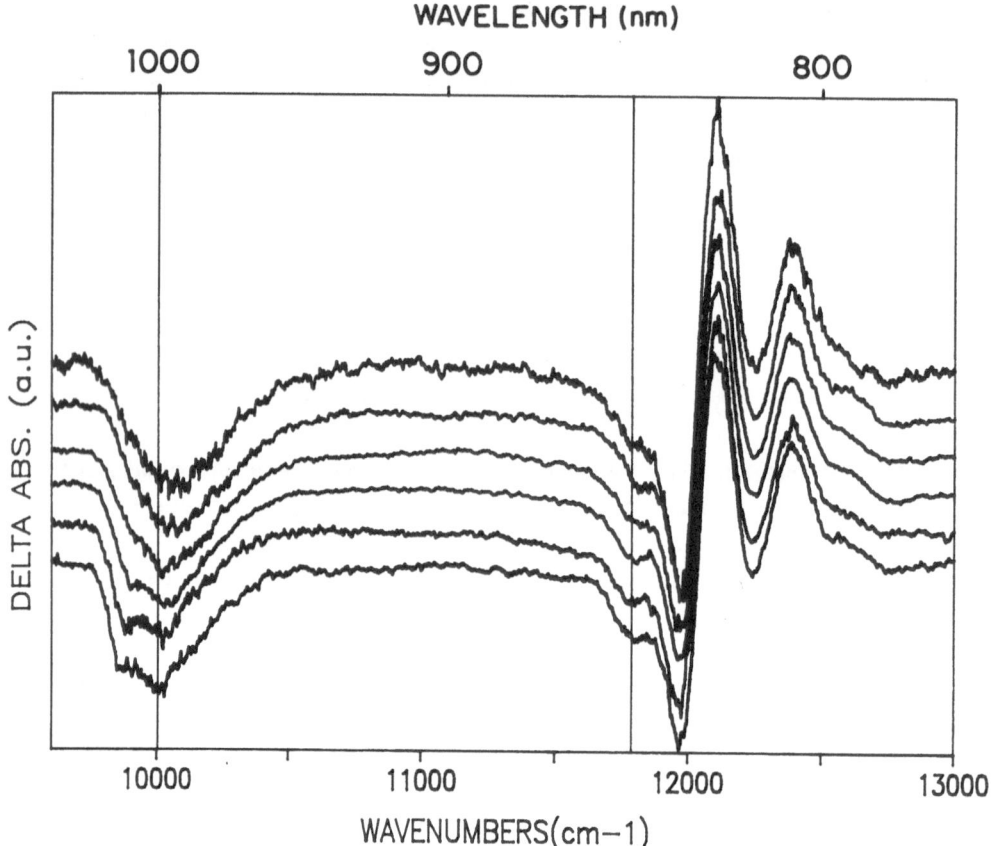

Fig. 9.   Hole burned spectra for *Rps. viridis* (PVOH/LDAO), T = 4.2 K, λ_B = 965, 980, 1000, 1010, 1012, 1015 nm (ordered from top to bottom). Resolution = 8 cm^{-1}.

cannot conclude that the apparent lack of correlation between the "accessory" states and P_ means that the accessory states are relatively pure, i.e. that they contain only a small contribution from the special pair. The reason is that the accessory bands (see Fig. 2) may, to a large extent, be homogeneously broadened due to ultra-fast downward electronic energy transfer processes [34,35]. Further PHB experiments are planned which should resolve this issue.

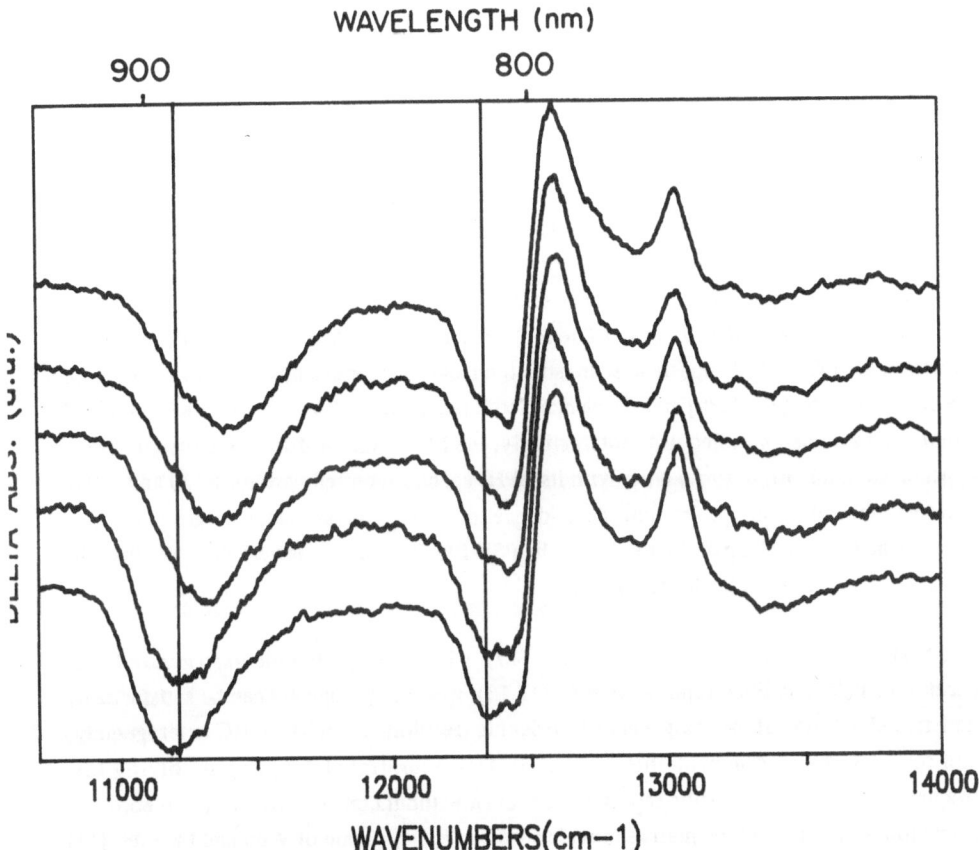

Fig. 10.   Hole burned spectra for *Rb. sphaeroides* (PVOH/LDAO), T = 4.2 K, λ_B = 860, 875, 885, 900, 905 nm (ordered from top to bottom). Resolution = 8 cm^{-1}.

IV. Further Discussion

We consider first the question of the nature of the ω_{sp} mode for P870 and P960. We assign it as a special pair intermolecular Franck-Condon <u>marker</u> mode for the following reasons: recent hole burned spectra for the antenna Chl *a* and *b* for photosystem I have shown that [23] the intramolecular S-factors for all modes are ≤ 0.04. With these data and the fluorescence excitation spectra of BChl *a* reported in ref. (24), the S-factors for BChl *a* modes possessing a frequency of ≤ 200 cm^{-1} can be estimated to be ≤ 0.02 in sharp contrast is the value of S_{sp} ~ 1.1 and 1.55 for ω_{sp} reported here. Furthermore, no intramolecular BChl *a*, Chl *a* and Chl *b* monomer modes with frequencies close to the ω_{sp} values reported here have been observed [23,24]. Since the ω_{sp} values reported are for P^* (P_-), it would be worthwhile to attempt resonance Raman or coherent 4-wave mixing experiments on P870 and P960 in order to determine the corresponding ground state frequencies. More important for future work, however, are the questions of the dynamical nature of the ω_{sp} mode associated with the special pair and a possible role for the geometry change associated with ω_{sp}, which accompanies electronic excitation of P, in the primary charge separation process.

We have recently reported structured hole spectra for P680, the primary electron donor state absorption of the RC of PSII [36]. The RC structure of PSII has not been determined. However, the RC of PSII appears to share structural and functional similarities with the RC of the purple bacteria [37,38,39] and the Nanba-Satoh preparation [40] of the PSII RC binds 4-5 Chl a and two Pheo a molecules. Interestingly, the P680 hole profiles are quite similar in appearance to the origin (ω_{sp}^0) hole (with its ZPH) spectra reported here for P870 and P960. However, the P680 hole spectra show *no* evidence for the ω_{sp}-marker mode progression. This suggests that if a special pair does exist for the PSII RC, its structure is very different than the special pair structure for the bacterial RC.

The basic conclusion from our earlier work [17], which interpreted the *unstructured* hole spectra for P870 and P960 reported in refs. (12,13), was that the spectra can be satisfactorily understood in terms of inhomogeneous broadening (resulting from RC to RC heterogeniety) and the Franck-Condon principle (linear electron-vibrational coupling involving low frequency modes). The results reported here, even without consideration of the theoretical calculations, prove that this interpretation is correct. The assertion of Won and Friesner [15] that P^* undergoes ultra-fast (≤ 100 fs) electronic relaxation prior to formation of P^+BH^- finds no support from our experimental results. The calculations of Won and Friesner do *not* yield the ZPH reported here because of the ultra-fast electronic dephasing imposed on the primary electron donor state.

The assertion in ref. (19) that a charge transfer (CT) state, lying ~ 300 cm^{-1} higher in energy than the low E shoulder of P960 (i.e. the origin (ω_{sp}^0) band), figures importantly in the PHB spectrum is brought into question by the present results. This assertion was mainly based on a ΔT PHB spectrum (Fig. 3) for which the OD of P960 was sufficiently high (0.6) to render the ΔT spectrum an unfaithful representation of the ΔA spectrum at higher energies (confirmed by measurements on the same but diluted sample) [41]. The present work establishes that the principal features of the PHB spectra can be accounted for without invoking a CT state although the possibility that a weakly absorbing CT state does contribute to the high energy side of the primary donor state absorption profile cannot be excluded.

In conclusion, the results presented here reveal the origin of the homogeneous broadening associated with the primary donor state absorption profile of bacterial RC, identify a special pair marker mode (the dynamical nature of which is yet to be determined), further establish [19,36] the utility of PHB for determining electron transfer dynamics from zero-point level and establish the potential of site excitation energy selection for the determination of correlation between different RC states.

Acknowledgement

Ames Laboratory is operated for the U.S. Department of Energy by Iowa State University under Contract No. W-7405-Eng-82. The research was supported by the Director for Energy Research, Office of Basic Energy Science. The research at Argonne Laboratory was supported by the Division of Chemical Sciences, Office of Basic Energy Sciences, U.S. Department of Energy under Contract No. W-31-109-Eng-38.

References

1.)   Hayes, J.M. and Small, G.J. (1986) 'Photochemical hole burning and strong electron-phonon coupling: primary donor states of reaction centers of photosynthetic bacteria', J. Phys. Chem., 90, 4928-4931.

2.)   Deisenhofer, J., Epp, O., Miki, K., Huber, R. and Michel, H. (1984) 'X-ray structure analysis of a membrane protein complex: electron density map at 3 Å resolution and a model of the chromophores of the photosynthetic reaction center from *Rhodopseudomonas viridis*', J. Mol. Biol., 180, 385-398.

3.)   Deisenhofer, J., Epp, O., Miki, K., Huber, R. and Michel, H. (1985) 'Structure of the protein subunits in the photosynthetic reaction center of *Rhodopseudomonas viridis*', Nature (London), 318, 618-624.

4.)   Michel, H., Epp, O. and Deisenhofer, J. (1986) 'Pigment-protein interactions in the photosynthetic reaction center from *Rhodopseudomonas viridis*', EMBO J., 5, 2445-2451.

5.)   Chang, C.H., Tiede, D., Tang, J., Smith, U. and Norris, J. (1986) 'Structure of *Rhodopseudomonas sphaeroides* R-26 reaction center', FEBS Lett., 205, 82-86.

6.)   Allen, J.P., Feher, G., Yeates, T.O., Rees, D.C., Deisenhofer, J., Michel, H. and Huber, R. (1986) 'Structural homology of reaction centers from *Rhodopseudomonas sphaeroides* and *Rhodopseudomonas viridis* as determined by x-ray diffraction', Proc. Natl. Acad. Sci. USA, 83, 8589-8593.

7.)   Yeates, T.O., Komiya, H., Chirino, A., Rees, D.C., Allen, J.P. and Feher, G. (1988) 'Structure of the reaction center from *Rhodobacter sphaeroides* R-26 and 2.4.1:protein-cofactor (bacteriochlorophyll, bacteriopheophytin, and carotenoid) interactions', Proc. Natl. Acad. Sci. USA, 85, 7993-7997.

8.)   Kirmaier, C. and Holten, D. (1987) 'Primary photochemistry of reaction centers from the photosynthetic purple bacteria', Photosyn. Res., 13, 225-260.

9.)   Budil, D.E., Gast, P., Chang, C.H., Schiffer, M. and Norris, J. (1987) 'Three-dimensional x-ray crystallography of membrane proteins: insight into electron transfer', Ann. Rev. Phys. Chem., 38, 561-583.

10.)   Warshel, A. and Parson, W.W. (1987) 'Spectroscopic properties of photosynthetic reaction centers. 1. Theory', J. Am. Chem. Soc., 109, 6143-6152.

11.)   Scherer, P.O.J. and Fischer, S. (1989) 'Long-range electron transfer within the hexamer of the photosynthetic reaction center *Rhodopseudomonas viridis*', J. Phys. Chem., 93, 1633-1637.

12.)   Boxer, S.G., Lockhart, D.J. and Middendorf, T.R. (1986) 'Photochemical hole burning in photosynthetic reaction centers', Chem. Phys. Lett., 123, 476-482.

13.)   Boxer, S.G., Middendorf, T.R. and Lockhart, D.J. (1986) 'Reversible photochemical hole burning in *Rhodopseudomonas viridis* reaction centers', FEBS Lett., 200, 237-241.

14.)   Meech, S.R., Hoff, A.J. and Wiersma, D.A. (1986) 'Role of charge-transfer states in bacterial photosynthesis', Proc. Natl. Acad. Sci. USA, 83, 9464-9468.

15).   Won, Y. and Friesner, R.A. (1988) 'Theoretical studies of photochemical hole burning in photosynthetic bacterial reaction centers', J. Phys. Chem., 92, 2214-2219.

16.)   Won, Y. and Friesner, R.A. (1989) 'Comment: Theoretical studies of photochemical hole burning in photosynthetic bacterial reaction centers', J. Phys. Chem., 93, 1007.

17.)   Hayes, J.M., Gillie, J.K., Tang, D. and Small, G.J. (1988) 'Theory for spectral hole burning of the primary electron donor state of photosynthetic reaction centers', Biochim. Biophys. Acta, 932, 287-305.

18.)   Tang, D., Jankowiak, R., Gillie, J.K., Small, G.J. and Tiede, D. (1988) 'Structured hole-burned spectra of reaction centers of *Rhodopseudomonas viridis*', J. Phys. Chem., 92, 4012-4015.

19.)   Tang, D., Jankowiak, R., Small, G.J. and Tiede, D. (1989) 'Structured hole burned spectra of the primary donor state absorption region of *Rhodopseudomonas viridis*', Chem. Phys., 131, 99-113.

20.) Vermeglio, A. and Paillotin, G. (1982) 'Structure of *Rhodopseudomonas viridis* reaction centers absorption and photoselection at low temperature', Biochim. Biophys. Acta, 681, 32-40.

21.) Fleming, G.R., Martin, J.L. and Breton, J. (1988) 'Rules of primary electron transfer in photosynthetic reaction centers and their mechanistic implications', Nature (London), 333, 190-193.

22.) Breton, J., Martin, J.L., Fleming, G.R. and Lambry, J.-C. (1988) 'Low-temperature femtosecond spectroscopy of the initial step of electron transfer in reaction centers from photosynthetic purple bacteria', Biochem., 27, 8276-8284.

23.) Gillie, J.K., Small, G.J. and Golbeck, J.H. (1989) 'Nonphotochemical hole burning of the native antenna complex of photosystem I (PSI-200)', J. Phys. Chem., 93, 1620-1627.

24.) Renge, I., Mauring, K. and Avarmaa, R. (1987) 'Site-selection optical spectra of bacteriochlorophyll and bacteriopheophytin in frozen solutions', J. Lumin., 37, 207-214.

25.) Lee, I.-J., Hayes, J.M. and Small, G.J. (1989) 'Hole and anti-hole profiles in non-photochemical hole burned spectra', J. Chem. Phys., to be published.

26.) Angerhofer, A. (1987) 'Optische und ODMR-untersuchungen an antennen und reaktionszentren photosynthetisierender bakterien', thesis (Ph.D.), Universitat Stuttgart, Stuttgart, West Germany.

27.) Maslov, V.G., Klevanik, A.V., Ismailov, M.A. and Shuvalov, V.A. (1983) 'Nature of long-wave absorption band of *Rhodopseudomonas viridis* reaction centers in relation to primary charge separation', Dokl. Akad. SSSR, 269, 1217-1221.

28.) Hayes, J.M., Fearey, B.L., Carter, T.P. and Small, G.J. (1986) 'Nonphotochemical hole burning – versatility and theoretical status', Int. Rev. Phys. Chem., 5, 175-184.

29.) Jortner, J. (1980) 'Dynamics of electron transfer in bacterial photosynthesis', Biochim. Biophys. Acta, 594, 193-230.

30.) Bixon, M. and Jortner, J. (1982) 'Quantum effects on electron-transfer processes', Faraday Discuss. Chem. Soc., 74, 17-29.

31.) Meech, S.R., Hoff, A.J. and Wiersma, D.A. (1985) 'Evidence for a very early intermediate in bacterial photosynthesis. A photon-echo and hole-burning study of the primary donor band in *Rhodopseudomonas sphaeroides*', Chem. Phys. Lett., 121, 287-292.

32.) Johnson, S.G. and Small, G.J. (1989) 'Spectral hole burning of a strongly exciton-coupled bacteriochlorophyll *a* antenna complex', Chem. Phys. Lett., 155, 371-375.

33.) Johnson, S.G., Tang, D., Jankowiak, R., Hayes, J.M., Small, G.J. and Tiede, D. (1989) 'Polarized photochemical hole burning of reaction centers of *Rhodopseudomonas viridis* and *Rhodobacter sphaeroides*', to be published.

34.) Breton, J., Martin, J.-L., Migus, A., Antonetti, A. and Orszag, A. (1986) 'Femtosecond spectroscopy of excitation energy transfer and initial charge separation in the reaction center of the photosynthetic bacterium *Rhodopseudomonas viridis*', Proc. Natl. Acad. Sci. USA, 83, 5121-5125.

35.) Martin, J.-L., Breton, J.,Hoff, A.J., Migus, A. and Antonetti, A. (1986) 'Femtosecond spectroscopy of electron transfer in the reaction center of the photosynthetic bacterium *Rhodopseudomonas sphaeroides* R-26: direct electron transfer from the dimeric bacteriochlorophyll primary donor to the bacteriopheophytin acceptor with a time constant of $2.8 \pm .2$ psec.', Proc. Natl. Acad. Sci. USA, 83, 957-961.

36.) Jankowiak, R., Tang, D., Small, G.J. and Seibert, M. (1989) 'Transient and persistent hole burning of the reaction center of photosystem II', J. Phys. Chem., 93, 1649-1654.

37.) Michel, H. and Deisenhofer, D. (1986) 'X-ray diffraction studies on a crystalline bacterial photosynthetic reaction center: a progress report and conclusions on the structure of photosystem II reaction centers', in A.C. Staehelin and C.J. Arntzen (eds.), Encyclopedia of Plant Physiology: Photosynthesis III, Springer-Verlag, Berlin, pp. 371-381.

38.) Michel, H. and Deisenhofer, J. (1987) 'The photosynthetic reaction center from the purple bacterium *Rhodopseudomonas viridis*', Chemica Scripta, 27B, 173-180.

39.) Trebst, A. (1986) 'The topology of the plastoquinone and herbicide binding peptides of photosystem II in the thylakoid membrane', Z. Naturforsch, 41C, 240-245.

40.) Nanba, O. and Satoh, K. (1987) 'Isolation of a photosystem II reaction center consisting of D-1 and D-2 polypeptides and cytochrome b-559', Proc. Natl. Acad. Sci. USA, 84, 109-112.

41.) Also, experiments performed on the undiluted sample yielded PHB spectra that agree with the ΔT spectra obtained earlier [18,19], yet result in ΔA spectra that are consistent with those presented here. The feature in Fig. 3 (which should be labeled as a ΔT spectrum) of ref. 19 assigned to a CT state is actually the ω_{sp}^2 satellite hole.

INFLUENCE OF CHROMOPHORES ON QUARTERNARY STRUCTURE OF
PHYCOBILIPROTEINS FROM THE CYANOBACTERIUM, *Mastigocladus
laminosus*

R. FISCHER, J. GOTTSTEIN, S. SIEBZEHNRÜBL,
H. SCHEER
Botanisches Institut der Universität
Menzinger Str. 67
D - 8000 München 19
FRG

ABSTRACT. Chromophores of C-phycocyanin and phycoerythrocyanin have been chemically modified by reduction to
rubins, bleaching, photoisomerization, or perturbation
with bulky substituents. Pigments containing modified
chromophores, or hybrids containing modified and unmodified chromophores in individual protomers have been prepared. All modifications inhibit the association of the
(αß)-protomers of these pigments to higher aggregates. The
results demonstrate a pronounced effect of the state of
the chromophores on biliprotein quaternary structure. It
may be important in phycobilisome assembly, and also in
the dual function of biliproteins as (i) antenna pigments
for photosynthesis and (ii) reaction centers for photomorphogenesis.

1. INTRODUCTION

Cyanobacteria, red algae and cryptophytes have specialized
antenna pigments which enable them to harvest light efficiently in the green spectral region where chlorophylls
have only poor absorption. These are the phycobiliproteins, extra-membraneous proteins carrying covalently
bound open-chain tetrapyrrolic chromophores (Gantt, 1986,
Glazer, 1983; MacColl and Guard-Friar, 1983; Scheer, 1982;
Schirmer *et al.*, 1987; Wehrmeyer, 1983; Zuber, 1986). They
are characterized by a remarkable degree of spectral adap-
tations covering the range from 480 to 670 nm. The spectral variation is based on the usage of only four different chromophore types, each of them being further mod-
ulated in its absorption properties by non-covalent interactions with the apoproteins.

In cyanobacteria and red algae, the phycobiliproteins show
a high degree of organisation: Together with the linker

J. Jortner and B. Pullman (eds.), Perspectives in Photosynthesis, 121–131.
© 1990 Kluwer Academic Publishers.

polypeptides, which are often devoid of chromophores, they form self-organizing microscopic structures, the phyco-bilisomes, which in turn form ordered arrays on the outer thylakoid membrane surface (Gantt, 1986; Glazer, 1983; Mörschel and Schatz, 1988; Wehrmeyer, 1983). By a combin-ation of energetic and spatial ordering, and low internal conversion rates of the chromophores, the phycobilisomes are an antenna system working with quantum efficiencies approaching 100%.

The factors controlling tuning of the absorption, photo-chemistry and aggregation in biliproteins are presently becoming understood in considerable detail. This is based on an increasing body of structural data (Bishop et al., 1987; Bryant, 1988; Dürring and Huber, 1989; Glazer, 1983; Schirmer et al., 1987; Tandeau de Marsac, Zuber, 1986) with theoretical (Sauer and Scheer, 1988; Schneider et al., 1988, Scheer, 1987) and spectroscopic studies (Fischer et al., 1988; Glazer, 1983; MacColl and Guard-Friar, 1987; Mimuro et al., 1986a,b, Scheer, 1982; Schirmer et al., 1987). They involve conformational changes of the chromophores imposed by non-covalent inter-actions with the apoproteins (e.g., point charges, and π-π-interactions, and -in higher aggregates- excitonic interactions among chromophores (MacColl and Guard-Friar, 1983; Sauer and Scheer, 1988; Schneider et al., 1988).

The assembly of phycobilisomes is a complex, and to a large extent autonomous process believed to be controlled by the presence of appropriate components. The factors controlling the aggregation of monomers ($\alpha\beta$) to oligomers ($\alpha\beta$) have hitherto mainly been located on the proteins. Here we want to present data which show a pronounced influence of the chromophores, too, on the aggregation. The results may also be important for an understanding of a the second function of biliprotein, e.g. as photomorpho-genetic reaction center pigments (Björn and Björn, 1980; Rüdiger and Scheer, 1983; Scheer, 1982; Song, 1988).

2. MATERIALS AND METHODS

C-phycocyanin (PC) and phycoerythrocyanin (PEC) were isol-ated from the cyanobacterium, *Mastigocladus laminosus*, as described before (Fischer et al., 1988). Subunits were prepared by isoelectric focusing (Schmidt et al., 1988).

Chromophores were bleached by irradiation with 350 nm light (Scheer, 1987).

Reduction of cyanin to rubin chromophores was done by a modification of the method of Kufer and Scheer, 1982). Denatured PC or isolated subunits ($8-25\mu M$) in potassium phoshphate buffer (0.9 M, pH = 7) were treated with NaBH (final concentration $\approx$170 mM) for 45 min at ambient tempé-rature. The reaction was followed spectrophotometrically (decrease at $\approx$600, increase at $\approx$420 nm). After completion,

excess reagent was destroyed by addition of glucose. Modified PC was renatured by dialysis against phosphate
buffer (100 mM, pH = 7) at ambient temperature. Hybrid-
ization to native protomers with modified chromophores on
only one of the subunits, was done by combining stoichio-
metric amount of the appropriate subunits in buffer containing urea (8M), followed by dialysis against decreasing
amounts of urea and finally urea-free buffer.

Cystein-111 of PC was modified by treatment with
p-chloro-mercurybenzenesulfonate (PCMS) (Siebzehnrübl *et
al.*, 1987).

Photochemistry was induced with a cold light source (150W)
equipped with light-guide and suitable interference fil-
ters (≈10 nm fwhh). Spectra were recorded with lambda2
(Perkin-Elmer) or ZWSII (sigma) spectrophotometers. Aggregation state of samples was determined by ultracentrifugation (Martin and Ames, 1961) with myoglobin and trimeric
phycocyanin as reference.

3. RESULTS AND DISCUSSION

3.1. Effect of reduced chromophores on aggregation

Chromophores of PC were completely reduced with NaBH in
the presence of urea (8M) to rubins, which show no absorption at wavelengths >500nm. After renaturation, the modified chromophores are unstable in the native protein
environment and are slowly oxidized to cyanin chromophores
within several days. The process can be followed quantitatively by monitoring the absorption increase at 620nm.

The aggregation state of phycocyanin containing reduced
chromophores on both subunits is monomeric. There is only
a single, yellow band present after centrifugation of
freshly prepared "phycorubin". Under uv-light, this band
is non-fluorescent. After prolonged standing of the samples, they become slowly green due to re-oxidation of the
chromophores. Ultracentrifugation of such samples shows
generally two bands: a greenish one at the position of
monomers, and an additional one at the position of the PC
trimer. The latter exhibits the characteristic absorption
and fluorescence of PC trimer, and lacks the 420 nm band
of the rubin chromophore(s) (λ ≈ 420 nm). It is at
first only visible by its red fluorescence, and at increasing re-oxidation times becomes concentrated enough to
become visible to the eye too. We, therefore, conclude
that the lower band contains trimeric PC containing the
re-oxidized native chromophores. If the aggregation pattern is followed through the reoxidation, the trimer band
is always fluorescent and blue (if visible), whereas the
monomer band is yellow or green. This shows clearly that
modified PC bearing rubin chromophores can no longer
aggregate.

In order to test if the inhibition of aggregation is due
to specific chromophore(s), similar experiments were
carried out with hybrids in which the α-subunit was
bearing a modified rubin chromophore and the ß-subunit
cyanin chromophores. For this purpose, isolated α-subunits
were modified, and then hybridized with the respective
complementary unmodified subunits. The results can be
summarized in one sentence: Whenever there was a rubin
chromophores present in the hybrid, the hybrids were monomeric; and they did only aggregate when the chromophores
became re-oxidized.

3.2. Effect of cystein-111 modification on aggregation

The single free cystein residue in PC, e.g. ß-111 in the
immediate vicinity of chromophore ß-84, was modified with
PCMS, and the aggregation studied at different protein
concentrations. At low concentrations (<0.3 μM), both
PCMS-modified and unmodified PC were monomeric. Increasing
concentrations shifted the equilibrium to trimers in unmodified PC, which at concentrations $\geq$3 μM PC was mostly
present as trimer. The PCMS-modified PC remained monomeric
up to the highest concentrations investigated, e.g. 10 μM.

This inhibition of aggregation was reversible, too. Treatment of PCMS-modified PC with an excess of dithiothreitol,
removes the mercurial reagent from the chromoprotein as
shown by the blue-shift of the spectrum. The difference
spectra are mirror-images of the reaction with the mercurial, and the final spectrum of the recovered PC showed
again the spectral characteristics of trimeric PC.

3.3. Effect of aggregation on photochemistry

3.3.1. Phycocyanin

Long-lived photoproducts have been observed in several
phycobiliproteins. In phycocyanin, a small but measurable
photochemistry is found only upon partial denaturation
and/or disaggregation of the native trimers. This photochemistry was originally studied in phycocyanin treated
with urea at concentrations of 3-5M, which are known to
change aggregation and to some extent also the tertiary
structure. The difference spectra are characterized by a
decrease around 622nm, and only a minor increase in the
500nm range (John et al., 1985). A similar photochemistry
(if judged from absorption difference spectra) has now
been observed under a variety of mildly denaturing conditions. Its magnitude (defined by the ratio of the amplitude
of the difference spectrum, to the maximum absorption) can
be as high as 60% (in the presence of 20% mercaptoethanol,
Schmidt et al., 1988). From comparison with the photoreac-
tions of phycocyanin in which the native 15Z-configured
(Schirmer et al., 1987) chromophores have been partially
converted to the 15E-isomers (Schmidt et al., 1988), it is
most likely due to a Z/E interconversion of the chromo-
phore(s) at the C-15,16 double-bond. Under such condi-

Figure 1: Difference spectrum of phycoerythrocyanin (pre-irradiated with green light, 500 nm), and the same sample after orange irradiation (600 nm).

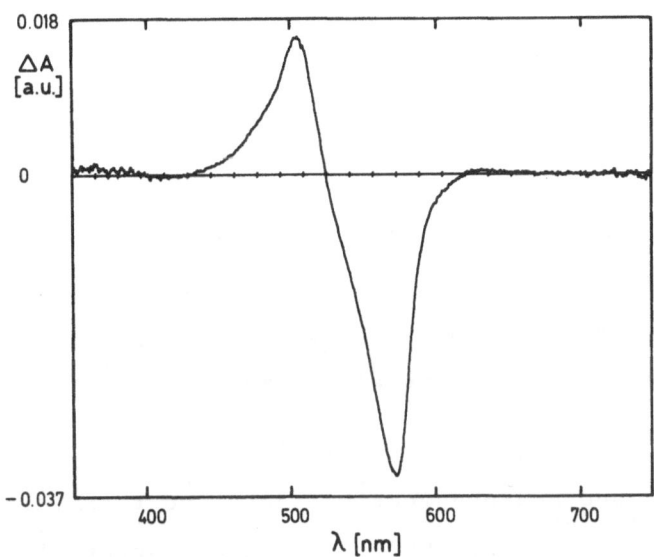

Table 1 : Absorption difference extrema and amplitudes of different PEC samples or samples subjected to different pre-treatments.

Sample	C_{urea} a)	pH	Difference extrema [nm]		$\Delta\Delta A$ b)
	[M]		orange	green	[%]
PEC	0	7.0	570	503	18
PEC	1	7.1	569	503	32
PEC	4	7.3	567	502	35
PEC	6	7.5	600	510	12
PEC	8	7.0	599	515	6
PEC	8	3.0	598	507	13
PEC-monomer	c)	7.0	565	504	36
α-Subunit	0	7.0	569	504	50
phycobilisome	d)	7.0	567	502	0.31
phycobilisome,diss	e)	7.0	567	504	2.82

a) Samples in potassium phosphate buffer (100 mM, pH = 7), if not indicated otherwise.
b) $\Delta\Delta A$ is the amplitude of the s-shaped difference spectrum from minimum to maximum, divided by the maximum absorbance of the sample (see Fig.1).
c) 1 M KSCN, no urea.
d) Coupled phycobilisomes, in phosphate buffer (0.9 M).
e) Dissociated phycobilisomes, in distilled water.

tions, the chromophores are apparently capable to perform
the same type of photochemistry as that of phytochrome,
but the subsequent changes -if there are any- must of
course be different (as are the proteins).

3.3.2. Phycoerythrocyanin

Phycocyanin shows photochemistry only under partially
denatured conditions and is photochemically "silent" in
the native as well as in the fully denatured state. A
closely related pigment, phycoerythrocyanin (PEC), shows a
pronounced photochemistry under all conditions from fully
native (Fig. 1) to fully denatured (Siebzehnrübl et al.,
1989). There is nonetheless a pronounced effect of aggregation on the amplitude of the photochemical difference
spectrum. It is very small in phycobilisomes, increases
upon dissociation to trimers and further to monomers, and
then decreases again in denatured PEC. Thus, the basic
pattern is similar in PC and PEC, but the amplitudes of
the difference spectra are always much larger in PEC
(Table 1) than in PC (John et al., 1985).

The α-subunit of PEC had been linked previously to
photochromic activities in cyanobacterial extracts, and
possibly to photomorphogenesis (Björn and Björn, 1980;
Kufer, 1988). This pigment, which is structurally very
similar to PC (Bryant, 1982; Dürring and Huber, 1989)
carries a rare phycoviolobilin chromophore (alternatively
also called PXB or cryptoviolin) at cystein α-84 (Bishop
et al., 1987), which replaces the common phycocyanobilin
chromophore present at the same location in PC. Being a
component of the phycobilisome, it is commonly regarded a
light-harvesting pigment. A distinct difference from other
phycobiliproteins, is however its pronounced photochemistry in the native state. It involves probably a Z/E-iso-
merization at the C-15 methine bridge similar to the
photoreaction of the chromphore in phytochrome (Thümmler
et al., 1983), but this still has to be demonstrated. This
reaction would require a decreased rigidity in the environment of α-84, which (contrary to preliminary indic-
ations) is not obvious in the crystal structure of (trimeric) PEC (Dürring and Huber, 1989).

3.4. Effect of chromophore α-84 photochemistry in PEC on aggregation

During photochemical studies with PEC, we noticed that it
shows not only increased photochemistry upon disaggregation, but that there is also a reciprocal dependence of
biliprotein aggregation on photochemistry. When PEC is
alternately irradiated with orange (600nm) and green light
(500nm), the α-subunit becomes enriched in one (15E) and
the other form (15Z), respectively. Ultracentrifugation
analysis showed, that at the same time there occurs a
photoreversible change in aggregation: The amount of
trimer increased each time the last irradiation was
performed with green light, and decreased each time it was

performed with orange light. This means, that the configuration of α-84 influences aggregation, a fact which can
be rationalized from the x-ray structure: α-84 is located
very close to the contact surface of monomers in trimers
(Dürring and Huber, 1989).

4. Discussion

In many cyanobacteria, profound alterations in phycobilisome structure (and other cyanobacterial activities) can
be induced by changes in the environmental light conditions. The photomorphogenetic receptors (adaptachromes,
photomophochromes) are hitherto unknown, but action
spectra suggest that they also belong to the biliproteins
(Björn and Björn, 1980; Scheer, 1982). Pigment(s) of the
biliprotein type, albeit with very different apoproteins,
are functional in higher plants for the same purpose. The
green-plant photomorphogenetic reaction center pigment,
phytochrome, has a chromophore which is very similar to
the ones of the light harvesting pigment, phycocyanin
(Rüdiger and Scheer, 1983). This multiple function of bile
pigment chromophores is an aspect of biliproteins which is
still poorly understood. The first event in photomorpho-
genetic pigments, is generally believed to be a structural
change of the chromophore(s). This signal can then be
propagated by induced structural changes of the protein,
and beyond. In phytochrome, the primary reaction has been
shown to involve a photoreversible Z/E-isomerization and
possibly a protonation/deprotonation of the chromophore,
while the induced changes on the protein are less understood (Rüdiger and Scheer, 1983; Song, 1988; Thümmler *et
al.*, 1983).

The multiple function of biliproteins containing very
similar chromophores, is reminescent of the multiple
functions performed by chlorophylls of the same molecular
structure in photosynthesis. In both cases, the photochemical and photophysical properties of the chromophores
are diverse. The relative contributions of different de-
excitation pathways must be regulated by specific interactions between the chromophores and the proteins. In
analogy to interactions among living systems, these interactions can be termed as "molecular ecology".

5. SUMMARY

The results obtained by different modifications of the
chromophores or their immediate vicinity, demonstrate an
involvment of the biliprotein chromophores not only in
energy transfer and photochemistry of biliproteins, but
also in their quaternary structure. This effect may well
be at the origin of a signal chain leading eventually to
photomorphogenesis. In a more general context, it is an
example for the intricate interplay of proteins with their
cofactors, which leads to the stunning variety of properties of pigments with the same or very similar molecular
structures.

References:

Bishop, J. E., H. Rapoport, A. V. Klotz, C. F. Chan, A. N. Glazer, P. Füglistaller, and H. Zuber, Chromopeptides from phycoerythrocyanin. Structure and linkage of the three bilin groups. *J. Am. Chem. Soc.* **109**,875-881.

Björn, G. S. (1979) Action spectra for *in vivo* and *in vitro* conversions of Phycochrome b, a reversibly photochromic pigment in a blue-green alga,and its separation from other pigments. *Physiol. Plant.* **46**,281-286.

Björn, G. S., N. Ekelund, and L. O. Björn (1984) Light-induced linear dichroism in photoreversibly photochromic sensor pigments .4.Lack of chromophore rotation in phy-cochrome-B immobilized *in vitro. Physiol. Plant.* **60**, 253-256.

Björn, L.O., and Björn, G.S. (1980), Photochromic Pigments and photoregulation in blue-green algae, *Photochem. Photobiol.*, **32**, 849-852

Braslavsky, S. E., A. R. Holzwarth, and K. Schaffner (1983) Solution conformations, photophysics and photochemi-stry of bili-pigments - bilirubin and biliverdin dimethyl esters and related linear tetrapyrroles. *Angew. Chem. Int. Ed.* **22**, 656-674.

Bryant, D. A. (1982) Phycoerythrocyanin and Phycoerythrin - properties and occurence in Cyanobacteria. *J. Gen. Microbiol.* **128**, 835-844.

Bryant, D. A. (1988) Phycobilisomes of *Synechococcus* sp. PCC 7002, *Pseudoanabaena* sp. PCC 7409, and *Cyanophora paradoxa*: An analysis by molecular genetics. In *Photosyn-thetic Light Harvesting Systems. (Eds. H. Scheer and S. Schneider)*, pp. 217-233, De Gruyter, Berlin.

Dürring, M. and R. Huber (1989) private communication

Fischer, R., S. Siebzehnrübl, and H. Scheer (1988) C-phy-cocyanin from *Mastigocladus laminosus*: Chromophore Assign-ment in higher Aggregates by Cystein Modification. In *Pho-tosynthetic Light Harvesting Systems:Organization and Function (Eds. H. Scheer and S. Schneider)*, pp. 71-76, De Gruyter, Berlin.

Gantt, E. (1986) Phycobilisomes. In *Photosynthesis III (Eds.* L. A. Staehelin and C. J. Arntzen), pp.260-268, Springer-Verlag, Berlin.

Glazer, A. N. (1983) Comperative biochemistry of photosyn-

thetic light-harvesting systems. *Ann. Rev. Biochem.*
52, 125-157.

John, W., R. Fischer, S. Siebzehnrübl, and H. Scheer
(1985) C-Phycocyanin from Mastigocladus laminosus.
Isolation and properties of subunits and small aggregates.
In *Antenna and Reaction Centers of Photosynthetic Bacteria
- Structure, Interactions and Dynamics (Ed. M. E.
Michel-Beyerle)*,pp. 17-25, Springer, Berlin.

Kufer, W. (1988) Concerning the relationship of light harvesting Biliproteins to Phycochromes in cyanobacteria. In
*Photosynthetic Light Harvesting Systems. (Eds. H. Scheer
and S. Schneider)*, pp. 89-93, De Gruyter, Berlin.

Kufer,W. and H. Scheer (1982) Rubins and rubinoid addition
products from phycocyanin. *Z. Naturforsch.* **37c**,179-192.

MacColl, R. and D. Guard-Friar (1987) *Phycobiliproteins.*
CRC Press, Boca Raton.

MacColl, R., K. Csatorday, and D. S. Berns (1981) The
relationship of the quaternary structure of Allophycocya-
nin to its spectrum. *Arch. Biochem. Biophys.* **208**, 42-48.

Martin, R. G. and B. N. Ames (1961) A method for
determining the sedimentation behavior of enzymes:
Application to protein mixtures. *J. Biol Chem.*
236,1372-1379.

Mimuro, M., P. Füglistaller, R. Rümbeli, and H. Zuber
(1986 a) Functional assignment of chromophores and energy
transfer in C-phycocyanin isolated from the thermophilic
cyanobacterium *Mastigocladus laminosus. Biochim. Biophys.*
848, 155-166.

Mimuro, M., R. Rümbeli, P. Füglistaller, and H. Zuber
(1986 b) The microenvironment around the chromophores and
its changes due to the association states in C-phycocyanin
isolated from the cyanobacterium *Mastigocladus laminosus.*
Biochim. Biophys. Acta **851**, 447-456.

Mörschel, E. and G.-H. Schatz (1988) On the structure of
photosystem II-Phycobilisome complexes of cyanobacteria.
In *Photosynthetic Light Harvesting Systems. (Eds. H. Scheer
and S. Schneider)*, pp. 21-35, De Gruyter, Berlin.

Rüdiger, W. and H. Scheer (1983) Chromophores in Photomor-
phogenesis. In *Photomorphogenesis - Encyclopedia of Plant
Physiology*, New Series. (Eds. W. Shropshire jr.and H.
Mohr), Vol. 16, pp. 119-151, Springer, Heidelberg.

Sauer, K., and H. Scheer (1988) Excitation transfer in
C-phycocyanin: Förster transfer rate and exciton calculations based on new crystal structure data for C-phycocyanins from Agmenellum quadruplicatum and Mastigocladus
laminosus, Biochim. Biophys. Acta.

Scheer, H. (1982) Light reaction Path of Photosynthesis.
In *Phycobiliproteins: Molecular Aspects of Photosynthetic
Antenna Systems (Ed. F. K. Fong)*, pp. 7-45, Springer,
Springer.

Scheer, H. (1987) Photochemistry and Photophysics of C-
phycocyanin. In *Progress in Photosynthesis Research
(Ed.* J. Biggins), pp. I.1.143-149, Martinus Nijhoff,
Doordrecht.

Schirmer, T. and M. G. Vincent (1986) Polarized absorption
and fluorescence spectra of single crystals of C-phycocyanin. *Biochim. Biophys. Acta* **893**, 379-385.

Schirmer, T., W. Bode, and R. Huber (1987) Refined
Three-dimensional structures of 2 cyanobacterial
C-phycocyanins at 2.1. and 2.5 A resolution - A common
principle of phycobilinprotein interaction. *J. Mol. Biol.*
196, 677-695.

Schmidt, G., S. Siebzehnrübl, R. Fischer, and H. Scheer
(1988) Photochromic properties of C-phycocyanin. In *Photosynthetic Light Harvesting Systems. Organization and Function* (Eds. H. Scheer and S. Schneider), pp. 77-89,
De Gruyter, Berlin.

Schneider, S., C. Scharnagl, M. Dürring, T. Schirmer, and
W. Bode (1988) Effect of protein environment and the exci-
tonic coupling on the excited-state properties of the
bilinchromophores in C-phycocyanin. In *Photosynthetic
Light Harvesting Systems*, (Eds. H. Scheer and S.
Schneider), pp. 483-491, De Gruyter, Berlin.

Schneider, S., P. Geiselhart, F. Baumann, S. Siebzehnrübl,
R. Fischer, and H. Scheer (1988) Energy transfer in "native" and chemically modified C-phycocyanin trimers and the
constituent subunits. In *Photosynthetic Light Harvesting
Systems. Organization and Function* (Eds. H. Scheer
and S. Schneider), pp. 469-483, De Gruyter. Berlin.

Siebzehnrübl, S., R. Fischer, and H. Scheer (1987) Chromophore Assignment in C-phycocyanin from *Mastigocladus laminosus. Z. Naturforsch.* **42c**, 258-262.

Siebzehnrübl, S., R. Fischer, W. Kufer, and H. Scheer
(1989) Photochemistry of phycobiliproteins: reciprocity of
reversible photochemistry and aggregation in
phycoerythrocyanin from Mastigocladus laminosus.
Photochem. Photobiol. **49**, 753 - 762.

Song, P. S. (1988) The molecular topography of Phytochrome
-Chromophore and Apoprotein. *J. Photochem. Photobiol.* 2,
43-57.

Tandeau de Marsac, N. and G. Cohen-Bazire (1977) Molecular
composition of bacterial Phycobilisomes. *Proc. Natl. Acad.*

Sci. USA **74**, 1635.

Thümmler, F., W. Rüdiger, E. Cmiel, and S. Schneider
(1983) Chromopeptides from phytochrome and phycocyanin.
Nuclear magnetic resonance studies of the P_r and P_{fr}
chromophore of phytochrome and E,Z isomeric chromophores
of phycocyanin. *Z. Naturforsch.* **38 c**, 359-368.

Wehrmeyer, W. (1983) Phycobiliproteins and phycobilipro-
tein organization in the photosynthetic apparatus of cyanobacteria, red algae,and cryptophytes. In *Proteins and
nucleic acids in plant systematics (Eds. U. Jensen
and D. E. Fairbrother)*, Springer, Berlin.

Zuber, H. (1986) Primary structure and function of the
light-harvesting polypeptides from cyanobacteria, red algae
and purple photosynthetic bacteria. In *Photosynthesis III
(Eds*. L. A. Staehelin and C. J. Arntzen), pp. 238-251,
Springer, Berlin.

ENVIRONMENTAL DYNAMICS AND ELECTRON TRANSFER REACTIONS

JAMES T. HYNES, EMILY A. CARTER,a) GIOVANNI CICCOTTI,b)
HYUNG J. KIM & DOMINIC A. ZICHI c)
Department of Chemistry and Biochemistry
University of Colorado
Boulder, CO 80309-0215, USA

MAURO FERRARIO
Istituto di Fisica Teorica
Universitá di Messina
Messina, ITALY

RAYMOND KAPRAL
Chemistry Physics Theory Group
Department of Chemistry
University of Toronto
Toronto, Ontario M5S 1A1, CANADA

ABSTRACT. Recent theoretical and computer simulation work on the dynamics associated with electron transfer processes in polar solvents is described. This includes solvent relaxation subsequent to photo-induced charge transfer, adiabatic electron transfer
rates, and the solvent influence on the electronic states relevant to electron transfers.

1. INTRODUCTION

A number of key issues that are evidently relevant for electron transfer (ET) dynamics in
the photosynthetic reaction center [1] also arise in the superficially remote context of ET
in solution, where the environment for the ET event is provided by the solvent. Among
these issues are the timescales for solvent relaxation subsequent to photo-induced ET,
the role of the solvent dynamics in influencing the ET rate, and the influence of the solvent on the nature of electronic states having charge transfer character. In the following,
we briefly summarize some of our recent theoretical and computer simulation results on

---

a) Permanent Address: Dept. of Chemistry and Biochemistry, Univ. of California, Los
   Angeles, CA 90024-1569.
b) Permanent address: Dip. di Fisica, Univ. 'La Sapienza', Pl. A. Moro, 2, 00185 Rome,
   ITALY.
c) Present Address: Agouron Pharmaceuticals, Inc. 11025 North Torrey Pines Road, La
   Jolla, CA 92037

J. Jortner and B. Pullman (eds.), Perspectives in Photosynthesis, 133–148.
© 1990 *Kluwer Academic Publishers.*

each of these questions for ET in polar solvents. One can hope that, modulo specific details, the central concepts emerging from these solution studies could be useful in the patently more complex photosynthetic arena.

2.   TIME DEPENDENT FLUORESCENCE AND SOLVENT RELAXATION

When a charge separation is induced in a solute by absorption of a photon in a Franck-Condon transition, the solvent--which is initially out of equilibrium with the new charge distribution--will ultimately relax to equilibrium with it (Fig. 1). These solvent dynamics can be probed in time dependent fluorescence (TDF) experiments. [2]

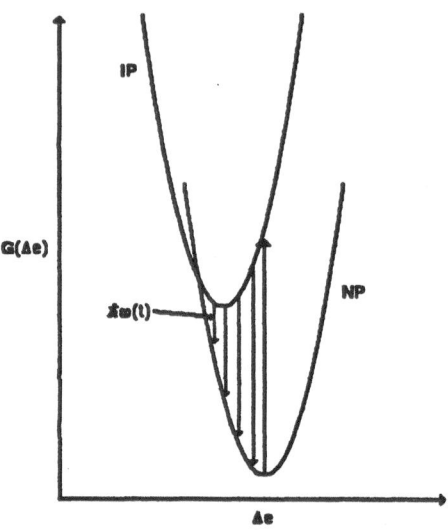

Figure 1.   Schematic diagram illustrating time dependent fluorescence transitions of frequency $\omega(t)$ between ion pair (IP) and neutral pair (NP) states, versus the numerical value Δe of the solvent coordinate ΔE. See the text.

While earlier work in this area focussed on creating analytic theory and models [3], there has been more recent activity in Molecular Dynamics (MD) computer simulation of the phenomenon [4,5]. Carter and Hynes [5] have studied TDF in a simulation of a neutral pair (NP) DA, photoexcited directly to a charge transfer ion pair (IP) state D^+A^-.
   The constituent members of the solute pair (SP) each have mass 40 amu; their centers are rigidly separated by 3.0Å, but the SP is free to translate and rotate. The solvent is composed of 342 rigid dipolar molecules with constituent atoms of mass 40 amu separated from each other by a fixed distance of 2.0Å and with partial charges such that the dipole moment is 2.4D. The number density is $0.012Å^{-3}$ and the temperature is 250K. This solvent [6], which is very roughly similar to methyl chloride, is akin to members of the class of dipolar aprotic solvents currently under experimental investigation [2].

The total potential energy consists of Lennard-Jones and Coulomb potentials between each atomic site. The LJ parameters are $\varepsilon/k_B = 200K$ and $\sigma = 3.5Å$ for each site in the SP and the solvent.

The solvent dynamics were monitored by following the dynamical collective variable ΔE. This is the difference, at fixed solvent configurations, between the SP-solvent interaction potential energy in the IP and NP states, i.e., an energy gap. For the present model, ΔE is just the Coulomb IP-solvent energy. Note that this variable is well-defined even in the absence of the IP, i.e., in the presence of the NP.

Constant temperature [7] MD simulations were carried out in a periodically replicated cubic box with side length 30.52Å. The equations of motion were integrated via the Verlet algorithm [8] with a time step of 10^{-2} ps. The long range forces were treated by the Ewald summation method [9] and the bond constants for the SP and solvent molecules were implemented with the SHAKE algorithm [10].

In the simulations, the solvent was initially equilibrated to the NP. 198 different initial configurations were then selected, the charges were instantly turned on to produce the IP, and then the ensuing dynamics were examined.

One important characteristic of the solvent subsequent to the FC transition is the normalized TDF shift [3-5]

$$S(t) = \frac{\bar{\omega}(t) - \bar{\omega}(\infty)}{\bar{\omega}(0) - \bar{\omega}(\infty)} = \frac{\Delta\bar{E}(t) - \Delta\bar{E}(\infty)}{\Delta\bar{E}(0) - \Delta\bar{E}(\infty)} , \qquad (1)$$

which is related to the average in the nonequilibrium ensemble of the TDF frequency ω and ΔE initially, at time t, and at "infinite" time $t=\infty$, when relaxation has concluded. Figure 2 shows that the relaxation is extensive and rapid, with a distinctly bimodal character. The celerity of the initial relaxation is especially to be noted.

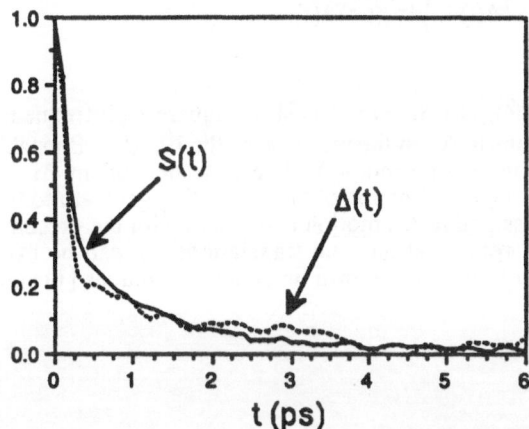

Figure 2.   MD results [5] for the TDF shift S(t), Eq. 1. The time correlation function Δ(t), Eq. 3, is also shown. The absolute magnitude of the shift $\Delta\bar{E}(0) - \Delta\bar{E}(\infty)$ is approximately 1.9×10^3 cm^{-1}

Carter and Hynes expressed the nonequilibrium average $\Delta\overline{E}(t)$--whose dynamics occur in the presence of the IP, but with initial conditions in the solvent determined by the NP--as the average [5]

$$\Delta\overline{E} = \left\langle e^{\beta\Delta E} \right\rangle_{IP}^{-1} \left\langle e^{\beta\Delta E} \, \Delta E(t) \right\rangle_{IP} \tag{2}$$

over an equilibrium IP ensemble. Here $\beta^{-1} = k_B T$. When developed to second order in ΔE, this leads to [5]

$$S(t) = \left\langle (\delta\Delta E)^2 \right\rangle_{IP}^{-1} \left\langle \delta\Delta E \, \delta\Delta E(t) \right\rangle_{IP} \equiv \Delta(t) \quad , \tag{3}$$

i.e., an equilibrium time correlation function of the type considered in a number of studies [3-5]. Here $\delta\Delta E = \Delta E - \langle\Delta E\rangle_{IP}$.

Figure 2 shows that this approximation is fairly accurate, so that, in the main, the nonequilibrium average TDF shift can in fact and somewhat remarkably be understood via the dynamics of solvent fluctuations at equilibrium; this was a key assumption of analytic approaches to TDF [3]. (This statement is not true for other measures of the TDF spectrum, the spectral width in particular [5].)

This being the case, further examination of the correlation function $\Delta(t)$ is in order. The most convenient formalism for this purpose is via a rigorous generalized Langevin equation (GLE) for ΔE developed by Zichi et al. [11], according to which $\Delta(t)$ satisfies

$$\ddot{\Delta}(t) = -\omega^2\Delta(t) - \int_0^t d\tau \zeta(t-\tau)\dot{\Delta}(\tau) \quad . \tag{4}$$

Here $\omega^2 = \langle(\delta\Delta E)^2\rangle_{IP}^{-1} \langle(\delta\dot{\Delta E})^2\rangle_{IP}$ is [12] the square well frequency for the free energy well for fluctuations in ΔE in the presence of the IP. (It was established previously [12] that this well is indeed harmonic.) The time dependent friction coefficient $\zeta(t)$ is essentially the correlation function of the fluctuating generalized force acting on ΔE [11]. It accounts for dissipative ΔE motion associated with the presumably complex solvent dipole librations, reorientations and translations. It can be extracted from the MD-generated $\Delta(t)$ via Fourier transform inversion techniques [11] and is displayed in Fig. 3.

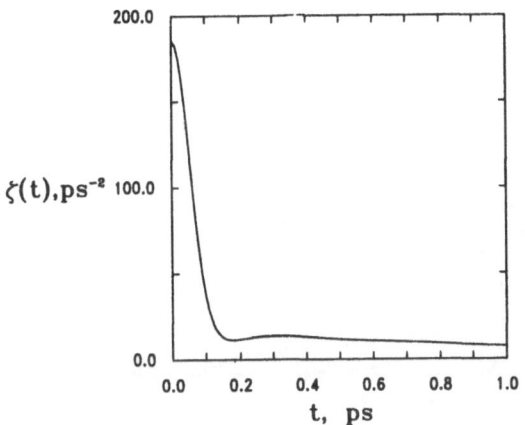

Figure 3.   The time dependent friction coefficient $\zeta(t)$ for the ion pair [11].

The substantial initial rapid drop in both the TDF shift S(t) and the tcf $\Delta(t)$ is governed [5] by the frequency ω according to the simple Gaussian behavior $\Delta^G(t) = \exp(-\omega^2 t^2/2)$, as shown in Fig. 4. This is an important observation, since ω^2 is an equilibrium quantity and one can hope to understand it in terms of the electrostatic forces and torques exerted by the solvent on the ion pair via the methods of modern equilibrium statistical mechanics. It can be further shown [11] that the longer time tails of S(t) and $\Delta(t)$ arise from the time dependent friction $\zeta(t)$ and thus contains information on dissipative solvent processes. Just what those processes are and how they can be approximately described analytically remain to be determined. But it is certainly clear from the present results that the popular continuum dielectric model--which would predict an exponential behavior for S(t) and $\Delta(t)$ [3]--fails significantly, a failure recently observed experimentally in femtosecond laser experiments [2]. Figure 4 shows the failure of the exponential decay predicted by a Langevin Equation (LE) simplification of the GLE, in which the time dependence of $\zeta(t)$ is approximated by a delta function. Obviously, a new initiative is required here to generate a useful molecular level description of the solvent dynamics observed in these simulations.

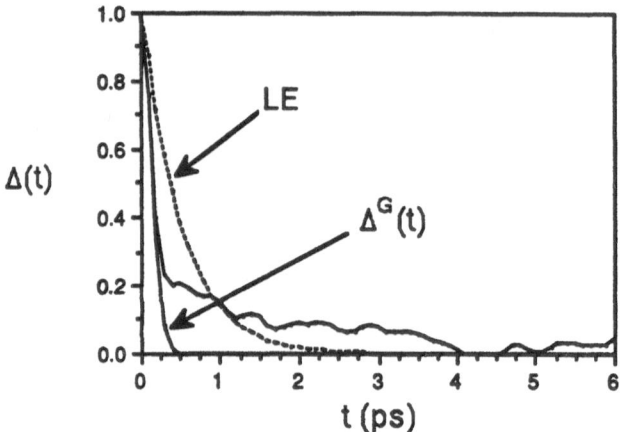

Figure 4.   The solvent time correlation function $\Delta(t)$ compared to the Gaussian description $\Delta^G(t)$ and a Langevin equation description [5]. $\Delta(t)$ is called $C(t)$ in many studies.

3.   ET REACTION RATES

The possible role of solvent dynamics in influencing charge transfer reaction rates in solution has received considerable recent scrutiny [3]. One particular focus has been the importance of solvent dynamics for electron transfer reaction rate constants k. In the standard Marcus Theory [14], which is a Transition State Theory (TST), the rate constant depends upon solvent free energetics but not upon the solvent dynamics. Recently, there has been an explosive analytical theoretical effort attempting to describe the influence of solvent dynamics in causing departures from the Marcus Theory predictions for k [15-17]. Parallel experimental efforts have indicated that some role is indeed played by the dynamics of the solvent in influencing e transfer rates [18]. Despite these intensive efforts, the picture nonetheless remains somewhat clouded; e.g., measures of solvent free energetics and dynamics employed in the theory and interpretations rely on continuum dielectric theory predictions of uncertain validity.

In response, Zichi, Ciccotti, Ferrario and Hynes (ZCFH) have undertaken [19] MD simulations for a well-defined homogeneous activated ET reaction in which deviations from the Marcus TST Theory due to solvent dynamical effects were examined and quantified [20].

ZCFH have considered the (artificial) model ET reaction

$$A^{-1/2}B^{1/2} \longrightarrow A^{1/2}B^{-1/2} \quad , \tag{5}$$

for a solute pair AB, with A = B, immersed in the solvent described in Sec. 2 (the rationale for the choice of fractional charges is given below). The reactant (R) and product (P) solute pair members with a fixed AB separation of 3Å interact via Coulomb and Lennard-Jones (LJ) potentials with the solvent.

The reaction coordinate adopted was the many-body solvent variable[21]

$$\Delta E = H_R - H_P = V_R^{coul} - V_P^{coul} \; ; \tag{6}$$

this energy gap is the difference, at fixed solvent configurations, between the R and P Hamiltonians, i.e., between the Coulomb potential energy of interaction between the solvent molecules and the R and P solute pairs respectively. By symmetry, the reaction transition state is obviously located at $\Delta E = 0$. This simple identification via symmetry considerations, which does not require extensive free energy simulations to establish it, together with the established character of the Ewald method applicable for overall charge-neutral systems, provided the rationale for the choice of the model reaction Eq. 5 for this initial study.

Only the electronically adiabatic limit was considered. Thus, the electronic coupling is sufficiently strong to provide a continuous electronic path between reactants and products (see below). This regime is applicable for many short range electron transfers and it is in this regime that solvent dynamical effects should be most pronounced [17].

The prescription for the adiabatic dynamics was the following. Let H_e formally represent the total system Hamiltonian including the electronic degree of freedom. At fixed solvent molecule configurations, the adiabatic Hamiltonian is determined by a straightforward variational calculation based on the trial two state electronic wave function [21] $\Psi = C_R \Psi_R + C_P \Psi_P$ in which Ψ_R and Ψ_P are the diabatic wave functions describing the R and P electronic distributions and the probability amplitudes C_R and C_P depend parametrically on the solvent configurations. The diabatic system Hamiltonians for the R and P charge distributions are $H_R = \langle \Psi_R | H_e | \Psi_R \rangle$ and $H_P = \langle \Psi_P | H_e | \Psi_P \rangle$ respectively. The resulting diagonalization then gives the (lowest) energy system adiabatic Hamiltonian for the solute pair as [19]

$$H = \frac{(H_R + H_P)}{2} - \frac{\left[(\Delta E)^2 + 4V_{el}^2 \right]^{1/2}}{2} \; ;$$

$$V_{el} = \frac{2H_{RP} - S(H_R + H_P)}{2(1 - S^2)} \; . \tag{7}$$

Here $(H_R + H_P)/2 = H_{NP}$ is just the classical system Hamiltonian for a solute neutral pair. V_{el} is a properly symmetrized electronic coupling, independent of the zero of energy, expressed in terms of the matrix element $H_{RP} = \langle \Psi_R | H_e | \Psi_P \rangle$ and $S = \langle \Psi_R | \Psi_P \rangle$, which is the overlap integral. No attempt was made to calculate V_{el} a priori, but rather it was regarded as an input (constant) parameter for the simulations. Note again that Eq. (7) governs the system dynamics.

In this adiabatic description, the quantum reactant occupation probability $C_R^2 = 1 - C_P^2$,

$$C_R^2(\Delta E) = 4V_{el}^2 \left[4V_{el}^2 + \left\{ \Delta E + \left[(\Delta E)^2 + 4V_{el}^2 \right]^{1/2} \right\}^2 \right]^{-1} , \qquad (8)$$

goes smoothly from unity to zero as ΔE goes from large negative values to large positive values. It can be directly established from the equations of motion associated with Eq. 7 that the dipolar solvent molecules experience the electric field of the apparent classical charge [19]

$$q(\Delta E) = \left[1 - 2C_R^2(\Delta E) \right](e/2) , \qquad (9)$$

on the solute pair member A and $-q(\Delta E)$ on member B. This apparent charge $q(\Delta E)$ proceeds smoothly from $q = -e/2$ at large negative ΔE, through $q = 0$ at $\Delta E = 0$, to $q = e/2$ at large positive ΔE. The adiabatic Hamiltonian Eq. (7) can in fact be written (after some taxing algebra) as [19]

$$H = H_{NP} - V_{el} - \int_0^{\Delta E} d\Delta E \left[q(\Delta E)/e \right] ,$$

whose integral contribution emphasizes the electronically "polarizable" character of the solvent-dependent solute pair charge distribution: the charge distribution is solvent-dependent.

The deviation from the Marcus TST Theory prediction k^{TST} was quantified by the transmission coefficient

$$\kappa = k/k^{TST} . \qquad (10)$$

This was calculated [19] by a flux time correlation function [22-25], as in other reaction simulations [13,26,27], based on trajectories sampled from an initial equilibrium distribution at the transition state [24,26], here located by $\Delta E = 0$. ZCFH attain the transition state in the simulation by imposing the coordinate and velocity constraints $\Delta E = 0$ and $\dot{\Delta E} = 0$ in a constant temperature simulation. However, this procedure introduces a distortion in the sampling compared to the desired initial equilibrium ensemble which is restricted to $\Delta E = 0$ but not $\dot{\Delta E} = 0$. But as described in detail by Carter, Ciccotti, Hynes and Kapral [28], this distortion can be analytically corrected for, and the transmission coefficient is given correctly by the formula

$$\kappa = \frac{\left\langle D^{-1/2} \dot{\Delta E} \theta[\Delta E(t)] \right\rangle_c}{\left\langle D^{-1/2} \dot{\Delta E} \theta(\dot{\Delta E}) \right\rangle_c} , \qquad (11)$$

in which D is [28] the sum over all molecules

$$D = m^{-1} \sum_i \nabla_i' \Delta E \cdot \nabla_i' \Delta E \quad ,$$

where ∇_i' denotes the spatial gradient subject to all bond length constraints, θ is the step function and t is a "plateau" time [22-24]. The average $\langle\langle...\rangle\rangle_c$ denotes the constrained reaction coordinate or "blue moon" ensemble due to Carter *et al.*[28], in which the initial conditions are prepared as described above and at t = 0, the constaint is released, with an appropriate equilibrium distribution of momenta sampled. We pause to note that this procedure can be straightforwardly applied in even more complex reaction problems, with an arbitrary many body, collective configuration-dependent reaction coordinate.

ZCFH first selected a moderate value of the electronic coupling, $V_{el} = 1$ kcal/mol. While no MD-simulated free energy curves are presented here, the approximate free energy curve displayed in Fig. 5 and described in the caption helps to provide perspective on the MD reaction simulation results. In particular, the barrier is cusped; the barrier frequency $\omega_b \sim 4\omega_R$ is estimated by the formula [19] $\omega_b = \omega_R [(2\Delta G^{\ddagger}/V_{el}) - 1]^{1/2}$, with ω_R the R well frequency.

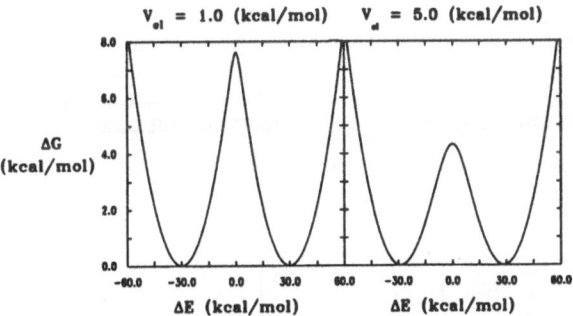

Figure 5. Schematic free energy curves for the two electronic coupling cases. These are generated [19] by a standard macroscopic description[14] with MD-simulated input parameters. See Ref. 19.

A representative example [19] out of 100 trajectories is shown in Fig. 6. There is a direct passage across the transition state, flanked by quite rapid equilibration within the wells. For the very few recrossing trajectories observed, only small excursions off the transition state and single recrossings are involved. The estimated transmission coefficient is $\kappa_{MD} = 0.95 \pm 0.04$. This proximity to unity reflects the feature that recrossings are rare, and Marcus Theory provides an excellent description for this case -- the basic TST assumption (see, e.g., [13]) of no recrossing of the barrier top is well satisfied.

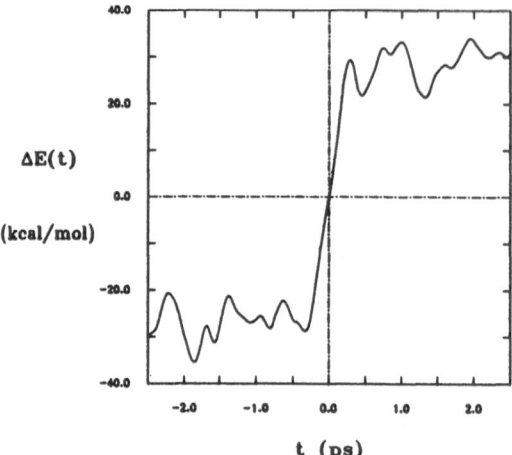

Figure 6.   Representative trajectory for the V_{el}=1 kcal/mol case [19].

In the second case studied by ZCFH, the electronic coupling was increased to $V_{el} = 5$ kcal/mol. The estimated barrier (Fig. 5) is somewhat broad: $\omega_b/\omega_R \sim 1.6$. A representative trajectory [19] is displayed in Fig. 7. Recrossing is now pronounced, with repeated recrossings occurring near the barrier top prior to ultimate rapid equilibration in the wells. The estimated transmission coefficient $\kappa_{MD} = 0.59 \pm 0.11$ represents a marked departure from Marcus TST Theory for this case.

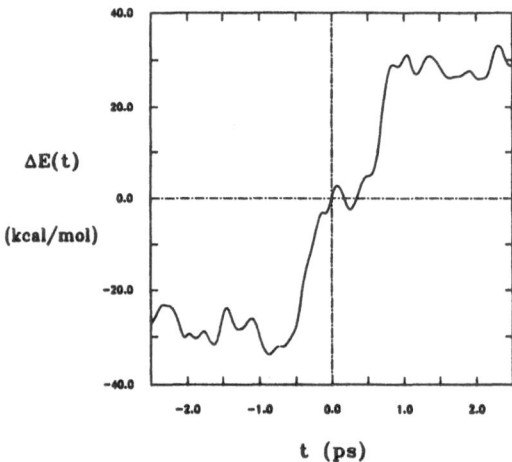

Figure 7.   Representative trajectory for the V_{el}=5 kcal/mol case [19].

Most current theories [15-17] for the solvent dynamical influence on κ values for sharp ("cusped") barrier ET reactions derive from the Zusman Theory [15]; this pictures the solvent dynamical effect as arising exclusively from slow overdamped solvent dynamics in the R and P wells and <u>not</u> at the barrier top. It was found that this description does not apply at all to the current simulations [19], essentially due to the vital and dominant importance of the barrier top dynamics.

A theory for activated barrier crossing, which has proved to be strikingly successful for a variety of reaction classes in solution [13,27,29] is Grote-Hynes Theory [17,30], according to which the transmission coefficient is given by the self-consistent relation

$$\kappa_{GH} = \left[\kappa_{GH} + \frac{1}{\omega_b} \int_0^\infty dt \, e^{-\omega_b \kappa_{GH} t} \, \zeta^\ddagger(t) \right]^{-1} , \qquad (12)$$

where $\zeta^\ddagger(t)$ is the time dependent friction (tdf) for the reaction system at the barrier top. (The GH theory focuses solely on events occurring in the barrier top region.) Evaluation of $\zeta^\ddagger(t)$ requires extensive special simulations at the transition state [13,27,29]. ZCFH followed [19] instead a simpler approximate exploratory route to $\zeta^\ddagger(t)$ and κ_{GH}, now described.

Note that at the transition state $\Delta E=0$, the solute pair charge distribution is that of a neutral pair (NP) (cf. Eq. (8) and Eq. (9)). An estimate of $\zeta^\ddagger(t)$ can then be obtained by determining the actual tdf $\zeta_{NP}(t)$ for a neutral solute pair. This can be determined [11] in precisely the same fashion as described in Sec. 2 for the ion pair friction, and the result is shown in Fig. 8.

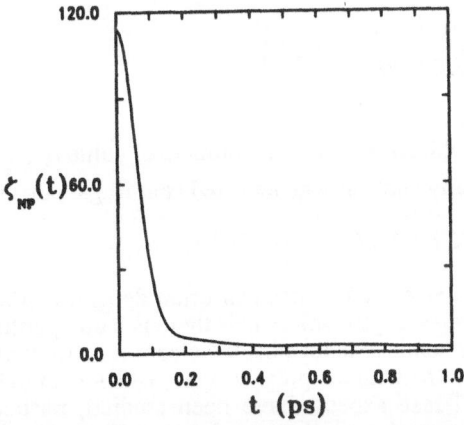

Figure 8.   The time dependent friction $\zeta_{NP}(t)$ for a neutral pair [11,19].

This was then used as an approximation for $\zeta^\ddagger(t)$ in the GH Eq. (12), with the results [19] listed in Table 1.

TABLE 1. ET Transmission Coefficients [19]

V_{el}(kcal/mol)	ω_b(ps^{-1})	κ_{MD}	κ_{GH}
1	26.6	0.95±0.04	0.94
5	10.2	0.59±0.11	0.74

The reasonable level of agreement obtained between these approximate κ_{GH} values and κ_{MD} for both the cusped and broad barrier cases examined is very encouraging. It strongly suggests (a) that the barrier region dynamics are the most important aspect of the solvent dynamics (also indicated by the detailed trajectories) and (b) that the shorter time scale solvent dynamics such as those responsible for the rapid decay of ζ_{NP} (t) in Fig. 8 are those most important in establishing the transmission coefficients and the departure from Marcus TST Theory. These aspects can now be explored for a range of ET reactions in the future.

Finally, the constrained reaction coordinate ensemble (CRCE) [28] provides a prescription for determining rigorous free energy profiles versus ΔE for general ET systems. In particular, the negative gradient $F(\Delta e)$ of the potential of mean force is given by

$$F = \left\langle D^{-1/2} \right\rangle_{c,\Delta e}^{-1} \left\langle D^{-1/2} \frac{\partial \underline{r}}{\partial \Delta E}\left[-\frac{\partial}{\partial \underline{r}}(V + C) \right] \right\rangle_{c,\Delta e} \quad ,$$

where $\langle\langle(...)\rangle\rangle_{c,\Delta e}$ means a CRCE with ΔE equal to the numerical value Δe, $\underline{r}$ denotes all coordinates, V is the potential energy and C is a generalized centrifugal potential [28].

4.   SOLVATION AND ELECTRONIC STRUCTURE

The charge distribution in solute electronic states can often depend markedly on the solvent. Thus in an electron donor-acceptor solute pair there is a competition between the electronic coupling V_{el}, which tends to delocalize the electron between the D and A sites, and the electrostatic interactions with the polar solvent, which tend to localize the electron on one of the sites. These aspects have been studied, particularly in a spectroscopic context, in early important work by several groups [31].

In all this work, however, it was assumed that the solvent was completely in equilibrium with the solute charge distribution. For charge transfer rate and relaxation problems, however, this equilibrium assumption perforce does not hold. Kim and Hynes have recently constructed a theoretical description for this nonequilibrium problem [32]. It is assumed that the solvent electronic polarization is equilibrated to the solute charge distribution, but that the solvent orientational polarization need not be. This leads to a nonlinear Schroedinger equation which is then solved to find the solute wave functions (and thus charge distributions) and the system free energies under nonequilibrium conditions.

Here we briefly describe just one result that emerges from this theory, for the activation free energy $\Delta G^{\ddagger}$ for electronically adiabatic ET reactions (Fig. 9). To place this in perspective,

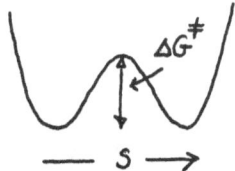

Figure 9.   Schematic free energy curve for an adiabatic ET reaction indicating the activation free energy $\Delta G^{\ddagger}$. Here s denotes a solvent coordinate.

we first recall that the activation free energy for a symmetric charge shift $D^- + A \rightarrow D + A^-$ harmonic reaction is conventionally constructed [14] by finding the intersection of the two diabatic free energy curves for the localized reactant $(D^- + A)$ and product $(D + A^-)$ states, and then subtracting the electronic coupling V_{el}. This gives, for a dielectric continuum solvent model, the standard and widely employed result

$$\Delta G^{\ddagger} = \frac{e^2}{4}\left(\frac{1}{\varepsilon_{\infty}} - \frac{1}{\varepsilon_0}\right)\left(\frac{1}{2R_D} + \frac{1}{2R_A} - \frac{1}{R_{AD}}\right) - V_{el} \ , \qquad (13)$$

where the R's are radii and R_{AD} is the AD separation, and ε_{∞} and ε_0 are the high frequency and static dielectric constants. By contrast, the Kim-Hynes theory gives the approximate result

$$\Delta G^{\ddagger} = \frac{e^2}{4}\left(1 - \frac{1}{\varepsilon_0}\right)\left(\frac{1}{2R_D} + \frac{1}{2R_A} - \frac{1}{R_{AD}}\right) - V_{el} \ . \qquad (14)$$

The origin of the difference in Eqs. (13) and (14) is the following [32]. In Eq. (13), the factor $(\varepsilon_{\infty}^{-1} - \varepsilon_0^{-1})$ represents the feature that the orientational polarization is fixed in the ET act, while the solvent electronic polarization keeps up. But, it is critical to note, the electronic polarization is that appropriate to the charge localized, nonadiabatic states. Eq. (14) instead refers to a transition state with a fixed orientational polarization, but with a solvent electronic polarization which is equilibrated to an adiabatic, charge delocalized symmetric transition state. In fact, the first term in Eq. (14) is essentially the difference in the equilibrium solvation free energies of the delocalized symmetric transition state and the localized reactant state [32]; this is reflected in the appearance of the $(1 - \varepsilon_0^{-1})^{-1}$ factor. Since in highly polar solvents $\varepsilon_0 \gg \varepsilon_{\infty} \approx 2$, there is roughly a factor of two difference in the predictions of Eqs. (13) and (14), which is a very large effect for the ET rate constant, which depends exponentially on $\Delta G^{\ddagger}$. It is clearly of considerable interest to assess the validity of Eq. (14) compared to Eq. (13).

The interplay between the solute quantum charge distribution and the solvent electronic and orientational polarization should prove to be quite important in the understanding of a wide array of dynamic spectroscopic and kinetic problems involving charge transfer.

ACKNOWLEDGMENTS

This work was supported in part by NSF grants CHE84-19830 and CHE88-07852 (JTH), the Natural Sciences and Engineering Research Council of Canada (RK), the donors of the Petroleum Research Fund, administered by the American Chemical Society (JTH, RK), EEC Contract No. ST2J-0094 (GC) and a NATO International Collaborative Grant. We acknowledge the Italian CNR for the allocation of computer time via the CRAY project on Statistical Mechanics, and through the P.F. "Sistemi Informatici Calcolo Parallelo", a grant from the Pittsburgh Supercomputer Center (JTH), and the Research Board of the University of Toronto and the Ontario Center for Large Scale Computation for grants of CRAY computer time. We thank Dr. M. Newton for suggesting the use of the adiabatic Hamiltonian Eq. (7). The work described in Sec. 3 commenced during a series of CECAM workshops in Orsay (1986-88), and we thank the other members of the workshops for useful discussions and Dr. C. Moser for his support and hospitality.

REFERENCES

1.   For a recent review, see R. A. Friesner and Y. Won, submitted to *Biochim. Biophys. Acta.*

2.   See, e.g., M. A. Kahlow, W. Jarzeba, T. J. Kang and P. F. Barbara, *J. Chem. Phys. 90*, 151(1989) and references therein, particularly to the work of the Fleming and Simon groups.

3.   B. Bagchi, D. W. Oxtoby and G. R. Fleming, *Chem. Phys. 86*, 259(1984); G. van der Zwan and J. T. Hynes, J. Phys. Chem. *89*, 4181(1985).

4.   D. A. Karim, D. J. Haymet, M. Banet and J. D. Simon, *J. Phys. Chem. 92*, 3391(1988); M. P. Maroncelli and G. R. Fleming, *J. Chem. Phys. 89*, 5044(1988); J. S. Bader and D. Chandler, *Chem. Phys. Let. 157*, 501(1989).

5.   E. A. Carter and J. T. Hynes, submitted to *J. Chem. Phys.*

6.   G. Ciccotti, M. Ferrario, J. T. Hynes and R. Kapral, *Chem. Phys. 129*, 241(1989).

7.   S. Nose, *J. Chem. Phys. 81*, 511(1984).

8.   L. Verlet, *Phys. Rev. 159*, 98(1967).

9. See, e.g., J. P. Hansen, in *Molecular Dynamics Simulation of Statistical Mechanical Systems*, G. Ciccotti and W. Hoover, Eds. (North Holland, New York, 1986).

10. G. Ciccotti and J. P. Ryckaert, *Comput. Phys. Rep. 4*, 345(1986).

11. D. A. Zichi, H. J. Kim, E. A. Carter and J. T. Hynes, submitted to *J. Chem. Phys.*

12. E. A. Carter and J. T. Hynes, *J. Phys. Chem. 93*, 2184(1989).

13. See, e.g., B. J. Gertner, K. R. Wilson and J. T. Hynes, *J. Chem. Phys. 90*, 3537(1989) and references therein.

14. R. A. Marcus, *J. Chem. Phys. 24*, 966, 979(1956); M. D. Newton and N. Sutin, *Ann. Rev. Phys. Chem. 35*, 437(1984).

15. L. D. Zusman, *Chem. Phys. 49*, 295(1980).

16. D. F. Calef and P. G. Wolynes, *J. Phys. Chem. 87*, 3387(1983); H. L. Friedman and M. D. Newton, *Faraday Discuss. Chem. Soc. 74*, 73(1982); H. Sumi and R. A. Marcus, *J. Chem. Phys. 84*, 4272(1986); I. Rips and J. Jortner, *J. Chem. Phys. 87*, 2090(1987).

17. J. T. Hynes, *J. Phys. Chem. 90*, 3701(1986).

18. See, e.g., G. E. McManis and M. J. Weaver, *J. Chem. Phys. 90*, 912(1989); S. G. Su and J. D. Simon, *J. Chem. Phys. 89*, 908(1988).

19. D. A. Zichi, G. Ciccotti, M. Ferrario and J. T. Hynes, *J. Phys. Chem*, in press.

20. For other simulation studies of different aspects of ET reactions, see e.g., J K. Hwang and A. Warshel, *J. Am. Chem. Soc., 109*, 715(1987); R. A. Kuharski, J. S. Bader, D. Chandler, M. Sprik, M. L. Klein and R. W. Impey, *J. Chem. Phys. 89*, 3248(1988); J. W. Halley and J. Hautman, *Phys. Rev. B, 38*, 11704(1988); C. Zheng, J. A. McCammon and P. G. Wolynes, 1989 preprint.

21. A. Warshel, *J. Phys. Chem. 86*, 2218(1982).

22. T. Yamamoto, *J. Chem. Phys. 33*, 281(1960).

23. R. Kapral, *J. Chem. Phys. 56*, 1842(1972).

24. D. Chandler, *J. Chem. Phys. 68*, 2959(1978).

25. S. H. Northrup and J. T. Hynes, *J. Chem. Phys. 73*, 2715(1980).

26.   R. O. Rosenberg, Jr., B. J. Berne, and D. Chandler, *Chem. Phys. 75,* 162(1980); J. S. Montgomery, Jr., D. Chandler and B. J. Berne, *J. Chem. Phys. 70,* 4056(1979).

27.   J. P. Bergsma, B. J. Gertner, K. R. Wilson and J. T. Hynes, *J. Chem. Phys. 86,* 1356(1987); J. P. Bergsma, J. R. Reimers, K. R. Wilson and J. T. Hynes, *J. Chem. Phys. 85,* 5625(1986).

28.   E. A. Carter, G. Ciccotti, J. T. Hynes and R. Kapral, *Chem. Phys. Lett. 156,* 472(1989).

29.   G. Ciccotti, M. Ferrario, J. T. Hynes and R. Kapral, submitted to *J. Chem. Phys.*

30.   R. F. Grote and J. T. Hynes, *J. Chem. Phys. 73,* 2715(1980); J. T. Hynes, in *The Theory of Chemical Reaction Dyanmics,* M. Baer, Ed. (CRC, Boca Raton, FL, 1985). Vol. IV, p. 171.

31.   See, e.g., H. Beens and A. Weller, *Chem. Phys. Lett. 3,* 666(1969); S. Yomosa, *J. Phys. Soc. Jpn. 35,* 1738(1973); *44,* 602(1978); N. Mataga and T. Kubota, *Molecular Interactions and Electronic Spectra,* (M. Dekker, New York, 1970).

32.   H. J. Kim and J. T. Hynes, to be submitted.

RELAXATION AND COHERENCE IN SIMPLE MODEL SYSTEMS

R. Silbey
Department of Chemistry
Massachusetts Institute of Technology
Cambridge, Massachusetts   02139

1. Introduction

In this paper, I will discuss various theoretical models for systems in which relaxation and coherent energy (or electron) transport occur simultaneously. In particular, I want to spend some time on the range of validity of certain models and approximations. I will do this by showing how unthoughtful applications of a model may lead to nonsense. Most of what I will say is known, so the paper will be more of a didactic exercise than a presentation of novel results.

2. Simple Bloch Equations

Consider a two level system interacting with a large heat bath at temperature T. This interaction causes relaxation of the populations and coherences in the two level system to their Boltzmann equilibrium values. I will designate the two states of the system by $|+>$ and $|->$ and the reduced density matrix for the system by σ. It is a reduced density matrix because the trace over the states of the heat bath has already been taken. A standard and simple model for the relaxation of the reduced density matrix to equilibrium is given (mathematically) by the Bloch equations [1]. I take the $|->$ state to have energy $+\Delta/2$ and the $|+>$ state to have energy $-\Delta/2$. Then I write (note that $\sigma_{++}+\sigma_{--}= 1$ since the total population in the 2 level system is conserved).

$$\dot{\sigma}_{++} - \dot{\sigma}_{--} = \frac{-1}{T_1} (\sigma_{++} - \sigma_{--}) + \frac{1}{T_1} (\sigma_{++}{}^{eq} - \sigma_{--}{}^{eq}) \qquad 1.$$

$$\dot{\sigma}_{+-} = \frac{-1}{T_2} \sigma_{+-} + i\Delta\sigma_{+-} \qquad 2.$$

$$\sigma_{-+} = \sigma_{+-}{}^{*} \qquad 3.$$

In these equations $1/T_1$ is the population relaxation rate and $1/T_2$ is the dephasing rate. These equations have been derived using various

149

J. Jortner and B. Pullman (eds.), Perspectives in Photosynthesis, 149–155.
© 1990 Kluwer Academic Publishers.

approximations by many authors and have been used to interpret experiments for forty years [1]. It is possible to get into trouble with these equations? Yes, and the simplest way is to assume (or find by some side calculation) that $\frac{1}{T_2} \lessgtr \frac{1}{2T_1}$ (or $T_2 > 2T_1$). Although it is

a simple exercise, (often given on physical chemistry exams for graduate students) to show that such an assumption leads to a density matrix which, for some times, has one eigenvalue greater than 1 and one eigenvalue less that 0, this error still crops up in the literature.

The simple exercise mentioned above can be exemplified by going to the limit that T_2/T_1 is infinite. Then σ_{++} and σ_{--} relax to their equilibrium values before σ_{+-} has changed from its initial value. The density matrix is then

$$\sigma = \begin{pmatrix} \sigma_{++}^{eq} & \sigma_{+-}(0) \\ \sigma_{+-}^{*}(0) & \sigma_{--}^{eq} \end{pmatrix} \qquad \qquad 4.$$

with eigenvalues

$$\mu_1 = \frac{1}{2} + \{(\frac{\sigma_{++}^{eq} - \sigma_{--}^{eq}}{2})\}^2 + |\sigma_{+-}(0)|^2\}^{1/2} \qquad \qquad 5.$$

and

$$\mu_2 = \frac{1}{2} - \{(\frac{\sigma_{++}^{eq} - \sigma_{--}^{eq}}{2})^2 + |\sigma_{+-}(0)|^2)^{1/2} \qquad \qquad 6.$$

Since we can easily arrange for the square root to be greater than 1/2, we find that σ is illegal.

3. Simple Two Level System: Spin-Boson Model

Let me now turn to a more difficult system, but one in which a lot of effort has been invested in recent years [2]. The Hamiltonian of the system and heat bath is given by

$$H = J(|L><R| + |R> <L|\} + \Delta_0\{|L><L| - |R><R|\}$$

$$+ H_{bath} + V_0\{|L><L| - |R><R|\} \qquad \qquad 7.$$

$$+ V_1 \{|L><R| + |R><L|\}$$

Here H_{bath} is the Hamiltonian of a set of independent harmonic oscillators (hence, bosons) and V_0 and V_1 are both operators, whereas Δ_0

and J are constants. To make it as simple as possible, I will take $\Delta_0 = 0$ and $V_1 = 0$ so that this represents a two level system with a static energy transfer matrix element (J) coupling the two states and a fluctuating term (V_0) which breaks the energy degeneracy of the two states. It is useful to rotate the states of the two level system, by defining

$$|\pm> = \frac{1}{\sqrt{2}} \{|L> \pm |R>\} \qquad\qquad 8.$$

In which case, H becomes ($\Delta_0 = 0$, $V_1 = 0$).

$$H = J\{|+><+1 - |-><-|\} + H_{bath} \qquad\qquad 9.$$
$$+ V_0 \{|+><-| + |-><+|\}$$

Now $|2J|$ is the energy splitting (I take J<0) and V_0 causes of population relaxation (and dephasing). If I now do the usual treatment for the dynamics of the density matrix, which is equivalent to second order time-dependent perturbation theory, I obtain the Redfield equations:

$$\dot{\sigma}_{++} - \dot{\sigma}_{--} = -(1 + A) \Gamma (\sigma_{++} - \sigma_{--}) - (1 - A) \qquad\qquad 10.$$

$$\dot{\sigma}_{+-} = -i2J \sigma_{+-} - \frac{1}{2} (1 + A) \Gamma (\sigma_{+-} - \sigma_{-+}) \qquad\qquad 11.$$

where A is the Boltzmann factor (exp $-2\beta|J|$) and Γ is the downward population relaxation rate (in the Golden Rule) from $|->$ to $|+>$

$$\Gamma = \int_{-\infty}^{\infty} dt \; e^{-i2J\tau} <V_0(0) V_0(\tau)> \qquad\qquad 12.$$

It is sometimes more suggestive to define $\sigma_x = \sigma_{+-} + \sigma_{-+}$, $\sigma_y = \sigma_{+-} - \sigma_{-+}$, and $\sigma_z = \sigma_{++} - \sigma_{--}$. Then ($\gamma = (1+A) \Gamma$)

$$\dot{\sigma}_z = -\gamma (\sigma_z - \sigma_z{}^{eq}) \qquad\qquad 13.$$
$$\dot{\sigma}_x = -i2J \sigma_y \qquad\qquad 14.$$
$$\dot{\sigma}_y = -i2J \sigma_x - \gamma \sigma_y \qquad\qquad 15.$$

These are similar, but not identical, to the Bloch equations. Note that had I thrown out the coupling between σ_{+-} and σ_{-+} in eq (11), the Bloch equations would have resulted. All of this is pretty standard. The equations for σ_x and σ_y are decoupled from that of σ_z in this model. The solutions to equ. (14-15) are of the form

$$a\, e^{\lambda_+ t} + b\, e^{\lambda_- t} \qquad\qquad 16.$$

where

$$\lambda_\pm = -\frac{\gamma}{2} \pm \left\{ \left(\frac{\gamma}{2}\right)^2 - 4J^2 \right\}^{1/2} \qquad\qquad 17.$$

If $4J^2 > \left(\frac{\gamma}{2}\right)^2$, λ are complex conjugate and σ_x and σ_y oscillate. If $4J^2 < \left(\frac{\gamma}{2}\right)^2$, then $\lambda_\pm$ are both real, and $|\lambda_+|$ is smaller than $|\lambda_-|$. If $\left(\frac{\gamma}{2}\right)^2 >> 4J^2$ then $|\lambda_+|$ is very small and $|\lambda_-|$ very large. The rate of energy transfer from L> to |R> corresponds to the rate of decay of the coherence in the $|\pm>$ representation, or the rate of decay of σ_x and σ_y. Thus, in the limit of $\left(\frac{\gamma}{2}\right)^2 >> 4J^2$ (large energy fluctuations in the LR representatior relative to the transfer matrix element J) the rate of energy transfer from L to R is slowed to $\sim 4J^2/\gamma$. This is the usual and expected result.

But, there is the same serious problem which we saw above: the rate of decay of σ_z (γ) is fast compared to $4J^2/\gamma$, so σ_z relaxes to its equilibrium distribution before σ_x has budged from its initial value. This leads once again to a density matrix which is illegal. The only way out is to have $\sigma_{++}^{eq} = \sigma_{--}^{eq}$ to order $|J|/\gamma$ which is the hallmark of high temperature theories, $2\beta|J| << 1$. The problem is that many of the interesting applications of this model are at very low T.

What has gone wrong with the theory? Clearly when $(\gamma/2)^2 >> 4J^2$, the upper state $(|->)$ has a width greater than the splitting between the states $1+>$ and $|->$. Therefore, the perturbation theory fails and the solutions are meaningless for $\gamma/2 >> 2J$. Can this be repaired by going to higher order perturbation theory? Probably, but note that the relationship given by eq. 14 is exact since it follows directly from the exact equations of motion. This, in addition to the L, R symmetry forces the exact equations of motion to be of the form of eq (13-15), except that <u>time dependent</u> coefficients are possible. If we accept the idea that a Redfield-like description [1,3] is possible, (i.e. time independent coefficients), then we <u>must</u> conclude that overdamped solutions is <u>impossible.</u> This raises some interesting questions about some of the approximate calculations of the symmetric spin-boson model. For example, what is the behavior of $\sigma_{++}(t)$ in the time regime, in which $\sigma_{+-}(t)$ is predicted to be overdamped?

In the asymmetric case, in which the state $|L>$ has a larger zero order energy than $|R>$ (i.e. $\Delta_0 > 0$ eq (7)), complex eigenvalues occur only if $\Gamma' > 2(\Delta_0^2 + J^2)^{1/2}$. Since in most physical systems, Δ_0 is large compared to the width of the states, this is unlikely. Note however that $\Gamma>>J$ is quite likely, and this leads to a decay rate of the population in the $|L>$ state proportional to $\dfrac{J^2}{\Delta_0^2}\Gamma$. By the way, in the limit $\Delta_0^2>>J$, Γ becomes a "pure dephasing" rate, rather that a population relaxation rate, since the exact eigenstates of the two level system are closer to $|L>$ and $|R>$ than to $|+>$ and $|->$. No unphysical results occur in this limit.

4. Three (and more) Level Systems

We have now seen that in certain limits, characterized by widths (γ) greater than splittings (J), the Redfield equations can lead to nonsense. Of course, this is not to be taken as a criticism of the Redfield approach, but of the application of the approach in limits in which it was never intended to be applied. I now discuss a 3 level model which, if care is not taken, leads to similar problems. This was pointed out to me by R. Friesner.

Consider a system of three levels, two of which are degenerate and coupled by a Hamiltonian matrix element J, and a third state below these

$$H_0 = E_0 \left(|a><a| + |b><b|\right) + J\left(|a><b|+|b><a|\right)$$

$$+ E_1\left(|c><c|\right) \qquad\qquad 19.$$

with $|E_1 - E_0| >> J$. We now allow a decay process from $|b>$ to $|c>$ via phonon (or photon) coupling. This model then represents a two molecule system in which molecule A has state $|a>$ and molecule B has states $|b>$ and $|c>$ with $|c>$ acting as a "sink". If $|a>$ is initially excited, then we expect this excitation to leak over to $|b>$ via the coherent (J) coupling and then to decay to $|c>$. If I organize the model so that the decay rate, Γ, from $|b>$ to $|c>$ is very fast compared to J, I expect to see slow irreversible decay of an initially prepared $|a>$ state. How can I go wrong?

Let me define coherent states

$$|\pm> = \frac{1}{\sqrt{2}}\left(|a> \pm |b>\right) \qquad\qquad 20.$$

Then (take $E_0 = 0$ for simplicity)

$$H_0 = J\big[\,|+\rangle\langle+| - |-\rangle\langle-|\,\big] + E_1\,|c\rangle\langle c| \qquad\qquad 21.$$

Now it <u>seems</u> that both the $|+\rangle$ and $|-\rangle$ states decay to $|c\rangle$ with the rate $\Gamma/2$. Therefore if I start in $|a\rangle = \frac{1}{\sqrt{2}}\,(|+\rangle + |-\rangle)$, I seem to predict <u>fast</u> ($\sim \Gamma/2$) decay, instead of slow ($\sim J^2/\Gamma$) decay. This is obviously absurd if we take $J=0$. What have I done wrong? I have neglected <u>correlations</u> which have to be present by my assumption that $|b\rangle$ is a decaying state. This can be seen either by writing the dynamical equations in the $|a\rangle$, $|b\rangle$, $|c\rangle$ representation and then transforming to the $|\pm\rangle$, $|c\rangle$ representation or by starting with a microscopic Hamiltonian including the terms which cause the decay and doing things carefully. I will briefly present the former method. In the $|a\rangle$, $|b\rangle$, $|c\rangle$ representation, the relevant reduced density matrix equations of motion are

$$\dot{\sigma}_{aa} = iJ(\sigma_{ab}-\sigma_{ba}) \qquad\qquad 22.$$

$$\dot{\sigma}_{bb} = iJ(\sigma_{ba}-\sigma_{ab}) - \Gamma\sigma_{bb} \qquad\qquad 23.$$

$$\dot{\sigma}_{ab} = iJ(\sigma_{aa}-\sigma_{bb}) - 1/2\,\Gamma\sigma_{ab} \qquad\qquad 24.$$

$$\dot{\sigma}_{cc} = \Gamma\sigma_{bb}, \quad \sigma_{ab}{}^* = \sigma_{ba} \qquad\qquad 25.$$

where I have assumed that back transfer from $|c\rangle$ to $|b\rangle$ is zero. Transforming to the $|\pm\rangle$ representation yields

$$\dot{\sigma}_{++} = \frac{-\Gamma}{2}\,\sigma_{++} + \frac{\Gamma}{2}\,(\sigma_{+-} + \sigma_{-+}) \qquad\qquad 26.$$

$$\dot{\sigma}_{--} = \frac{-\Gamma}{2}\,\sigma_{--} + \frac{\Gamma}{2}\,(\sigma_{+-} + \sigma_{+-}) \qquad\qquad 27.$$

$$\dot{\sigma}_{+-} = -i2J\sigma_{+-} - \frac{\Gamma}{2}\,\sigma_{+-} + \frac{\Gamma}{2}\,(\sigma_{+-} + \sigma_{-+}) \qquad\qquad 28.$$

$$\sigma_{-+} = \sigma_{+-}{}^* , \quad \dot{\sigma}_{cc} = \frac{\Gamma}{2}\,(\sigma_{++} + \sigma_{--} - \sigma_{+-} - \sigma_{-+}) \qquad\qquad 29.$$

The important point to note is that the coupling between population modes (σ_{++} and σ_{--}) and coherences (σ_{+-} and σ_{-}) <u>cannot</u> be ignored, because the coupling coefficient $(\Gamma/2)$ is of the same order of magnitude as the decay coefficient $(\Gamma/2)$. Thus the usual procedure of decoupling these variables fails. It was incorrect of me to say (above) that $|+>$ and $|->$ decay; in this case the dynamics are more complex. If equ (26-29) are solved, it is easy to see that $\sigma_{aa}(t)$ decays with a rate proportional to J^2/Γ and all is well. If I had decoupled the populations and coherences, I would have found the nonsensical result that σ_{aa} decays with a rate Γ even if $J = 0$.

I can see this in a direct way from the Hamiltonian by adding to H_0, the term coupling $|b>$ and $|c>$ through the phonons:

$$\hat{V} = \phi \ (|b><c| + |c><b|)$$

and using the standard Redfield method (see for example [3]). I find that the coefficient coupling σ_{++} to σ_{+-} is

$$R_{++,+-} = +\int_0^\infty < \ \hat{V}_{-c} \ V_{c+}(t) \ > \ e^{-i(E_- - E_c)t} dt$$

$$= + \Gamma/4 \qquad\qquad\qquad\qquad\qquad 31.$$

just as in eg (26). Thus, the standard methods work, as long as I do not neglect the coupling between modes.

5. Conclusions

As I stated in the introduction, all of this is well known, but it is important to be reminded of the pitfalls of simple approaches to dynamical problems when coherence and relaxation are intertwined, as they certainly are in the short time dynamics in electronically excited systems.

Acknowledgements: I am indebted to Rich Friesner and Bob Harris for advice and questions, and to the NSF for support.

References

1. A. Redfield, Adv. Mag. Res. 1, 1 (1965); K. Blum, Density Matrix Theory and Applications (Plenum, New York, 1981).

2. See for example, A. Caldeira and A. Leggett, Phys. Rev. Lett. 46, 211 (1981); A. Leggett et al., Rev. Mod. Phys. 59, 1 (1987); H. Grabert, U. Weiss and P. Hanggi, Phys. Rev. Lett. 52, 2193 (1984); D. Carmeli and D. Chandler J. Chem. Phys. 84, 3400 (1985); E. Pollak, Phys. Rev A33, 4244 (1986); R.A. Harris and R. Silbey, J. Chem. Phys. 80, 2615 (1984), ibid 83, 1069 (1985).

3. R. Wertheimer and R. Silbey, Chem. Phys. Lett. 75, 243 (1980).

SUPEREXCHANGE COUPLING MECHANISMS FOR ELECTRON TRANSFER PROCESSES

M. D. NEWTON
Chemistry Department
Brookhaven National Laboratory
Upton, NY 11973 USA

ABSTRACT. Electron transfer matrix elements for electron exchange between various pairs of transition metal complexes in close contact have been calculated and analyzed for a variety of approach geometries for the two reactants. The coupling between the nominal metal ion donor/acceptor sites is achieved by superexchange of the "hole" type arising from ligand-to-metal charge transfer (LMCT), the dominant ligand-field interaction for the electron-donor ligands considered (H_2O, NH_3, and the cyclopentadienide anion). The pronounced variations of H_{if} with geometry are not correlated with the separation distance of the metal ions (between which the direct overlap is negligible) and span the range from non-adiabatic to strongly adiabatic electronic coupling. The values for metallocene/metallocinium redox pairs bracket recently reported experimental values. Analysis of the results using the method of corresponding orbitals demonstrates the validity of an effective 1-electron model for the electron transfer process to within about 10% for the class of systems considered. A higher-order superexchange mechanism was encountered for the $Co(NH_3)_6^{2+/3+}$ exchange process, in which the LMCT-driven hole-transport mechanism couples excited local states of the metal ions, which in turn are connected to the corresponding ground states by spin-orbit mixing. This mechanism yields on electronic transmission factor within two orders of magnitude of unity.

1. INTRODUCTION

The kinetics of electron transfer processes are controlled by a number of energetic and dynamical factors involving both nuclear and electronic degrees of freedom [1-4]. While traditional approaches have focused most attention on the activation energy [1,2], with primary application to small molecular reactants in close contact, more recent interest in transfer between widely separated donor and acceptor sites has increasingly focussed attention on the electronic structural aspects of the process [5-9]. Since direct orbital overlap between local donor and acceptor sites becomes negligible if their

J. Jortner and B. Pullman (eds.), Perspectives in Photosynthesis, 157–170.

separation exceeds a few angstroms, the electronic coupling is
typically formulated perturbatively as an indirect process involving
virtual intermediate states [2,8,10,11]. In spite of the fact that
such superexchange mechanisms are in principle many-electron
phenomena, one in general expects a one-electron model to provide a
viable approximation for electron transfer processes [6c], in contrast
to analogous two-electron superexchange coupling between localized
spin sites [12].

While the formulation of superexchange coupling is fairly
straight-forward, the multiiplicity of possible "pathways" which arise
in complex systems and which may interfere with each other
constructively or destructively, often makes quantitative
implementation of superexchange schemes quite difficult.
Nevertheless, the availability of sophisticated computational
techniques for determining electronic structural features of complex
molecular systems has made it possible to attempt detailed
superexchange analyses for certain redox processes [6c].

In the present paper we illustrate two different types of
superexchange coupling associated with electron exchange processes
involving coordinated transition metal ions:

$$ML_n^{z+} + ML'_n^{(z+1)+} \longrightarrow ML_n^{(z+1)+} + ML'_n^{z+} \tag{1}$$

When the redox partners are in contact, good overlap will exist
between one or more pairs of ligands (L,L'). Indirect coupling of the
nominal donor and acceptor orbitals (taken as the 3d orbitals of the
metal ions M,M') can then be facilitated by ligand-metal covalent
mixing [6c]. In some cases, this type of supercharge coupling is
supplemented by a second type arising from spin-orbit coupling
[13,14]. This latter situation will be illustrated for the case of
M = Co, L = NH$_3$ (z=2, n=6).

As noted above, one expects to be able to formulate an electron
transfer process within a "one-electron" framework, a consideration of
no small importance when it comes to modeling very complex processes
as, for example, occur in protein-based systems [8,9] or in
photosynthesis [15,16]. In the present paper we will address the
validity of the one-electron model from the perspective of results
obtained from a computational model which includes all the valence
electrons of the two redox partners. In particular, we shall evaluate
the electron transfer matrix elements,

$$H_{if} = \int \psi_i H \psi_f d\tau \tag{2}$$

which couple the initial and final states in the electron transfer
process, where H is the full (many-electron) Schroedinger electronic
Hamiltonian associated with the reaction partners in their
close-contact encounter complex (this precursor complex will be
treated as a supermolecule complex in the calculations reported
below).

We wish to understand the sensitivity of H_{if} values to the
detailed electronic structure of the reactants and to structural

variations within the encounter complex, especially the relative orientation of the redox partners and the separation of the nominal donor/acceptor sites (M,M'). Relative to the simple model in which H_{if} is taken as a one-electron resonance integral between spherical donor and acceptor orbitals, so that the only structural dependence is an exponential dependence on donor/acceptor separation, we shall find a much richer pattern in which through-bond (TB) and through-space (TS) factors control the variations of H_{if} with orientation. These variations can be sufficiently pronounced to span the limits of adiabatic and non-adiabatic behavior for a given redox pair.

Any polarized solvent surrounding the encounter complexes dealt with in the present study is assumed to have a minor influence on H_{if} magnitudes, as supported by recent calculations [16] in which the supermolecule complex was placed in a cavity within a polarized continuum.

2. THEORETICAL AND COMPUTATIONAL DETAILS

2.1. Kinetic Model

In order to establish a concrete link between the rate constant for electron transfer, k_{et}, and the electron transfer matrix element, H_{if}, we consider the following generalized transition state theory (TST) expression [2,6c]:

$$k_{et}^{TST} = (K_{pre-eq})(exp[-\beta E^+])(v_n \kappa_{el}) \qquad (3)$$

where the three factors in parentheses correspond respectively to the formation of the precursor complex (K is the pre-equilibrium constant), the activation of the precursor complex, and the passage from initial to final state at the transition state. Our interest in the present work lies in the attenuation of the rate of the latter process due to the electronic transmission coefficient κ_{el}. With the help of the Landau-Zener model [18] we may represent κ_{el} as

$$\kappa_{el} = 2P_0/(1+P_0) \qquad (4)$$

where the probability P_0 for hopping between diabatic surfaces, H_{ii} and H_{ff}, is given by

$$P_0 = 1 - exp(-2\pi\gamma) \qquad (5)$$

and where, for a harmonic oscillator model,

$$2\pi\gamma = |H_{if}|^2 \pi^{3/2}/2hv_n(k_B TE^+)^{1/2} \qquad (6)$$

In eq 5, v_n is the harmonic frequency associated with the initial state (H_{ii}) and final state (H_{ff}) wells, and E^+ is the activation energy (the energy at which H_{ii} and H_{ff} cross relative to the minimum of the H_{ii} well). Thus the magnitude of H_{if} is a crucial factor in determining where a given process lies relative to the

non-adiabatic ($\kappa_{e\ell} \ll 1$) and adiabatic limits ($\kappa_{e\ell} \sim 1$). For
typical activated electron processes, one is near the adiabatic limit
when $|H_{if}| > k_BT$ at room temperature (i.e., ~ 200 cm^{-1}) [2].

2.2. Wavefunctions and Matrix Elements

The most straightforward approach to evaluating H_{if} (eq 2) is to
obtain it directly, using calculated wavefunctions ψ_i and ψ_f
[6a,7]. In general, ψ_i and ψ_f, which can be thought of as
charge-localized valence bond structures, corresponding, respectively,
to the left and right hand side of eq 1, are non-orthogonal ($S_{if} \equiv
\int \psi_i \psi_f d\tau \neq 0$), As a result, eq 2 is generalized as follows:

$$H'_{if} = (H_{if} - S_{if}H_{ii})/(1 - S_{if}^2) \tag{7}$$

The states ψ_i and ψ_f are represented as single configuration
(i.e., single determinant), wavefunctions and are determined
variationally using the self-consistent field method (SCF) [6]. In
previous studies of small model coordination complexes, we performed
these calculations using <u>ab initio</u> methods. For the larger molecular
complexes treated in the present study we have, for the most part,
employed a version of the INDO method developed transition metal
complexes by Zerner et al. [19]. In cases where comparison is
possible, the <u>ab initio</u> and INDO approaches are found to yield
comparable estimates of H'_{if} (generally within 20 or 30 percent of
each other) [6c].

As an alternative to eq 6, in cases of symmetric electron
exchange it is convenient to estimate H'_{if} values as the splittings
of charge-delocalized state energies [6],

$$H'_{if} = (H^+ - H^-)/2 \tag{8}$$

where H^+ and H^- are the expectation values of the Hamiltonian with
respect to the symmetric (+) and antisymmetric (−) charge-delocalized
SCF wavefunctions, respectively. While the constraint of
charge-localization suppresses a certain amount of electronic
relaxation (a many-electron effect) relative to the fully-relaxed
functions ψ_i and ψ_f, H'_{if} values obtained from eq 8 are generally
quite close to those obtained from eq 7 [6c]. Most of the results
presented below are based on eq 8, although comparisons with eq 7, are
also included.

The method of corresponding orbitals [20] plays an important role
not only in implementing eq 7, but also in casting the calculated
results in a form which provides a straightforward definition of the
effective donor and acceptor orbitals and allows a quantitative
assessment of departures from a simple one-electron model for H'_{if}.
The corresponding orbital transformations among the orbitals of the
n-electron wavefunctions, ψ_i and ψ_f, define sets of n molecular
orbitals, $\{\phi^i\}$ and $\{\phi^f\}$, respectively, which correspond maximally
in pairs (ϕ_1^i, ϕ_1^f; ϕ_2^i, ϕ_2^f; ... ϕ_n^i, ϕ_n^f). A
one-electron model is valid to the extent that a nearly invariant

$(n-1)$-electron core can be identified $(\phi_k{}^i \sim \phi_k{}^f, \; k \leq n-1)$, and the remaining pair of orbitals, $\phi_n{}^i$ and $\phi_n{}^f$, provide the natural definition of the donor acceptor orbitals [6c]. Departures from a purely one-electron model are identified with departures from unity of the overlap integrals, $S_k{}^{i,f} \equiv \int \phi_k{}^i \phi_k{}^f d\tau, \; k \leq n$.

For the $Co(NH_3)_6{}^{2+/3+}$ complexes, multiplet splittings were calculated from __ab__ initio wavefunctions employing large basis sets (s,p,d, and f orbitals) and including electron correlation at the level of 2nd-order perturbation theory (MP2 [21]) relative to the SCF reference [14].

3. SUPEREXCHANGE COUPLING

3.1. The Role of Ligand Field Mixing

The role of ligand-field mixing in indirect superexchange coupling of metal-ion donor orbitals can be understood in terms of the following simplified scheme based on a single contact ligand-pair, in which the ligands are taken as electron donors (the case relevant to the water, ammine, and cyclopentadienide ligands considered below):

$$
\begin{array}{ccccccccc}
\text{reactants:} & M \xrightarrow{z+} L\cdots L & \longrightarrow & M \xrightarrow{(z+1)+} & \longleftrightarrow & M \xrightarrow{z+} L\cdots L & \longrightarrow & M \xrightarrow{1+} & z+ \\
\end{array}
$$

reactants: $M \overset{z+}{\rule{2cm}{0.4pt}} L\cdots L \rule{1cm}{0.4pt} M \overset{(z+1)+}{\longleftrightarrow} M \overset{z+}{\rule{1cm}{0.4pt}} L\cdots L \rule{1cm}{0.4pt} M^{z+}$

direct   indirect
(first-order)   (third-order)

products: $M \overset{(z+1)+}{\rule{2cm}{0.4pt}} L\cdots L \rule{1cm}{0.4pt} M^{z+} \longleftrightarrow M \overset{z+}{\rule{1cm}{0.4pt}} L\cdots L \overset{1+}{\rule{1cm}{0.4pt}} M^{z+}$

The reactant (ψ_i) and product (ψ_f) states are each represented as a resonance mixture of a primary valence bond structure (left-hand-side) and a valence structure corresponding to ligand-to-metal charge transfer (LMCT). Direct (first-order) electron transfer between metal ion sites (left-hand side) is of minor importance in comparison with indirect (third-order superexchange) electron transfer which exploits direct (first-order) coupling of adjacent ligands in conjunction with first order LMCT within each redox partner (right-hand side).

The above superexchange mechanism is of the "hole" type, involving electron-deficient virtual states of the intervening ligands [2]. For cases of electron-acceptor ligands, one would expect an analogous superexchange mechanism of the "electron" type in which the indirect donor/acceptor coupling is established via intervening electron-attachment states. The third-order "hole" mechanism for electron exchange between various aquo- and ammine complexes has been demonstrated quantitatively for encounter geometries with an apex-to-apex contact (i.e., a common four-fold axis), using ligand-field covalency parameters inferred from the calculated wavefunctions as a measure of the LMCT [6c].

3.2. Dependence of H'_{if} on Encounter Geometry

The magnitude of electron coupling by super-exchange of the type
illustrated above will depend on the relative orientation of reactants
in the encounter complex, since the nature of the ligand-ligand
contacts will depend on the details of the encounter geometry. Thus
for ligands involving peripheral hydrogen atoms, the primary contact
may involve overlap of the hydrogen orbitals on the respective
ligands, even in the case of π-type electron transfer, in which case
the hydrogenic orbitals are coupled to the heavy atoms via
hyper-conjugation. In addition, direct overlap between the orbitals
of the heavy atoms of the ligands on the two reactants may be
appreciable for certain orientations [6c]. On the other hand, one
does not expect any particular correlation of overall coupling
strength (and hence H'_{if} magnitude) with the distance between the
nominal redox sites ($r_{MM'}$).

3.2.1. Hexa-aquo and Hexa-ammine Complexes. In Table 1 are displayed
calcualted H'_{if} magnitude for three different encounter geometries,
illustrating cases of both π-type (t_{2g}, $Fe(H_2O)_6^{2+/3+}$) and σ-type
(e_g, $Co(NH_3)_6^{2+/3+}$) electron transfer. The selected geometries --
apex-to-apex, edge-to-edge, and face-to-face -- correspond to
van-der-Waals contact between counter pairs sharing common four-fold,
two-fold, and three-fold axes, respectively.
    The H'_{if} values for π-electron transfer vary rather little
inspite of appreciable changes in the number of ligand-ligand contacts
and the metal-metal distance. The small increase which is observed in
proceeding from the head-on apex-to-apex approach to the rather
oblique ligand-ligand contacts in the edge-to-edge and face-to-face
approaches probably reflects the increased role of direct overlap
between reactant oxygen π orbitals (antisymmetric in the water planes)
as the angle between FeO-FeO bonds in contact changes from 180°
(apex-to-apex) to 90° (edge-to-edge and face-to-face) [6c].
    The coupling for sigma electron transfer in the $Co(NH_3)_6^{2+/3+}$
system is both stronger and more sensitive to orientation than for the
above case of π-electron transfer (a qualitatively similar conclusion
was reached on the basis of extended Hückel calculations [8c,22]).
Furthermore, the strong decrease of H'_{if} values is seen to occur even
though the number of ligand-ligand contacts is increasing and the
metal-metal separation is decreasing. The increased bulkiness of the
NH_3 ligand relative to water leads to less variation in $r_{MM'}$
compared with that exhibited by the hexa-aquo complexes, and the
directionality of the ammine lone pairs, leads to less effective
direct inter-reactant overlap involving these orbitals than for the
analogous case of the water pi-orbitals.
    While all the displayed cases for $Fe(H_2O)_6^{2+/3+}$ correspond to
an appreciable degree of non-adiabatic behavior ($\kappa_{el} < 1$), the
various $Co(NH_3)_6^{2+/3+}$ values are seen to span the range from weak
adiabatic to strong adiabatic coupling. Of course, an overall
assessment of the rate constant would require an estimate of the
relative energies of the different precursor states in solution. For

TABLE 1.   Orientation Dependence of H'_{if}

Orientation	H'_{if} $(cm^{-1})^a$	Number of L,L' Contacts	$r_{M,M'} (\text{Å})$
π/t_{2g} transfer $(FeH_2O)_6^{2+/3+}$			
apex-to-apex[b]	18	1	7.4
edge-to-edge	30	2	6.4
face-to-face	40[c]	6	5.3
σ/e_g transfer $(Co(NH_3)_6)^{2+/3+}$			
apex-to-apex[d]	700	1	7.0
edge-to-edge[b]	120	2	6.9
face-to-face	80	6	5.8

a) Based on eq 8, using the lowest energy supermolecule
states of g and u symmetry. For each of the three
orientation, the redox partners are related by inversion
through the symmetry center of the supermolecule
complex. The geometrical parameters of the ML_6
complexes are the same as those given in ref. [6].
b) The adjacent "octahedral edges" are parallel.
c) A very similar result (38 cm^{-1}) was obtained using
eq 7.
d) The process involves an excited spin state (^2E_g;
t_{2g}^6/e_g); this state is coupled to the ground state
($^4T_{1g}$; t_{2g}^5/e_g^2) by spin-orbit mixing, as discussed
in Section 3.3.

$Fe(H_2O)_6^{2+/3+}$ the face-to-face approach is likely to be dominant
[23,24].

3.2.2. Metallocene Systems. We turn now from the saturated H_2O and
NH_3 ligands to the unsaturated cyclopentadienide (Cp^-) ligand. When a
metallocene (Cp_2M) is ionized to form the metallocinium ion (Cp_2M^{1+}),
the hole formed in one of the metal d-orbitals is strongly screened by
LMCT into the vacancies in the other 3d orbitals [19a]. Accordingly,
it is of interest to investigate the effect of such screening on H'_{if}
values for metallocene/metallocinium redox pairs, especially since
experimental estimates of H'_{if} have been obtained by Weaver et al.
for the cases M = Fe and Co [25]. Theoretical and experimental
results are presented in Table 2.
     The Fe and Co results offer an interesting comparison since the
transferred electron in the case of Fe has "delta" symmetry (i.e.,
transforming as x^2-y^2, or xy, where the reference direction is the
five-fold symmetric z-axis), while the analogous symmetry for Co is
"pi" (transforming as xz or yz). Thus for cobaltocene, we expect
better metal-ligand overlap, and hence, more delocalization in the
effective donor and acceptor orbitals.
     The data in Table 2 are presented in order of increasing
interreactant coupling, and include coaxial (D_{5h}) and side-by-side
(D_{2h}) approaches, as well as an intermediate (C_{2h}) structure. We
have also included the covalently-bound bi-metallocene system [27].
As in Table 1, we find no apparent correlation between H'_{if} and

TABLE 2. H_{if}' Values for Metallocene Redox Pairs (cm^{-1})[a]

Structure	M = Fe[b] calc[d]	exp	M = Co[c] calc[d]	exp	$r_{MM'}$ (Å)
(D$_{2h}$) ...1+	20		40		5.92
(C$_{2h}$) ...1+	50	35[e]	350	175[e]	5.29
(D$_{5h}$) ...1+	140		870		6.58
(C$_{2h}$) ...1+	1050 ≳ 500[f]		2240 ≳ 1100[f]		5.14

a) (CpMCp)$^{0/1+}$ redox pairs, where Cp ≡ C$_5$H$_5$. In the present model studies, a common geometry was used for all metallocene species: D$_{5h}$ symmetry; r_{CC} = 1.42 Å; r_{CM} = 1.62 Å; r_{CH} = 1.10 Å. Experimental geometrical data and pertinent references are summarized by Weaver et al. [26]. The first three structures represent van-der-Waals contact (edge-to-edge, with H···H ~ 2.2 Å for D$_{2h}$, C$_{2h}$), and C···C = 3.5 Å for D$_{2h}$). In the last case (bimetallocene), the metallocenes are linked by a single covalent bond (r_{CC} = 1.64 Å). For D$_{2h}$ and C$_{2h}$ cases, results pertain to lower energy component of split degenerate states (see footnote a, Table 3).
b) The zeroth-order picture of the ferrocinium cation involves removal of a 3dδ electron (defined with respect to the five-fold σ axis of the metallocene).
c) The zeroth order picture of the cobaltocinium cation involves removal of a 3dπ electron (defined as in footnote b).
d) Calculated using eq 8.
e) Ref. [25].
f) Obtained as lower limits from analysis of intervalence charge-transfer spectra [27].

$r_{MM'}$, and once again, the orientational variation is seen to span the range from non-adiabatic to strong adiabatic coupling. The theoretical values for the non-bonded metallocene-metallocinium contact pairs bracket the experimental estimates [25], while the theoretical values for the bi-metallocene systems, assuming trans conformations, are consistent with the experimental lower limits [27]. More precise contact with experiment will require estimates of relative energies of the different structures.

The effect of electronic relaxation on H'_{if} is illustrated in Table 3 for the case of the C_{2h} encounter geometry. Relaxing the constraint of change-delocalization entailed in the use of eq 8 is seen to reduce H'_{if} magnitude by only 10-30%, and we thus find quantitative support for a one-electron model of the electron transfer process. The details of the calculations also reveal that non-unit overlap factors associated with the n-1 electron core (see Section 2) attenuate the one-electron contribution to H'_{if} in the localized representation (eq 7) by only 10%, thus providing further support for the one-electron model. The effective donor and acceptor orbitals in this model are found from the corresponding orbital analysis to be primarily localized on the metal atom (~ 75%) for the case of Fe, while for Co the orbitals are about evenly shared by metal and ligands.

The robustness of the one-electron model, as assessed by the preceding criteria, is maintained in spite of the appreciable screening (an intrinsically many-electron effect) which accompanies the ionization of metallocenes, as noted above. For both ferrocene and cobaltocene, more than 70% of the d-orbital depletion of the donor molecular orbital is screened by LMCT involving the other molecular orbitals.

TABLE 3. Effect of Relaxation on H'_{if} (cm^{-1})[a]

		Fe[b]		Co[c]	
		deloc[d]	loc[e]	deloc[d]	loc[e]
(C$_{2h}$)	xy	50	35	yz	350	280
	x^2-y^2	13	10	xz	40	35

a) The 2-fold degeneracy associated with the metallocene five-fold axis is broken in the encounter complex. The first row of H'_{if} values refers to the lower-energy component in each case (xy and yz symmetry, respectively, for the donor/acceptor orbitals in the Fe and Co systems).
b) See footnote b, Table 2.
c) See footnote c, Table 2.
d) Calculated using eq 8.
e) Calculated using eq 7.

3.3 A Role for Spin-Orbit Coupling

Situations may arise in which ground-state spin multiplets do not by
themselves serve as useful localized donor and acceptor states in
superexchange schemes of the type considered above. Such a situation
may well pertain to the cease of the $Co(NH_3)_6^{2+/3+}$ exchange process
[13,14]. As indicated in the following scheme,

$$Co(NH_3)_6^{2+} + Co(NH_3)_6^{3+} \longrightarrow Co(NH_3)_6^{3+} + Co(NH_3)_6^{2+}$$

reactant 1 (2+)	reactant 2 (3+)

$$^4T_{1g} + C\ ^2E_g \qquad\qquad ^1A_{1g} + C'\ ^3T_{1g}$$

$$(t_{2g}^5/e_g^2)\quad (t_{2g}^6/e_g^1) \qquad\qquad (t_{2g}^6)\quad (t_{2g}^5/e_g^1)$$

$$^1A_{1g} + C'\ ^3T_{1g} \qquad\qquad ^4T_{1g} + C\ ^2E_g$$

$$(t_{2g}^6)\quad (t_{2g}^5/e_g^1) \qquad\qquad (t_{2g}^5/e_g^2)\quad (t_{2g}^6/e_g^1)$$

product 1 (3+)	product 2 (2+)

the pathway connecting ground spin states (dashed vertical arrows) is
formally a "three-electron" process (inter-reactant transfer of an
electron and an $e_g \leftrightarrow t_{2g}$ rearrangement within each redox partner)
and is thus expected to have very low probability.

   On the other hand, the observed reaction in aqueous solution
exhibits "normal" electron transfer kinetics [28] based on the Marcus
transition state model [1], thus appearing to rule out the existence
of an unusually small κ_{el} factor. Accordingly we are led to
consider various "one-electron" pathways (solid diagonal arrows) which
become accessible to the extent that spin-orbit coupling occurs in the
ground state species. It is these one-electron pathways which are
governed by the LMCT-driven superexchange mechanism illustrated in
Section 3.2.1, but the overall superexchange process here will be
higher order (formally fifth order) since first-order spin-orbit
coupling is required in each reaction partner.

   A critical ingredient for perturbatively estimating the strength
of the spin-orbit coupling is the high-spin/low-spin energy splitting
for the 2+ and 3+ ions. While the value for the 3+ ion is known from
spectroscopy [29], the most likely smaller, and hence more important,
value for the 2+ ion is not available experimentally, although INDO

calculations [22] have suggested that it may be quite small (< 5000 cm^{-1}). Accordingly, we have attempted to calculate this quantity [14], using correlated electronic structure techniques with very large basis sets (see section 2), and empirically correcting the calculated splitting by exploiting the close analogy between the electronic state differences for the two different charge states: i.e., for each charge state the low-spin →high-spin process corresponds to the breaking of a t_{2g}^2 pair accompanied by a $t_{2g} \rightarrow e_g$ excitation. The sensitivity of the calculated splittings to various levels of calculation is indicated in Table 4.

TABLE 4. Sensitivity of Calculated high-spin/low-spin energy splittings [14], (ΔE^{z+}, 10^3 cm^{-1})

$$\Delta E^{3+} \equiv E(^3T_{1g}) - E(^1A_{1g})$$
$$\Delta E^{2+} \equiv E(^2E_g) - E(^4T_{1g})$$

Computational Level	ΔE^{3+} [a]	ΔE^{2+} [a]
SCF	1.2	16.2
2nd-order correlation (MP2)[b]		
s,p,d basis	---	13.9
s,p,d,f basis[c]	7.7	12.4
empirically corrected[c,d]	13.7	6.4

a) Based on calculated equilibrium Co-N bond lengths: 2.07 Å(3+) and 2.29 Å(2+).
b) Ref. [21].
c) Includes basis functions of $\underline{f}$ symmetry on Co.
d) The positive correction term, E_{corr}, which brings the 3+ value into exact agreement with the known experimental value [29], was also employed to correct the 2+ value. Given the definitions of ΔE^{2+}, the correction of ΔE^{3+} by $+E_{corr}$ implies correction of ΔE^{2+} by $-E_{corr}$.

Table 5 reveals that for the geometry most pertinent to the kinetics (i.e., the transition state value of the Co-N bond length), the splittings are significantly less than for the respective equilibrium geometries, and hence the spin-orbit coupling is expected to be correspondingly greater. Using atomic ion values for the necessary spin-orbit matrix elements, we finally obtain [14] estimates of ~ 0.10 and ~ 0.30 for the spin-orbit mixing coefficients in the two low lying states of the 2+ ion, and ~ 0.15 for the 3+ ion (these are the coefficients C,C' in the above scheme.

Applying these results to the case of the apex-to-apex approach of reactants, we obtain an effective H'_{if} value of ~ 25 cm^{-1} (scaling the "one-electron" H'_{if} value of 700 cm^{-1} listed in Table 1 by a spin orbit attenuation factor of ~ 0.03) and hence via (eqs 4-6) an overall $\kappa_{e\ell}$ value of ~ 10^{-2}. Thus the spin-orbit

TABLE 5.   Variation of ΔE^{2+} values with Co-N bond length[a]

species	equilibrium		transition state	
	r_{CoN} (Å)	$\Delta E(10^3\ cm^{-1})$	r_{CoN} (Å)	$\Delta E(10^3\ cm^{-1})$
3+ ion	2.07	13.7	2.15	10.0
2+ ion	2.29	6.4	2.15	2.2

a) All ΔE value have been corrected (last entry in
Table 4). The following general linear relationships have
been obtained [14]: $\Delta E^{3+} = 109.3 - 46.2\ r_{CoN}$ and $\Delta E^{2+} =$
$-61.6 + 29.7\ r_{CoN}$. The calculated r_{CoN} values are
uniformly about 0.1 Å larger than the experimental values.
Compensation for errors in calculated ΔE values arising from
the systematic shift of bond lengths is included in the
empirical correction (see Table 4).

coupling mechanism seems capable of accounting for the absence of very
strong departures from the adiabatic limit in the observed kinetics
[28].   Further calculations suggest that the activation energy for the
spin-orbit ground state pathway is somewhat smaller than that for the
pathway involving the thermally excited ^2E state of $Co(NH_3)_6^{2+}$ [30].
     As a final comment it is worth noting that the sensitivity of
high-spin/low-spin splitting energies to variations in r_{CoN}
(Table 5) implies a corresponding variation in the H'_{if} value along
the reaction coordinate (since the r_{CoN} values control the
inner-sphere component of this coordinate), and hence, an interesting
departure from the Condon approximation [2].

4. SUMMARY

Electron transfer matrix elements for electron exchange between
various pairs of transition metal complexes in close contact have been
calculated and analyzed for a variety of approach geometries for the
two reactants. The coupling between the nominal metal ion
donor/acceptor sites is achieved by superexchange of the "hole" type
arising from ligand-to-metal charge transfer (LMCT), the dominant
ligand-field interaction for the electron-donor ligands considered
(H_2O, NH_3, and the cyclopentadienide anion). The pronounced
variations of H'_{if} with geometry are not correlated with the
separation distance of the metal ions (between which the direct
overlap is negligible) and span the range from non-adiabatic to
strongly adiabatic electronic coupling. The values for
metallocene/metallocinium redox pairs bracket recently reported
experimental values. Analysis of the results using the method of
corresponding orbitals demonstrates the validity of an effective
1-electron model for the electron transfer process to within about 10%
for the class of systems considered. A higher-order superexchange
mechanism was encountered for the $Co(NH_3)_6^{2+/3+}$ exchange process, in
which the LMCT-driven hole-transport mechanism couples excited local

states of the metal ions, which in turn are connected to the corresponding ground states by spin-orbit mixing. This mechanism yields on electronic transmission factor within two orders of magnitude of unity.

5. ACKNOWLEDGEMENT

This research was carried out at Brookhaven National Laboratory under contract DE-AC02-76CH00016 with the U.S. Department of Energy and supported by its Division of Chemical Sciences, Office of Basic Energy Sciences.

6. REFERENCES

1. (a) Marcus, R. A. J. Chem. Phys. (1956) **24**, 966; (b) 1956, **24**, 979.
2. Newton, M. D.; Sutin, N. Ann. Rev. Phys. Chem. (1984) **35**, 437–480.
3. Calef, D. F.; Wolynes, P. G. J. Phys. Chem. (1983) **87**, 3387–3400.
4. Hynes, J. T. J. Phys. Chem. (1986) **90**, 3701–3706.
5. Marcus, R. A.; Sutin, N. Biochim. Biophys. Acta (1985) **811**, 265–322.
6. (a) Logan, J.; Newton, M. D. J. Chem. Phys. (1983) **78**, 4086; (b) Newton, M. D. J. Phys. Chem. (1988) **92**, 3049; (c) Newton, M. D. J. Phys. Chem. (1988) **92**, 3049.
7. Ohta, K.; Closs, G. L.; Morokuma, K.; Green, N. J. J. Am. Chem. Soc. (1986) **108**, 1319–1320; Ohta, K.; Morokuma, K. J. Phys. Chem. (1987) **91**, 401.
8. (a) Larsson, S. J. Am. Chem. Soc. (1981) **103**, 4034; (b) Larsson, S. J. Chem. Soc. Faraday Trans. (1983) **279**, 1375–1388; (c) Larsson, S. J. Phys. Chem. (1984) **88**, 1321; (d) Larsson, S. Chemica Scripta (1988) **28A**, 15–20.
9. (a) Beratan, D. N.; Onuchic, J. N.; Hopfield, J. J. J. Chem. Phys. (1987) **86**, 4488; (b) Kuki, A.; Wolynes, P. Science (1987) **236**, 1647.
10. McConnell, H. M. J. Chem. Phys. (1961) **35**, 508.
11. Kuznetsov, A. M.; Ulstrup, J. J. Chem. Phys. (1981) **75**, 2047.
12. Goodenough, J. B. (1963) Magnetism and the Chemical Bond, John Wiley & Sons, New York.
13. Buhks, E.; Bixon, M.; Jortner, J.; Navon, G. Inorg. Chem. (1979) **18**, 2014.
14. Newton, M. D. ACS Symp. Ser., in press.
15. Plato, M.; Möbius, K.; Michel-Beyerle, M. E.; Bixon, M.; Jortner, J. J. Am. Chem. Soc. (1988) **110**, 7279–7285.
16. Scherer, P. O. J.; Fischer, S. F. J. Phys. Chem. (1989) **93**, 1633–1637.
17. Mikkelsen, K. V.; Dalgaard, E.; Swanstøm, P. J. Phys. Chem. (1987) **91**, 3081–3092.

18. Landau, L.; Phys. Z. Sowjet. (1932) **2**, 46; Zener, C.
     Proc. Roy. Soc. London (1932) **Ser. A**, 696.
19. (a) Zerner, M. C.; Loew, G. H.; Kirchner, R. F.;
     Mueller-Westerhoff, U. T. J. Am. Chem. Soc. (1980) **102**, 589;
     (b) Anderson, W. P.; Edwards, W. D.; Zerner, M. C. Inorg. Chem.
     (1986) **25**, 2728-2732.
20. King, H. F.; Stanton, R. E.; Kim, H.; Wyatt, R. E.; Parr, R. G.
     J. Chem. Phys. (1967) **47**, 1936.
21. Krishnan, R.; Firsch, M. J.; Pople, J. A. J. Chem. Phys. (1980)
     72, 4244.
22. Larsson, S.; Stahl, K.; Zerner, M. C. Inorg. Chem. (1986) **25**,
     3033-3037.
23. (a)Tembe, B. L.; Friedman, H. L.; Newton, M. D. J. Chem. Phys.
     (1982) **76**, 1490; (b) Friedman, H. L.; Newton, M. D. Faraday
     Discuss. Chem. Soc. (1982) **74**, 73-81.
24. Kuharski, R. A.; Bader, J. S.; Chandler, D.; Sprik, M.;
     Klein, M. L. J. Chem. Phys. (1988) **89**, 3248-3257.
25. Weaver, M. J.; Gennett, T. Chem. Phys. Lett. (1985) **113**, 213;
     Gennett, T.; Milner, D. F.; Weaver, M. J. J. Phys. Chem. (1985)
     89, 2787-2794.
26. McManis, G. E.; Nielson, R. M.; Gochev, A.; Weaver, M. J.
     J. Am. Chem. Soc. (1989) in press.
27. McManis, G. E.; Nielson, R. M.; Weaver, M. J. Inorg. Chem.
     (1988) **27**, 1827-1829.
28. Hammershoi, A.; Geselowitz, D.; Taube, H. Inorg. Chem. (1984)
     23, 979.
29. Wilson, R. B.; Solomon, E. I. J. Am. Chem. Soc. (1980) **102**,
     4085.
30. Newton, M. D., to be published.

SEQUENTIAL VERSUS SUPEREXCHANGE ELECTRON TRANSFER IN THE PHOTOSYNTHETIC REACTION CENTER

Yuming Hu and Shaul Mukamel
Chemistry Department
Rochester, New York 14627, USA

ABSTRACT. We present a novel expression for the complete electron transfer rate matrix in a three level system, which holds for an arbitrary value of the system free energies and reorganization energies. When the intermediate energy level is sufficiently high we recover the conventional superexchange rate. The relative contribution of the superexchange and the sequential mechanisms in the primary electron transfer process in the photosynthetic reaction center is discussed.

1. Introduction

The primary electron transfer process in the photosynthetic reaction center (RC) involves a transition from a photoexcited Bacteriochlorophyll dimer (P) to Bacteriopheophytin (H). In this process the electron transfers over a distance of 17Å in 2.8 psec. Since a direct (through space) coupling between P and H is excluded by their large separation, it has been suggested that the Bacteriochlorophyll monomer (B), which is located between P and H plays an active role in this event.[1-3]

This system can be described using the following adiabatic (Born-Oppenheimer) model Hamiltonian

$$H = \sum_{j=1}^{3} |j> H_j <j| + V_{12}(|1><2| + |2><1|) + V_{23}(|2><3| + |3><2|).$$

$$(1)$$

State $|1>$ is the optically excited chlorophyll dimer state (P*BH), state $|3>$ is the charge transfer state observed after 2.8 psec (P+BH-) and state $|2>$ is a possible intermediate state (P+B-H). $H_j(Q)$ denotes the adiabatic Hamiltonian of the polar medium whose degrees of freedom are denoted collectively by Q. V_{jn} represents the electronic coupling between states $|n>$ and $|j>$. The direct coupling V_{13} between states $|1>$ and $|3>$ was neglected in Eq.(1). The complete kinetic scheme for this system involves a 3 x 3 rate matrix whose matrix element K_{jn} denotes the transition rate from state $|n>$ to state $|j>$; $n,j = 1,2,3$. There are two basic mechanisms that have been suggested for the electron transfer process.[2,3] The first is <u>superexchange</u>, whereby $|2>$ serves as an intermediate virtual state. In the superexchange mechanism[2-4] level $|2>$ is never populated and the electron tunnels from level $|1>$ to $|3>$. Level $|2>$ simply contributes to the tunneling matrix element. The other possible mechanism involves a <u>sequential</u> process whereby the electron transfers from level

J. Jortner and B. Pullman (eds.), Perspectives in Photosynthesis, 171–184.

I1> to I2> and then from I2> to I3>. This mechanism is described by the rates K_{21} and K_{32} and does not involve K_{31}. Since in the sequential mechanism level I2> is intermediate in the kinetic scheme $|1>\rightarrow|2>\rightarrow|3>$ and it could be populated in the course of the reaction. Ultrafast optical measurements[5] have shown that the rate of appearance of level I3> following the preparation of level I1> by photoexcitation is 3.6×10^{11} sec^{-1}. These measurements have failed to detect a transient population of level I2>. This observation could support the superexchange mechanism. However, a sequential scheme with $K_{32}/K_{21} \geq 70$ will be consistent with this as well.[6] Extensive additional linear and nonlinear spectroscopic information (such as hole-burning and photon echo) is available in this system.[7,8] A major obstacle in resolving this issue is the lack of precise information regarding many of the relevant energetic parameters.[2,3,9] In particular the vertical transition energy $\Delta G_{12}^0 - \lambda_{12}$ is not known. ΔG_{12}^0 is the free energy difference between states I1> and I2>, λ_{12} is the corresponding reorganization energy (Figure 1). The common superexchange expression is valid only when $\Delta G_{12}^0 - \lambda_{12}$ is large enough and it diverges when $\Delta G_{12}^0 - \lambda_{12} = 0$. In this paper we report a microscopic calculation of the rate matrix K_{jn} which is obtained by formulating the problem using the density matrix in Liouville space.[10] This formulation was applied earlier to nonadiabatic and adiabatic electron transfer in a two level system and was used to explore the role of the dynamics of solvation in electron transfer processes.[11] The present formulation provides a general connection between the dynamics of electron transfer and nonlinear optical processes such as four wave mixing.[12] We have shown that the interplay between the sequential and the superexchange mechanisms is completely analogous to the interplay between the Fluorescence and the Raman components in an optical measurement. This profound connection is apparent when the problem is formulated using the density matrix in Liouville space. Our expression is valid for any value of $\Delta G_{12}^0 - \lambda_{12}$ and can therefore be used to explore the relative contribution of the superexchange and the sequential electron transfer even when all three levels are degenerate.

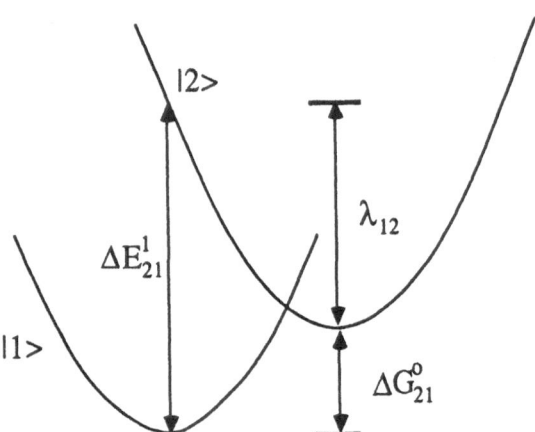

Figure 1: Configuration coordinate scheme showing the transition free energy ΔG_{21}^0 the reorganization energy λ_{12} and the vertical transition energy ΔE_{21}^1. Similar schemes apply for the other pairs of levels (13 and 23).

2. Sequential and Superexchange Electron Transfer

We start our analysis by partitioning the nuclear Hamiltonian $H_j(Q)$ into the following terms

$$H_j(Q) = E_j + H_B + U_j \tag{2}$$

where E_j represents the electronic energy of the unsolvated state $|j>$ and H_B represents the Hamiltonian of the bath, i.e. the nuclei of the RC which form a polar medium. U_j denotes the interaction between the electronic system and the medium, which may be expressed in terms of the polarization of the bath $P(r)$, and the electrostatic field $D_j(r)$ produced by the charge distribution of the system in the $|j>$ state, i.e.

$$U_j = -\int dr\ D_j(r) \cdot P(r). \tag{3}$$

It should be emphasized that U_j depends on a macroscopic number of polarization degrees of freedom $P(r)$, which undergo complicated motions resulting from thermal fluctuations. The statistical properties of U_j contain all the relevant information for our problem. We shall model U_j as Gaussian random variables. This is a common assumption in electron transfer theories.[2,3,13] It has been recently verified by an extensive numerical simulation of outer sphere electron transfer in water.[14] Eqs.(1) - (3) constitute our basic model Hamiltonian for the RC and they can be used to evaluate the electron transfer rates. We shall now introduce the following equilibrium density matrices for the bath

$$\rho_j = \exp(-H_j/k_BT)/\mathrm{Tr}\ \exp(-H_j/k_BT) \qquad j=B,1,2,3 \tag{4}$$

where Tr denotes a trace over the bath degrees of freedom. ρ_B is the density matrix of the bath in the absence of the electron transfer system. ρ_1, ρ_2 and ρ_3 are the density matrices of the bath when the electron transfer system is in states 1, 2 and 3, respectively. For the sake of simplicity we assume that the energy fluctuation U_j of the $|j>$ state is totally uncorrelated with the energy fluctuation U_n on any other state $|n>$, i.e.

$$\mathrm{Tr}\ (U_j\ U_n\ \rho_B) = 0 \qquad\qquad j \neq n \tag{5}$$

This is a reasonable assumption for the RC electron transfer, because the different electronic states correspond to the electron residing on different parts of RC, which have a different local environment, and their energy fluctuations are expected to be uncorrelated. The present calculation of the rate is based on formulating the dynamics in terms of the density matrix in Liouville space.[11] The formalism allows a perturbative expansion of the rate matrix K_{jn} in a power series in the nonadiabatic coupling V. This formalism was developed for electron transfer in a two level system resulting in general expressions for K_{21} and K_{32}.[11] We have extended the calculation to a three level system.[10] The calculation is based on expanding the rate matrix to fourth order V, followed by a partial resummation of the series to infinite order via a Pade approximant. It should be noted that the lowest order contribution to K_{21} and K_{32} is second order whereas for K_{31} it is fourth order. The bookkeeping of the various contributions is best illustrated using a Liouville-space coupling diagram. In Figure 2a we show the eight Liouville-space pathways contributing to K_{21} and in Figure 2b we show the three pathways contributing to K_{31}.

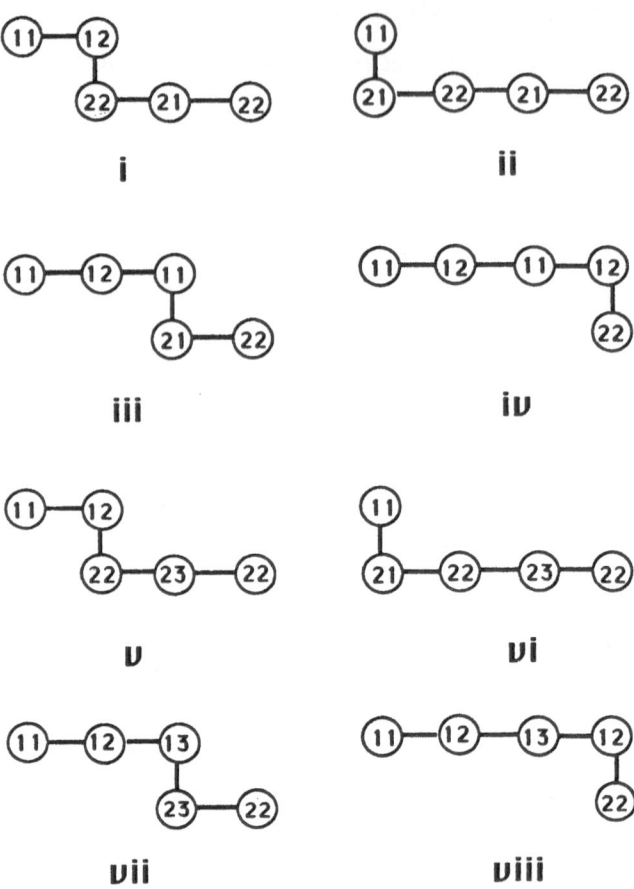

Figure 2a: The Liouville-space pathways[11] contributing to the rate K_{21} to fourth order in the nonadiabatic coupling. Each pair of indexes jn implies that the system is in the state |j><n|, where j=n stands for a population and j≠n for a coherence. Each bond represents a nonadiabatic coupling V. There are sixteen pathways which can lead from 11 to 22 in fourth order (four bonds). These pathways come in complex conjugate pairs so that we need consider only the eight pathways shown. Pathways (i) - (vi) represent a sequential nonequilibrium process whereby the system passes through an intermediate population. Pathways (vii) and (viii) represent a tunneling process.

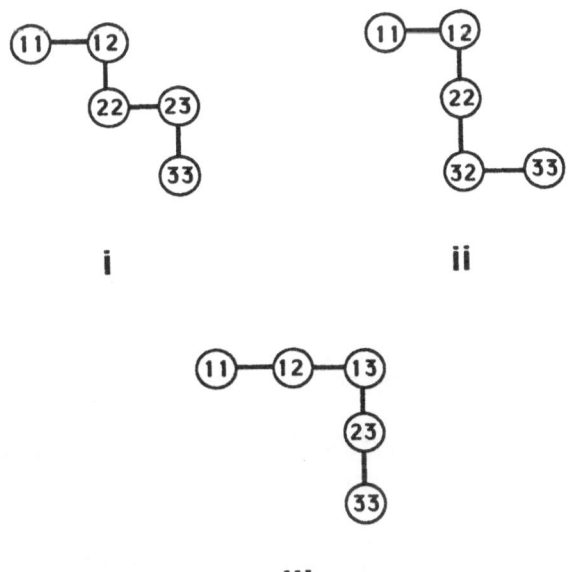

Figure 2b: The Liouville space pathways[11] contributing to the rate K_{31} to fourth order in the nonadiabatic coupling. There are six pathways which lead from 11 to 33 in fourth order. We need to consider only the three pathways shown (the others are their complex conjugates). Pathways (i) and (ii) represent a sequential nonequilibrium process and pathway (iii) is the tunneling (superexchange) term.

The rate matrix is given in terms of several physical quantities. We first define the free energy change for the $|j>$ to $|n>$ transition (Figure 1)

$$\Delta G_{jn}^0 = Tr(H_j\rho_j) - Tr(H_n\rho_n), \tag{6}$$

and the vertical transition energy for this transition when the system is in state $|m>$

$$\Delta E_{jn}^m \equiv Tr[(H_j - H_n)\,\rho_m]. \tag{7}$$

We further define the solvent reorganization energy in the $|j>$ state

$$\lambda_j \equiv Tr\,(U_j^2\,\rho_B)/2k_BT \tag{8}$$

and the reorganization energy for the $|j>$ to $|n>$ transition

$$\lambda_{jn} \equiv \lambda_j + \lambda_n. \tag{9}$$

Note that $\Delta E_{jn}^m = -\Delta E_{nj}^m$, $\Delta G_{jn}^0 = -\Delta G_{nj}^0$ and $\lambda_{jn} = \lambda_{nj}$.

We next introduce the solvent correlation function

$$M_{jk}(t) = \frac{<\exp(i\ H_B t)\ U_{jk}\ \exp(-i\ H_B t)\ U_{jk}\ \rho_B>}{<U_{jk}^2\ \rho_B>}$$

(10a)

$$M(t) = \frac{<\exp(i\ H_B t)\ U_{23}\ \exp(-i\ H_B t)\ U_{12}\ \rho_B>}{\sqrt{<U_{23}^2\ \rho_B><U_{12}^2\ \rho_B>}}.$$

(10b)

The rate matrix was calculated by assuming a Gaussian statistics for U_j and using condition (Eq.(5)). The details of these calculations are given elsewhere.[10] We thus have

$$K_{21} = \frac{\sigma_{21}}{1+\sigma_{12}\tau_a+\sigma_{21}\tau_b+\sigma_{32}\tau_c} + \frac{2\pi}{\hbar} V_{12}^2 V_{23}^2 [R_{12}I(\eta_{12}^1)S_{12}(\Delta E_{12}^1)+R_{23}I(\eta_{23}^1)S_{23}(\Delta E_{23}^1)]$$

(11a)

$$K_{12} = \frac{\sigma_{12}}{1+\sigma_{12}\tau_a+\sigma_{21}\tau_b+\sigma_{32}\tau_c} + \frac{2\pi}{\hbar} V_{12}^2 V_{23}^2 [R_{12}I(\eta_{12}^2)S_{12}(\Delta E_{12}^2)+R_{23}I(\eta_{23}^2)S_{23}(\Delta E_{23}^2)]$$

(11b)

$$K_{32} = \frac{\sigma_{32}}{1+\sigma_{32}\tau_a'+\sigma_{23}\tau_b'+\sigma_{12}\tau_c} + \frac{2\pi}{\hbar} V_{12}^2 V_{23}^2 [R_{12}I(\eta_{12}^2)S_{12}(\Delta E_{12}^2)+R_{23}I(\eta_{23}^2)S_{23}(\Delta E_{23}^2)]$$

(11c)

$$K_{23} = \frac{\sigma_{23}}{1+\sigma_{32}\tau_a'+\sigma_{23}\tau_b'+\sigma_{12}\tau_c} + \frac{2\pi}{\hbar} V_{12}^2 V_{23}^2 [R_{12}I(\eta_{12}^3)S_{12}(\Delta E_{12}^3)+R_{23}I(\eta_{23}^3)S_{23}(\Delta E_{23}^3)]$$

(11d)

$$K_{31} = \sigma_{21}\sigma_{32}\tau_c$$
$$+ \frac{2\pi}{\hbar}V_{12}^2 V_{23}^2 [R_{13}I(\eta_{13}^1)S_{13}(\Delta E_{13}^1)-R_{12}I(\eta_{12}^1)S_{12}(\Delta E_{12}^1)-R_{23}I(\eta_{23}^1)S_{23}(\Delta E_{23}^1)]$$

(11e)

$$K_{13} = \sigma_{23}\sigma_{12}\tau_c$$
$$+ \frac{2\pi}{\hbar}V_{12}^2 V_{23}^2 [R_{13}I(\eta_{13}^3)S_{13}(\Delta E_{13}^3)-R_{12}I(\eta_{12}^3)S_{12}(\Delta E_{12}^3)-R_{23}I(\eta_{23}^3)S_{23}(\Delta E_{23}^3)]$$

(11f)

The diagonal elements of the rate matrix are given by

$$- K_{11} = K_{21} + K_{31}$$

$$- K_{22} = K_{12} + K_{32}$$

$$- K_{33} = K_{13} + K_{23}$$

(12)

σ_{jn} is the nonadiabatic transition rate from state n to j,

$$\sigma_{jn} \equiv \frac{2\pi}{\hbar} V_{jn}^2 \, S_{jn} \, (\Delta E_{jn}^n) \tag{13a}$$

where S_{jn} is the Franck Condon weighted density of states

$$S_{jn}(x) = \frac{1}{\sqrt{4\pi \, k_B T \, \lambda_{jn}}} \exp[- \frac{x^2}{4 k_B T \, \lambda_{jn}}]. \tag{13b}$$

τ_j where j = a, b, c, a', b' are characteristic solvation timescales. They are defined as follows. We first introduce two timescale functions

$$\tau_{jk}(q) = \int_0^\infty dt \left\{ \frac{1}{\sqrt{1-M_{jk}^2(t)}} \exp[\frac{2M_{jk}(t)q^2}{1+M_{jk}(t)}] - 1 \right\} \tag{14a}$$

$$\tau(q,q') = \int_0^\infty dt \left\{ \frac{1}{\sqrt{1-M^2(t)}} \exp[\frac{2qq'M(t)-(q^2+q'^2)M^2(t)}{1-M^2(t)}] - 1 \right\}. \tag{14b}$$

Here q and q' are dimensionless parameters. We further define

$$q_{nm} \equiv \frac{\Delta E_{nm}^m}{\sqrt{4 k_B T \, \lambda_{nm}}}. \tag{15}$$

In terms of these quantities we have $\tau_a = \tau_{12}(q_{12})$, $\tau_b = \tau_{12}(q_{21})$, $\tau_a' = \tau_{23}(q_{32})$, $\tau_b' = \tau_{23}(q_{23})$, $\tau_c = \tau(q_{32}, q_{12})$. Note that $q_{nm} \neq q_{mn}$ and that τ_a' and τ_b' are obtained from τ_a and τ_b, respectively by interchanging the indexes 1 and 3. $\tau_{jk}(q)$ was calculated and analyzed previously in detail.[11] We further have

$$\Delta E_{nm}^n = \Delta G_{nm}^0 - \lambda_{nm} \tag{16a}$$

$$\Delta E_{nm}^m = \Delta G_{nm}^0 + \lambda_{nm} \tag{16b}$$

$$\Delta E_{nm}^k = \Delta G_{nm}^0 + \lambda_n - \lambda_m \qquad k \neq n,m \tag{16c}$$

$$\eta_{jn}^k = \sqrt{R_{jn}} \frac{\Delta E_{jm}^k \lambda_n + \Delta E_{nm}^k \lambda_j}{\lambda_{jn}} \tag{17}$$

$$R_{nm} = \frac{\lambda_{nm}}{[\lambda_1\lambda_2 + \lambda_1\lambda_3 + \lambda_2\lambda_3] \, 2 k_B T} \tag{18}$$

where j, n, m are permutations of 1, 2, 3 and k = 1, 2, 3.
The function I(x) is given by

$$I(x) = -\int_0^\infty t \cos(x\,t)\,\exp(-t^2/2)\,dt\,.$$

(19)

I(x) is displayed in Figure 3. We have constructed the following Pade approximant for I(x),

$$I(x) \cong \begin{cases} \dfrac{-1 + 0.5851x^2}{1 + 0.4149x^2 + 0.1235x^4} & |x| < 2.1 \\[4mm] \dfrac{-1 + 0.5994x^2}{11.2551 - 3.8396x^2 + 0.5994x^4} & |x| \geq 2.1 \end{cases}$$

(20)

The dashed line in Figure 3 shows the Pade approximant. The fit is excellent. We further note the following limiting behavior of I(x)

$$I(x) = \begin{cases} \dfrac{1}{x^2} & x \gg 1 \\[3mm] -1 + x^2 & x \ll 1 \end{cases}.$$

(21)

3. Discussion

We shall now analyze our rate matrix (Eqs.(11)). We first note that to second order in V, Eqs.(11) reduce to the well known nonadiabatic rates[2,3] $K_{21} = \sigma_{21}$, $K_{32} = \sigma_{32}$, $K_{31} = 0$. When setting $V_{23} = 0$ all rates vanish except for K_{12} and K_{21},

$$K_{21} = \frac{\sigma_{21}}{1 + \sigma_{12}\tau_a + \sigma_{21}\tau_b}$$

(22a)

$$K_{12} = \frac{\sigma_{12}}{1 + \sigma_{12}\tau_a + \sigma_{21}\tau_b}$$

(22b)

Eqs.(22) have been derived and analyzed previously.[11] τ_a and τ_b are typical solvent timescales. Eqs.(22) interpolate from the nonadiabatic limit ($\tau_a, \tau_b \to 0$) to the adiabatic limit whereby τ_a and τ_b are long and the rate becomes equal to an inverse solvent timescale. The first term in K_{12}, K_{21}, K_{32} and K_{23} (Eqs.(11a) - (11d)) is an extension of this expression for a three level system. It depends on five different solvent timescales τ_a, τ_b, τ_c, τ'_a, τ'_b. Like Eqs.(22), the first term interpolates between the nonadiabatic rate (for $\tau_j \to 0$) to an inverse average timescale (as $\tau_j \to \infty$). The second term in K_{12}, K_{21}, K_{32}

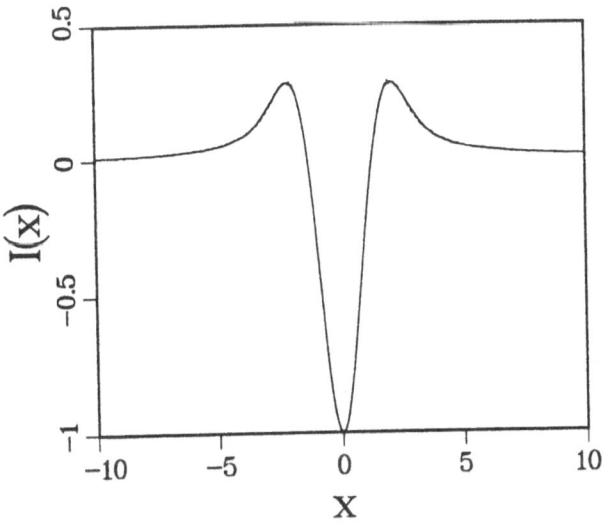

Figure 3: The auxiliary function I(x) (Eq.(19)) (solid line) and its Pade approximant (Eq.(20)) (dashed line).

and K_{23} comes from a coherent process in which the system goes between the states via coherences without passing through a population. A graphical illustration of the terms contributing to K_{21} is given in Figure 2a. All the diagrams lead from population of level 1 (11) to level 2 (22) in fourth order (four bonds). Diagrams (i) - (vi) have a population as an intermediate after two interactions (either 11 or 22). These diagrams contribute to the first term in the expression of K_{21} (Eq.(11a)). Diagrams (vii) and (viii) pass through a coherence (13) after two interactions and they contribute to the second (tunneling) term in Eq.(11a). Similar diagrams can be drawn for K_{12}, K_{32} and K_{23}. K_{31} (Eq.(11e)) has two terms. The first represents the process in which the electron does go through the population of the state |2>, but before it equilibriates with the bath at the state |2>, the electron transfers to state |3>. This term is represented by diagrams (i) and (ii) in Figure 2b. The second term in Eq.(11e) represents the process in which the electron transfers from state |1> to state |3> without actually passing through the state |2>. This is therefore a tunneling process. This term is represented by diagram (iii) in Figure 2b. We shall now consider K_{31} more closely. For simplicity we assume that the relaxation of the system (RC and environment) is sufficiently rapid, $\tau_c \rightarrow 0$, so that the only quantity we need to consider in K_{31} is the second term in Eq.(11e).
We shall now examine a few limiting cases of Eq.(11e). We first consider the case whereby the energy level |2> is sufficiently far from the energy levels |1> and |3>, i.e.

$$\Delta E_{12}^1 \gg \sqrt{\lambda_{12} k_B T}, \quad \Delta E_{23}^1 \gg \sqrt{\lambda_{23} k_B T}. \tag{23}$$

Condition (23) implies that the activation energies $\Delta G_{12}^* \equiv (\Delta G_{12}^0 - \lambda_{12})^2/4\lambda_{12}$ and ΔG_{23}^* $\equiv (\Delta G_{23}^0 + \lambda_2 - \lambda_3)^2/4\lambda_{23}$ are much larger than $k_B T$. This limit is where the conventional superexchange theory is usually formulated.[2-4] In this case Eq.(11e) assumes the form

$$K_{31} = \frac{2\pi}{\hbar} |V_{SE}|^2 S_{13}(\Delta E_{13}^1)$$

(24a)

with

$$V_{SE} = \frac{V_{12} V_{23}}{\Delta E_{12}^1 (\lambda_3/\lambda_{13}) + \Delta E_{32}^1 (\lambda_1/\lambda_{13})} .$$

(24b)

We next consider another limiting case where all the three states are completely degenerate, i.e. $\Delta E_{12}^1 = \Delta E_{23}^1 = \Delta E_{13}^1 = 0$. We further assume $\lambda_1 = \lambda_2 = \lambda_3 \equiv \lambda$. We then obtain from Eq.(11e)

$$K_{31} = \frac{2\pi}{\hbar} |V_{SE}|^2 S_{13}(0)$$

(25a)

where

$$V_{SE} = \frac{V_{12} V_{23}}{\sqrt{3\lambda k_B T}} .$$

(25b)

Eqs.(25) may be rationalized using a simple physical argument. The variance of the energy level fluctuations is $\sim \sqrt{\lambda k_B T}$. V_{SE} is thus given by $V_{12} V_{23}$ divided by a typical energy fluctuation.

We have calculated K_{31} using Eq.(11e) and the results are displayed in Figure 4 as a function of $\Delta E_{21}^1 = \Delta G_{21}^0 + \lambda_{12}$ for different values of ΔE_{13}^1. The dashed line in Figure 4 represents the asymptotic rate expression (Eqs.(24)). Assuming that $K_{32} \gg K_{21}$, as suggested by the ultrafast measurements,[5,6] then the sequential rate is equal to K_{21}. The following dimensionless parameter

$$R \equiv \frac{K_{31}}{K_{31} + K_{21}}$$

(26)

provides a measure of the relative contribution of the superexchange and the sequential mechanisms. R is bounded between 0 and 1. $R \rightarrow 0$ when the sequential transfer dominates, and $R \rightarrow 1$ when the superexchange transfer dominates. Figures 5 - 7 display the dependence of R on the energy ΔE_{21}^1 for different values of ΔE_{13}^1, coupling V_{12} and reorganization energy λ_{12}.

Our rate expression (Eqs.(11)) was obtained in the high temperature limit. It may be extended to low temperatures by adopting a more specific model for H_B and U_j. A common model in electron transfer theories assumes that U_j is proportional to a single harmonic coordinate representing an intramolecular or intermolecular vibration. In this case Eqs.(11) should be modified by replacing all $k_B T$ factors by the average oscillator energy $<\varepsilon>$[6,13]

$$<\varepsilon> = \frac{\hbar\omega}{2} \coth\left(\frac{\hbar\omega}{2k_B T}\right)$$

(27)

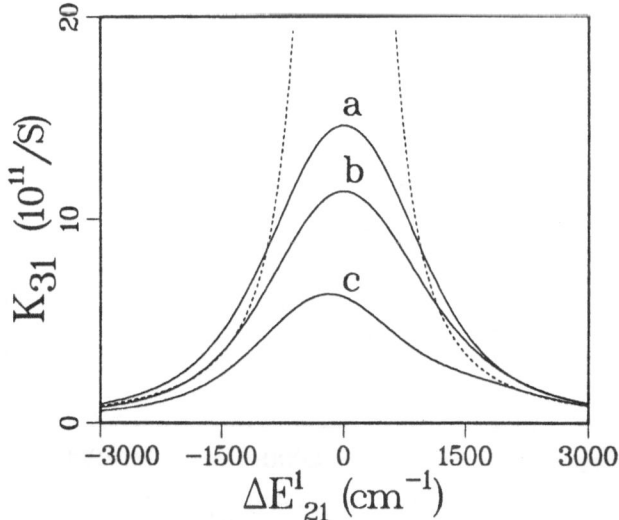

Figure 4: The dependence of the superexchange rate K_{31} on ΔE^1_{21} for different values of ΔE^1_{13}. Curves a, b and c correspond to the $\Delta E^1_{13} = 0$, 400 cm^{-1} and 700 cm^{-1}, respectively. The other parameters in this calculation are: $\lambda_{12} = 1000$ cm^{-1}, $\lambda_{23} = 1500$ cm^{-1}, $\lambda_{13} = 2000$ cm^{-1}, $V_{12} = 80$ cm^{-1}, $V_{23} = 6V_{12}$, T = 300K. The dashed line represents the conventional superexchange rate (Eqs.(24)) with $\Delta E^1_{13} = 0$.

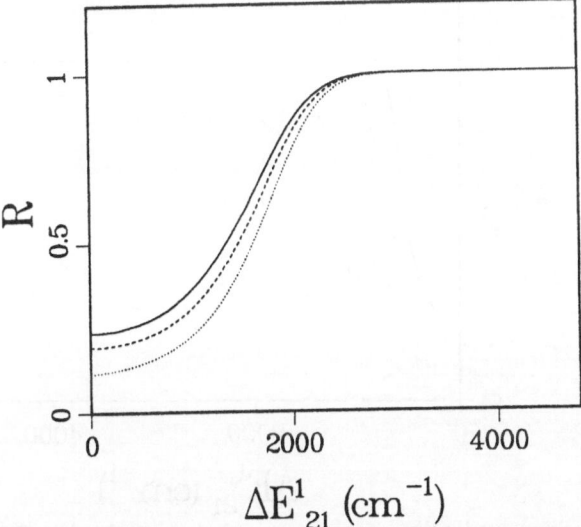

Figure 5: The relative contribution of the superexchange mechanism R is displayed versus ΔE^1_{21} for different values of ΔE^1_{13}. The solid, dashed and dotted lines correspond to $\Delta E^1_{13} = 0$ cm^{-1}, 400 cm^{-1} and 700 cm^{-1}, respectively. The other parameters are: $\lambda_{13} = 2000$ cm^{-1}, $\lambda_{23} = 1500$ cm^{-1}, $\lambda_{12} = 1000$ cm^{-1}, $V_{12} = 80$ cm^{-1}, $V_{23} = 6V_{12}$, T = 300K.

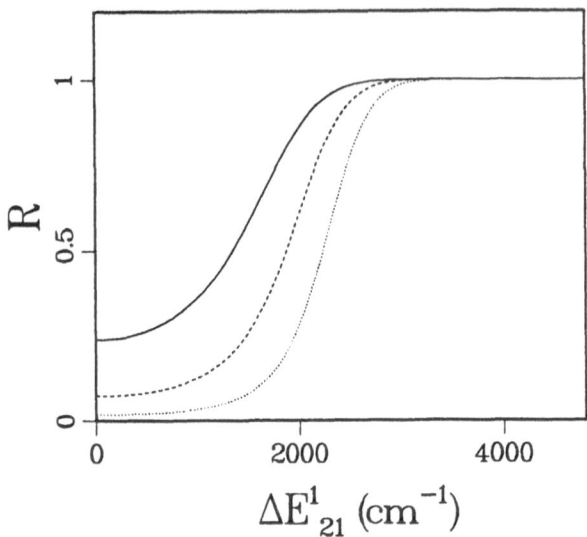

Figure 6: The relative contribution of the superexchange mechanism R is displayed versus ΔE^1_{21} for different values of the coupling V_{12}. These curves correspond to $V_{12} = 80$ cm^{-1} (solid line), 40 cm^{-1} (dashed line) and 20 cm^{-1} (dotted line), respectively. The other parameters are: $\lambda_{13} = 2000$ cm^{-1}, $\lambda_{23} = 1500$ cm^{-1}, $\lambda_{12} = 1000$ cm^{-1}, $\Delta E^1_{13} = 0$, $V_{23} = 6\ V_{12}$, T = 300K.

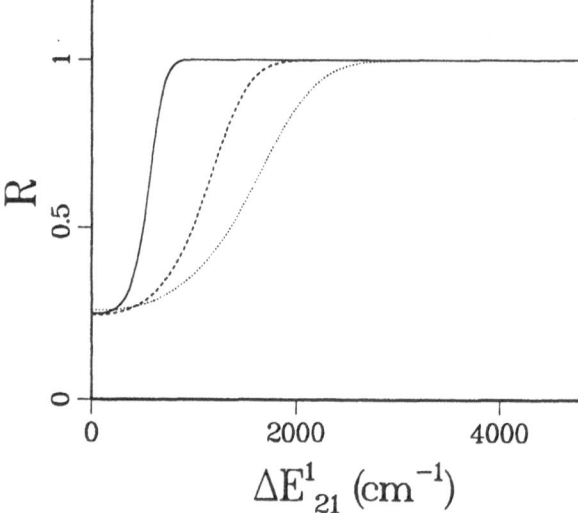

Figure 7: The relative contribution of the superexchange mechanism R is displayed versus ΔE^1_{21} for different values of the reorganization energy λ_{12}. The solid, dashed and dotted lines correspond to $\lambda_{12} = 100$ cm^{-1}, 500 cm^{-1} and 1000 cm^{-1}, respectively. The other parameters are: $\Delta E^1_{13} = 0$, $\lambda_{13} = 2000$ cm^{-1}, $\lambda_{23} = 2000$ cm^{-1}, $V_{12} = 80$ cm^{-1}, $V_{23} = 6\ V_{12}$, T = 300K.

Note that in the high temperature ($k_BT \gg \hbar\omega$) limit $\langle\varepsilon\rangle = k_BT$. When k_BT in Eqs.(11) is replaced by $\langle\varepsilon\rangle$ (Eq.(27)) we obtain a rate expression which is not restricted to the high temperature limit. Martin, et.al.[6] have measured the temperature dependence of the RC electron transfer in the range 10K to 300K. For Rb. sphaeroides the variation of the rate was found to be in agreement with the conventional superexchange rate (Eqs.(24)) assuming a single mode with $\omega = 80$ cm^{-1}. For Rps. viridis on the other hand the conventional expression is inadequate and the best fit obtained[6] (using $\omega = 25$ cm^{-1}) is shown in Figure 8 (dashed line). The temperature dependence predicted by our present theory for K_{31} (Eq.(11e) together with Eq.(27)) is found to be in a much better agreement with experiment. This is illustrated in Figure 8.

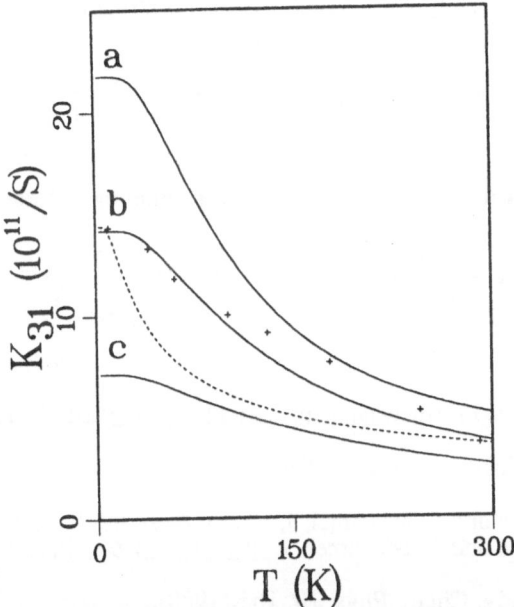

Figure 8: Temperature dependence of the superexchange rate K_{31}. A coupling to a single vibrational mode with $\omega = 90$ cm^{-1} is assumed and the calculation was made using Eq.(11e) with k_BT replaced by $\langle\varepsilon\rangle$ (Eq.(27)). Curves a, b and c correspond to $\Delta E_{21}^1 = 800$ cm^{-1}, $\Delta E_{21}^1 = 1050$ cm^{-1} and $\Delta E_{21}^1 = 1500$ cm^{-1}, respectively. The dashed curve represents the best fit with the conventional superexchange rate using $\omega = 25$ cm^{-1} (Eq.(2) in Reference 6). The points (+) denote the experimental results of electron transfer rate in Rps. viridis from Martin, et.al.[6] Other parameters in this calculation are $\lambda_{13} = 2000$ cm^{-1}, $\lambda_{12} = 200$ cm^{-1}, $\lambda_{23} = 2000$ cm^{-1}, $V_{12} = 79$ cm^{-1}, $V_{23} = 6 V_{12}$, $\Delta E_{13} = 0$. This figure shows that our superexchange expression (curve b) provides a better fit with experiment, compared with the conventional superexchange expression (dashed line).

Acknowledgement

The support of the Office of Naval Research, the National Science Foundation and the Petroleum Research Fund, administered by the American Chemical Society, is gratefully acknowledged.

References

1.   See papers in "The Photosynthetic Bacterial Reaction Center" edited by Jacques Breton and Andre Vermeglio, Plenum, NY (1988).

2.   R.A. Marcus, Chem. Phys. Letters **133**, 471 (1987); in Reference 1, p.389.

3.   M. Bixon, J. Jortner, M. Plato and M.E. Michel-Beyerle in Reference 1, p.399.

4.   A. Beretan and J.J. Hopfield, Journ. Am. Chem. Soc. **106**, 1584 (1984); M. Redi and J.J. Hopfield, J. Chem. Phys. **72**, 6651 (1980); J.R. Miller and J.V. Beitz, J. Chem. Phys. **74**, 6746 (1981).

5.   J. Breton, J.L. Martin, A. Migus, A. Antonetti and A. Orszag, Proc. Nat. Acad. Sci. USA **83**, 5121 (1986); G.R. Fleming, J.L. Martin and J. Breton, Nature **333**, 190 (1988).

6.   J.L. Martin, J. Breton, J.C. Lambry and G.R. Fleming in Reference 1, p. 195.

7.   J.M. Hayes and G.S. Small, J. Phys. Chem. **90**, 4928 (1986); S.R. Meech, A.J. Hoff and D.A. Wiersma, Chem. Phys. Lett. **121**, 287 (1985); S.J. Boxer, D.S. Gottfried, D.J. Lockhart and T.R. Midderdorf, J. Chem. Phys. **86**, 2439 (1987).

8.   Y. Won and R.A. Friesner, J. Phys. Chem. **92**, 2214 (1988).

9.   Y. Won and R.A. Friesner, Biochem. Biophys. Acta **935**, 9 (1988).

10. Y. Hu and S. Mukamel (to be published).

11. M. Sparpaglione and S. Mukamel, J. Chem. Phys. **88**, 5 (1988); Y.J. Yan, M. Sparpaglione and S. Mukamel, J. Phys. Chem. **92**, 4842 (1988).

12. S. Mukamel, Adv. Chem. Phys. **70**, 165 (1988).

13. M. Bixon and J. Jortner, J. Phys. Chem. **90**, 3795 (1986).

14. J.S. Bader and D. Chandler, Chem. Phys. Lett. (in press).

NON-ADIABATIC ELECTRON TRANSFER: SOME DYNAMICAL AND ELECTRONIC
EXTENSIONS OF STANDARD RATE EXPRESSIONS

Mark A. Ratner
Department of Chemistry
Northwestern University
Evanston, IL 60208

ABSTRACT: The standard Marcus-Hush approach for computing rates of
non-adiabatic electron transfer processes is based upon a combination
of perturbation theory and transition state theory. While the standard
approach is both extremely powerful and widely applicable, there are
certain experimental situations in which the transition state theory
argument, in particular, is no longer valid. We discuss a number of
situations in which extensions of Marcus-Hush theory are required to
deal with electron transfer rate phenomena. These situations include:
1) nuclear tunneling reactions, in which a vibronic treatment, related
to standard small polaron theory, can be used; 2) control of fast
reactions by solvent dynamics, in which case the relaxation time
spectrum of the solvent must be included in the rate discussion; 3)
secondary stable minima on the potential energy surface, in which case
gating phenomena may occur because alternative barriers and steepest
descent pathways can be found for trajectories moving from reactant to
product; 4) highly anisotropic diffusion or friction along different
coordinates, in which case trajectories wander far from the steepest
descents pathway and nonexponential transient behavior may be observed;
5) specific electronic effects, including modified initial states, the
importance of coincidence events, choice of the initial state for
electron transfer, breakdown of perturbation theory and generalized
superexchange behavior.

I. Introduction

Non-adiabatic electron transfer processes, occurring over relatively
long distances and in situations involving bridge assistance between
donor and acceptor sites, is usually analyzed using the standard
Marcus-Hush formalism based on transition state theory.[1-12] This
theory is, in principle, simply the study of intersecting parabolas:
the activation energy involved is that required to reach the parabolic
intersection, and electron tunnelling terms are added to permit the
system to transfer from reactant to product geometries. This picture
is relatively easily extended to a full polaron type vibronic picture,

185

J. Jortner and B. Pullman (eds.), Perspectives in Photosynthesis, 185–210.

with quantitative inclusion of nuclear tunnelling effects.[3,5-15] This vibronic picture of electron transfer in the non- adiabatic limit has been extensively developed over the last two decades, and provides very useful interpretations of the great majority of solution phase electron transfer reactions.

Recent experimental advances, both in the time scales over which electron transfer (ET) reactions can be studied[17-26] and in the sorts of systems amenable to experimental study, require extension of the standard non-adiabatic polaron-type treatment. These experiments include extremely fast (picosecond or femtosecond) electron transfer processes, which investigate dynamical relaxation effects not included in the vibrational averaging that underlies standard vibronic theory. Extension of ET studies to protein environments can entail complications arising from protein conformational motions, that can be slow and overdamped and therefore not easily describable in simple harmonic oscillator terms.[16,27-35] Studies of intramolecular electron transfer in bridge assisted situations may complicate a simple vibronic picture because the initial state may not be fully localized on the donor,[9,36,37] as is assumed in the standard treatments. Effects such as electron transfer resonances, interferences, and dephasing processes can then enter, and may invalidate the simple perturbation theory upon which a standard electron transfer formalism is based.[36] Finally, while the Kramers formulation of diffusion-type reactions in solution provides important additional insights into the electron transfer process occurring essentially along a given reaction coordinate, anisotropic friction, arising from differential damping of conformational and vibrational modes, can result in an anisotropic diffusion coefficient in protein type ET systems; this in turn can result in very different behavior[28,34,38,39] from that predicted by standard vibronic theory.

In this manuscript we will touch on some of the extensions to standard electron transfer theory that have been necessitated by these new experimental interests. We will stress the utility of direct time dependent study of electron transfer situations,[36,40-48] to follow the evolution of electron transfer events, and to provide tests for, and insight into, useful formal schemes derived from perturbation theory, vibronic coupling models, density matrix formulations, and so forth. While direct time dependent calculations for slow electron transfer processes are clearly impractical, nevertheless they can yield important insights into the role played by superexchange-type coupling, the relative importance of particular geometries to the electron transfer process, and a possible recurrences, power dependences, and interference effects.

Most of this manuscript is intended as a conceptual overview of progress described extensively elsewhere. Section II briefly recaps the standard formal theory of non-adiabatic electron transfer reactions and its vibronic coupling extension. In Section III, a very brief discussion is given of the extensions required by fast experiments measuring solvent relaxation effects. Section IV deals with the important extensions that arise from the existence of multiple stable minima on the potential energy surface. This can lead to gating of the ET reactions, and to completely different activation behaviors.[49,50,51] Section V considers anisotropic diffusion in systems with diffusive type ET processes; here important transient effects can be observed, and the evolution of the system point on the potential surface may stray exceedingly far from the simple reaction coordinate. Finally, section VI considers specific electronic structure effects, including some important insights gained from recent dynamical studies of electron transfer in model systems.

II. Non-adiabatic electron transfer theory: the standard formulation.

Standard theory of non-adiabatic ET reactions is based upon the transition state theory approach: one computes the overall rate constant, k_{NA}, for the reaction from precursor to successor complex, or from donor to acceptor, or from reactant to product, as:

$$k_{NA} = k_o \, \kappa e^{-\Delta G^{\ddagger}/RT} . \tag{1}$$

The three terms on the right hand side each can be interpreted in simple physical terms. The activation free energy is given as:

$$\Delta G^{\ddagger} = (\Delta - \lambda)^2/4\lambda, \tag{2}$$

where Δ and λ are respectively the reaction exoergicity and the reorganization energy or free energy.[1-16,52] These quantities are indicated schematically on the configuration coordinate diagram of figure 1. While the exoergicity Δ is simply the energy or free energy difference between reactant and product, the reorganization λ consists of contributions both from the solvent (generally referred to as the outer sphere reorganization energy) and from the coupled harmonic vibrations of the reacting species (generally called the inner sphere reorganization). As is clear from the construction in figure 1, $\Delta G^{\ddagger}$ is simply the barrier height that the system must overcome in passing from the reactant geometry over the activation barrier to the product geometry.

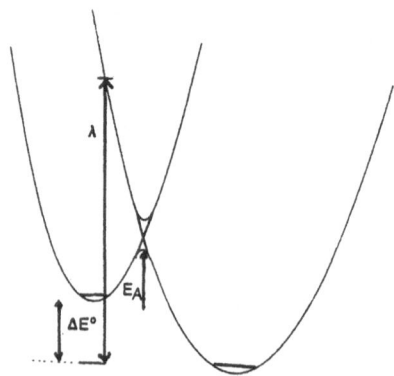

Figure 1.   Schematic one dimensional potential energy surface for nonadiabatic electron transfer. The transition state is the maximum in the lower curve, and the activation energy is the energy difference between the minimum on the left and that maximum. The reorganization energy λ and the overall exoergicity are indicated.

The intrinsic rate constant at barrier top, k_o, is given by:

$$k_o = \frac{2\pi}{\hbar} \frac{|V_{DA}|^2}{(4\pi\lambda kT)^{\frac{1}{2}}} .$$
(3)

Here V_{DA} is the electronic-structure-determined electron coupling matrix element mixing donor and acceptor states. This matrix element is simply half the splitting of the non-crossing curves at the saddle point of figure 1. Its determination, magnitude, variation with geometry, and unique definition, comprise very important components of the study of non-adiabatic electron transfer.[3,9,53-57]   Verbally, however, the interpretation of equation (3) is simple: this is just the intrinsic rate at which the system passes over the saddle point top.

Finally, the transmission coefficient κ of equation (1) determines the relative probability that the system will remain in the product state once in fact it passes over the transition region. While many factors can contribute to κ (such as final state scattering, spin orbit interactions, reaction coordinate curvature, and relative slopes at the crossing point), dominant recent attention has been devoted to the effect of relaxation on κ: if the system is simply not allowed to relax, the geometry will remain that of the reactant, the system cannot evolve into the final state geometry, recrossing effects will be dominant and κ will be very small.[19,47,58-61]

The above is, scandalously simplified, the essential picture of standard non-adiabatic electron-transfer rate theory. More careful and complete expositions are found in several excellent review articles.[3-12]   Several important assumptions underlie the development of these formulas: It is the failure of these assumptions under particular experimental conditions that requires extension of the results of equations (1-3). It is just these experimental conditions which are the focus of our study here. The great majority of electron transfer reactions, in the nonadiabatic limit, are straightforwardly and accurately described in terms of equations (1-3).

The first important extensions of this simple nonadiabatic transition-state theory picture are based on polaron type models, involving the simple hamiltonian[37] of equation (4).

$$H = |\alpha\rangle\langle\alpha| \; [\sum_{n}^{2} P_n^2/2M_n + \tfrac{1}{2} M_n\omega_n(q_n - \lambda_n)^2]$$

$$+ |\beta\rangle\langle\beta| \; [\sum_{n} P_n^2/2M_n + \tfrac{1}{2} M_n\omega_n^2 q_n^2]$$

$$+ V(|\alpha\rangle\langle\beta| + |\beta\rangle\langle\alpha|) \tag{4}$$

Here the projection operators are onto the two electronic states, with α, β, representing for example donor and acceptor. The sum is over all the coupled vibrations in the system, with P_n the vibrational momentum, M_n the vibrational reduced mass, ω_n the vibrational frequency, and q_n the vibrational displacement. The oscillator displacement λ_n is simply the difference in equilibrium geometry of the situations with the electron localized on the donor, α and on the acceptor β. This hamiltonian can be treated to infinite order in the linear coupling arising in the first bracket on the right hand side of equation (4) and to second order in the electronic mixing parameter V. The resulting vibronic formalism has been extended broadly[3,5-16] to the study of the temperature dependence of ET, tunnelling effects, the importance of the inverted region, etc. Intrinsically, however, this vibronic formalism simply extends the standard nonadiabatic result to include nuclear tunnelling below the barrier; when such nuclear tunnelling effects are unimportant, the result of the vibronic theory and the picture based on activated complex theory, and summarized in equations (1-3), are essentially identical. Nevertheless, vibronic theory does provide important extensions of the standard picture since particular approximations (such as the steepest descents approximation)[13] are easily expressed using this vibronic approach.

Formally, the vibronic coupling hamiltonian (4) couples a boson field (the vibrations) with a fermion field (the two-state electronic structure manifold). It has been known for some time that if the relaxation and correlation times within the phonon or boson manifold are relatively short, the essential dynamics of the situation can be re-expressed in terms of the electronic manifold coupled only to a set of stochastic bath variables.[62] This has been investigated fairly extensively in recent ET studies utilizing path integral formulations.[63,64] This approach is an interesting one; it lays primary focus on the electronic manifold, and permits study of extended systems. On the other hand, by ignoring the details of the vibrational mode that are coupled to the electronic system, it cannot really be used to examine particular molecular situations, deuteron effects, or small changes in structure.

While the three standard pictures just discussed (standard activated complex picture, vibronic coupling picture, and stochastic bath treatment) are all adequate for most electron transfer situations, we will now turn our attention to situations in which they require substantial extension.

III.   Solvent Relaxation Effects on Electron Transfer Dynamics

Extensive recent discussion of the role of solvent dynamics in adiabatic and nonadiabatic electron transfer reactions has made clear that, especially for rapid reactions, solvent dynamics will generally contribute importantly to the overall rate.[17,19-26,58-61,65] The subject has been discussed extensively, including important contributions by authors within this volume, therefore we limit ourselves here to some qualitative remarks. Firstly, for situations in which the inner sphere reorganization energy λ_i is much greater than the outer sphere reorganization λ_o, one does not expect a major contribution from solvent dynamics. In these cases, the overall vibrational contribution to the ET behavior (the dominant contribution to the boson baths, in the statistical model), as well as the dominant energy acceptance, are due to the inner sphere vibrations. Solvent dynamics is then essentially irrelevant, and one would not expect to see substantial differences even in different solvents. That is eminently reasonable on a physical basis: if both the activation terms and the energy acceptance are due to internal vibrations, external solvents should be unimportant. Secondly, depending on the actual shape of the potential energy surface in the neighborhood of the peak, different components of the solvent relaxation can in fact contribute to relaxation trapping of the product. Hynes and his collaborators,[32,33,47,61] as well as Zusman,[58,66] have pointed out that no single solvent relaxation time parameter is then capable of describing the overall response. Rather, frequency dependent friction, corresponding to the entire spectrum of relaxation times, is required for quantitative understanding. Under these conditions, the electron transfer rate may exceed the average relaxation time, since only faster components of solvent response are then important in trapping the product state.[47]

Sumi and Marcus have discussed[38] situations in which both λ_i and λ_o are important. Under these conditions, the electron transfer kinetics can become nonexponential, since two different relaxation processes, one corresponding to solvent domination, the other to inner sphere domination, can contribute to the overall reaction.

Finally, while one expects in general that adiabatic electron transfer reactions will be more sensitive to solvent relaxation effects than will be nonadiabatic, in fact both processes can mirror the effects of solvent relaxation. In particular, if solvent relaxation effects are dominant (slow relaxation time spectrum), then the system can spend a great deal of time at the peak in the potential energy diagram of figure 1; under these conditions, the rate is no longer necessarily determined by the factors of equations (1-3), but rather by the solvent relaxation rate itself. In a very simplified picture, Rips and Jortner have shown that, assuming that only the longitudinal solvent relaxation time is relevant for characterizing the relaxation, the equations (1-3) can be replaced by:

$$k = \frac{k_{NA}}{1 + K} \tag{5}$$

$$k_{NA} = \frac{2\pi}{\hbar} \cdot \frac{T_{ab}^2}{(4\pi\lambda k_B T)^{\frac{1}{2}}} \, e^{-E_A/RT} \tag{6}$$

$$K = \frac{4\pi \, T_{ab}^2}{\lambda\hbar} \, \tau_\ell \tag{7}$$

$$\tau_\ell = (\epsilon_\infty/\epsilon_0)\tau_D \tag{8}$$

$$k \to \begin{cases} k & K \ll 1 \\[2mm] \left(\dfrac{NA_\lambda}{16\pi k_B T}\right)^{\frac{1}{2}} \exp(-E_A/RT) \cdot \dfrac{1}{\tau_\ell} & K \gg 1 \end{cases} \tag{9}$$

Here τ_D, ϵ_0 and ϵ_∞ are respectively the Debye relaxation time, and static and optical dielectric constants. Thus, when longitudinal solvent relaxation is very slow, it indeed becomes the rate determining process.

A number of elegant, fast-time experiments in several different solvents have been devoted to the study of solvent relaxation effects on electron transfer.[18,20-26] This work is conveniently discussed elsewhere.[19] It is thus abundantly clear that solvent relaxation processes require extension of the simple forms of equations (1-3). The remainder of our discussion here will be devoted to less well characterized situations in which extensions are also required.

IV. Secondary Stable Minima on Potential Surfaces: Gating.

The Born Oppenheimer potential energy surface for any chemical reaction is in fact a multidimensional one, with the coordinates corresponding to the various nuclear motions. For electron transfer reactions, there is a large number of such nuclear motions, generally broken up into inner and outer sphere, corresponding to the vibrations of the super molecule and the solvent motions, respectively. Simple extensions of the one dimensional coordinate diagram of figure 1 to several dimensions, such as figure 2a, are then straightforward.[3] In this case, the two coordinates might be, say, an inner sphere and outer sphere (note

the higher frequency for the horizontal, inner sphere motions). The
reaction coordinate is then simply the path of steepest descent from
the minimum at the upper left to the minimum at the lower right, and
the conceptual generalization is relatively minor.

More interesting situations, requiring substantial extensions of
the simple formulas of equations (1-3) and picture of figure 1, arises
when multiple stable minima exist on the potential energy surface. For
instance, one can consider a substituted cyclohexane in which intramolecular electron transfer from a photo excited donor to an
acceptor takes place, as sketched in figure 2b. In this case, the
vertical axis corresponds to configurational change, and the horizontal
axis corresponds to inner shell vibrational changes. Notice that in
this case in going from the reactant at upper left to the stable
product at lower right, there are three possible pathways: the first
is the direct, concerted pathway from chair structure with excited
donor to boat structure with separated charges, while the other two
correspond to indirect pathways either through the upper right (chair
structure with charge transfer) or through the lower left (boat
structure before electron motion). A schematic potential energy
diagram for this situation is shown in figure 2c.

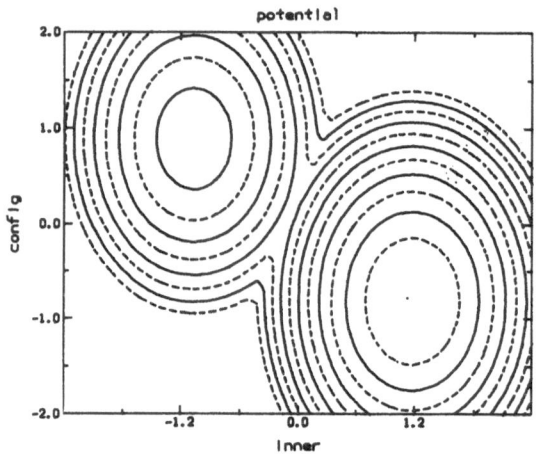

Figure 2a. Schematic potential energy surface for ordinary electron
transfer without conformational state change. The abscissa is the
inner sphere vibrational coordinate, the ordinate is either the outer
sphere coordinate or an internal low frequency motion. The reaction
proceeds over a saddle point located on a line joining precursor
minimum (upper left) to successor minimum (lower right).

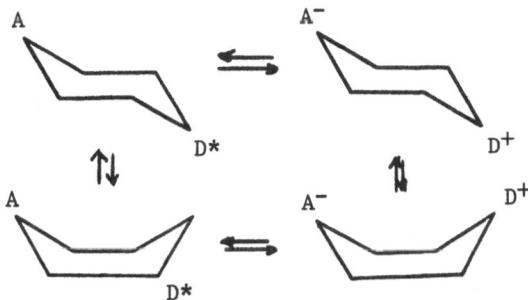

Figure 2b. Schematic pathway from photoexcited disubstituted cyclohexane to electron transferred, conformationally changed successor state.
The configurational changes (vertical motion) and electron transfer
changes (horizontal motions, corresponding to inner sphere motion)
produce four separate stable minima on the surface.

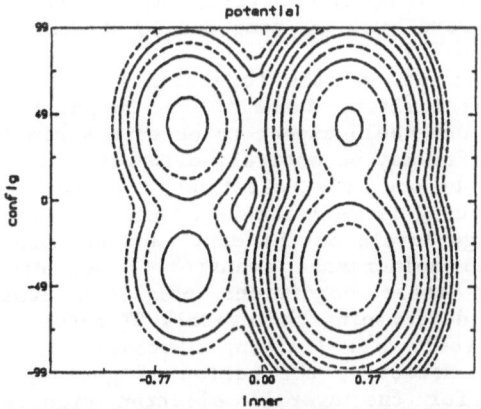

Figure 2c. Schematic potential energy surface for the electron
transfer with conformation equilibrium, such as that in figure 2b. The
ordinate is a conformational coordinate, the surface describes processes such as the coupled electron transfer/conformational change
described in the text. Transfer from the conformationally favored
precursor state at the upper left to the stable successor conformer on
the lower right occurs by a two step process (either across and then or
down and then across), rather than by a combined upper left to lower
right path, which has a higher barrier. From ref. 49.

An important observation is, then, that in general[49,50] the
activation barrier corresponding to the direct concerted process from

upper left to lower right in figure 2c has a higher activation barrier than either of the two indirect processes, through the secondary stable minimum in the upper right or the lower left. Since the increased height of the direct activation barrier would not easily be overcome by pre-exponential factors, it is then clear that in these situations the actual rate process will correspond not to concerted reaction, but to stepwise reaction. The system then wanders a very substantial distance from the simple concerted pathway that is relevant to the 2-minimum situation either of figure 1 or figure 2a.

Such indirect electron transfer processes, reminiscent of the "square scheme" used in electrochemical situations,[67] have been considered in several recent publications, especially with regard to electron transfer in protein systems.[49,50,51] The configuration coordinate, in these protein systems, is no longer simple "chair/boat" interconversion but rather some more complex configurational motion of the protein host itself. Nevertheless, the conclusions reached remain the same: the overall rate must be computed by the indirect, dogleg pathways through the intermediates in figure 2c, rather than by the direct route. Although along each of the two steps in the indirect process the nonadiabatic or adiabatic standard electron transfer formulations, such as those in equations (1-3), may still be relevant, nevertheless the overall rate must be computed by a concatenation of the two separate rate processes.

Simple rate theory analysis of the resulting kinetic scheme then yields several interesting limiting cases.[49-51] In particular, under rather general conditions one would expect to observe gating: that is, the rate of electron transfer may be entirely determined by a conformational motion required before the electron transfer itself can occur.[68,69] While this gating is reminiscent of the activation process required to make the energy levels on the donor and acceptor in figure 1 degenerate (a so-called coincidence event;[70,71] see discussion in section 6 below), under these conditions only the conformational motion may be relevant in determining the overall ET rates.

The gating process itself has limiting subcases, including fast gating and slow gating. Moreover, the notion of gating, and of two separate steps required for the overall electron transfer, may be highly relevant to disentangling a number of ET reactions in protein systems, in which one can in fact separately alter the electronic mixing and the nature of conformational change.[49-51]

Limitations on the arguments just made, and their extension to considerations involving the abnormal regime of highly exoergic electron transfer, are presented in an elegant study by Brunschwig and Sutin.[50] Once again, as in the discussion of solvent relaxation, the major point to be made here is simply that such reactions require extension of the simple arguments of equations (1-3), and figure 1.

V. Dynamical Corrections in the Diffusive Limit: Anisotropic Friction.

In a recent discussion of the effects of solvent dynamics on electron transfer reactions, Fleming and collaborators note[19] that "it is worth pointing out ... that in some cases attainment of the (transition) state requires large amplitude motions of the solute. This motion may involve viscous aspects of solvent-reaction coupling of a sort not considered in present theories." Large amplitude motions, particularly of modes with large masses and small force constants such as conformational motions in proteins, cannot reasonably be described by the simple harmonic oscillator picture generally used for coupled vibrations in polaron type models such as that of equation (4). Situations in which overdamped, slowly relaxed motion, corresponding to large masses and displacements, enters into the dynamics involving electron transfer processes, (or other dynamical processes) may then be the rule.

Perhaps the most studied situation involves ligand rebinding in myoglobin. An extensive series of investigations by Frauenfelder and his collaborators has shown[16,69,72] that ligand recombination kinetics, on the microsecond time scale, is generally not of single exponential type. The observed nonexponential kinetics in this and in other structural reorganization processes is not simply resolved into the sum of a few exponentials. This suggests that non-exponential time evolution may be related to the multitude of degrees of freedom associated with the large molecule and its configurational motions. At very high temperatures, the system explores configuration space efficiently, and transition-state arguments hold. On the other hand, at lower temperatures, other relaxation times may fall within the experimental time scale, then even though the long time behavior will be dominated by the lowest barrier crossing process, shorter time nonexponential kinetics may be observed.

In situations of this type, deviations from simple transition state rate theory arise not from strange features of the potential surface, but rather from strongly anisotropic diffusion coefficients. In the regime in which electron transfer motions are strongly coupled, both to molecular vibrations and to the solvent, diffusive type chemical kinetics can be used to describe the reaction rate, as first discussed by Kramers.[70] Even for so simple a potential surface as that in figure 2a, with simple minima corresponding to reactant and product, the actual evolution of the electron transfer system may be very different from the transition state choice moving along the steepest descents reaction coordinate. Agmon and Hopfield[27] first pointed out that if the diffusion coefficient, or effective friction, in the vertical coordinate of figure 2a (the configurational coordinate of the protein) is extremely slow, then the chemical reaction might be considered to occur along horizontal lines on the surface. Under these conditions, trajectories running from the reactant region of the surface to the product region of the surface would pass over barriers all along the so-called ridge line, separating reactant and product. There would then be no single exponential behavior either in time or in inverse temperature, since the overall rate would be an average, over the initial distribution in the configurational coordinate, of the

rates determined for motion along the inner sphere coordinate at each
value of the (fixed) configurational coordinate.

While the Agmon/Hopfield picture is intriguing and important, and
contains very significant physical insight, it is also substantially
oversimplified. Actual trajectories will in fact not follow horizontal
lines, but rather will fan out throughout the two dimensional surface,
and may very well cross the ridge line several times. The resulting
anisotropic diffusive dynamics can exhibit a multitude of interesting
rate phenomena, and have been recently investigated by Dygas et al.[34]

The analysis of the reaction process with strongly overdamped
diffusion along the configurational coordinate is based upon the
Smoluchowski system of equations corresponding to overdamped motion in
the plane. These are:

$$\dot{X} = -\frac{1}{m\gamma_1} \frac{\partial V}{\partial X_1} + (2kT/m\gamma_1)^{\frac{1}{2}} \dot{w}_1 \qquad (10)$$

$$\dot{X}_2 = \frac{-1}{m\gamma_2} \frac{\partial V}{\partial X_2} + (2kT/m\gamma_2)^{\frac{1}{2}} \dot{w}_2 \qquad (11)$$

Where $V(X_1, X_2)$ is the potential surface of equation 2a, γ_1 and γ_2 are
the friction coefficients for motion along the x and y coordinates
respectively (and are assumed large relative to all characteristic
frequencies), and w_1 and w_2 are independent Gaussian white noise terms.
Anisotropy in the motion then arises from inequivalence of γ_1 and γ_2,
corresponding to stronger friction, or stronger damping, in the large
amplitude conformational mode than in the high frequency, small
amplitude vibrations. The analysis of the coupled equation set (10,
11) leads to some general statements about trajectories on the surface
which, in turn, describe the observed kinetics. The analysis is
presented elsewhere,[34] but several critical results, with important
experimental implications, emerge. These include:

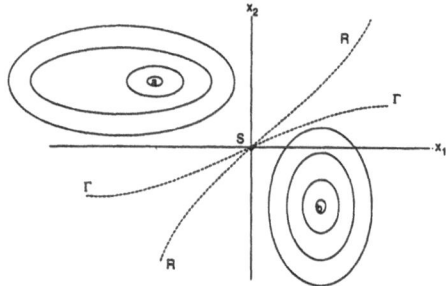

Figure 3a. A schematic sketch of the potential surface for the ligand
and protein rebinding reaction. R denotes the ridge, Γ the separatrix,
and S the saddle point. The stable reactant and product conformations
are denoted by a and b, respectively. Note that the shapes and
orientations of R and Γ are schematic and other possibilities exist as
discussed in the text.

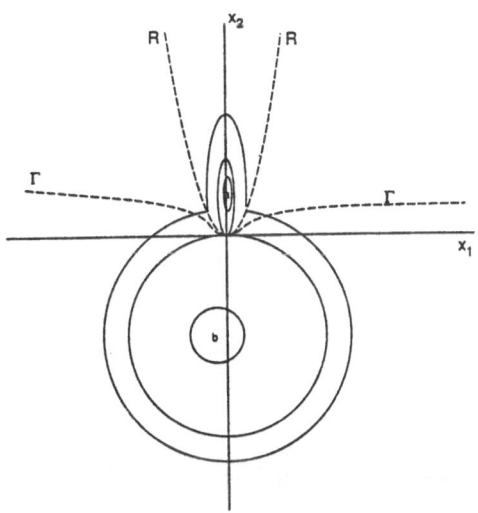

Figure 3b. A plot of the potential corresponding to the lower eigen
values 2x2 vibronic coupling problem.   The ratio of the diffusion
coefficients is δ - 0.1. The ridge and separatrix shown in the figure
have been calculated for this ratio. Note that unlike the situation in
the schematic picture (a), R and Γ touch (at the saddle point) but do
not cross one another. From ref. 34.

1. The separatrix, the locus of points that separate stability
   regions in the trajectories, is no longer coincident with the
   ridgeline, which is the geometric separation between the attraction
   region of the reactant well and the product well. This is indicated in figures 3a and 3b, for two different situations. Figure
   3a is a schematic representation of the ligand association process
   in myoglobin, whereas figure 3b is the potential surface corresponding to a 2x2 Jahn-Teller surface. This is a a highly
   exaggerated situation, where δ - 0.1. Here the ridge and separatrix touch one another, precisely at the saddle point, but never in
   fact cross.
2. Long time kinetics either from reactant or into the product fits a
   single exponential, with rate constant given by k - 1/2τ, where τ
   is the Kramers time. This is the mean first passage time for
   crossing the separatrix, and analytic forms are available in
   particular limits.
3. Multiple transients occur at shorter times, corresponding to
   crossing and recrossing of the ridge line, but not of the separatrix. If a given experiment is sensitive to the geometric structure of the system (such as a vibrational spectrum, or an EPR line
   shape, or a fluorescence frequency), then it will exhibit highly
   nonexponential behavior on this transient time scale, as the system

crosses and recrosses from product side of the ridge line to the reactant side.

4. So long as the activated complex energy is large compared to the thermal energy, the long time Kramers rate indeed corresponds to motion along the reaction coordinate, passing over the saddle point. This is in agreement with the ordinary understanding of Kramers theory, but quite different from the results of Agmon and Hopfield,[27] who assume essentially infinitely slow motion along the conformational coordinate.

5. When the ratio $\delta = \gamma_1/\gamma_2$ of the damping coefficients (or the xx and yy components of the diffusion tensor) approaches unity, the results of standard Kramers theory are obtained.

6. Since the evolution of the system probes substantially different parts of the reaction surface on a transient time scale, different experimental probes might well result in the observation of different kinetics.

When the diffusion tensor is isotropic, and $\delta = 1$, the ridge line and the separatrix are identical. Under these conditions, one expects the dynamics, even in the strongly diffusive, overdamped regime, to be dominated entirely by trajectories that cross at the saddle point, with no transient behaviors, since no recrossing of the ridge line will occur. When the diffusion coefficient is no longer isotropic, ridge line and separatrix will be different, and recrossing, transients, and nonexponential behavior will occur. The classical treatment discussed here bears some similarity to the quantum treatment of ET reactions in which solvent dynamics are taken into account, but the dynamics of the vibrational coordinates is assumed instantaneous.[32,33,38,39] Investigations in ET processes, and other dynamical processes in macromolecular systems, in which overdamping of a protein coordinate might well lead to strongly anisotropic diffusion has not our knowledge been extensively pursued in the experimental literature other than the ligand rebinding kinetics already referred to. Such investigations might well lead to improved concepts of how large amplitude, overdamped protein motions lead to dynamical processes in these biological systems.

VI. Electronic Effects: Time Dependent Analysis of Electron Transfer Rates

The transition-state based, nonadiabatic electron transfer rate theory of equation (1) has the great advantage that it is not explicitly time dependent: it therefore can be used to describe both very slow and very fast electron transfer processes. Time independent calculations of rates always have this advantage; they also have a concomitant disadvantage, which is that the actual time scale of the process is not entirely explicit from perturbation theoretic, time independent studies. It is therefore occasionally of real use to employ time dependent dynamic calculations to study chemical kinetic processes. Such methods have become extremely popular in the past decade, first for spectroscopic studies[73] and later, using fft techniques,[74] for more general dynamic situations.

In the particular case of electron transfer, there are several reasons why an explicitly time dependent calculation is useful. These include:

A.   The role of relaxation processes can be directly monitored, and the efficacy of the adiabatic separation between electronic and nuclear motion can be pursued.

B.   Generalized questions of electronic motion, including phenomena such as superexchange, the role of barrier heights, the exoergic gap law behavior and so on, can be directly observed.

C.   The role of dephasing, and of statistical fluctuations, can be directly monitored in time.

D.   The validity of continuum approximations, versus semi-continuum or microscopic descriptions, can be assessed.

E.   The validity of perturbation theory can be studied, by comparison with correct time dependent behavior.

Several recent studies of dynamic electron transfer have appeared.[40-48,75] They have illuminated all of the points above, and have led to interesting suggestions for changing the nature of particular electron transfer reactions. We will discuss two situations: first, we will examine a time-dependent self-consistent field treatment of intramolecular ET, using a full electronic structure description for the electronic manifold, and a phonon type description for the vibrations and solvent degrees of freedom. Second, we will use a very simplified, small basis, independent electron model to suggest the importance of superexchange, dephasing, gap laws and interference processes in bridge assisted ET.

Mikkelsen and his collaborators[41-43] have presented both a formalism and a number of examples for direct study of ET from a prepared precursor state to a final product. Their approach is to prepare the open shell, odd electron species by localizing an electron on the donor at initial time, and then follow the evolution of the system density matrix as that electronic density evolves. They then write the hamiltonian as

$$H = H_e + H_{vib} + W_{el-vib} + H_{pol} + W_{pol-el} \qquad (12)$$

where the terms are, schematically, the electronic hamiltonian of fixed nuclear geometry, the vibrational hamiltonian in the harmonic approximation, the coupling between electrons and vibrations (assumed linear in the displacement, the usual small polaron form[3,5-11,13-15]), the polarization hamiltonian of the continuum dielectric solvent outside the studied molecular complex, and finally the interaction between the solvent polarization and the electronic system. The explicitly considered electronic and vibrational structures include the donor species, the acceptor species and any solvent or bridge molecules that exist in the cavity, within the dielectric continuum, in which the electron transfer process occurs. The electronic hamiltonian, in particular, is taken as the full all-electron non-relativistic hamiltonian, within a given finite basis.

The solution to the electron transfer dynamics then is given by studying the expectation value of a number of operators. In particular, Mikkelsen et al.[43] consider the operator manifold

$$\{q\} = \{a_r^+ a_s, \ B_n^+, \ B_n, \ b_k^+, \ b_k\} \tag{13}$$

where a_r^+ is a creation operator for the electron in atomic basis function r, B_n^+ creates a vibrational excitation in the nth normal mode, and b_k^+ creates a phonon of the continuum solvent. The Ehrenfest limit of the coupled equations of motion for this operator manifold is then solved according to

$$i\hbar \frac{\partial \langle q_i \rangle}{\partial t} = \langle [q_i, \ H] \rangle \tag{14}$$

The expectation values are calculated self consistently, using a generalized Fock operator, which contains both the usual electronic structure Fock components and extra components due to the interaction of vibrations and polarization with the electrons.

Given an initial condition with the electron localized on the donor, the solution for the Fock elements self consistently yields the time dependence of the electron amplitude for being on donor, bridge, or acceptor sites. This calculation has several important advantages: by use of well defined, ab-initio electronic structure methodology it avoids some of the inherent simplifications of independent electron models or semiempirical hamiltonians. By bringing in all vibrations of the supermolecular complex, plus the phonons in the dielectric continuum approximation, it can evaluate all the important dynamic effects. It can be pursued at any level of complication, from the transfer complex in vacuo, through a continuum solvent, into a semi continuum picture in which the dielectric cavity contains a number of solvent molecules. It can be used to evaluate the relative roles of superexchange, bridge assistance, energy gaps and so forth.

Some typical results of these calculations[42] are contained in figure 4. The calculations are performed for a five molecule complex in the cavity, corresponding to a benzene radical ion as donor, benzene as acceptor, and water molecules as bridge/solvent species. Some features of these calculations include the fact that the dielectric medium reduces the maximum rate by about a factor of four, the rate is very sensitive to the angle of interaction, largely because the matrix elements, both direct and indirect, depend on overlap factors which in turn strongly depend on angle, bridge assistance is absolutely necessary for transfer over relatively long distances (in this case, in interplanar transfer distance of 5.0 Å), the effective transfer matrix element is itself strongly angle dependent, and (perhaps most relevant for comparison with standard analytic theories) the maximum in the transfer rate occurs at precisely the angular geometry (roughly 105°) at which the energy difference between the electronic levels on donor and acceptor goes essentially to zero. This so called "coincidence events" geometry[9,70,71] is that at which the two potential curves of figure 1 undergo their avoided crossing; it is the geometry at which the electronic levels on donor and acceptor become degenerate, and therefore transfer is optimal; it is the geometry at which no energy transfer to the boson system is necessary, and therefore no Franck Condon modulation of the transfer rate occurs. It is also the rate at which one generally expects the transfer rate to be maximized.

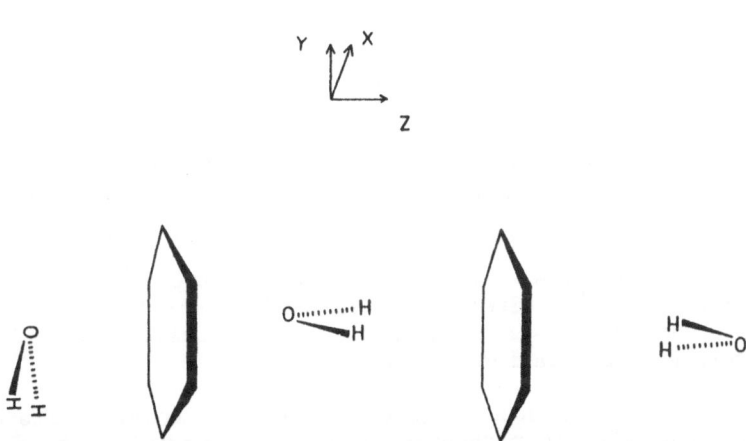

Figure 4.   Some results for dynamic calculation of intramolecular electron transfer. The computations are for the five molecule system of figure 4a, imbedded in a dielectric continuum. Figure 4b shows the computed electron transfer rates in the vacuum (+) and with a dielectric medium surrounding the encounter complex (x).

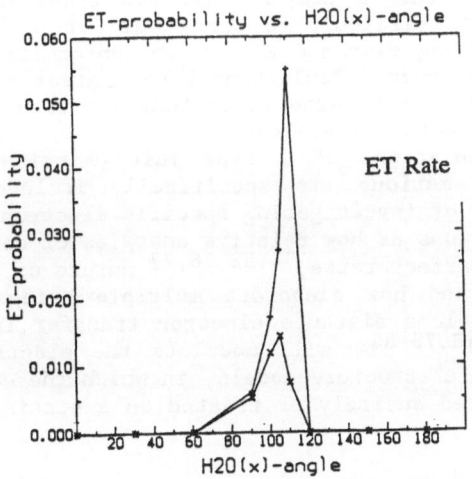

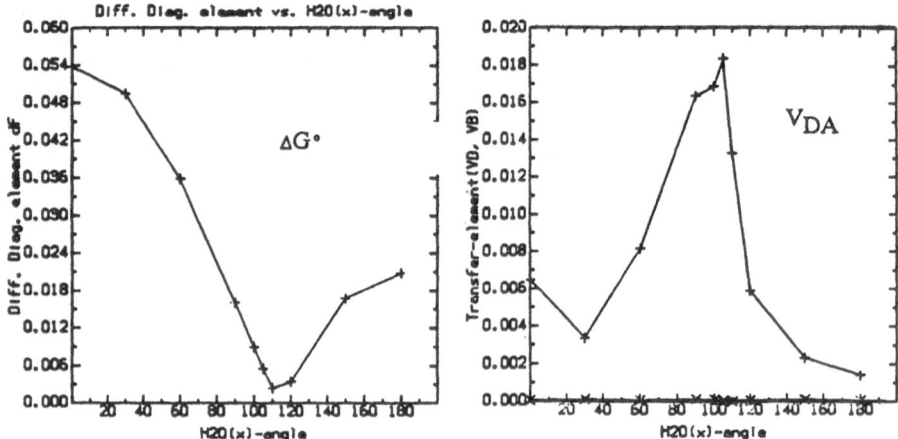

Figure 4c. Angular dependence of the electron transfer matrix elements corresponding to direct (x) and bridge assisted (+) transfer. Note that the direct transfer route is essentially unavailable, for an interplanar separation of 5 Å.

Figure 4d. The donor-acceptor electronic energy difference, df, for the five molecule encounter complex in vacuum. Note that the energy levels become essentially degenerate (coincidence event) at an angle near 105°, precisely where the average ET probability of figure 4a maximizes. From ref. 42.

Realistic calculations such as those in references 40-43, in which the electronic structure is treated carefully and the interactions both with the vibrations and with the continuum solvent are included, yield substantial insight into the nature of electron transfer reactions. Such calculations will be difficult to carry out for slow reactions, such as those found in long distance ET and in conformationally gated processes. Nevertheless, such calculations have a great deal to teach us about the nature of transfer, especially indirect transfer, and how the electronic structure actually evolves.

Time dependent calculations of a type just described, in which electronic and nuclear motions are specifically included, are not necessarily the best way of investigating specific electronic structure effects. For such questions as how relative energies of donor/acceptor and superexchange sites affect rates,[36,52,76,77] nature of coherence or interference events,[36] and how elaborate multiple bridge structures (such as those found in long distance electron transfer in natural or synthetic systems)[9,11,12,78-84] will modulate the electron transfer process, simple electronic structure models, in which the coupled boson systems are either ignored entirely or treated on a statistical basis,

are relevant.[36,44-46]   In particular, use of the model hamiltonian 15
permits study of transfer of non-interacting electrons.

$$H = \sum_i \epsilon_i a_i^+ a_i + \sum_{i \neq j} t_{ij} a_i^+ a_j \tag{15}$$

This tight binding, or generalized extended Huckel, hamiltonian has
been used by several investigators to examine[36,44-46] some of the
specific electronic structure effects just mentioned.   Formally, one
can solve the evolution of an electronic system described by equation
(15) coupled to a boson continuum; utilizing semigroup methods,[85-88]
the equations of motion for any operator Ω can then be written

$$\frac{d\Omega}{dt} = \frac{i}{\hbar} [H,\Omega] + (\dot{\Omega})_R \tag{16}$$

where the first term on the right is the hamiltonian evolution and the
second term is the relaxation contribution arising from coupling to the
boson continuum.   A typical result for a three site model (donor,
superexchange bridge, acceptor) is shown in figure 5.

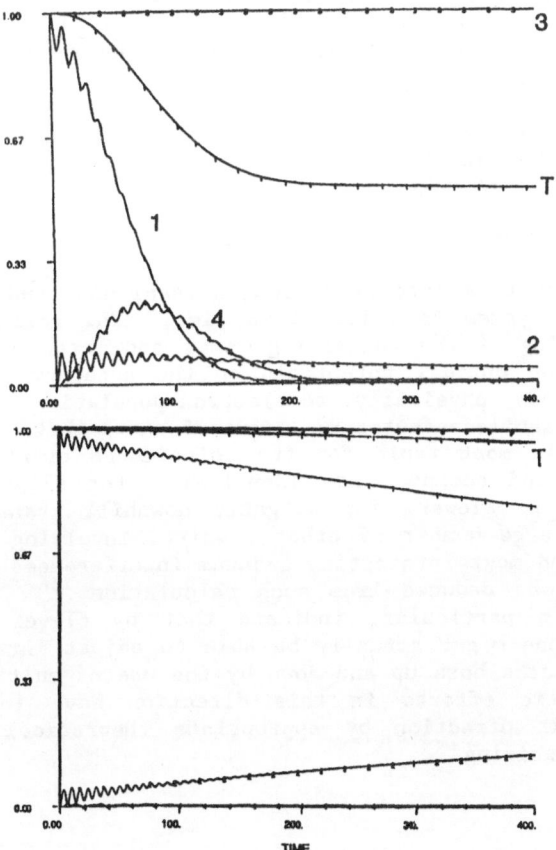

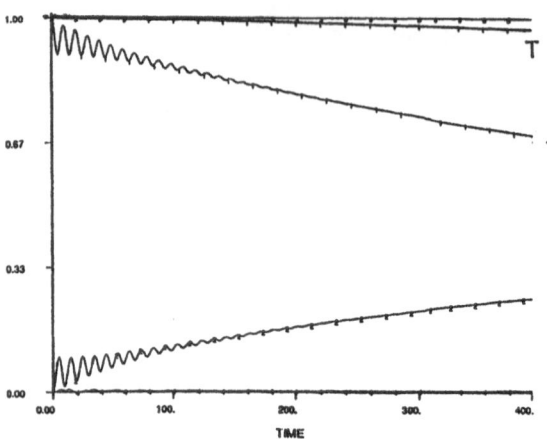

Figure 5.   Dynamic results of eq. (15,16), a model bridge-assisted ET system.   The parameters: $t_{12} = t_{24} = t_{13} = t_{34} = .1$, $\epsilon_3 = -14$, $\epsilon_2 = 1.6$.   The lines show the populations of the four levels and half the total population (T) as a function of time: level 1 is the donor, level 4 is the acceptor whose population decays due to the semigroup algebra.   The energy parameters:

     (5a): $\epsilon_1 = \epsilon_4 = 1$
     (5b): $\epsilon_1 = .7$, $\epsilon_4 = 1$
     (5c): $\epsilon_1 = 1$, $\epsilon_4 = .7$

Note fastest transfer for the degenerate case.

Here the electron at time zero is placed in level one, and the evolution of the total system is followed in time.   The relaxation term consists of dephasing of the tunnelling terms and loss of population (corresponding to an optical potential) on the acceptor state; this final term corresponds, physically, to electron population loss such as might occur from a labile acceptor.[36]   Figure 5 shows that, once again, electron transfer is most rapid for the coincidence event; when the energies of donor and acceptor are identical.   For slightly uphill transfer the rate is slower, for slightly downhill transfer it is intermediate.   A large number of other results, involving dephasing, power dependence and most interesting quantum interferences of several pathways, can also be deduced from such calculations.[36]   The interference effects, in particular, indicate that by clever design of molecular bridges one might actually be able to adjust intramolecular electron transfer rates both up and down by the use of multiple bridge structures; synthetic efforts in this direction have in fact appeared,[89] and their direction by appropriate theoretical modelling should prove very exciting.

VII. Remarks

Use of rapid spectroscopic techniques, elegant synthetic systems, and advanced theoretical methods, has greatly extended the range of systems amenable to, and challenging for, ET rate studies. We have described here several theoretical directions in which traditional transition state theory of nonadiabatic electron transfer reactions has been extended. These directions include use of a full vibronic model (section II), effects of solvent relaxation (section III), the role of substructures or stable secondary minima on the potential surface in providing gating reactions, especially important in protein systems (section IV), result of anisotropy in the diffusion coefficient in completely changing the trajectories on the potential surface (section V), and how electronic structure effects can change the assumed transition state dynamics, and the role of explicitly time dependent studies in understanding the transfer process (section VI). A number of other very important effects, such as the selection of initial state for electron transfer,[36,37] the role of high frequency optical phonons in providing non-coincidence events (energetically inelastic) efficient transfer pathways,[71] the role of applied fields and ion charges in the neighborhood of the ET species, and how continuum levels (electrodes) would change the whole process have been ignored. Clearly this is a very dynamic and vibrant field, indeed it is one of the most vibrant research in chemistry at present.

There is a rapidly growing, and very important, role for dynamical calculations in understanding the intimate details of how ET processes proceed. Such explicitly time dependent calculations, by illuminating the pathway taken by ET reactions, should help both in the understanding of reactions as they occur, and the design of ET reactions for particular purposes, such as facilitated rates over long distances, non-exponential profiles of the electron transfer, control of ET processes by external fields for possible device applications, photovoltaic charge separation processes and micromodulation of electron transfer processes.[45,71]

Acknowledgements: This work was supported by the Chemistry Division of the National Science Foundation (Grant CHE 8805585). I am very grateful to a number of colleagues, including P. Hale, R. Kosloff, B. Hoffman, A. Nitzan, N.S. Hush, S.F. Fischer, A. Aviram and particularly K.V. Mikkelsen for close collaborations and important insights.

References

1. Marcus, R.A. (1963) J. Phys. Chem., 67, 853, 2889; (1956) J. Chem. Phys., 24, 966, 979; (1965) Ann. Revs. Phys. Chem., 16, 155.

2. Hush, N.S. (1961) Trans. Far. Soc., 57, 557; (1968) Electrochem. Acta, 13, 1005.

3. Newton, M.D. and Sutin, N. (1984) Ann. Revs. Phys. Chem., 35, 437; Sutin, N. (1983) Prog. Inorg. Chem., 30, 441.

4. Cannon, R.D. (1980) Electron Transfer Reactions, Butterworths, London.

5. De Vault, D. (1984) Quantum-Mechanical Tunneling in Biological Systems, Cambridge.

6. Marcus, R.A. and Sutin, N. (1985) Biochem. Biophys. Acta 811, 265.

7. Ulstrup, J. (1979) Charge Transfer Processes in Condensed Media, Springer, Berlin.

8. Cristov, S. (1980) Collision Theory and Statistical Theory of Chemical Reactions, Springer, Berlin.

9. Mikkelsen, K.V. and Ratner, M.A. (1987) Chem. Revs., 87, 113.

10. (1983) Prog. Inorg. Chem. vol. 30.

11. (1979) Tunneling in Biological Systems, Chance, B., DeVault, D., Frauenfelder, H., Marcus, R.A., Schrieffer, J.R. and Sutin, N. (eds), Academic, New York, 1979.

12. Guarr, T. and McLendon, G., (1985) Coord. Chem. Revs., 68, 1.

13. Fischer, S.F. and Van Duyne, R.P. (1974) Chem. Phys., 5, 183); (1977) 26, 9.

14. Jortner, J. (1976) J. Chem. Phys., 64, 4860.

15. Scher, H. and Holstein, T. (1981) Phyl. Mag., B44, 343.

16. Frauenfelder, H. and Wolynes, P.G. (1985) Science, 229, 337.

17. Fleming, G.R. (1986) Ann. Revs. Phys. Chem., 37, 81.

18. Fleming, G.R. (1986) Chemical Applications of Ultrafast Spectroscopy, Oxford, New York.

19. Maroncelli, M., Mac Innis, J. and Fleming, G.R. (1989) Science, 243, 1674.

20. Opallo, M. (1986) J. Chem. Soc. Far. Trans., 1, 82, 339.

21. McGuire, M. and McLendon, G. (1986) J. Phys. Chem., 90, 2549.

22. Kosower, E.M. and Huppert, D. (1986) Ann. Revs. Phys. Chem., 37, 127.

23. Simon, J.D. (1988) Accts. Chem. Res., 21, 128; Simon, J.D. and Su, S.-G. (1988) J. Phys. Chem., 92, 2395; (1988) J. Chem. Phys., 89, 908.

24. Kahlow, M.A., Kang, T.J. and Barbara, P.F. (1987) J. Phys. Chem., 91, 6452.

25. Heitele, H., Michel-Beyerle M.E. and Finckh, P. (1987) Chem. Phys. Lett., 38, 237.

26. McManis, G.E. and Weaver, M.J. (1988) Chem. Phys. Lett., 145, 55; J. Chem. Phys. in press; McManis, G.E., Golovin, M.N. and Weaver, M.J. (1986) J. Phys. Chem., 90, 6563.

27. Agmon, N. and Hopfield, J.J. (1983) J. Chem. Phys., 78, 6947; 79, 2042.

28. Agmon, N. (1988) Biochemistry 27, 3507.

29. Agmon, N. and Kosloff, R. (1988) J. Phys. Chem., 91.

30. Kramers, H.A. (1940) Physica, 7, 284.

31. Schuss, Z. and Matkowski, B.J. (1979) SIAM J. Appl. Math., 35, 604.

32. Grote, A.F. and Hynes, J.T. (1981) J. Chem. Phys., 74, 4465; (1981) 75, 2171.

33. van der Zwan, G. and Hynes, J.T. (1982) J. Chem. Phys., 77, 1295.

34. Klosek-Dygas, M.M., Hoffman, B.M., Matkowsky, B.J., Nitzan, A., Ratner, M.A. and Schuss, Z. (1989) J. Chem. Phys., 90, 1141.

35. Frauenfelder, H. and Young, R.D. (1986) Comments Mol. Cell. Biophys., 3, 347.

36. Kosloff, R. and Ratner, M.A. Isr. J. Chem., in press.

37. Knapp, E.W. and Fischer, S.F. (1989); J. Chem. Phys., 90, 354; Davydov, A.S. (1978) Phys. Stat. Sol., B90, 457; Beratan, D.N. Onuchic, J.J. and Hopfield, J.J. (1987) J. Chem. Phys., 86, 4489; Nussbaum, I. and Fischer, S.F. (1986) Phys. Lett., 115, 268.

38. Nadler, W. and Marcus, R.A. (1987) J. Chem. Phys., 86, 3906; Sumi, H. and Marcus, R.A. (1986) J. Chem. Phys. 84, 4272.

39. Onuchic, J.N. (1987) J. Chem. Phys., 86, 392.

40. Deumens, E. and Ohrn, Y. (1988) J. Phys. Chem. 92, 3181; Deumens, E., Lathouwers, L. and Ohrn, Y. (1987) Int. J. Quant. Chem. Symp., 21, 321.

41. Mikkelsen, K.V., Dalgaard, E. and Swanstrom, P. (1987) J. Phys. Chem., 91, 3081.

42. Mikkelsen, K.V. and Ratner, M.A. (1987) Int. J. Quant. Chem., S21, 341; (1988) S22, 707.

43. Mikkelsen, K.V. and Ratner, M.A. (1989) J. Phys. Chem., 93, 1759; (1989) J. Chem. Phys. 90, 4237.

44. Ondrechen, M.J., Ratner, M.A. and Sabin, J.R. (1979) J. Chem. Phys. 71, 2244.

45. Reimers, J.R. and Hush, N.S. Chem. Phys., in press; Riemers, J.R. and Hush, N.S. (1989) in Aviram, A. (ed) Molecular Electronics, The Engineering Society, New York.

46. Joachim, C. (1987) Chem. Phys., 116, 339; Joachim, C. (1989) in Aviram, A. (ed) Molecular Electronics, The Engineering Society, New York.

47. Zichi, D.A., Ciccotti G. and Hynes, J.T. J. Phys. Chem. submitted.

48. Chandler, D. and Kuharski, R.A. (1988) Faraday Disc., 85, 329; Hwang, J.K. and Warshel, A. (1987) J. Am. Chem. Soc., 109, 715.

49. Hoffman, B.M. and Ratner, M.A. (1987) J. Am. Chem. Soc., 109, 6237; (1988) erratum ibid., 110, 8267.

50. Brunschwig, B. and Sutin, N. J. Am. Chem. Soc., in press.

51. Hoffman, B.M., Ratner, M.A. and Wallin, S. Adv. Chem. Ser., King, R.B. (ed.) in press.

52. Marcus, R.A. (1988) Chem. Phys. Lett., 133, 47; (1988) 146, 13.

53. McManis, G.E., Nielson, R.M., Gochev, A. and Weaver, M.F., J. Am. Chem. Soc., in press; Mayo, S.L., Ellis, W.R, Crutchley, R.J. and Gray, H.B. (1986) Science, 233, 948.

54. Balaji, S., Ng, L., Jordan, K.D., Paddon-Row, M.N. and Patney, H.K. (1987) J. Am. Chem. Soc., 109, 6957; Paddon-Row, M.N. and Jordan, K.D. (1988) in Modern Methods of Bonding and Delocalization, J.F. Liebman, J.F. and Greenberg, A. (ed), VCH Publishers, New York, p. 116.

55. Ondrechen, M.J., Ellis, D.E. and Ratner, M.A. (1984) Chem. Phys. Lett., 109, 50.

56. Newton, M.D. (1986) J. Phys. Chem., 90, 3437; (1988) ibid., 92, 3049; (1982) ACS Symp. Ser., 198, 255.

57. Beratan, D.N. and Hopfield, J.J. (1984) J. Am. Chem. Soc., 106, 1584.

58. Zusman, L.D. (1980) Chem. Phys. 49, 295; Calef. D.F. and Wolynes, P.G. (1983) J. Phys. Chem., 87, 3387.

59. Rips, I. and Jortner, J. (1987) J. Chem. Phys., 87, 2090, 6513; (1988) 88, 818.

60. Sparpaglione, M. and Mukamel, S. (1988) J. Chem. Phys., 88, 1465; (1988) 88, 3263.

61. Hynes, J.T. (1986) J. Phys. Chem., 90, 3701.

62. Sewell, G.L. (1963) Phys. Rev., 129, 597; (1963) in Polarons and Excitons, Kuper, C. and Whitfield, G. (eds.) Plenum, New York, 1963).

63. Carmeli, B. and Chandler, D. (1985) J. Chem. Phys., 82, 3400; Kuki, A. and Wolynes, P. (1987) Science, 236, 1647.

64. Allinger, K. and Ratner, M.A. (1989) Phys. Rev., A39, 864.

65. Murillo, M. and Cukier, R.I. (1985) J. Chem. Phys., 89, 6736.

66. Zusman, L.D. (1988) Chem. Phys., 51, 119.

67. Bond, A.M. and Oldham, K.B., (1983) J. Phys. Chem., 87, 2472; O'Connell, K.M. and Evans, D.H. (1983) J. Am. Chem. Soc., 105, 1473.

68. Leiber, L.M., Karas, J.L. and Gray, H.B. (1987) J. Am. Chem. Soc., 109, 3778; McLendon, G., Pardue, K. and Bak, P. (1987) J. Am. Chem. Soc., 109, 7540; Bernardo, M.M., Robandt, P.V., Schroeder, R.R. and Rorabacher, D.B. (1989) ibid., 111, 1224; Bechtold, R., Kuehn, C., Lepre, C. and Isied, S. (1986) Nature, 322, 286; ref. 18, p. 179; Northrup, S.H. and McCammon, J.A. (1986) J. Am. Chem. Soc., 106, 930.

69. Debrunner P.G and Frauenfelder, H. (1982) Ann. Revs. Phys. Chem. 33, 283.

70. Emin, D. (1971) Phys. Rev. B4, 3639.

71. Ratner, M.A. (1989) in Aviram, A. (ed) Molecular Electronics, The Engineering Society, New York.

72. Frauenfelder, H. and Young, R.D. (1986) Comm. Mol. Cell. Biophys., 3, 347.

73. Tannor, D. and Heller, E.J. (1982) J. Chem. Phys., 77, 202; Heller, E.J. (1981) Accts. Chem. Res., 14, 368.

74. Kosloff, R. (1988) J. Phys. Chem. 92, 2087.

75. Kosloff, R. and Ratner, M.A. Israel J. Chem., in press.

76. McConnell, H.M. (1961) J. Chem. Phys., 35, 508.

77. Plato, M., Mobius, K., Michel-Beyerle, M.-E., Bixon, M and Jortner, J. (1988) J. Am. Chem. Soc., 110, 7279.

78. Closs, G. and Miller, J.R. (1988) Science, 240, 440.

79. Oevering, H. Paddon-Row, M.N., Heppener, M. Oliver, A.M. Cotsaris, E., Verhoeven, J.W. and Hush, N.S. (1987) J. Am. Chem. Soc., 109, 3258.

80. Peterson-Kennedy, S.E., McGourty, J.L., Kalweit, J.A., and Hoffman, B.M. (1986) J. Am. Chem. Soc., 108, 1939.

81. Nocera, D.G., Winkler, J.R., Yocom, K.M., Bordignon, E. and Gray, H.B. (1984) J. Am. Chem. Soc., 106, 5145.

82. Isied, S.S., and Vassilian, A. (1984) J. Am. Chem. Soc., 106, 1732.

83. Wasielewski, M.R., Niemczyk, M.P., Svec, W.A. and Pewitt, E.B. (1985) J. Am. Chem. Soc. 107, 1080.

84. Kong, J.L.Y., Spears, K.G., and Loach, P.A. (1982) Photochem. Photobiol., 35, 565.

85. Kosloff, R. and Ratner, M.A. (1984) J. Chem. Phys. 80, 2352.

86. Kosloff, R. and Rice, S.A. (1980) J. Chem. Phys. 72, 4591.

87. Spohn, H. and Lebowitz, J. (1978) Adv. Chem. Phys. 38, 1096.

88. Gorini, V., Kossakowski, A. and Sudarshan, E.C.G. (1976) J. Math. Phys., 17, 821; Lindblad, G. (1978) Comm. Math Phys., 48, 119.

89. Stein, C.A., Lewis, N.A., and Seitz, G. (1982) J. Am. Chem. Soc., 104, 2596.

BIOMOLECULAR DYNAMICS — QUANTUM OR CLASSICAL? RESULTS FOR PHOTOSYNTHETIC ELECTRON TRANSFER

José Nelson Onuchic,[1] Robert F. Goldstein,[2] and William Bialek[3]
[1] Instituto de Física e Química de São Carlos
Universidade de São Paulo, 13560 São Carlos, SP Brazil
[2] Department of Chemistry and University Computer Center
University of Illinois at Chicago Circle, Chicago, Illinois
[3] Departments of Physics and Biophysics
University of California, Berkeley, California 94720

ABSTRACT. We discuss the relations between quantum and classical descriptions of the atomic motions which accompany electron transfer reactions. Rather than arguing that a specific model is required to account for some particular observations, we show that a broad class of models can be qualitatively understood, and that is possible to understand the mechanism of quantum/classical crossover without discussing details of a particular model. We then motivate the hypothesis that the photosynthetic reactions operate in the quantum regime of these models. This hypothesis has several robust consequences which are confirmed by recent experiments.

1. Introduction

Since the inception of quantum mechanics there have been several suggestions that quantum effects may be important for the dynamics of various biological proceses. The more conventional view was forcefully stated by Pauling [1,2] at the outset — quantum mechanics is important because it determines the rules of chemical bonding, but once these rules are established biology operates in the classical limit. This outlook is repeatedly echoed in recent molecular dynamics simulations of proteins, where it is argued that internal motions of biomolecules are dominated by modes with characteristic quantum energies $\hbar\omega$ much smaller than the thermal energy k_BT, so that quantum mechanics is irrelevant [3,4]. All of these discussions, however, are phrased in rather general terms, often without regard to the specifics of some particular molecule or reaction [5]. In this paper we show how the primary events of photosynthesis offer a unique opportunity to address the role of quantum mechanics in biology.

In trying to understand whether the dynamics of some particular reactions are classical or quantum we face several fundamental physical issues. Most textbook examples focus on the application of classical mechanics to objects on a macroscopic, human scale, while quantum mechanics is introduced as a theory appropriate to the microscopic domain of atoms and sub-atomic particles. We are taught that quantum mechanics gives way to classical mechanics as masses, energies, length scales, ... become large. In fact the border between the microscopic classical world and the macroscopic classical world is notoriously difficult to define. It is at this border, for example, that many of the conceptual difficulties

J. Jortner and B. Pullman (eds.), Perspectives in Photosynthesis, 211–226.

in the quantum theory of measurement are to be found. Recent theoretical work has emphasized and clarified the subtleties of the classical/quantum interface, showing both how quantum effects can manifest themselves at a macroscopic level and how semi-classical ideas can be used to understand the seemingly strongly quantum behavior of microscopic systems. Applications of these ideas have been pursued in a number of experiments, ranging from the behavior of electrons in disordered metals to the dynamics of superconducting circuits. Some of the theoretical developments may be found in [6,7,8,9], while relevant experiments include those discussed in [10,11,12].

Persistent experimental inspiration for the idea that quantum effects are important in biology has come from studies on the primary events of photosynthesis, both in green plants and in bacteria. In particular, it was discovered many years ago that some photo-induced charge separation reactions can occur quite efficiently even at temperatures approaching absolute zero, suggesting that these reactions may proceed by quantum "tunneling" rather than by classical activation [13]. More quantitative studies on the rate of a single electron transfer step in the bacterium *Chromatium vinosum* demonstrated a crossover from classical activation, with its associated Arrhenius temperature dependence, near room temperature to apparent tunneling — with essentially no temperature dependence — below 200 K [14]. More recently it has been found that essentially all of the electron transfer steps in the reaction centers of photosynthetic bacteria deviate from the classical Arrhenius temperature dependence of reaction rates (see Fig. 1 and several other articles in this book); in two cases which have been checked this apparently non-classical behavior appears robust to changes in the thermodynamic driving force for the reaction [13,15,16].

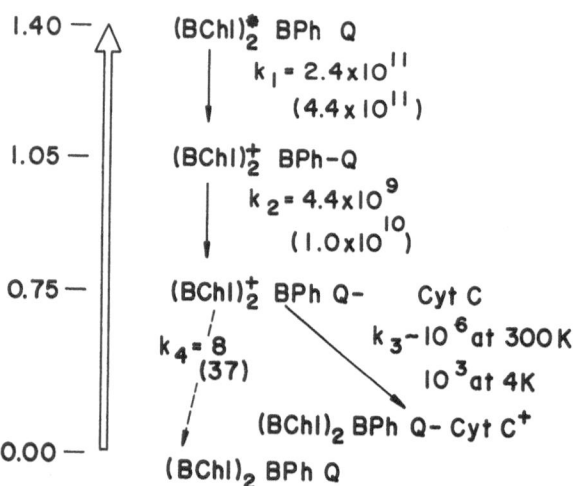

Figure 1. Schematic representation of the primary electron transfer processes in the photosynthetic bacteria. Rates are given in s^{-1}. Rates were measured at room temperature and rates in parentheses were determined below 100K [13,15,16].

Theoretical ideas regarding the interpretation of these observations can be a bit confusing. To begin, it now seems completely clear that electron transfer across biologically realistic distances involves tunneling of the transferred electron [13,17,18]. This means that one must think about the electronic dynamics in terms of quantum mechanics, but this is hardly surprising; it is of course the quantum dynamics of electrons which determine the rules of chemical bonding, so presumably even Pauling would concede the relevance of quantum mechanics at this level. But electron transfer is accompanied by atomic rearrangements at the donor/acceptor sites and in the surrounding protein or solvent — are the dynamics of these rearrangements classical or quantum? This question, whose answer is far from clear in looking at the current literature, is the focus of this paper. We emphasize three central issues:

(a) One must be very careful to define what one means by "behaving classically" or "quantum effects are important." It is usually argued that quantum mechanics gives way to the classical limit as Planck's constant $\hbar$ becomes small. But even in the simplest of models for photosynthetic electron transfer we shall see that there are several different energy and time scales which govern the reactive dynamics, so there are many quantities to which $\hbar$ can be compared. We present mathematical methods which allow us to identify these different scales and identify these mathematical constructs with several different physical senses in which the reactive event may be termed "classical" or "quantum." We hope to make this more complex notion of classical *vs.* quantum dynamics both mathematically rigorous and intuitively appealing.

(b) Simple models for photosynthetic electron transfer have a very broad range of parameters in which a "classical" approximation fails miserably. This mathematical failure can be traced physically to a breakdown of the notion that the reaction occurs at a unique "transition state." Specific results on models in this "strongly quantum" regime have been presented by several groups, and indeed models in this regime have been used to analyze recent experiments on electron transfer in photosynthesis and synthetic model compounds [15,16,19]. Our contribution is the demonstration that this breakdown of the classical theory is to a large extent generic, independent of many (unknown) details in the reactive dynamics.

(c) We provide detailed estimates of the key parameters controlling the crossover between classical and quantum behavior, making plausible the hypothesis that the electron transfer reactions in the bacterial reaction center operate in the strongly quantum regime *at room temperature*. Correspondingly we suggest that many of the generic experimental signatures of this regime have correlates in recent observations on the reaction center. This comparison of theory and experiment provides a strong case for believing that biologically significant features of photosynthetic electron transfer are indeed understandable only in terms of quantum mechanics.

We should like to emphasize at the outset that our goal in this paper is very different from that in most previous theoretical discussions of photosynthetic electon transfer. Specifically, we are *not* attempting to construct a model which accounts for some particular observations. On the contrary, we are trying to argue that a broad class of models can be qualitatively understood, and that in particular it is possible to understand the mechanism of quantum/classical crossover without discussing many details of a particular model. Once we can motivate the hypothesis that the photosynthetic reactions operate in the quantum regime of these models, a number of predictons follow generically — rather than trying to fit data we are claiming that certain features of the data are inevitable consequences of the theory.

2. A Simple Model for Electron Transfer

We now try to construct a model within which we can address all of the issues raised in the previous section. At the outset it is crucial to understand what we do not intend to do: (1) We do not intend to argue that particular models are necessary to account for certain data. As noted above we hope eventually to argue just the opposite — a broad class of models generically predicts certain qualitative features of the relevant experiments. (2) We do not intend to compare "the classical model" for electron transfer with "the quantum model" for electron transfer. On the contrary we want to study a family of models in which the reactive dynamics become progressively more classical or more quantum as some fundamental molecular parameters are varied. (3) We do not intend to advocate a radically new family of models; in fact we shall arrive at a model which is quite familiar and well studied. The point of the exercise is to understand what we have assumed (or not assumed) just in writing down the model before we make any approximations (such as classical mechanics. A more complete discussion of these ideas is given elsewhere [20].

To start it is necessary to include at least two electronic states, corresponding to the electron being localized on the donor or acceptor sites. In the simplest case *only* two states will be important. For this to be a good approximation we must have a system where the energy difference between donor and acceptor states is much smaller than the other electronic excitation energies — we must make sure that all other localized states are inaccessible for the transferring electron. This is clear for the molecules that are closed shell in the "oxidized" state where the core electrons act as an effective potential for the tunneling electron; note that if we represent the reaction by $D + A \rightarrow D^+ + A^-$, the oxidized states are D^+ and A. This closed shell picture holds true for acceptor molecules such as the quinones. It may not be valid for excited state donors such as the chlorophyll dimer. In this case the "oxidized" state is open shell and more than one reduced state may have to be included. It will turn out that the addition of these extra electronic states does not affect the main conclusions of subsequent sections, as will be discussed elsewhere.

Once we have isolated two electronic states we can write the Hamiltonian for the system in terms of a spin one-half, identifying spin up with the reactant state (electron localized on donor) and spin down with the product (electron on acceptor). The most general Hamiltonian is then

$$\mathbf{H} = \frac{1}{2}\hat{\epsilon}\sigma_z + \hat{\Delta}_{\mathrm{el}}\sigma_x + \mathbf{H}_{\mathrm{vib}} \tag{2.1},$$

where $\hat{\epsilon}$ is the instantaneous energy difference between reactant and product states and $\hat{\Delta}_{\mathrm{el}}$ is the matrix element for tunneling between the donor and acceptor sites. We have written these quantities with hats to emphasize that they may in principle depend on the trajectories of all the atoms in the system, which are in turn governed by the vibrational Hamiltonian $\mathbf{H}_{\mathrm{vib}}$. We would like to say that ϵ and Δ_{el} depend only on the positions of the atoms, which is a version of the Born-Oppenheimer approximation. For such approximations to be valid, two conditions should be satisfied. The first and trivial condition is that vibrational excitations should not be to able access any other electronic state. This is a reasonable assumption since excitations of valence electrons go from near infrared to ultraviolet, 1–4 eV. The thermodynamic driving force for most chemical reactions are smaller than 1 eV, and the highest energy vibrational modes are in the range of 0.2–0.35 eV (C=O or C=C, for example).

The second condition for the validity of the Born-Oppenheimer approximation in this context is that when the electron tunnels from donor to acceptor, the "time" spent in the forbidden region ("traversal" time) should be much shorter than any vibrational frequency

[20,21,22]. Modes that are much faster than the "traversal" time would adiabatically follow the electron and simply renormalize the electronic matrix element in the two level system; these very fast modes should simply be excluded from the vibrational dynamics in the same sense that we exclude highly excited electronic states. For typical distances in biolgical electron transfer this cutoff would be at $\sim 10,000\,cm^{-1}$, well above any modes of interest.

In defining our model we will make a further approximation, and this is the "one-electron" approximation. Rigorously speaking all we need are two states separated from all other states, but it is very convenient to be able to talk about the transferred electron. This is possible only if the motion of this electron is not strongly correlated with that of other electrons on the donor and acceptor sites. The one electron approximation is widely applied in the computation of tunneling matrix elements, and it is also very useful when discussing questions of time scale. Qualitatively this approximation is justified by the fact that the distance tunneled by the electron between donor and acceptor is much larger than any typical distance for the "bonding" electrons. Another way of putting it is that a one electron wave function is a good approximation "far" from the nuclei — electronic correlation with the core electron strongly affects the orbital "effective" energies, but the hopping matrix element between sites is basically a one electron matrix element [21,23]; the "local" correlation just serves to renormalize the site energy. The ideas behind this qualitative argument have been borne out in careful quantum chemistry calculations by Logan and Newton [24]. More generally, if the matrix element coupling donor and acceptor states is dominated by medium orbitals (through bond mechanism [23]), the approximations descrbed above are the same with the only difference that now every medium orbital should be included in the effective potential.

The fact that we have a two level system does not mean that we can calculate a well defined rate for electron transfer. Indeed, if the paramaters of the two level system (ϵ and Δ_{el}) are fixed for all time, the electron will just oscillate back and forth between the donor and acceptor sites. Fluctuations in ϵ or Δ_{el} are required to destroy this coherence and make the transfer irreversible. It is this dynamics which determines the activation energy for the reaction, the adiabaticity or non-adiabaticity of the transfer, and so on (see refs. 13, 25, 26, and 27, for example). We will assume that the atomic motions which give rise to fluctuations in ϵ and Δ_{el} can be described as superpositions of harmonic oscillator motions at different frequencies, that is in terms of vibrational modes with some spectrum.

Clearly there is a broad range for such modes, going all the way from fast local interatomic vibrations like $C=O$ down to protein breathing. In fact we want to distinguish explicitly between local, high frequency motions and the more global low frequency modes. Since the high frequency modes are strongly localized, motions along these modes couple to electron transfer only if the transferred electron spends significant time on or near the bonds whose vibrations participate in the mode. This allows us to make some semi-quantitative estimates of the couplings, as discussed below. Because of this localization, the high frequency modes which couple to the transfer reaction are typically discrete and well resolved from any continuum background; one may treat them as if they were perfect normal modes of the molecule, with (approximately) no damping. These local vibrational motions are not exactly harmonic, but on the energy scales of relevance to electron transfer one never stretches an individual bond enough to see these anharmonicities.

Besides the localized high frequency modes, there are also modes describing the dynamics of all the other degrees of freedom in the protein and solvent environment. For simplicity we consider these low frequency motions as a continuum of of harmonic oscillators linearly coupled to our two level system. There are several justifications for such a major simplification:

(1) The harmonic approximation is much less stringent than one might imagine. The bath of harmonic oscillators that we consider here is *not* equivalent to the set of normal

modes obtained by expanding the full potential to second order in the displacements about their equilibrium coordinates. Rather they represent a model for excitations in protein–solvent continuum [26,28,29]. The harmonic approximation is the assumption that these elementary excitations do not interact [30].

(2) One can think of the low frequency modes as the 'dielectric' response of the protein to the charge transfer. The harmonic approximation is equivalent to the assumtpion that this dielectric response is linear — the electric field is proportional to the polarization. Certainly we know that macroscopic dielectric responses can be linear even though the microscopic degrees of freedom are far from harmonic or weakly interacting. We are actually trying to do concrete demonstrations of this point in semi-realistic models for protein dynamics, but this will be discussed elsewhere; see also ref. 20.

(3) It will turn out that many of our crucial conclusions are not sensitive to the details of the low frequency dynamics.

We are now prepared to write our model Hamiltonian. We assume for simplicity that the tunneling matrix element Δ_{el} is constant (the Condon approximation), and we think about the coupling between the oscillators and the electronic state by assuming that the transfer of the electron applies a force to each mode, shifting its equilibrium displacement. Then we have

$$\mathbf{H} = \Delta_{el}\sigma_x + \frac{E}{2}\sigma_z + \sum_h \left\{ \frac{p_h^2}{2m_h} + \frac{1}{2}m_h\Omega_h^2 \left(Q_h + \frac{\delta Q_h\sigma_z}{2} \right)^2 \right\} +$$

$$+ \sum_l \left\{ \frac{p_l^2}{2m_l} + \frac{1}{2}m_l\Omega_l^2 \left(Q_l + \frac{\delta Q_l\sigma_z}{2} \right)^2 \right\} , \qquad (2.2)$$

where E is the energy gap or reaction driving force, and δQ_i is the separation of the equilibrium positions for the ith mode when electron is in the donor or in the acceptor. The reorganization energy for the ith mode is then defined as $\lambda_i = 1/2\ m_i\Omega_i^2\delta Q_i^2$. The notation h and l stands for the high and low frequency modes. A more detailed discussion of the applicability of this class of models is given in ref. 20.

3. Methods for Non–Adiabatic Rate Calculation

Having defined a model that includes a two level system (donor and acceptor states) coupled to high and low frequency modes the problem now is to compute reaction rates. To simplify the discussion we rewrite Eq. 2.2 as

$$\mathbf{H} = \Delta_{el}\sigma_x + \frac{1}{2}\epsilon(Q)\sigma_z + \mathbf{H}_Q \qquad (3.1)$$

where $\epsilon(Q)$ is the instantaneous energy gap between donor and acceptor states. Because the biological electron transfers of interest here are in the weak electronic coupling (small Δ_{el}) limit we confine ourselves to the non–adiabatic regime. Further discussion about validity of the non–adiabatic approximation can be found, for example, in refs. 25, 28, 32, and 33. Since the electronic matrix element is considered constant, in this limit the only important dynamics are fluctuations in $\epsilon(Q)$ [20,31]. Since we described ϵ in terms of a collection of modes, we must understand how the rate depends on the spectrum of these modes and on the strength of coupling to these modes.

To simplify the discussion still further we will assume that the entire collection of low frequency modes can be described in terms of a single mode a frequency Ω_L with a frequency-dependent damping constant $\gamma(\omega)$. This model actually describes a continuum of modes, since the damping γ corresponds to coupling between the "mode" at Ω_L and a continuum "heat bath" [22]. For $\gamma \ll \Omega_L$ one has a well-resolved resonance, while for $\gamma \gg \Omega_L$ this resonance dissolves and one should think of the dynamics as overdamped or diffusive. Our discussion here focuses on the underdamped case, although we shall see that there is a broad regime in which the precise value of the damping constant is irrelevant; for a discussion of the strongly damped case see [28]. Note that describing the system in terms of a single damped mode is just a convenience; by choosing different functions $\gamma(\omega)$ we can synthesize almost any spectrum we choose, so that we are not really sacrificing generality. The ability to trade the spectral density for a description in terms of damped modes is discussed, for example, in ref. 28.

As discussed above, the high frequency modes are at least approximately undamped. To simplify once more we imagine adding just one high frequency mode to the low-frequency continuum. We start our discussion by ignoring this mode, solving the low-frequency problem, and then examining the effects of the high-frequency mode. This relatively simple calculation captures most of the essential physics in the problem.

If we have only one low frequency mode, Ω_L, the rate can be written as an integral over a time dependent correlation function [31]

$$
k_L \sim \frac{\Delta_{\text{el}}^2}{\hbar^2} \int d\tau \; \exp \left\{ \frac{iE\tau}{\hbar} - i\frac{(\delta Q_L)^2}{\hbar} \int \frac{d\omega}{2\pi} \sin(\omega\tau) \frac{\gamma(\omega)}{\omega} \times \frac{\Omega_L^4}{|-\omega^2 - i\omega\gamma(\omega) + \Omega_L^2|^2} \right.
$$

$$
\left. - i\frac{(\delta Q_L)^2}{\hbar} \int \frac{d\omega}{2\pi} [1 - \cos(\omega\tau)] \frac{\gamma(\omega)}{\omega} \times \frac{\Omega_L^4 \coth(\hbar\omega/2k_BT)}{|-\omega^2 - i\omega\gamma(\omega) + \Omega_L^2|^2} \right\} , \qquad (3.2)
$$

where $\gamma(\omega)$ is the damping constant. If γ is constant we are in the "ohmic" limit. The integral we must do is of the general form $\int d\tau \exp[S(E,\tau)/\hbar]$, by which we mean that as $\hbar \to 0$, S approaches some non-zero function which depends on the classical energies E, λ_L, k_BT. This suggests that we apply the saddle point, or steepest descent method. For underdamped modes there are multiple saddle points, and one must sum their contributions. We obtain [31]

$$
k = A \exp[-G] \sum_{n=0}^{\infty} \cos(2\pi n E/\hbar\Omega_L) \; \exp[-n\gamma/\Gamma] \qquad (3.3a)
$$

where

$$
\Gamma^{-1}(T = 0) = \frac{2\pi E}{\hbar\Omega_L^2}. \qquad (3.3b)
$$

$$
\Gamma^{-1}(T \gg \hbar\Omega_L/k_B) \sim \frac{2\pi E}{\hbar\Omega_L^2} \cdot \frac{k_BT}{\hbar\Omega_L} \cdot \left[1 + \mathcal{O}\left(\frac{\Delta E - S_L\hbar\Omega_L}{S_L\hbar\Omega_L} \right)^2 \right], \qquad (3.3c)
$$

and

$$
A \sim \frac{2\sqrt{\pi}\Delta_{\text{el}}^2}{\hbar} \left\{ 2S_L(\hbar\Omega_L)^2(2\bar{n}_L + 1) + \cdots \right\}^{-1/2} , \qquad (3.3d)
$$

$$
G \sim \frac{(\Delta E - S_L\hbar\Omega_L)^2}{2S_L(\hbar\Omega_L)^2(2\bar{n}_L + 1)} - \frac{(\Delta E - S_L\hbar\Omega_L)^3}{6S_L(\hbar\Omega_L)^3(2\bar{n}_L + 1)^5}[3 + (2\bar{n}_L + 1)^2] + \cdots, \qquad (3.3e)
$$

where $\bar{n}_L = (e^{\hbar\Omega_L/k_BT} - 1)^{-1}$ and $\lambda_L = S_L\hbar\Omega_L$. We have expanded G around the term which dominates when $\hbar \to 0$ to emphasize that we are looking for quantum corrections to an the classical or semi-classical result: more general expressions are available.

Eq. 3.3 has everything we need: we just have to take it apart. Suppose that $\hbar \to 0$. Then all the $n \neq 0$ terms are forced to zero, since $e^{-n\gamma/\Gamma} \sim \exp[-\text{stuff}/\hbar]$, where stuff is non-zero in the classical limit. In addition, $G \to E_{act}/k_B T$, where the activation energy is just $E_{act} = (E - \lambda_L)^2/4\lambda_L$. Clearly this is just the non-adiabatic version of Marcus theory [13,34], as we would hope.

We can derive non-adiabatic Marcus theory much more directly by thinking about the motion along Q_L as the motion of a classical particle, and using the Landau-Zener picture of curve-crossing to compute the probability of a reaction each time the particle passes near the transition state defined by having the instantenous energy gap equal to zero. Using this reaction probability as a weighting factor one computes the thermally-averaged flux in the spirit of transition state theory. The result is precisely $k = A e^{-E_{act}/k_B T}$ with A and E_{act} as above. Physically what this means is that, in the non-adiabatic limit, the rate is proportional to the probability of being at the transition state independent of the dynamics around the transition state. We emphasize that this is a classical ($\hbar \to 0$) result.

We can make this argument still more precise. The non-adiabatic rate can be written formally in terms of a time-ordered expectation value,

$$k = \frac{\Delta_{el}^2}{\hbar^2} \int dt \langle \mathbf{T} \exp\left[-i \int_0^t d\tau \hat{\epsilon}(\tau)/\hbar\right]\rangle, \qquad (3.4)$$

where $\hat{\epsilon}$ is the Heisenberg operator corresponding to the instantaneous energy gap and $\langle ... \rangle$ denotes both the quantum expectation value and the thermal average. It is then easy to show that as $\hbar \to 0$,

$$k \to \Delta_{el}^2 \hbar^{-1} \langle \delta(\hat{\epsilon}) \rangle, \qquad (3.5)$$

i.e. the rate is proportional to the probability of being on the crossing surface $\hat{\epsilon} = 0$. If one evaluates this probability in quantum statistical mechanics rather than classical statistical mechanics one finds that the Boltzmann factor of Marcus is replaced by the result of Hopfield [13,27], $k = A e^{-E_{act}/k_B T_{\text{eff}}}$, where $k_B T_{\text{eff}} = (\hbar\Omega_L/2)\coth(\hbar\Omega_L/2k_B T)$. Of course the Hopfield result is identical to that of Marcus if $k_B T >> \hbar\Omega_L$.

It is clear, then, that the Hopfield-Marcus theory can be obtained as an $\hbar \to 0$ limit of an exact quantum theory for electron transfer rates. This is a "semi-classical" limit of some sort. Since we can derive this result by looking at the passage of a classical particle through the crossing surface, the result is classical in the sense that trajectories in real time are well defined. Of course in quantum mechanics we have no such notion of trajectory, and it can be recovered only approximately as $\hbar \to 0$.

The picture of classical trajectories as dominant does not mean that we are neglecting quantum effects such as tunneling. On the contrary, we know that the Hopfield picture corresponds to tunneling of the reaction coordinate from its ground vibrational state if one extrapolates to zero temperature. This tunneling, however, just means that the quantum system has some non-zero probability to be at a classically forbidden point, namely the crossing. The dynamics of the traverse through the crossing — that is, the dynamics of the reactive event itself — are assumed to be classical. This is our first indication that the distinction between quantum and classical behavior can be subtle. In particular, a reaction can be classical in the sense that Hopfield's result is correct, but quantum in the sense that the Marcus result is not correct. Corrections to the Hopfield/Marcus theory arise from the deeply quantum mechanical fact that particles do not have well-defined trajectories. We hope it is clear that giving up on a mental image of trajectories is "more quantum mechanical" than simply admitting that these trajectories continue into classically forbidden regions (tunneling).

Eq. 3.3 allows us to explore quantitatively the corrections to the semi-classical limit. Intuitively we expect that the notion of well defined trajectories through the crossing region will break down because of the uncertainty principle. The crossing is defined by demanding that the energies of the intial and final states be equal at a single value of the coordinate. But fixing the coordinate implies a large uncertainty in momentum and hence kinetic energy, so the uncertainty principle forbids us from simultaneously specifying energy and position. Another way of saying this is that a wavepacket initially localized at the crossing will spread during the time it takes the molecule to "decide" if it will react; this spreading blurs the transition state and allows the reaction to occur at lower energies. We emphasize once more that this quantum effect (spreading of the wavepacket) is in addition to the effect of tunneling. Blurring of the transition state through the uncertainty principle is captured by keeping just the $n = 0$ term of Eq. 3.3 , which is the contribution from the saddle point closest to $\tau = 0$.

A different quantum effect is related to phenomenon of interference, which is normally manifest simply as the quantization of energy levels. If an electronic transition with energy gap E is coupled to a single vibrational mode with frequency Ω_L, we expect that the rate will be maximal only if E can be evenly divided into an integer number of phonons, $E \sim N\hbar\Omega_L$. As discussed in ref. 25 one can look at these quantum resonances as arising from constructive interference of transition amplitudes along paths which make multiple passes through the crossing region. Intuitively we expect that the resonances are blurred by damping, but sometimes this is unclear in the literature. If one ignores damping altogether it is clear both mathematically and physically that there is no well defined rate; formally if one computes the rate in perturbation theory one finds a discrete sum of delta functions, which is meaningless. Jortner [35], for example, found himself with this problem but did not give an adequate resolution — in the end his rate expression depends on an arbitrary energy scale which is introduced to smooth the singularities of the rate $vs.$ driving force.

Eq. 3.3 provides a nice picture of how damping destroys the quantum resonances even in a single mode model. If we set $\gamma = 0$ the sum of cosines becomes a series of delta functions, as it must. As γ is increased these singular terms are smoothed into an approximately sinusoidal ripple in a plot of rate $vs.$ driving force, and the amplitude of these ripples declines exponentially as either γ increases or $\hbar$ decreases. One crucial result is that, for reasonable parameters, the ripple is essentially obliterated even when the mode is underdamped — the resonances disappear when $\gamma \gg \Gamma$, which is a much less stringent condition than $\gamma \gg \Omega_L$. In fact as long as $\Omega_L > \gamma > \Gamma$, the $only$ effect of the damping is to destroy quantum resonances, leaving the contribution from one saddle point essentially equal to its value at zero damping.

To summarize, we have identified at least three different senses in which a reaction can be classical or quantum, and we have connected these different physical effects with the mathematical structure of our saddle point calculation:

1. Quantum systems display resonances when the driving force for the reaction can be broken into an integer number of vibrational quanta. This effect is washed out by damping, which restores a smooth dependence of rate on driving force, as expected classically. This aspect of the quantum/classical crossover is captured by examing the contribution of secondary saddle points.

2. In quantum systems the reaction does not occur at a well-defined "curve crossing," since this is forbidden by the uncertainty principle, while in the classical limit one can identify the non-adiabatic rate rigorously with the probability for sitting at the crossing point. This uncertainty-induced blurring is controlled by the dominant saddle point; in particular the position of the saddle point in the complex τ plane determines the importance of wavepacket spreading.

3. Finally, rates are dominated by tunneling if vibrational quanta are much larger than the thermal energy, while activation over the barrier dominates if vibrational quanta

are small. As in previous discussions [27,35] this can be seen from the explicit factors of $\bar{n}$ in the rate expressions.

We suspect that most biological electron transfers are classical in the first sense (no resonances), although this is not crucial for what follows. Hopfield's [27] suggestion that biological transfers can be quantum in third sense, at least at low temperatures, is of course what sparked interest in this class of theories. Our contribution is to show that physically reasonable couplings to high frequency modes makes all of the primary photosynthetic transfers quantum in the second sense, at least at large driving forces.

To explore the effects of high frequency modes we note that since the high frequency modes of interest are well-resolved from the continuum and have quanta much larger than $k_B T$, the rate can be computed by assuming that we have solved the low frequency problem $k_L(E)$, and then include the possibility of emitting n high frequency phonons with the probability $e^{-S_H} S_H^n / n!$; for a discussion of this idea see refs. 36 and 37. This leads to

$$k(E) \sim \sum_{n=0}^{\infty} e^{-S_H} \frac{S_H^n}{n!} k_L(E - n\hbar\Omega_H). \qquad (3.6a)$$

It is interesting to rewrite this as

$$k(E) \sim \sum_{n} k_L[E^{\text{eff}}(n); \Delta_{\text{el}}^{\text{eff}}(n)], \qquad (3.6b)$$

where we have introduced renormalized energy gaps and matrix elements.

$$\Delta_{\text{el}}^{\text{eff}}(n) = \Delta_{\text{el}} \sqrt{\frac{\exp(-S_H) S_H^n}{n!}} \quad \text{and} \quad E^{\text{eff}}(n) = E - n\hbar\Omega_H. \qquad (3.6c)$$

Fig. 2 shows a plot of Eq. 3.6 for typical parameters. In this figure the low frequency modes are assumed to be totally classical, but from the discussion above we need not restrict ourselves in this way. Parameter values will be carefully discussed in the next section. It is interesting to notice that even in the case of $S_H < 1$ where emission of high frequency phonons is unfavorable, the strong dependence of $k_L(E)$ on the energy gap compensates, so that extremely small couplings can have (exponentially) larger effects on the rate.

From Eq. 3.6 and Fig. 2 we find that there are two distinct regions for the dependence of the rate on energy gap and temperature. At small energy gaps the rate is extremely dependent on temperature and E, while at high energy gaps the high frequency modes wipe out these variations. From Fig. 2 we notice that this qualitative behavior obtains even at weak coupling, $S_H \ll 1$ — rates are strongly temperature and driving force dependent for $E < \lambda_L$ and weakly temperature and driving force dependent otherwise.

If we now calculate the rate in our two mode model by the analog of Eq's. 3.2 and 3.3 and keep only the dominant saddle point we obtain essentially the result shown in Fig. 2 but without the quantum resonances. This shows the strength of the saddle point methods, since this is exactly what we expect if instead of one we include a few high modes of incommensurate frequencies. Thus our approximate treatment of a model with one high frequency mode approaches the exact answer for a multi-mode problem. In particular, the main results regarding temperature and energy gap dependence are captured by this simple calculation. If we do a truly semi-classical calculation, however, computing the probability that the system is one the crossing surface (a line in the two-mode problem), we grossly underestimate the rate at large energy gaps if S_H is not much larger than 1. In addition,

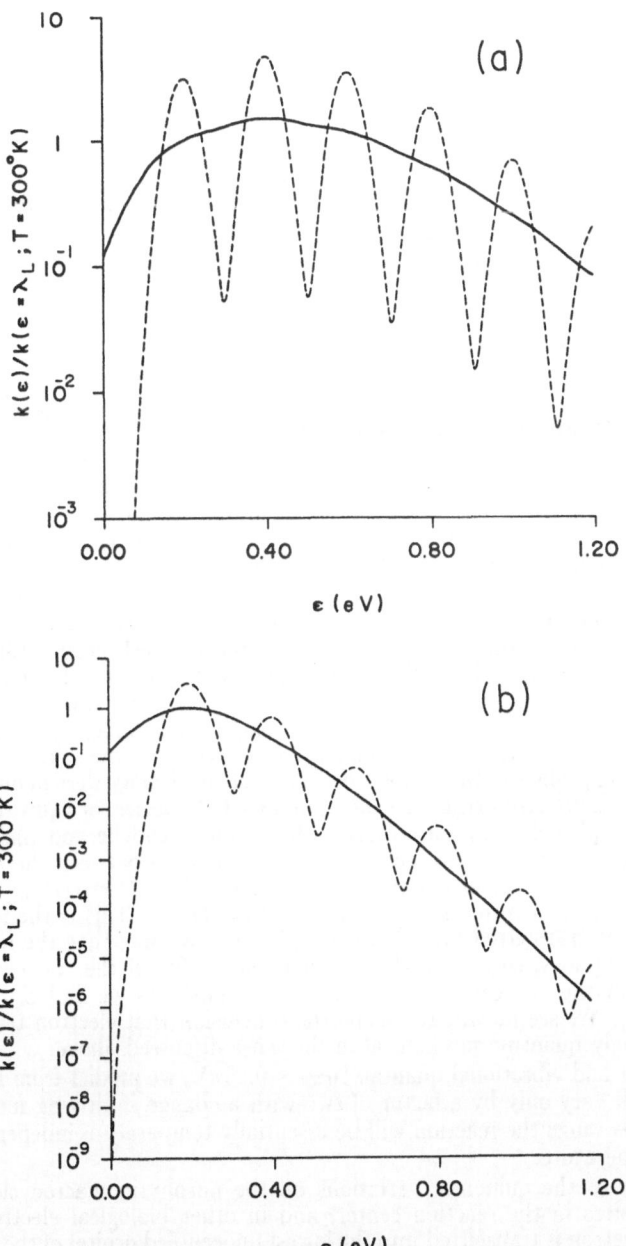

Figure 2. Plot of rate vs. driving force for the case of one high frequency and one low frequency mode. The low frequency mode is assumed to be totally classical. $\lambda_L = 0.2\text{eV}$ and $\hbar\Omega_H = 0.2\text{eV}$. The solid lines show the rate for $T = 300\text{K}$ and the dashed ones for $T = 30\text{K}$. The plots (a) consider $S_H = 1.5$ and the plots (b) consider $S_H = 0.2$.

the semi-classical calculation predicts the strong energy gap dependence of the rate will continue at large energy gaps, in contrast to the correct quantum result.

The comparison of the exact rate, the single saddle point rate, and the semi-classical rate leads us to our central conclusion:

> *Small to moderate coupling of an electron transfer reaction to localized, high frequency vibrational modes wipes out the temperature and energy gap dependence at large energy gaps that one expects from a classical analysis. This experimentally observable behavior is a signature of a true quantum effect on the reactive dynamics, namely the uncertainty-induced blurring of the reactive trajectories. These reactions are thus quantum in the second sense defined above.*

4. Comparison with experiment

Are the arguments of the previous section relevant to biological electron transfer? The key issue is clearly the strength of coupling to high frequency vibrational modes. If the typical couplings to C=O or C-C stretching modes are even of order 0.1, then we can see from Fig. 2 and Eq's. 3.2, 3.3 and 3.6 that quantum effects are significant at room temperature, at least for large energy gaps. We believe that there is sufficient data from model compunds to motivate semi-quantitative estimates of the relevant coupling constants; here we outline the arguments, saving the details for a longer paper [38].

Two of the electron donor/acceptor sites in the bacterial reaction center are quinones, and the transfers into and out of these sites are well studied (cf. Fig. 1). There also exist synthetic model systems, such as that studied by the CalTech group [19], which involve transfer between quinone and porphyrin moieties. EPR studies of semiquinone radical anions [39] suggest that the extra electron spends $\sim 40\%$ of its time in the quinone ring (therefore 60% of its time in the two carbonyl bonds); more precisely $\sim 40\%$ of the spin density is delocalized onto the quinone ring, but we expect that the spin and charge density have similar spatial profiles in these molecules. If all of the density were in one C=O bond, it would mean that the electron transfer generated a C=O$^-$ species, or equivalently converted a carbonyl into a phenol. Crystal structure data from carboylic and phenolic molecules allows us to estimate rather accurately the structural change between these two electronic states, and knowing the typical vibrational frequencies of C=O stretching motion we can convert this displacement into a force on the bond [40,41,42]. This is the force applied by one electron if it is transferred into the bond. We then assume that the force scales with the fractional occupancy, so that in the reaction center 30% of the "one-electron" force is applied to each of the two carbonyls. From these arguments we find $S_{CO} = 1$ within a factor of two [43]. We see no way to escape the conclusion that electron transfer involving quinones is strongly quantum mechanical in the sense discussed above.

With $S \sim 1$ and vibrational quanta $\hbar\Omega_{CO} \sim 0.25\,\mathrm{eV}$, we predict from Eq. 3.6 that the rate constant will vary only by a factor of two with a change in driving force of $\sim 0.5\,\mathrm{eV}$, and that over this range the reaction will be essentially temperature independent until well above room temperature.

In addition to the quinones, variations on the porphyrin macrocycle also serve as donor/acceptor sites in the reaction center, and in other biological electron transfres as well. Here the electron is transferred into the lowest unoccupied orbital of the π−conjugation system, which means that it applies forces to the C–C and C=C bonds of the ring. Again we can imagine the case in which all of the electron were placed in one bond, which converts (for example) a double bond to a single bond since the extra electron goes into an anti-bonding orbital. The coupling to the C=C stretching mode in this extreme case would be

$S_1 \sim 4$. This is a linear extrapolation but it is a good number to work with because we never actually have the entire charge in one bond.

What happens when the electron is distributed over N bonds which participate in the conjugation? If all of the bonds are equivalent — which corresponds to a "metallic" or pure molecular orbital view of the conjugation — then the C–C stretches become a band of acoustic phonons and the added electron applies all its force to the lowest acoustic mode, expanding or contracting the ring uniformly. If, on the other hand, one can identify alternating short and long bonds around the ring, the extra electron applies a force to the nearly dispersionless band of optical modes which involve relative motion of the short and long bonds; in this case there is a gap between two bands of electronic states. These pictures give rise to very different predictions for the scaling of the total high-frequency reorganization energy with the size of the ring, and it is a difficult problem to decide which is correct for the biologically interesting examples. We favor the optical mode picture because we can think of at least two seemingly different scenarios in which it is correct, each of which has some experimental support from studies of model systems:

i. The electrons in the conjugation system are highly correlated and the correct description of the electronic states is in terms of resonance among valence bond configurations. In this case each valence bond configuration has the bond-alternating structure, and the description in terms of optical modes is approximately valid.

ii. The molecular orbital picture is approximately correct but the metallic state is unstable to a Peierls distortion, as in the Su-Schrieffer-Heeger (SSH) model for polyacetylene [44].

If the optical mode picture is correct we can think of the force applied by the extra electron as being distributed uniformly among the N bonds, which leads to $S_N \sim 4/N$ if we assume the band gap is roughly N-independent. In the SSH model one can explictly calculate the dependence of the reorganization energy on the size of the ring, and one finds that this prediction is correct for intermediate size rings; the decline in coupling saturates at large N, leaving some finite reorganization energy in this limit. For example, in the case of porphyrin where $N = 16$, $S \sim 0.25$. If the transfer is between two such molecules, then because there is coupling to the both donor and acceptor this number should be doubled, $S \sim 0.5$. Also, even if the bond-alternation picture is not fully correct, so that we overestimate the coupling, we recall that $S \sim 0.1$ is enough for quantum effects to dominate the large energy gap regime (see Fig. 2b).

Before discussing experiments we present some evidence of why we believe most of these reactions are non-adiabatic, i.e., that we can calculate rates perturbatively in the matrix element. Since the distances that the electron tunnels is of the order of 10 Å, the matrix elements of interest are in the range of $10^{-4} - 10^{-6}$ eV, and we strongly believe that such matrix elements are sufficiently small to justify perturbation theory (see refs. 25, 27, 28, and 31, for example). Also, recent results for electron transfer in chemically modified myoglobin, showing an exponential distance dependence of the rate as theoretically predicted, provides clear evidence for non-adiabaticity [18].

We want to make clear, however, that the rate does not have to be non-adiabatic for our conclusions to be valid. Actually our arguments are even stronger in the adiabatic limit. Since we can think of the high-frequency modes as just renormalizing the matrix element and energy gap [cf. Eq. 3.6], it is clear that in the adiabatic limit — where the rate is independent of the matrix element — all that counts is the energy gap renormalization, and even the gradual fall off at large energy gaps will disappear [36].

To summarize, we expect that all the electron transfer reactions in the bacterial reaction center will be dominated by quantum effects at large energy gaps, since the expected couplings to high frequency modes of chlorophylls, pheophytins, and quinones are sufficient to push the system into the strongly quantum regime which we have identified in the discussion above. This means that we should observe and approximate temperature and driving force independence of the transfer rates. In fact, as noted in Fig. 1, most of these rates, both forward and reverse, are weakly only temperature dependent. The old idea of temperature independence due to activationless reactions [13,45] does not sound plausible for so many rates. To believe that nature has optimized forward rates to optimize charge separation (k_1 and k_2 are weakly temperature dependent) may be fine, but to do the same with reverse rates (k_4 is weakly temperature dependent) looks totally improbable. We have shown that such behavior is a generic consequence of the fact that these reactions are dominated by quantum effects.

To understand this temperature independence in several rates, Gunner and Dutton performed a series of experiments where they changed E for the rates k_4 and k_2 of Fig. 1 [15,16]. For large E they observe exactly what we see in Fig. 2, not only weak temperature dependence but also weak energy gap dependence. For the small E regime they observe the expected strong energy gap dependence but the temperature dependence still needs some understanding; specifically the prediction in this regime depends on the spectrum of the low frequency modes. Although further work is necessary the available data is enough to confirm that quantum effects associated with the high frequency modes are responsible for the weak temperature dependence of the rate. The idea of activationless reactions can not explain the weak dependence of the rate on the energy gap.

The photosynthetic rate k_4 exhibits very similar temperature and energy gap behavior as found for the CalTech model system [19]. These two systems probably have completely different low frequecy modes but both of them have strong coupling to the CO in the quinone. This leaves us with a nice picture: Chemical and biological systems show similar behavior for electron transfer rates when we study temperature and driving force dependence because high frequency local modes dominate even though the low frequency dynamics are presumably very different. The differences between the absolute rates in these two systems is due to different tunneling matrix elements, and indeed scaling of the rates with distance between two systems is in reasonable agreement with the studies performed by the Gray group for distance dependence of the rate [18].

Taken together, these observations strongly suggest that Gunner and Dutton have observed a clear signature of quantum effects in a room temperature biochemical reaction. We believe that several other observations on the reaction center may also be generic consequences of the fact that photosynthetic electron transfer operates in a strongly quantum regime, as will be discussed elsewhere [38].

Acknowledgements

We thank M. Gunner and P.L. Dutton for helpful discussions. This work was supported by the Brazilian Agencies CNPq and FINEP, and by the National Science Foundation through a Presidential Young Investigator Award, supplemented by funds from Sun Microsystems and Cray Research. Part of this work was done during a visit of W.B. to Brazil, sponsored by CNPq. Additional support for the early stages of this work came from the Miller Institute for Basic Research in Science and from the USPHS through a Biomedical Research Support Grant.

References

1. Pauling, L. and Delbrück, M. (1940) Science 92, 77.
2. Pauling, L. (1948) Nature 161, 707.
3. McCammon, J.A., and Harvey, S.C. (1987) 'Dynamics of Proteins and Nucleic Acids', Cambridge Press, New York.
4. Warshel, A., Sussman, F., and Hwang. J.-K. (1988) J. Mol. Biol. 201, 139.
5. Bialek, W. (1983) Ph.D. Thesis, University of California at Berkeley.
6. Kivelson, S., Kallin, C., Arovas, D.P., and Schrieffer, J.R. (1987) Phys. Rev. B 36, 1620.
7. Miller, W.H. (1986) Science 233, 171.
8. Chakravarty, S., and Schmid, A. (1986) Physics Reports 140, 193.
9. Leggett, A.J., Chakravarty, S., Dorsey, A., Fisher, M.P.A., Garg, A., and Zwerger, W. (1987) Revs. Mod. Phys. 59, 1.
10. Imry, Y. (1986) in 'Directions in Condensed Matter Physics', G. Grinstein and G. Masenko (eds.), World Scientific, Singapore.
11. Webb, R.A., and Washburn, S. (1988) Physics Today 41(12), 46.
12. Martinis, J.M., Devoret, M.H., and Clarke,J. (1985) Phys. Rev. Lett. 55, 1543.
13. Chance, B., DeVault, D.C., Frauenfelder, H., Marcus, R.A., Schrieffer, J.R., and Sutin, N. (eds.) (1979) 'Tunneling in Biological Systems', Academic Press, New York.
14. DeVault, D.C., and Chance, B. (1966) Biophys. J. 6, 826.
15. Gunner, M.R., Robertson, D.E., and Dutton, P.L. (1986) J. Phys. Chem. 90, 3783.
16. Gunner, M.R., and Dutton, P.L., submitted.
17. J. Jortner and B. Pullman (eds.) (1986) 'Tunneling', D. Reidel Publishing Company, Dordrecht.
18. Cowan, J.A., Upmacis, R.K., Beratan, D.N., Onuchic, J.N., and Gray, H.B. (1988) Ann. N.Y. Acad. Sci 550, 68.
19. Joran, A.D., Leland, B.A., Felker, P.M., Zewail, A.H., Hopfield, J.J.,and Dervan, P.B. (1987) Nature 327, 508.
20. Bialek, W., Bruno, W.J., Joseph, J., and Onuchic, J.N. (1989) Photosyn. Res., to be published.
21. Onuchic, J.N., Beratan, D.N., and Hopfield, J.J. (1986) J. Phys. Chem. 90, 3707.
22. Caldeira, A.O., and Leggett, A.J. (1983) Ann. Phys. 149, 374.
23. Onuchic, J.N., and Beratan, D.N. (1987) J. Am. Chem. Soc. 109, 6771.
24. Logan, J., and Newton, M.D. (1983) J. Chem. Phys 78, 4086.
25. Onuchic, J.N., and Wolynes, P.G. (1988) J. Phys. Chem. 92, 6495.
26. Bialek, W., and Onuchic, J.N. (1988) Proc. Nat. Acad. Sci USA 85, 5908.
27. Hopfield, J.J. (1974) Proc. Nat. Acad. Sci USA 71, 3640.
28. Garg, A., Onuchic, J.N., and Ambegaokar, V. (1985) J. Chem. Phys. 83, 1491).
29. Goldstein, R.F, and Bialek, W. (1986) Comments Mol. Cell. Biophys. 5, 407.
30. Landau, L.D. and Lifshitz, E.M. (1977) 'Statistical Physics', Pergamon, Oxford.
31. Bialek, W., and Goldstein, R.F. (1986) Phys. Scr. 34, 273.
32. Zusman, L.D. (1980) Chem. Phys. 49, 295.
33. Fruenfelder, H., and Wolynes, P.G. (1985) Science 229, 337.
34. Marcus, R.A. (1956) J. Chem. Phys. 24, 966.
35. Jortner, J. (1976) J. Chem. Phys. 64, 4860.
36. Onuchic, J.N. (1987) J. Chem. Phys. 86, 3925.
37. Bialek, W., Goldstein, R.F., and Kivelson, S. (1987) in 'Structure, Dynamics and Function of Biomolecules', A. Ehrenberg, R. Rigler, A. Cräslund, and L. Nilsson (eds.), Springer-Verlag, Berlin.
38. Bialek, W., Goldstein, R.F., and Onuchic, J.N. (1989) to be submitted.

39. Higasi, K., Baba, H., and Rembaum, A. (1965) 'Quantum Organic Chemistry', Inter-
    science Publishers, New York.
40. Anno, T., and Sadô, A. (1958) Bull. Chem. Soc. Japan 31, 734.
41. Harris, D.C., and Bortolucci, M.D. (1978) 'Symmetry and Spectroscopy', Oxford
    Press, New York.
42. Swingle, S.M. (1954) J. Am. Chem. Soc. 76, 1409.
43. Leland, B. (1987) Ph.D. Thesis, California Institute of Technology.
44. Heeger, A.J., Kivelson, S., Schrieffer. J.R., and Su, W.-P. (1988) Rev. Mod. Phys.
    60, 781.
45. Jortner, J. (1980) Biochim. Biophys. Acta 594, 193.

FEMTOSECOND-PICOSECOND LASER PHOTOLYSIS STUDIES ON ELECTRON TRANSFER DYNAMICS AND MECHANISMS IN SOME MODEL SYSTEMS

N. MATAGA
Osaka University
Faculty of Engineering Science
Department of Chemistry
Toyonaka, Osaka 560
Japan

ABSTRACT. For the more satisfactory understanding of the dynamics and mechanisms underlying the photoinduced CS (charge separation) and CR (charge recombination) of produced CT (charge transfer) or IP (ion pair) state, we have conducted femtosecond-picosecond laser photolysis and time-resolved transient absorption spectral studies on various D (donor)-A (acceptor) composite systems such as those combined by spacers $D-(CH_2)_n-A$ (n=1,2,3) or directly D-A, and on the uncombined fluorescer-quencher pairs as well as CT complexes in polar solvents. On the basis of the obtained results concerning the effects of D, A electronic interaction, energy gap for the reaction and solvent dynamics on the photoinduced CS and CR of produced IP states, a discussion is given on the photoinduced CS mechanism in biological photosynthetic reaction center.

1. INTRODUCTION

The electron transfer (ET) dynamics and mechanisms regulating the photoinduced charge separation (CS) and charge recombination (CR) of the produced charge transfer (CT) or ion pair (IP) state are the most important fundamental problems in the photochemical and photobiological primary processes. The elucidation of factors underlying these processes are of crucial importance and directly related to the mechanisms of the highly efficient ultrafast CS taking place in the biological photosynthetic reaction center (RC) [1,2].

It is believed in general that the rate of ET in the photoinduced CS and CR of produced CT or IP state is regulated by the following factors [1-4]: (a) the magnitude of electronic interaction responsible for the ET between donor (D) and acceptor (A) groups, (b) the Franck-Condon (FC) factor for the ET which is related to the energy gap between the initial and final state of the reaction, (c) reorganization energies of D, A as well as surrounding solvent (d) interactions between the solute D, A and solvent in the course of ET including the orientation dynamics of polar solvent or polar groups in the environment. However, our understanding of the ET mechanisms and dynamics in relation to these problems is quite unsatisfactory yet not only in photobiological reactions but also in photochemical reactions in condensed phase in

J. Jortner and B. Pullman (eds.), Perspectives in Photosynthesis, 227–240.
© 1990 *Kluwer Academic Publishers.*

general.
    Since the fundamental dynamical processes related to the above
photoinduced ET reactions are very rapid in general, the picosecond–
femtosecond laser photolysis and time-resolved absorption spectral
studies as well as fluorescence dynamics measurements are very important
for the elucidation of the above factors governing the photoinduced ET
and related phenomena. For the more satisfactory understanding of the
above problems, we are conducting such ultrafast laser photolysis
studies on CS at encounter between fluorescer and quencher [5] as well
as CR of produced geminate IP [6] and on various D, A composite systems
such as the model systems combined by spacers [7,8] and molecular
aggregates due to CT interactions [9,10]. In the following, we discuss
recent results of our femtosecond-picosecond laser photolysis studies
related to the above problems.

2. FEMTOSECOND–PICOSECOND LASER PHOTOLYSIS STUDIES ON PHOTOINDUCED CS
    AND CR OF PRODUCED IP STATE OF D, A COMBINED SYSTEMS AND SOME CT
    COMPLEXES

    In Figure 1, conceptual diagrams for ET or CT in the excited
electronic state are indicated.

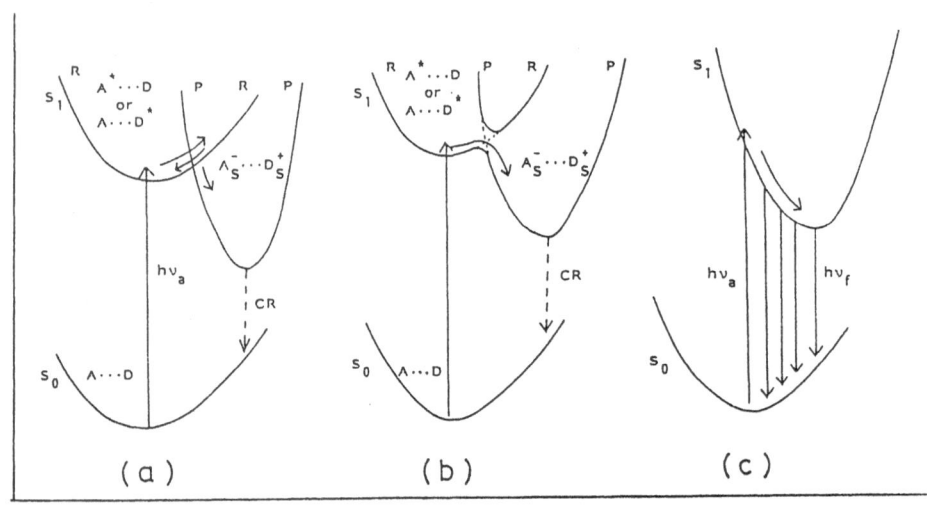

Reaction Coordinate

Figure 1. Conceptual diagrams of potential energy surfaces for ET or CT
in the excited electronic state, (a) nonadiabatic CS process, (b)
adiabatic CS process, (c) fluorescence Stokes shift due to solvation of
polar excited state.

When the electronic interaction between D and A is very weak, the ET is
considered nonadiabatic (a). When the interaction is fairly strong, the
reaction will be adiabatic (b). If the electronic interaction becomes

sufficiently strong in (b) and the energy gap relation is also favorable the ET will become almost barrierless (b'). In such case it is believed that the R→P process is governed mainly by orientational motions of polar solvent and the longitudinal dielectric relaxation time τ_L will be important as a factor controlling the ET rate, $\tau_L = (\varepsilon_\infty/\varepsilon_S)\tau_D$, where ε_∞ and ε_S are optical and static dielectric constant, respectively, and τ_D the Debye relaxation time.

In a limit of strong interaction, we can regard the combined system as a very polar single molecule and we can observe a large fluorescence Stokes shift due to solvation in polar solvents, for which the first theoretical formula (eq 1) was given by the present author [11] and Lippert [12] and has been extended recently by Bagchi et al. [13] and others to take into account its dynamical aspects.

$$h(\nu_a - \nu_f) = \text{Const.} + \frac{2(\vec{\mu}_e - \vec{\mu}_g)^2}{a^3}\left[\frac{\varepsilon_S - 1}{2\varepsilon_S + 1} - \frac{\varepsilon_\infty - 1}{2\varepsilon_\infty + 1}\right] \tag{1}$$

It should be noted that the nonequilibrium polarization with respect to the solute–solvent interaction in the FC state in the light absorption and emission involved in the calculation of eq 1 is essentially similar to that involved in the calculation of ET rate proposed [14] almost the same time as eq 1.

2.1. D, A Systems Combined by Spacer or Directly by Single Bond

For the elucidation of the above mechanisms, especially the case (b') of the ET process controlled by the solvent dynamics, femtosecond–picosecond laser photolysis studies on D, A combined systems with different degrees of electronic interactions is of crucial importance. However such investigations on appropriate systems are quite few, and the discrimination among above cases (a)-(c) does not seem very clear in the systems examined up to now [15]. From such viewpoint, we have investigated following systems with the femtosecond–picosecond laser photolysis and time–resolved transient absorption spectral measurements: $p-(CH_3)_2N-\phi-(CH_2)_n-(1-pyrenyl)$ (Pn, n=1,2,3), $p-(CH_3)_2N-\phi-(CH_2)_n-(9-anthryl)$ (An, n=1,2,3) [8], A_n and its derivative [16], 9,9'-bianthryl (BA) and its derivative [7] in alkylnitrile and viscous alcohol solvents. The time–resolved absorption spectral measurements give direct information on the electronic structures of the system undergoing ET, which is extremely important for discriminating various cases of ET processes. It is clear from such measurements that, in Pn and An with n=1,2 and 3, the photoexcitation in polar solvent like acetonitrile induced R→P process from the locally excited (LE) state of pyrene or anthracene part to the intramolecular IP state.

Time–resolved spectra of An in acetonitrile solution are indicated in Figure 2. These systems show rapid rise of the characteristic absorption band at 480 nm which is ascribed to the DMA (N,N–dimethylaniline) cation of the IP state. The rise curves converge respectively to constant values within several ps and do not show any

decrease up to several 10 ps. The latter fact is ascribed to the much
slower CR decay of the IP state compared with the photoinduced CS
process. Similar results have been obtained also in the case of Pn in
acetonitrile. From the observed rise curves the time constant τ_{CS} of

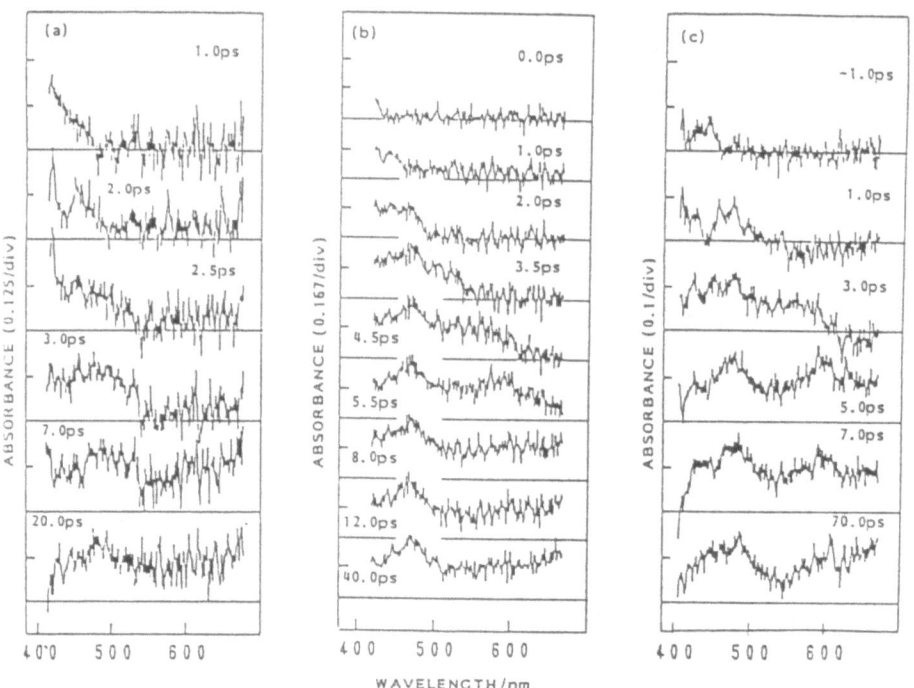

Figure 2. Time-resolved transient absorption spectra of A_1(a), A_2(b)
and A_3(c) in acetonitrile measured with femtosecond laser[1] photolysis
method.

the photoinduced IP state formation in acetonitrile has been obtained
as: A_1; ~0.7 ps, A_2; 2.1 ps, A_3; 2.7 ps, P_1; 2.6 ps, P_2; 6.1 ps, P_3; 11
ps. These τ_{CS} values are much longer than the τ_l=0.2 ps of
acetonitrile. Therefore, the photoinduced CS of these systems in
acetonitrile solution is not controlled by the solvent reorientation
dynamics. Even if we use the solvation time τ_s estimated from the time
dependent Stokes shift of the polar fluorescence probe molecule, this
conclusion is not altered except the case of A_1 where τ_{CS} is rather
close to τ_s suggesting the possibility of control by solvation dynamics.
On the other hand, the time constant τ_{CR} of CR decay of the IP state in
acetonitrile has been obtained as: A_1; 4.0 ns, A_2; 1.0 ns, A_3; 0.7 ns,
P_1; 7.0 ns, P_2; 3.3 ns, P_3; 1.1 ns. These τ_{CR} values are almost three
orders of magnitude longer than τ_{CS} and become shorter with increase of
the intervening chain number n, contrary to the case of τ_{CS}.
It is possible in general to give a satisfactory account of the
above results by considering the magnitude of the electronic tunneling

matrix element and the FC factor (the energy gap) of the reaction
without resorting to the solvent dynamics. The energy gap for CS,
$-\Delta G^{\circ}_{CS}$, of the D, A pairs of these systems in acetonitrile is around ~ 0.5
eV, where $k_{CS}(=\tau^{-1}_{CS})$ vs. $-\Delta G^{\circ}_{CS}$ relation is rather dull according to the
theoretical prediction [17]. Therefore, the difference of k_{CS} among
compounds with different number of methylene chains may be ascribed not
only to small difference in $-\Delta G^{\circ}_{CS}$ but also to the factor $\bar{A}$ containing
the tunneling matrix element $<i|\mathcal{H}'|f>$ and the average angular frequency
of intramolecular mode $<\omega>$.

$$\bar{A} = 2\pi |<i|\mathcal{H}'|f>|^2/\hbar^2 <\omega> \tag{2}$$

$\bar{A}$ for An will be larger than that of Pn due to the larger AO coefficient
at the 9-position where DMA is attached. This fact together with the
slightly larger $-\Delta G^{\circ}_{CS}$ value for anthracene-DMA pair results in the
shorter τ_{CS} of An.
     On the other hand, much smaller rate of CR can be interpreted
satisfactorily as due to the overwhelming effect of the FC factor.   As
it is discussed in 3., our experimental investigation by the ultrafast
laser photolysis and transient absorption spectral measurements has
established the bell-shaped $k_{CR}(=\tau^{-1}_{CR})$ vs. $-\Delta G^{\circ}_{IP}$ relation which can be
reproduced also by the theoretical calculation. $-\Delta G^{\circ}_{IP}$ is the free
energy difference between the IP state and the ground state.   According
to this result, the k_{CR} vs. $-\Delta G^{\circ}_{IP}$ relations for both pyrene$^-$-DMA$^+$ and
anthracene$^-$-DMA$^+$ pairs are in the inverted region at large energy gap
around $-\Delta G^{\circ}_{IP} \sim 2.8$ eV and the energy gap for the latter pair is a little
smaller.   From the bell-shaped relation, we obtain $k_{CR} \sim 10^8$ s^{-1} by
taking $\bar{A}=10^{13}$ s^{-1} for n=1 compounds in agreement with observation.   For
n=3 compounds, configurational change to sandwich type can take place
immediately after CS, which increases $\bar{A}$ and decreases $-\Delta G_{IP}$ leading to
the faster CR with $k_{CR} \sim 10^9$ s^{-1}.   In the n=2 compounds, the freedom for
such configuration change is smaller, giving the intermediate τ_{CR}
values.
     According to the discussion given concerning Figure 1, it will be
possible to observe the typical case of the photoinduced CS controlled
by the solvent dynamics by examining the more strongly interacting D, A
system combined directly by single bond. TM-A$_0$ (4-(9-anthryl)-2,3,5,6-
tetramethyl-DMA) shows time-resolved absorption spectra corresponding to
the simple CS process, $R(A^*-D) \rightarrow P(A^--D^+)$, according to our previous study
in the viscous alcohol solution [16a]. Our preliminary study of the
time-resolved absorption spectra of TM-A$_0$ in acetonitrile with the
femtosecond laser photolysis shows also the simple CS process and gives
$\tau_{CS} \sim 6$ ps [16b] which is much longer than τ_l and τ_S. Therefore, in this
case too, CS takes place by very weak interaction mechanism probably
from the dimethylamino group to the anthracene ring because the benzene
plane is perpendicular to both the anthracene plane and the
dimethylamino group and through bond interaction via benzene ring does
not seem very effective.
     In A$_0$, where the D, A electronic interaction is much stronger   due

to the more co-planar structure, our previous picosecond time-resolved absorption spectral studies in 1-butanol solution [16a] cannot be interpreted by the simple CS process but we must assume the gradual change of the electronic structure over delay times more than 200 ps from less polar to strongly polar ones via many states with different degrees of solvation and intramolecular structural rearrangements. Our preliminary femtosecond transient absorption spectral studies on this molecule in acetonitrile also indicate some gradual spectral changes in ps time regions. These results on A_0 imply that, for such system with strong D, A electronic interaction, the photoinduced CS is not simply determined by the solvent dynamics and also cannot be interpreted simply by the usual ET theories assuming weak interaction but proceeds via many states with different electronic structures. This is due to the fact that for the CS in such strongly interacting system, extensive solvations and some geometrical rearrangements are necessary in order to prevent the electronic delocalization interaction in the CS states.

In the case of 9,9'-bianthryl (BA), according to our investigation on the 1-pentanol solution [7], the photoinduced CS takes place with time constant of $\tau_{CS} \sim 170-180$ ps at 23°C, which is very close to $\tau_L \sim 174$ ps. However, if we slightly perturb the symmetry of BA by substituting a Cl atom at 10-position, the photoinduced CS becomes considerably faster than τ_L ($\tau_{CS} \sim 140$ ps) [7]. This result means that the intramolecular ET process is not simply determined by the solvent reorientation but the light absorption of those compounds projects the ground state equilibrium distribution onto a nonzero gradient of the excited state potential or the slightly pre-solvated state for those symmetry disturbed compounds will facilitate the CS process [7].

It should be noted here that the photoinduced CS of BA and 10-Cl-BA in polar solvents is not complete but a little delocalization interaction between the cation and the anion in the pair or some admixture of nonpolar electronic configuration into the intramolecular ET configuration should be taken into consideration because of the rather strong electronic interaction between two groups combined directly by single bond even though they are perpendicular to each other. In this respect, the results of our picosecond time-resolved absorption and fluorescence studies on the solvation induced electronic symmetry breaking in 1,2-dianthrylethanes in various polar solvents are very interesting [18].

In the case of 1,2-di(1-anthryl)ethane in acetonitrile solution, the formation of the completely charge separated state within ca. 10 ps has been observed by the time-resolve absorption spectral measurements. The $-\Delta G^\circ_{CS}$ of this system has been estimated to be 0.33 eV, which gives $k_{CS} \sim 10^{11}$ s^{-1} according to our theoretical calculation [17], in agreement with experiment observation. On the other hand, $-\Delta G^\circ_{IP}$ is estimated to be 2.85 eV, which leads to $k_{CR} \sim 10$ s^{-1} according to the bell-shaped energy gap relation [6]. However, observed k_{CR} was 400-500 ps, which can be ascribed to the rapid conversion of the IP state into the intramolecular excimer state by the diffusional rearrangements in the IP state, restoring the electronic symmetry. This intramolecular diffusional rearrangement is somewhat similar to that observed for P_3

and A_3 in acetonitrile solution [8].

2.2. CS Processes in the Excited CT Complexes in Polar Solutions

Another extreme case of the photoinduced CS due to the strong D, A electronic interaction is provided by excited CT complexes [8a,9]. We give here a brief discussion on the results of femtosecond laser photolysis and time-resolved absorption spectral studies on TCNB (1,2,4,5-tetracyanobenzene)-toluene (Tol) complex in acetonitrile in comparison with the case of D, A systems combined by spacer or directly by single bond discussed in 2.1. Results of our measurements show that it takes ca. 20 ps for the IP formation by the complete CS from the excited FC state [9]. This CS, however, is not a simple one step process but involves more complex processes of rearrangements in the solvation states as well as intracomplex structures as follows.

$$(A^{-\delta} \cdot D^{+\delta})_S \xrightarrow{h\nu_a} (A^{-\delta'} \cdot D^{+\delta'})_S^* \xrightarrow[\substack{\text{solvation and intracomplex} \\ \text{rearrangements}}]{\leqslant 1 \text{ ps}}$$

$$(3)$$

$$(A^{-\delta''} \cdot D^{+\delta''})_{S'}^* \xrightarrow[\substack{\text{further rearrangements in solvation} \\ \text{and intracomplex structure}}]{\sim 20 \text{ ps}} (A^- \ldots D^+)_{S''}$$

A solvent reorientation and a slight intracomplex rearrangement in the course of relaxation from FC state induce CS within 1 ps to a considerable extent but not completely. Further structural rearrangements of a large extent which take ca. 20 ps are necessary to realize complete CS by cutting off the D, A electronic delocalization interaction. This circumstance is quite different from that of A_1 and P_1 in acetonitrile where photoinduced CS is completed very rapidly within 1~3 ps [8b], but rather similar to the case of A_0. In general, the rate of the CS in the excited CT complexes will depend on the geometries of D and A in the complex in the ground state, excited FC state and relaxed IP state, and strength of D, A interactions as well as nature of the environment. In the case of TCNB-durene and TCNB-hexamethylbenzene (HMB) complex in acetonitrile, τ_{CS} has been determined to be ca. 8 ps and to be smaller than 4 ps, respectively [19]. Probably, the extent of the structural rearrangements necessary for complete CS may be much smaller for the TCNB-HMB complex compared to the TCNB-Tol complex. We have examined photoinduced CS processes of many other CT complexes with various strengths of D, A interactions, such as PMDA (pyromellitic dianhydride) (A), TCNE (tetracyanoethylene) (A), and TCNQ (tetracyanoquinodimethane) (A)-aromatic hydrocarbon (D) complexes. We have obtained results similar to the case of TCNB complexes [20].

After the CS in the course of the rapid rearrangements from the excited FC state, the produced IP state in acetonitrile undergoes the CR deactivation and the dissociation into solvated free ions. The rate of the CR deactivation depends strongly upon the energy gap between the IP and the ground state. We discuss this problem together with the energy gap dependence of the CR of IP produced by CS at encounter between fluorescer and quencher in acetonitrile solution in 3.

3. ULTRAFAST LASER PHOTOLYSIS AND TIME-RESOLVED ABSORPTION SPECTRAL STUDIES ON THE CR DEACTIVATION OF THE GEMINATE IP STATE

We haves discussed in 2. the CR dynamics and its energy gap dependence of the intramolecular IP state of Pn, An and also 1,2-dianthrylethane in the acetonitrile solution. Contrary to the rapid photoinduced CS in these systems, the CR of produced IP state is much slower and the relevant energy gap for the CR is much larger than that for the photoinduced CS in these systems. Actually, the CR processes in those systems are in the inverted region and quite slow due to the large energy gap, $-\Delta G_{IP}^{\circ}$.

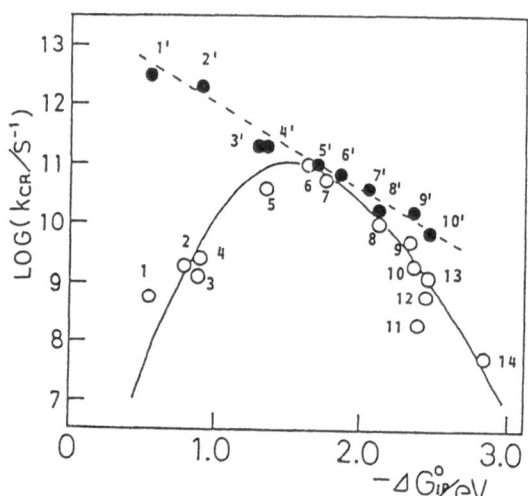

Figure 3. Energy gap dependences of CR rate constant of IP in acetonitrile solution.
o: IP formed by CS at encounter between fluorescer and quencher.
●: IP formed by excitation of ground state CT complex.
1. Per^{+}-$TCNE^{-}$, 2. $BPer^{+}$-$TCNE^{-}$, 3. DPA^{+}-$TCNE^{-}$, 4. Py^{+}-$TCNE^{-}$,
5. Per^{+}-$PMDA^{-}$, 6. Per^{+}-MA^{-}, 7. $TMPD^{+}$-Per^{-}, 8. Per^{+}-PA^{-}, 9. $BPer^{+}$-PA^{-},
10. An^{+}-PA^{-}, 11. DMA^{+}-Per^{-}, 12. o-DMT^{+}-Per^{-}, 13. Py^{+}-PA^{-}, 14. DMA^{+}-Py^{-}.
1'. Per^{+}-$TCNE^{-}$, 2'. Py^{+}-$TCNE^{-}$, 3'. $Naph^{+}$-$TCNQ^{-}$, 4'. Per^{+}-$PMDA^{-}$,
5'. Py^{+}-$PMDA^{-}$, 6'. Chr^{+}-$PMDA^{-}$, 7'. $Naph^{+}$-$PMDA^{-}$, 8'. Per^{+}-PA^{-},
9'. An^{+}-PA^{-}, 10'. Py^{+}-PA^{-}.
Per; perylene, BPer; 1,12-benzperylene, DPA; 9,10-diphenylanthracene, Py; pyrene, TMPD; N,N,N',N'-tetramethyl-p-phenylenediamine, An; anthracene, o-DMT; N,N-dimethyl-o-toluidine, PA; phthalic anhydride,

MA; maleic anhydride, Chr.; chrysene, Naph; naphthalene.

With respect to the energy gap dependence of the CR rates of geminate IP produced by CS at encounter between fluorescer and quencher in strongly polar solvent, no result for the normal region but only the results for the inverted region were available. We have made systematic studies on this problem by directly observing the CR deactivation of geminate IP with ultrafast laser spectroscopy and have obtained results not only for the inverted region but also for the normal as well as top regions, establishing the theoretically predicted bell-shaped relation as indicated in Figure 3 [6]. The observed results can be well fitted by theoretical calculation with appropriate parameter values.

In Figure 3, the energy gap dependence of the CR rate constant of IP formed by excitation of ground state CT complexes of the D, A system similar to the case of the CS at encounter in the same solvent is demonstrated [10], where the $-\Delta G^{\circ}_{IP}$ values are evaluated in both cases by the same conventional way. The k_{CR} vs. $-\Delta G^{\circ}_{IP}$ relation for the IP produced by the CT complex excitation is essentially different from that of the IP formed by CS at encounter. The k_{CR} values of the former are much larger than those of the latter in both the inverted and the normal regions. We can see from Figure 3 an enormous difference (three to four orders of magnitude) in k_{CR} between two kinds of IP at small $-\Delta G^{\circ}_{IP}$ values. As an example, the transient absorption spectrum of Py^{+}-$TCNE^{-}$ IP in acetonitrile observed immediately after excitation at CT absorption band (710 nm) by the femtosecond laser pulse and the time profile of the absorbance at 450 nm (Py^{+} absorption peak) are indicated in Figure 4. Compared with the very short lifetime (500 fs) of this IP, which is equal to k_{CR}^{-1}, the k_{CR}^{-1} of the IP formed by CS at encounter between the same pair is equal to 400 ps.

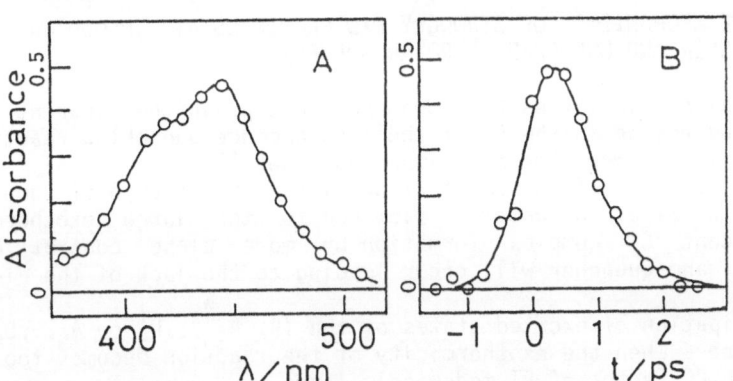

Figure 4. A. The transient absorption spectrum of the Py^{+}-$TCNE^{-}$ IP in acetonitrile immediately after excitation of the CT complex with the femtosecond laser pulse. B. The time profile of absorbance at 450 nm (Py^{+} absorption peak) of the Py^{+}-$TCNE^{-}$ IP. The solid line represents the simulation curve calculated with τ(lifetime of IP)=500 fs.

The observed results for the former IP can be given approximately by the relation,

$$k_{CR} \sim \alpha \cdot \exp[-\beta \cdot |\Delta G_{IP}^{\circ}|] \tag{4}$$

where α and β are constants independent of ΔG_{IP}°. This remarkable difference in the energy gap dependence of CR between the two kinds of IP may be ascribed to the difference in the structures including D^+, A^- configurations as well as solvation. The IP formed by CS at encounter in the fluorescence quenching reaction in acetonitrile will have a loose structure with some intervening solvent molecules between D^+ and A^- ions (loose IP, abbreviated as LIP), while the IP formed by CT complex excitation will have a more compact structure (compact IP, CIP) probably without intervening solvent molecules between the ions and with stronger electronic interaction between D^+ and A^-. It is interesting that the energy gap dependence of eq 4 is qualitatively analogous to that of the radiationless transition probability in the so-called "weak coupling" limit [21]. Anyhow, it is difficult to give a reasonable interpretation for the relation of eq 4 covering such wide energy gap range on the basis of the conventional ET theories [1].

It is well-known that, contrary to the results of the CR of the geminate IP formed by the CS at encounter between fluorescer and quencher, the CS itself in polar solvent does not show such bell-shaped energy gap dependence. The rate constant shows a steep rise around $-\Delta G_{CS}^{\circ} \sim 0$ and after that a constant diffusion-controlled value even at strongly exothermic region. It is difficult to explain this energy gap dependence of the photoinduced CS and the large difference between the energy gap dependences of the CR of IP and the photoinduced CS processes on the basis of usual ET theories [5].

4. ON THE MECHANISMS OF STRONGLY EXOTHERMIC CS AT ENCOUNTER BETWEEN FLUORESCER AND QUENCHER IN POLAR SOLVENTS

Several possibilities were proposed to explain the apparent lack of the inverted region in the CS of the fluorescence quenching reaction and the following two were the most important ones [5].
(a) Formation of nonfluorescent CT complex in the course of quenching. Even if the ET at encounter is slow due to too large exothermicity, nonfluorescent CT complex formation by more close contact between fluorescer and quencher will occur leading to the lack of the inverted region.
(b) Participation of excited states of the IP, $A_S^{-*} \ldots D_S^+$ or $A_S^- \ldots D_S^{+*}$, in the ET process when the exothermicity of the reaction becomes too large, keeping the FC factor of ET moderately large.

However, the results discussed in 3. concerning the quite different behaviors of the compact and loose IP reject the mechanism (a). If the nonfluorescent CT complex is formed in the course of the strongly exothermic CS in the fluorescence quenching reaction of the Py-TCNE system in acetonitrile, we should not observe the formation of the geminate IP by the picosecond laser photolysis method since the CIP of this system undergoes ultrafast deactivation with decay time of 500 fs. Actually, however, we have observed the formation of rather long-lived LIP with lifetime of 200 ps. We have obtained analogous results also in the case of the Py-PMDA system in acetonitrile.

On the other hand, we have examined the energetics of the photoinduced CS of the Per-PMDA (i) and Py-PMDA (ii) systems in acetonitrile concerning the mechanism (b). The exothermicities $-\Delta G^{\circ}_{CS}$ are estimated to be 1.5 eV (i) and 1.63 eV (ii), respectively. The usual ET theories predict the rate constant of CS should be smaller than 10^6 M^{-1} s^{-1} at $-\Delta G^{\circ}_{CS} \gtrsim 1.5$ eV contrary to the observed result of the fluorescence quenching rate constant $k_q > 2 \times 10^{10}$ M^{-1} s^{-1}. Even if we take into account the contribution of the excited states of ions in the course of CS, $-\Delta G^{\circ}_{CS}$ is almost zero or slightly negative, which leads to $k_q < 10^9$ M^{-1} s^{-1}, according to the usual ET theories as well as the experimental results of the energy gap dependence of k_q in the normal region. These results reject also the mechanism (b).

We need new concept which can interpret the fact that the observation of the inverted region in the photoinduced CS reaction is difficult and that the energy gap dependences of the CR of geminate IP and the photoinduced CS are quite different from each other. In this respect, we have proposed a new model which takes into consideration the possibility of different potential curvatures for the solute-solvent interactions between the charged state and neutral state of solute in polar solvents, which leads to the different energy gap dependences of photoinduced CS and CR of the produced IP state [17]. Some calculations based on this idea have actually demonstrated that the different energy gap dependences of observed rates of photoinduced CS and CR of produced IP can be well reproduced by this model [17]. Nevertheless, natures of the different potential curvatures of solute-solvent interaction between the charged state and neutral state of solute must be further elucidated.

5. CONCLUDING REMARKS

The results of the femtosecond-picosecond laser photolysis and the time-resolved transient absorption spectral studies on various combined and uncombined D, A systems discussed above seem to provide useful informations in considering the mechanisms of very fast and high efficiency photoinduced CS in biological photosynthetic reaction center (RC).

For example, the bell-shaped energy gap dependence of k_{CR} in Figure 3 indicates that the CR rate of porphyrin$^+$(P^+)-quinone$^-$(Q^-) geminate IP in polar solvent is around the top region in view of its energy gap ($-\Delta G^{\circ}_{IP} \sim 1.5$ eV) and actually ultrafast CR deactivation of P^+-Q^- geminate IP was confirmed [22]. Therefore, direct photoinduced ET between P and Q is not favorable for efficient CS. However, in the bacterial photosynthetic RC, the primary ET does not take place between P and Q, but among porphyrins, i.e. special pair (SP) of bacteriochlorophylls and bacteriopheophytin. The first CS process seems to take place at SP and extend to pheophytin and then to quinone. The photoinduced CS in the dimer model compounds such as 1,2-dianthrylethane discussed above can take place very rapidly in some suitable intramolecular conformation and environment, and CR deactivation of the produced IP state is much slower because k_{CR} vs. $-\Delta G^\circ_{IP}$ relation is in the inverted region and $-\Delta G^\circ_{IP}$ is very large. Similar circumstance will be prevailing also in the SP and between chlorophyll and pheophytin. Moreover, we have recently confirmed in the case of hybrid metal porphyrin dimer models that the back ET deactivation is much slower compared to the photoinduced forward ET [23].

Integrating the above described results and discussions, we can understand very well the highly efficient and very rapid photoinduced ET from SP to pheophytin and then to quinone. When electron has moved to quinone, CR decay will be prevented by very small tunneling matrix element due to the long distance between SP and quinone.

The works, the results of which are discussed in this article, are supported by a Grant-in-Aid (No. 62065006) from the Japanese Ministry of Education, Science and Culture to the author.

References
1. Marcus, R. A. and Sutin, N. (1985) 'Electron Transfers in Chemistry and Biology', Biochim. Biophys. Acta 811, 265-322.
2. Mataga, N. (1988) 'Dynamics and Energy Gap Dependences of Photoinduced Charge Separation and Recombination of the Produced Ion Pair State', in J. R. Norris and D. Meisel (eds.), Photochemical Energy Conversion, Elsevier, New York, pp. 32-46.
3. (a)Sumi, H. and Marcus, R. (1986) 'Dynamical Effects in Electron Transfer Reactions', J. Chem. Phys. 84, 4894-4914.
    (b) Rips, I. and Jortner, J. (1987) 'Dynamic Solvent Effects on Outer-Sphere Electron Transfer', ibid. 87, 2090-2104.
    (c) Sparpaglione, M. and Mukamel, S. (1988) 'Dielectric Friction and the Transition from Adiabatic to Nonadiabatic Electron Transfer. I. Solvation Dynamics in Liouville Space', ibid. 88, 3263-3280.
4. Rips, I., Klafter , J., and Jortner, J. (1988) 'Solvation Dynamics and Solvent-Controlled Electron Transfer', in J. R. Norris and D. Meisel (eds.), Photochemical Energy Conversion, Elsevier, New York, pp. 1-22.
5. Mataga, N., Kanda, Y., Asahi, T., Miyaska, H., Okada, T., and Kakitani, T. (1988) 'Mechanisms of the Strongly Exothermic Charge Separation Reaction in the Excited Singlet State. Picosecond Laser Photolysis Studies on Aromatic Hydrocarbon-Tetracyanoethylene and Aromatic Hydrocarbon-Pyromellitic Dianhydride Systems in Polar Solutions', Chem. Phys. 127, 239-248.
6. Mataga, N. Asahi, T. Kanda, Y., Okada, T., and Kakitani, T. (1988) 'The Bell-Shaped Energy Gap Dependence of the Charge Recombination Reaction of Geminate Radical Ion Pairs Produced by

Fluorescence Quenching Reaction in Acetonitrile Solution', Chem. Phys. 127, 249-261.

7. Mataga, N. Yao, H., Okada, T., and Rettig, W. (1989) 'Charge Transfer Rates in Symmetric and Symmetry-Disturbed Derivatives of 9,9'-Bianthryl', J. Phys. Chem. 93, 3383-3386.

8. (a) Mataga, N., Miyasaka, H., Asahi, T., Ojima, S., and Okada, T. (1988) 'Ultrafast Laser Spectroscopy of Transient Ion Pair States in Solution', Ultrafast Phenomena VI, 511-516.
    (b) Mataga, N. Nishikawa, S., Asahi, T., and Okada, T. (1989) 'Femtosecond-Picosecond Laser Photolysis Studies on the Photoinduced Charge Separation and Charge Recombination of Produced Ion Pair State of Some Typical Intramolecular Exciplex Compounds in Alkylnitrile Solvents', J. Phys. Chem. 93, submitted.

9. Miyasaka, H., Ojima, S., and Mataga, N. (1989) 'Femtosecond-Picosecond Laser Photolysis Studies on Ion Pair Formation Process in the Excited State of Charge Transfer Complex in Solution', J. Phys. Chem. 93, 3380-3382.

10. Asahi, T. and Mataga, N. (1989) 'Charge Recombination Process of Ion Pair State Produced by Excitation of CT Complex in Acetonitrile Solution – Essentially Different Character of Its Energy Gap Dependence from That of Geminate Ion Pair Formed by Encounter between Fluorescer and Quencher', J. Phys. Chem. 93, submitted.

11. Mataga, N. Kaifu, Y. and Koizumi, M. (a) (1955) 'The Solvent Effect on Fluorescence Spectrum. Change of Solute-Solvent Interaction during the Lifetime of Excited Solute Molecule', Bull. Chem. Soc. Jpn. 28, 690-691, (b) (1956) 'Solvent Effects upon Fluorescence Spectra and the Dipolemoments of Excited Molecules', ibid. 29, 465-470.

12. Lippert, E. (a) (1955) 'Dipolmoment und Elektronenstruktur von angeregten Molekulen', Z. Naturforsch. 10a, 541-545, (b) (1957) 'Spektroskopische Bestimmung des Dipolmomentes aromatischer Verbindungen im ersten angeregten Singulettzustand', Ber. Bunsenges. Phys. Chem. 61, 962-975.

13. Bagchi, B., Oxtoby, D. W., and Fleming, G. (1984) 'Theory of the Time Development of the Stokes Shift in Polar Media', Chem. Phys. 86, 257-267.

14. Marcus, R. A. (1956) 'On the Theory of Oxidation-Reduction Reactions Involving Electron Transfer. I.', J. Chem. Phys. 24, 966-978.

15. Kosower, E. M. and Huppert, D. (1986) 'Excited State Electron and Proton Transfers', Ann. Rev. Phys. Chem. 37, 127-156.

16. (a) Okada, T., Mataga, N., Baumann, W., and Siemiarczuk, A. (1987) 'Picosecond Laser Spectroscopy of 4-(9-Anthryl)-N,N-dimethylaniline and Related Compounds', J. Phys. Chem. 91, 4490-4495, (b) Mataga, N., Kanaji, K., Hagihara, M., Okada, T., and Baumann, W., unpublished work.

17. Kakitani,. T. and Mataga, N. (a) (1985) 'Photoinduced Electron Transfer in Polar Solutions. I. New Aspects of the Role of the Solvent Mode in Electron Transfer Processes in Charge-Separation

Reactions', Chem. Phys. 93, 381–397, (b) (1985) 'New Energy Gap Laws for the Charge Separation Process in the Fluorescence Quenching Reaction and the Charge Recombination Process of Ion Pairs Produced in Polar Solvents', J. Phys. Chem. 89, 8–10, (c) (1986) 'Different Energy Gap Laws for the Three Types of Electron–Transfer Reactions in Polar Solvents', ibid. 90, 993–995, (d) (1987) 'Comprehensive Study on the Role of Coordinated Solvent Mode Played in Electron–Transfer Reactions in Polar Solutions', ibid. 91, 6277–6285.

18. Yao, H., Okada, T. and Mataga, N. (1989) 'Solvation Induced Charge Separation in the Excited State of Composite Systems with Identical Halves and Intramolecular Excimer Formation by Recombination – Picosecond Laser Photolysis Studies on 1,2-Dianthrylethanes', J. Phys. Chem. 93, in press.

19. Ojima, S. Miyasaka, H. and Mataga, N. (1989) 'Femtosecond-Picosecond Spectroscopy of Excited Tetracyanobenzene–Aromatic Hydrocarbon Charge Transfer Complexes', to be published.

20. Asahi, T. and Mataga, N. (1989) 'Ultrafast Laser Photolysis Studies on the Dynamics of Excited Charge Transfer Complexes', to be published.

21. Englman, R. and Jortner, J. (1970) 'The Energy Gap Law for Radiationless Transitions in Large Molecules', Mol. Phys. 18, 145–164.

22. Mataga, N., Karen, A., Okada, T., Nishitani, S., Sakata, Y., and Misumi, S. (1984) 'Picosecond Laser Photolysis Studies of Photoinduced Electron Transfer in Porphyrin–Quinone Systems in Solution. Detection of the Short–Lived Exciplex State and Its Solvation–Induced Ultrafast Deactivation', J. Phys. Chem. 88, 4650–4655.

23. Mataga, N., Yao, H., Okada, T. Kanda, Y., and Harriman, A. (1989) 'Picosecond Dynamics of Intramolecular Electron and Energy Transfer in Porphyrin Dimer Model Compounds', Chem. Phys. 131, 473–480.

PROTEIN CONFIGURATIONAL FLUCTUATION DEPENDENCE OF THE ELECTRONIC TUNNEL FACTOR OF MODIFIED METALLOPROTEIN ELECTRON TRANSFER SYSTEMS AND IN FAST DIRECT AND SUPEREXCHANGE SEPARATION AND RECOMBINATION IN BACTERIAL PHOTOSYNTHESIS

Aleksandr. M. Kuznetsov
The A.N. Frumkin Institute of Electrochemistry of the
Academy of Sciences of the USSR
Leninskij Prospect 31, Moscow V-71, USSR

Jens Ulstrup and Merab G. Zakaraya *)
Chemistry Department A, Building 207
The Technical University of Denmark
2800 Lyngby, Denmark

1. INTRODUCTION

Electron transfer (ET) reactions involving solute biological macromolecules such as metalloproteins, or electron exchanging pigment molecules fixed in transmembrane protein environments possess a number of distinctive features compared with ET reactions between "small" molecules. Some of these features are related to specific properties of the environmental reorganization terms or the inter-reactant work terms which for protein systems include diffusion-like, large amplitude protein conformational motion[1-4] and nonlocal or image dielectric forces in both protein and external solvent[5,6]. A feature of fundamental importance is, however, that biological ET distances frequently exceed the geometric extension of the donor and acceptor groups. While inherent in common diabatic ET theory, this long-range electron tunnel feature raises several questions relating not only to orientation and distance between donor and acceptor groups but also to the relative importance of direct and superexchange ET, and to the effect of environmental configurational fluctuations on the electron tunnel factor.

Biological "long-range" ET in the above sense is encountered at least in the following classes of ET systems:

(A) ET involving small metalloproteins in which the metal centre is located "inside" the protein. Examples are the "east-site" ET channel in blue copper proteins[8] and, ET reactions involving high-potential iron sulphur proteins[9,10].

(B) ET in modified metalloproteins where ET groups have been attached to suitable surface sites, providing a pathway for intramolecular ET. Such investigations have been reported for cytochromes[11-13], blue copper proteins, iron-sulphur proteins and myoglobin[16].

(C) Intramolecular ET in native and modified multi-site redox proteins, for example metal-substituted hemoglobins. ET in metalloprotein complexes belong to this category.[18]

*)Permanent address: Institute of Inorganic Chemistry and Electrochemistry, Georgian Academy of Sciences, Tbilisi, 380086, USSR.

J. Jortner and B. Pullman (eds.), Perspectives in Photosynthesis, 241–259.
© 1990 Kluwer Academic Publishers.

(D) Electrochemical ET involving small ET metalloproteins[19] or redox enzymes such as glucose oxidase[20] or laccase[21]. Electrochemical ET can be "assisted" by internal attachment of suitable ET groups such as ferrocene[20].

(E) ET in membrane-bound systems. The availability of detailed crystallographic data[22] combined with a vast variety of spectroscopic, magnetic, and kinetic data over broad temperature ranges [23] makes the reaction centre ET reactions in bacterial photosynthesis the most important representative in this class.

The recent availability of precisely characterized modified metalloprotein ET systems and the structural and kinetic characterization of the bacterial photosynthetic reaction centre have placed biological ET theory in a new perspective and offered clues to subtler and less known elements of ET theory. In the present work we illustrate this by a brief analysis of ET systems in the classes (B), (C) and (E) listed above, with particular emphasis on environmental frequency dispersion effects and on the effects of nuclear motion on the electronic tunnel factor in both direct and superexchange-mediated long-range ET.

2. LONG-RANGE ELECTRON TRANSFER IN VIBRATIONALLY DISPERSIVE MEDIA

The diabatic rate constant in the form commonly applied, is[7]

$$W_{fi} = \frac{2\pi}{\hbar} Av_v \sum_w |(<\Psi_f|V|\Psi_i>)|^2 \delta(\epsilon_{fw} - \epsilon_{iv}) = \frac{2\pi}{\hbar}(V_{fi})^2 S_{fi}(\Delta G_0; T) \ (1)$$

where Ψ_i and Ψ_f are the overall Born-Oppenheimer wave functions in initial (i) and final (f) state, ϵ_{iv} and ϵ_{fw} the vibrational energies, V the perturbation which couples the two electronic states, and Av_v indicates averaging with respect to all the vibrational states v. The Condon approximation has been invoked in the second part of eq.(1), V_{fi} here being the nondiagonal part of the electronic transition matrix element, S_{fi} the thermally averaged nuclear Franck Condon factor and $\hbar$ Plancks constant divided by 2π. Eq.(1) most commonly refers all reaction free energy (ΔG_0) and temperature (T) dependence to the nuclear factor. However, as noted elsewhere[24], eq.(1) conceals in particular two features of importance to low temperatures and long ET distances, both appropriate to ET in membrane-bound protein systems. First, although the overall kinetics may be understood in terms of nuclear tunnelling and thermal activation of suitable finite frequency modes, the rate constant at low temperatures is crucially determined by the behaviour of the environmental permittivity function at the lowest frequencies ω[25]. Finite rate constants only emerge when the permittivity approaches zero no faster than ω, the prevalence of the lowest frequencies being what ensures process irreversibility[26]. The rate constant divergency for discrete and cut-off dispersions is often formally annihilated by "coarse graining"[27]. Vibrational dispersion is, however, well documented for water[28], ice[29], and proteins[30], and both physically and formally it can be adequately simulated by Debye, resonance or similar vibrational distributions[31-33]. We therefore explicitly incorporate vibrational dispersion in the rate constants. No divergency problems arise in this way, and the resulting rate constant form is adequate at all temperatures.

The second feature concealed by eq.(1) is that the "tails" of the electronic wave functions in long-range ET are strongly exposed to fluctuations in the environmental polarization field. The particular nuclear configuration of importance is the saddle point of suitable intersecting potential surfaces, spanned by all the molecular and environmental nuclear coordinates. This configuration is always quite different from the equilibrium configuration in either the reactant or product state and depends strongly on both ΔG_0 and T. This dependence must be reflected in the electronic wave functions of either the donor, the acceptor or both. We can therefore expect that not only the nuclear factor in eq.(1), but also the electronic factor depends on temperature and reaction free energy[24]. We shall show below that this can have profound effects on long-range ET parameters extracted from experimental ET data for modified metalloproteins[34,35] and fast separation and recombination reactions in bacterial photosynthesis[35].

These expectations extend to long-range ET effected by superexchange. The electronic factor in this mechanism can either be viewed as composed of separate two-level electronic transition matrix elements[36], each exposed to non-equilibrium environmental polarization fields and therefore depending on ΔG_0 and T. Alternatively, in the limit of long-range ET the effect of the intermediate group levels can be incorporated into a single, approximately exponential electronic factor the orbital exponent now being smaller than for direct ET. This observation is important for the initial charge separation from excited special pair chlorophyll to pheophytin and back, in the presence of auxiliary chlocophyll monomer in bacterial photosynthesis.

2.1 ET in vibrationally dispersive media

The diabatic rate constant for a linear, vibrationally dispersive medium, extending to the lowest frequencies and strongly coupled to the ET centre, is[7,25]

$$W_{fi} = \frac{1}{k_B T \hbar}(V_{fi})^2 \left\{ \frac{1}{2\pi} |F''(\theta^*)| \right\}^{-\frac{1}{2}} \exp[-\beta\theta^*\Delta G_0 - F(\theta^*)]; \beta = (k_B T)^{-1} \quad (2)$$

where k_B is Bolzmann's constant. $F(\theta*)$ Contains all information about the vibrational dispersion

$$F(\theta^*) = \frac{1}{\hbar}\int_{-\infty}^{\infty}\frac{d\omega}{\omega} \; \mathcal{E}_s(\omega)\frac{sh\left[\frac{1}{2}\beta\hbar\omega\theta^*\right]sh\left[\frac{1}{2}\beta\hbar\omega(1-\theta^*)\right]}{sh\left(\frac{1}{2}\beta\hbar\omega\right)} \quad (3)$$

where $\mathcal{E}_s(\omega)$ is the environmental reorganization free energy density, related to the dielectric permittivity function, $\epsilon(\omega)$, and overall reorganization free energy, E_s, by

$$\mathcal{E}_s(\omega) = \frac{2E_s}{\pi\omega c}\frac{Im\epsilon(\omega)}{|\epsilon(\omega)|^2}; C = \epsilon_0^{-1} - \epsilon_s^{-1}; \int_0^{\infty}d\omega \; \mathcal{E}_s(\omega) = E_s \quad (4)$$

ϵ_0 being the optical and ϵ_s the static dielectric constant. Provided that $|\Delta G_0| < E_s$ (the "normal" free energy range) or $(|\Delta G_0 - E_s|)/E_s$ (the "weakly inverted" range), $\theta*$ is finally determined by the saddle point equation

$$\Delta G_0 = \int_0^\infty d\omega \quad \mathcal{E}_s(\omega) \frac{sh\left[\frac{1}{2}\beta\hbar\omega(2\theta^* - 1)\right]}{sh\left(\frac{1}{2}\beta\hbar\omega\right)} \tag{5}$$

and coincides with the transfer coefficient of ET theory in the strong-coupling limit of solute-solvent interaction, i.e. $\theta^* = -k_B T d \ln W_{fi}/d\Delta G_0$.

Eqs.(2)-(5) provide a procedure for broad classes of dielectric dispersion functions. In the following we specifically exploit a broad resonance form[31]

$$\frac{Im\epsilon(\omega)}{|\epsilon(\omega)|^2} = (\epsilon_0^{-1} - \epsilon_s^{-1}) \frac{1 + \omega_R^2 \tau_R^2}{2\omega_R \tau_R} \left[\frac{1}{1 + (\omega - \omega_R)^2 \tau_R^2} - \frac{1}{1 + (\omega + \omega_R)^2 \tau_R^2} \right] \tag{6}$$

$(\omega_R \tau_R \approx 1)$ where ω_R is the resonance frequency and τ_R^{-1} the damping period. Other forms of W_{fi} and $\epsilon(\omega)$ relating to narrow resonance media[7], the weak-coupling limit and relaxing the saddle point approximation[38] are also available. Eqs.(2)-(6) can be supplemented by discrete modes and spatial dielectric dispersion by reported procedures[39].

2.2 Environmental modulation of the electronic tunnel factor

Theoretical approaches to the effects of environmental "matter" on the electron tunnel factor have rested on higher order perturbation (superexchange) theory,[40] quantum chemical methods,[41-45] tunnel theory,[46] and on theory of environmental continuum dynamics including vibrational and spatial dielectric dispersion[24]. Only the latter incorporates dynamic environmental polarization in detail, the electronic factor being elsewhere regarded as independent of the instantaneous nonequilibrium polarization and therefore of the temperature and reaction free energy.

In previous reports we have shown[24], however, that the ΔG_0- and T-variation of the polarization configuration at the saddle point between the reactants' and products' potential surfaces are strongly reflected in the electronic wave functions. The effects have been analyzed explicitly for single-parameter exponential wave functions

$$\psi_i(\rho) = (\lambda_i^3/\pi)^{\frac{1}{2}} \exp\left(-\lambda_i |\vec{\rho}|\right); \psi_f(\rho) = (\lambda_f^3/\pi)^{\frac{1}{2}} \exp\left(-\lambda_f |\vec{\rho} - \vec{R}|\right) \tag{7}$$

where R is the donor-acceptor distance. This form emerges self-consistently for central symmetric fields including both a dispersive continuum and molecular trapping fields, but is broadly representative for chemical, electrochemical and biological systems.

The dynamic environmental effects are represented by the temperature and reaction free energy dependence of λ_i and λ_f. The dependence is obtained by minimization of suitably constructed free energy functionals[24] and takes the form $\lambda_i = \lambda_i(\theta^*)$ and $\lambda_f = \lambda_f(\theta^*)$. Representative results are shown in fig.1 which reveals that the electron clouds expand, i.e. λ_s (s = i,f) decreases compared to equilibrium, for system motion along the potential surface in the "normal" free energy range, whereas contraction emerges in the inverted region. The figure and reported calculations[24] also suggest that the effect may be significant and comparable to the variation of the nuclear factor for ET over long distances.

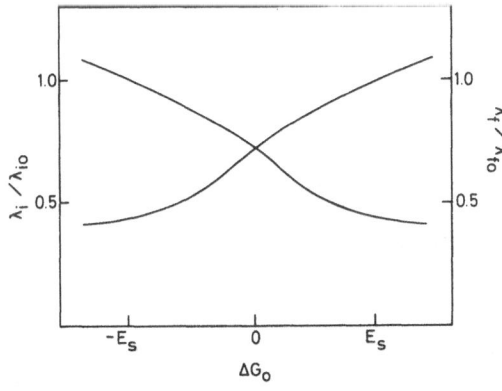

Fig. 1 Polarization fluctuation induced variation of λ_i/λ_{io} and λ_f/λ_{fo} with ΔG_0. $\epsilon(\omega)$ given by eq.(6) with $\omega_R = \tau_R^{-1} = 150$ cm^{-1}. R = 10 Å. T = 298 K. No excess core charges.

The effect takes a simple analytical form close to the activationless region, i.e. when $\theta^* < 0.3$., Provided that $\lambda_i(\theta^*) < \lambda_f(\theta^*)$, i.e. when a loosely bound electron is transferred to a deep trap. This approximation is representative for recombination to the bacteriochlorophyll dimer cation radical. The electron factor is then approximately

$$\kappa_{fi}(\theta^*) \approx \kappa_{fi}^0 \exp(2\lambda_{io} \upsilon \xi R \theta^*)$$

$$\xi = \frac{5/11}{1 + (16/11)(Z_0/c\epsilon_s)}; \quad \upsilon = \frac{2}{\pi c} \int_0^\infty \frac{d\omega}{\omega} \frac{Im\epsilon(\omega)}{|\epsilon(\omega)|^2} \frac{\frac{1}{2}\beta\hbar\omega}{th\left(\frac{1}{2}\beta\hbar\omega\right)} \quad (8)$$

where $\lambda_{io} = \lambda_i(\theta^* \approx 0)$ and $V_{fi}^0 = V_{fi}(\theta^* \approx 0)$ are the equilibrium values of λ_i and V_{fi}, respectively. It is noted that $V_{fi} > V_{fi}^0$ for $\theta^* > 0$, i.e. in the "normal" free energy range, whereas $V_{fi} < V_{fi}^0$ in the inverted region where $\theta^* > 0$. It is also noted that $\upsilon \to 1$ in the high-T limit, while $\upsilon > 1$ at lower T. Even in the Condon approximation the reaction free energy and temperature variation of the electron tunnel factor can thus be substantial. Generalizations of the procedure both to variation of the electronic-vibrational coupling and distortion of the potential surfaces as the electron density varies along the latter (the "improved" Condon approximation), and relaxation of the Condon approximation altogether are also available[7,47].

3. THE SUPEREXCHANGE MECHANISM INCLUDING ELECTRONIC-VIBRATIONAL COUPLING AND ENVIRONMENTALLY MODULATED ELECTRON TUNNEL FACTORS.

Three-level transitions, rather than the two-level transitions in the preceding section are inherent in superexchange effected ET. This mechanism is believed to operate the first rapid ET from the special pair to pheophytin - and perhaps the corresponding back reactions- in bacterial photosynthesis, the intermediate state being the lowest antibonding orbital in the L-branch accessory chlorophyll.

The superexchange mechanism rests on second order perturbation theory. The rate constant, corresponding to eq.(1), is[7,40]

$$W_{fi}^{(2)} = \frac{2\pi}{\hbar} A v_v \sum_w |(<\Psi_f | V^{(2)} | \Psi_i>)|^2 \delta(\epsilon_{fw} - \epsilon_{iv}); \quad V^{(2)} = \sum_{d,u} \frac{V |\Psi_d>)(<\Psi_d| V}{\epsilon_{iv} - \epsilon_{du} + i\gamma}$$

$$\gamma \to 0 \qquad (9)$$

where Ψ_d is the Born-Oppenheimer electronic-vibrational wave function in the intermediate electronic state "d", and u the vibrational summation index in this electronic state. The energy denominator means, however, that eq.(9) cannot in general be recast into the simple form of eq.(1). The following two cases must be distinguished[7,40]:

3.1 High-energy intermediate states

This limit corresponds to the location of potential surfaces as fig. 2a and is representative of the common use of the superexhange concept in ET theory, for example in relation to charge separation in photosynthesis or the bridge-group assisted molecular ET. The rate constant then takes approximately the same form as in eq.(1), i.e.

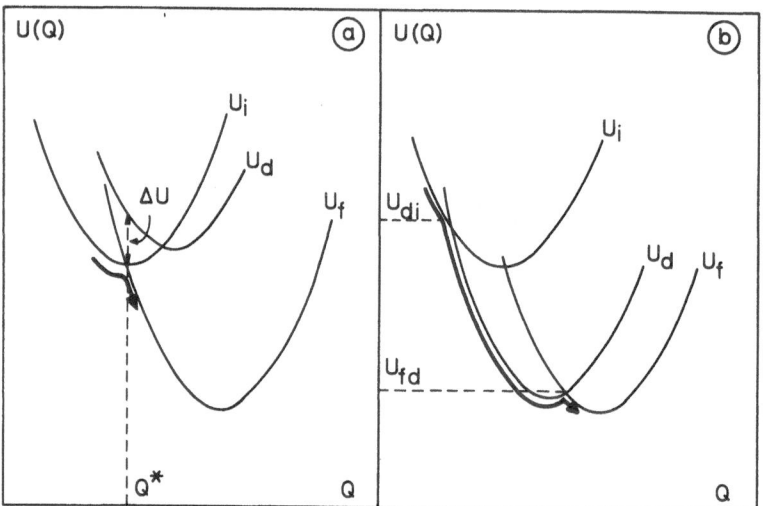

Fig. 2. Potential energy surfaces for the superexchange mechanism. (a): high-energy intermediate state. (b): low-energy intermediate state. The arrows indicate the trajectory for the chemical process. The relative potential surface location in fig.2a corresponds approximately to the primary charge separation in bacterial photosynthesis.

$$W_{fi}^{(2)} \approx \frac{2\pi}{\hbar} (V_{fi}^{(2)})^2 S_{fi}(\Delta G_0; \tau); \quad V_{fi}^{(2)} = \frac{1}{\Delta U(Q^*)} (V_{fd}^{(1)})(V_{di}^{(1)}) \qquad (10)$$

where the energy gap $\Delta U(Q^*)$ is indicated in fig.2a, while $V_{fd}^{(1)}$ and $V_{di}^{(1)}$ are the first order nondiagonal electronic transition matrix elements involving the intermediate electronic state.

The distance dependence of $V_{fi}^{(2)}$ can be illustrated in the following way. If both first order transition matrix elements in eq.(10) depend exponentially on the ET distance, with the same orbital exponent α, i.e.

$$V_{fd}^{(1)} \approx V_0^{(1)} \exp(-\alpha R)_{fd}; V_{di}^{(1)} \approx V_0^{(1)} \exp(-\alpha R_{di}) \qquad (11)$$

where the subscripts on R refer to the particular transition, then

$$V_{fi}^{(2)} \approx V_0^{(1)} \exp(-\alpha_{eff} R_{fi}); \alpha_{eff} \approx \alpha[1 - \frac{1}{\alpha R_{fi}} ln \frac{V_0^{(1)}}{\Delta U}] < \alpha \qquad (12)$$

In comparison, for direct ET between donor and acceptor states

$$V_{fi}^{(1)} \approx V_0^{(1)} \exp(-\alpha R_{fi}) \qquad (13)$$

Eq.(12) exhibits approximately exponential distance variation, but with a lower orbital exponent than for "direct" ET. For example, $\alpha R_{fi} \approx 7$ and $\Delta U \approx 0.1$ ev for the $(bch)_2^+/bph$ forward reaction. If $V_0^{(1)} \approx 10$ ev as for electronic mobility in molecular crystals of aromatic hydrocarbons,[27] then $\alpha_{eff}^0 \approx (0.3-0.4)\alpha$, and the superexchange mechanism is clearly more favourable than direct ET unless the intermediate state energy is very high.

An alternative view of the superexchange mechanism can be based on tunnel theory (fig.3)[48]. The high-energy intermediate state represents a local "indentation" or "hole" in the electron tunnel barrier which corresponds to direct ET. A deep hole leads to temporary electron localization in the region of the intermediate level and a corresponding decrease of the barrier width, endowing the ET process with the character "resonance tunnelling", quantitative frames for which in relation to electrochemical and surface ET reactions are given elsewhere.[48,49]

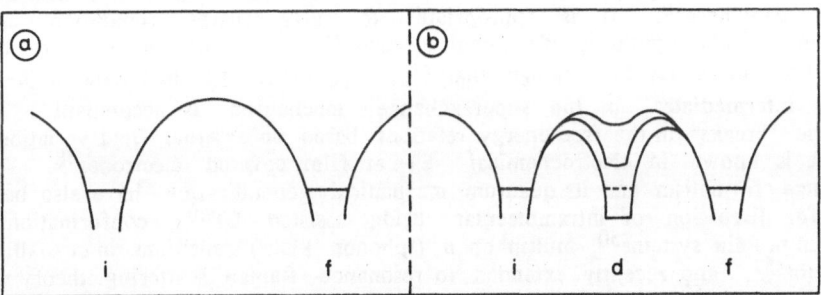

Fig. 3. Qualitative electron tunnel barrier view of the superexchange mechanism. (a): direct two-level ET. (b): superexchange mediated three-level ET for different intermediate state energies. The lowest level corresponds to resonance tunnelling.

3.2 Low-energy intermediate states

When the potential surfaces are located as in fig.2b, ET is a two-step process. Full evaluation of the rate expressions in eq.(9) by means of quasiclassical,[40a,b] trajectory[40d] or vibrational density matrix theory reveals,[40c] however, that the overall ET process cannot be regarded simply as two independent two-level transitions. The second order nature of the process implies that the vibrational modes do not relax in the intermediate state, and the system proceeds to the final state prior to nuclear relaxation. This is reflected in the following high-temperature form of the rate constant for motion along a single harmonic displaced mode of frequency Ω[40a]

$$ W_{fi}^{(2)} = \frac{2\pi^{3/2}(V_{fd}^{(1)})^2(V_{di}^{(1)})^2}{\hbar(\hbar\Omega)(E_r^{di}E_r^{fd})^{\frac{1}{2}}(|U_{di}-U_{fd}|k_BT)^{\frac{1}{2}}}\exp\left[-\frac{1}{k_BT}(U^*-U_{io})\right] \quad (14) $$

where U^* is the highest of the two intersection points, E_r^{di} and E_r^{fd} the reorganization energies for the $i\to d$ and $d\to f$ transitions, respectively, and $|U_{di}-U_{fd}|>k_BT$ the energy difference between the two intersection points.

The unrelaxed character of the nuclear motion in the intermediate electronic state can be appreciated by noting that the pre-exponential factor in eq.(14) takes the form of the product of two Landau-Zener factors in the diabatic limit, i.e.[40]

$$ W_{fi}^{(2)} = \frac{2}{\sqrt{\pi}}\Omega\gamma_{di}\gamma_{fd}\exp\left[-\frac{1}{k_BT}(U^*-U_{io})\right] \quad (15) $$

$$ \gamma_{di}=2\pi(V_{di}^{(1)})^2/\hbar\upsilon_{di}\sqrt{2\hbar\Omega E_r^{di}}; \quad \gamma_{fd}=2\pi(V_{fd}^{(1)})^2/\hbar\upsilon_{fd}\sqrt{2\hbar\Omega E_r^{fd}} \quad (16) $$

where υ_{di} and υ_{fd} are the two nuclear velocities at the appropriate intersection points. The velocity at the highest intersection point corresponds to thermal velocity, say $\upsilon_{di}=\sqrt{2k_BT\Omega/\hbar}$ whereas the lower point has a higher velocity arising from the conversion of $|U_{di}-U_{fd}|$ to kinetic energy, i.e. $\upsilon_{fd}=\sqrt{2|U_{di}-U_{fd}|\Omega/\hbar}$. This form is clearly different from the rate constant for two independent transitions.

This formalism has been generalized to quantum mechanical terms and discussed extensively elsewhere[40]. It is appropriate to early charge separation and recombination in photosynthesis if external electric fields can shift the accessory bch monomer level to values low enough that $U_{di}, U_{fd} < U_{fi}$. The shift from high- to low-energy intermediates in the superexchange mechanism is accompanied by characteristic "breaks" in the free energy relations based on external field variations. This effect is known in electrochemical ET at film covered electrodes[48]. The superexchange formalism and its quantum mechanical generalization have also been the basis for discussion of intramolecular bridge-assisted ET[40e], conformational relaxation in protein systems[50], multiphonon ("phonon kick") transitions in crystalline semiconductors[51], and recently extended to resonance Raman scattering theory for molecules in solution[52].

3.3 Environmental modulation effects on the superexchange electronic factor

The dynamic environmental modulation effects on the electronic transition matrix elements in superexchange through high-energy intermidiate states can be approached in two ways. The intermediate groups can be regarded solely as lowering the orbital exponent in an exponential representation, and the overall process otherwise as following the two-level process pattern. This would rightway extend the procedure in section 2.2 to superexchange.

A more detailed view emerges if the two transition matrix elements are considered separately. Modulation is then determined by the two "local" two-level transfer coefficients which can be obtained once the relative location and the topology of the potential sufaces are known. A general outcome, apparent from fig.2a is that characteristically one of the two transitions involving the intermediate state, corresponds to the "normal" and the other one to the "inverted region". If the obital exponent of the high-energy intermediate wave function is substantially smaller than those of the initial and final states, then the superexchange mechanism exhibits weaker modulation than direct ET. On the other hand, when the intermediate state orbital exponent is comparable to or higher than those of the initial and final states, modulation effects can be significant since ET occurs in strongly distorted nuclear configurations for both states.

The same considerations apply to superexchange via low-lying intermediate states, where each of the two electronic transition matrix elements can correspond either to the "normal" or the "inverted" region depending on the relative location of the three surfaces.

4. ENVIRONMENTAL ELECTRONIC FLUCTUATION EFFECTS IN MODIFIED METALLOPROTEINS

Environmental fluctuation effects are reflected in characteristic shifts and distortions of the free energy relations of long-range ET. In this section we shall estimate the effects for two metalloprotein systems, i.e. intramolecular ET in myoglobin (Mb) modified by attachment of redox active ruthenium complex fragments to surface histidines[16], and intermolecular ET in non-covalently bound modified cyt c/cyt b5 (cb5) complexes[18]. The estimated effects are significant but hard to substantiate experimentally. This is due to the limited quantitative diagnostic validity of free energy relations, to which the effects primarily can be referred.

4.1 Electronic tunnel modulation in intramolecular ET of modified myoglobins

In the expanding class of biological long-range intramolecular ET processes, data for distance and free energy variation of the reaction[16]

$$Ru^{3+} - Mb \rightarrow Ru^{3+} - Mb^* \xrightarrow{k_r} Ru^{2+} - Mb^+ \rightarrow Ru^{3+} - Mb \qquad (17)$$

have recently become available. Mb^* is the first excited singlet state of the Mb heme or substituted (Zn, Pd) heme and k_r the ET constant to be investigated. The distance variation is obtained by Ru-modification at the his-12, -48, -81 and -116 positions and the free energy variation by modification with different Ru-fragments, or metal substitution in the mesoporphyrin IX group. In these ways the centre-to-centre ET distance can be varied from 12 to 22 Å and the reaction free energy from zero to -1 ev.

The myoglobin systems possess the following additional important properties:

(a) The ET reactions belong to the "normal" free energy range, with a total reorganization free energy of about 2 ev. The transfer coefficient varies from 0.5 for $\Delta G_0 \approx 0 ((NH_3)_5 Ru(48)MbFe)$ to 0.25 for $\Delta G_0 \approx 1 \, ev ((NH_3)_4 py Ru(48)MbPd)$.

(b) The electron donor is an excited heme group and substantially less localized than the strongly confined acceptor state at the Ru-site, and in all cases the condition $\lambda_i(\theta^*) < \lambda_f(\theta^*)$ can be regarded as valid. this has been substantiated by ab initio calculations[34] and also implies that the maximum of the overlap is close to the acceptor group.

(c) The experimental distance variation can be approximated by the relation $exp[-0.9(R - R_{min})](R \, in \, \AA)$. By the procedure in section 2.2 this gives $\lambda_{io} = 0.6 \, \AA^{-1}$ for the donor orbital equilibrated with the environmental polarization, in line with several activationless processes where the electron configuration is that of the initial state equilibrium.[34]

(d) The free energy variation takes the approximate high-temperature form[24]

$$W_{fi}(\theta^*) \quad \alpha \quad exp[2\lambda_{io}\xi\theta^* R - \beta E_s \theta^{*2}] \tag{18}$$

where the first term in the exponent arises from the modulated electronic factor and the second term from the nuclear factor. Eq.(18) and fig. 1a show that the dominating effect of the modulation is a red-shift of the whole free energy relation by 0.15-0.2 ev for R = 10-15 Å. In addition, the value of E_s extracted from experimental data is larger than in the absence of electronic modulation and the relation broader because the maximum in the modulation-corrected relation is at a larger $|\Delta G_0|$-value than appears directly from the data.

(e) The estimated effects are important but hard to substantiate in other ways than from free energy relations or corresponding optical charge transfer bandshapes. Their distance and temperature dependence could, however, be the basis for investigation of T-dependent shifts of free energy relations or corresponding shifts on surface-attachment of families of Ru-fragments at different surface sites.

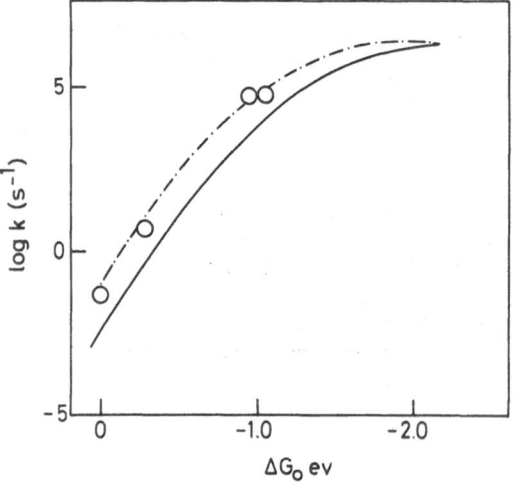

Fig. 4. Free energy relation, normalized to maximum at ΔG_0 =-2 ev for ET in the modified myoglobin system. (----): modulation of the electronic factor included to fit the data points of ref.16. ( ): corresponding plot when the modulation effects are omitted.

4.2 Electronic modulation in intermolecular ET in modified cb₅ complexes

A similar but less accurate analysis has been undertaken for intermolecular ET in non-covalently bound cb$_5$, of the kind[18]

$$^3(Zn-porph)^*c/Fe^{III}b_5 \rightarrow (Zn-porph^+)c/Fe^{II}b_5 \quad etc. \quad (19)$$

and involve coplanar heme groups, a centre-to centre ET distance of 16 Å and in most cases an electronically excited donor obital and a strongly confined acceptor orbital. The free energy range spans nearly an ev and extends into the "inverted" range. A similar analysis as in section 4.1 reveals the following[35]:

(a) The orbital exponent of the electronic transmission coefficient emerges as $\alpha_{app} = 1.5 Å^{-1}$ from the estimated rate constant at edge-to-edge contact. This gives for the orbital exponents of the wave functions $\lambda_{io} \approx 0.75 Å^{-1}$ and $\lambda_{fo} \geq 1.5 Å_{io}$ if $\lambda_{io} < \lambda_f(\theta^*)$ whereas $\lambda_{fo} \approx 1.5 Å^{-1}$ is the result that emerges if $\lambda_{io} > \lambda_f(\theta^*)$. Neither of these can a priori be ruled out.

(b) The former alternative leads to a shift and a corresponding broadening of the free energy relation can thus be significant also for this system.

5. ENVIRONMENTAL ELECTRON TUNNEL FLUCTUATION EFFECTS IN FAST CHARGE SEPARATION AND RECOMBINATION IN BACTERIAL PHOTOSYNTHESIS

5.1 Superexchange ET involving pheophytin

The rate constant of the primary fast activationless charge separation from excited chlorophyll dimer, $(bch)_2^*$ to pheophytin (reaction free energy $\Delta G_0 \approx -0.24 ev$[53])

$$(bch)_2^* bchbph \rightarrow (bch)_2^+ bchbph^- \quad (20)$$

at room temperature is 3.6 10^{11} s^{-1} for both Rps. Sphaeroides and Viridis[54]. This value rises to 8.4 10^{11} and 14.3 10^{11} s^{-1}, respectively in the low-T limit[54]. Most recent femto- and pico-second time-resolved spectroscopy favours a single superexchange mediated ET step via a virtual accessory bch anion radical state[55-58]. This is substantiated by the long (17 Å) direct ET distance and the low energy of the bch radical state (0.09 ev above the $(bch)_2^*$ state[53]). In comparison, the (bch)$_2$/bch and bch/bph centre-to-centre distances are 11 and 10 Å, respectively. The energetics of eq.(20) is illustrated in fig.2a where it is also assumed that the two environmental reorganization energies are approximately equal. In general E_r^{fi} would then be smaller than in fig.2a and a full graphic representation should rest on higher-dimensional surfaces

Recombination occurs when forward ET is blocked by quinone reduction or depletion[53,59,60]. The dominating path is singlet-triplet radical pair mixing induced by hyperfine interactions, followed by triplet decay to the ground state, either directly or via the lowest (bch)$_2$ triplet state. The latter is located 0.15 ev below the radical pair triplet state and proceeds to the ground state by intersystem crossing. In addition, direct singlet recombination to the ground state carries about 10% of the overall decay[53,59,60].

The room temperature rate constant for triplet and singlet decay is $6 \cdot 10^8$ and $6 \cdot 10^7$ s^{-1}, respectively. The former is independent of the T in the range 90-290°K, while the latter has dropped to $1.5 \cdot 10^7$ s^{-1} at 90°K[53]. If the reorganization free energy of the forward ET estimated from the activationless behaviour of this reaction is representative also for the back reactions, then triplet decay is approximately activationless, while singlet decay should belong to the inverted free energy range, $-\Delta G_0$ being 1.1-1.2 ev.

The analysis below appears to reveal a more entangled ET pattern. We shall attend particularly to the following elements:

(a) The effect of environmental modulation of the electronic factors.

(b) Relation between superexchange ET and magnetic interactions.

(c) The T-dependence of both the dispersive nuclear factor and the electronic factor.

We consider in turn forward and recombination ET.

5.1.1 The forward superexchange and direct ET

The activationless rate constant for the dispersive medium is

$$ W_{fi}^{act.1.} = \frac{1}{\hbar}(V_{fi})^2 \left\{ \frac{\hbar E_s}{\pi c} \int_0^\infty d\omega \frac{Im\epsilon(\omega)}{|\epsilon(\omega)|^2} cth\left(\frac{1}{2}\beta\hbar\omega\right) \right\}^{-1/2} \quad (21) $$

Comparison of eqs.(6) and (21) with the data gives $V_{fi} = 5.6 \cdot 10^{-3}$. Using the exponential representation of V_{fi} in eqs.(12) and (13) gives $\alpha_{eff} = 0.38$ A^{-1} if $V_0^{(1)} \approx$ 10 ev and $R_{fi} = 17$ Å. This value has the following implications:

(A) α_{eff} is very close to the value obtained by direct insertion of the equilibrium value of α from the bph/Q_A reaction into eq.(11) (cf. section 5.2 and estimates in section 3.1).
This formal consistency leaves little room for environmental fluctuation effects which would lead to significantly smaller α_{eff}-values than 0.38 A^{-1} in the superexchange mode.

(B) $\alpha_{eff} = 0.38$ A^{-1} would both be consistent with the bph$^-$/Q_A parameters and concur with expectations for strong environmental fluctuation effects if the process is viewed as direct, two-level ET, with no role for accessory bch. This view would, however, disregard several other observations concerning the role of this molecule[53,55-58].

(C) In separate consideration of the two-level transition matrix elements in the superexchange electronic factor (eq.(10)) the i → d transition corresponds to the activationless and the d → f transitions to the inverted region. Both acceptor orbitals would be strongly distorted at the ET moment if environmental fluctuation were important. Such effects would give larger superexchange ET transition matrix elements than obtained from magnetic interactions, such as observed recently[61]. However, since the superexchange mode leaves little room for fluctuation effects the difference between superexchange transition matrix elements estimated from thermal ET and magnetic data must be accounted for by the different nuclear configurations and corresponding energy denominators in eq.(11)[55].

5.1.2 The (bch) $\frac{+}{2}$/bph⁻ recombination

Presently we attend only to singlet decay which we shall regard as direct or superexchange mediated ET from bph⁻ to the (bch) $\frac{+}{2}$ ground state. The values of E_s ≈ 0.24 ev and $\Delta G_0 = -(1.1-1.2)$ ev estimated from the forward ET suggest that the process is strongly exothermic. Reactions which are not too strongly exothermic can still be handled by the saddle point method, giving an approximately Gaussian free energy dependence, (cf. eq. (21))

$$W_{fi} = 2\frac{\sqrt{\pi}}{\hbar\Delta}(V_{fi})^2 \exp[-(E_s + \Delta G_0)^2/\Delta^2]$$

$$\Delta = 2\left[\frac{\hbar E_s}{\pi c}\int_0^\infty d\omega \frac{Im\epsilon(\omega)}{|\epsilon(\omega)|^2} cth\left(\frac{1}{2}\beta\hbar\omega\right)\right]^{\frac{1}{2}} \qquad (22)$$

where $\Delta \to 2\sqrt{E_s k_B T}$ in the high-T limit. Cubic and quartic terms must be added to the exponent for still more strongly exothermic processes.[62,63] Procedures for relaxing the saddle point approximation altogether are also available[38].

In the following we shall use eq.(22). The transfer coefficient is $\theta^* = (2k_B T/\Delta^2)(E_s + \Delta G_0)$. $|\theta^*|$ increases significantly with increasing T in in the range where the nuclear motion is totally or partially quantum mechanically frozen but assumes a constant, larger value at higher T. We can therefore expect that the nuclear factor in eq.(22) increases with increasing T in the inverted region, wereas the electronic factor decreases if environmental electronic modulation is important (cf. eq.(8)).

The small value of E_s ≈ 0.24 ev gives both a far too small nuclear factor in W_{fi} and far too strong T-variation, so that V_{fi} has to assume unphysically large values in order to fit the data. Physically reasonable, i.e. small values of V_{fi} can only be obtained either if E_s is significantly larger than 0.24 ev or if $|\Delta G_0|$ is significantly smaller than 1.1-1.2 ev.

If ΔG_0 is assumed to be fixed, infinitely many combinations of V_{fi} and E_s can still be brought to fit the data, but constrictions can be invoked by the following considerations:

(A) V_{fi} must be smaller than for the forward reaction since the acceptor orbital now represents a strongly confined ground state and since the energy denominator in the superexchange matrix element is larger. This contriction gives a lower limit of $\Delta = \Delta^{low}$ ≈ 0.143 ev, or E_s ≈ 0.41 ev at 120 °K and Δ^{low} ≈ 0.171 ev or E_s ≈ 0.28 ev at T = 298 °K, calculated from eqs.(6) and (28) for $\omega_R = \tau_R^{-1} = 150 cm^{-1}$. In either case both Δ, θ^* and the nuclear part of W_{fi}, $W_{fi}^{nucl} = W_{fi}/(V_{fi})^2$ increase strongly with increasing T, as shown in table 1. The only way the overall rate constant can remain approximately constant over the whole T-range is then that the electronic factor rises by a similar factor. This requires both that the electronic factor is subject to strong-environmental polarization fluctuations and that the bph radical orbital exponent, λ_0^{bph}, is significantly smaller than that of the (bch)$_2^+$, λ_0^{bch}. However, using the values in table 1, the transmission coefficient can increase by no more than an order of magnitude unless the orbital exponent assumes unphysically large values.

Table 1. Nuclear rate parameters from eqs.(6) and (22) based on the upper limit of $V_{fi} = 5 \cdot 10^{-3}$ ev at 120°K or 298°K

T^0K	Δ^{low} ev	E_s ev	$\Delta(298)/\Delta(120)$	$W_{fi}^{nucl}(298)/W_{fi}^{nucl}(120)$	$\theta^*(298)/\theta^*(120)$
120	0.143	0.41	1.45	$1.2 \cdot 10^6$	1.181
298	0.171	0.28	1.45	$1.5 \cdot 10^{13}$	1.181

(B) A lower limit of V_{fi} can be obtained if an upper limit of E_s is invoked. This limit is suitably chosen as the value for recombination from quinone, i.e. 0.5-0.6 ev (cf. section 5.2). Variation of the nuclear factor over the 90-120°k T-range is now three orders of magnitude (table 2) while the observed variation of the overall rate constant is a factor of four. Using the data in table 2 the ratio of the transmission coefficients at the two temperatures calculated from eq.(8) is about $\exp(2.83 \lambda_0^{bph})$, which would leave a discrepancy of two orders of magnitude between observed and calculated temperature variations if λ_0^{bph} is the same as in forward ET, or smaller due to the superexchange effect.

Table 2. Rate parameters estimated from ET data[53] and $E_s = 0.55$ ev.

T^0K	Δ	W_{fi}^{nucl}	θ^*	V_{fi} from data	V_{fi}^{dir} calc.	V_{fi} Superex calc.
90	0.162	$6 \cdot 10^{-6}$	-0.35	$1.2 \cdot 10^{-6}$ ev	$1.0 \cdot 10^{-5}$ ev	$5.3 \cdot 10^{-4}$ ev
120	0.171	$3.5 \cdot 10^{-5}$	-0.41	$3 \cdot 10^{-6}$	$7.4 \cdot 10^{-6}$ ev	$4.2 \cdot 10^{-4}$ ev
298	0.250	$1.3 \cdot 10^{-2}$	-0.48	$3 \cdot 10^{-7}$	$5.0 \cdot 10^{-6}$ ev	$3.2 \cdot 10^{-4}$ ev

(C) Alternatively, Δ or E_s can be estimated from the experimental data if V_{fi} is known. If we use the λ_0^{bph} value estimated from the forward bph$^-$/Q$_A$ reaction, i.e. $\lambda_0^{bph} \approx 0.7$ Å^{-1}, then $\lambda_0^{bph}(\theta^*) \approx 0.85$ Å^{-1} for direct and $\lambda_0^{bph}(\theta^*) \approx 0.7$ Å^{-1} for superexchange recombination at 298 °K. This gives the value of V_{fi} in last two columns in table 2, while E_s would be 0.45 ev and 0.36 ev respectively. This is still not far from the values compatible with the overall T-dependence of the rate constant.

It can be noted in conclusion that if singlet decay is regarded as a two- or three-level electronic transition in the inverted free energy range, then the approximate T-independence of the rate constant conceals strong opposite temperature effects in the electronic and nuclear parts. However even if E_s is somewhat higher or $|\Delta G_0|$ somewhat lower than estimated on the basis of forward ET data, the observed

T-dependence is still far weaker than estimated from the T-dependence of the electronic and nuclear factor, unless other quantities (E_S, ΔG_0) also exhibit T-dependence.

5.2 The forward and back reactions involving the primary quinone

The nearest intermediate group in these reactions are environmental amino acids the lowest antibonding orbital energies of which are much higher than that of the bch monomer. We shall therefore regard these processes as direct ET where, however, the distance variation of the electronic factor may still reflect the nature of the environments and superexchange effects.

The halflives ($\tau_{1/2}$), reaction free energies and approximate centre-to-centre distances of these two reactions are[23]

	$\tau_{\frac{1}{2}}$	$\Delta G_0 \, ev$	$R_{fi} \, \AA$
$bph^- + Q_A \rightarrow bph + Q_A^-$	0.2 ns	-0.6	13
$Q_A^- + (bch)_2^+ \rightarrow Q_A + (bch)_2$	0.05 s	-0.5	28

$$(23)$$

The T-dependence suggest that both reactions are close to the activationless free energy range and the rate constant represented by eq.(22). Discussion elsewhere[35,64] also shows that quantitative agreement between eq.(22) and experimental data in the whole T-range can only be obtained if additional T-dependent effects such as thermal expansion, relative donor-acceptor group motion, conformational reorganization, phase transitions etc., are combined with eq.(22).

From the low-T limit of eq.(22) and the data of refs. 64 and 65 the diabatic transition matrix elements are $V_{fi} = 6.3 \ 10^{-4}$ ev for forward ET and $4.6 \ 10^{-8}$ for recombination. Assuming that V_{fi} has the same exponential form as above, practically the same orbital exponent of $\alpha \approx 0.7 \ \AA^{-1}$ is obtained for both reactions. The widely different values of V_{fi} thus primarily reflect the different ET distances and suggest that dynamic environmental effects on the electronic factor might be important.

In estimating the environmental fluctuation effects, we invoke the constriction that the quinone orbital exponent at equilibrium, λ_0^{quin}, must be the same for the two processes (even though the ET directions are slightly different). It is then convenient to distinguish between the following two limits:

5.2.1 Weak environmental fluctuation effects

The constriction and the data are compatible with all three equilibrium orbital exponents taking approximately the same value, i.e. $\lambda_0^{bph} \approx \lambda_0^{quin} \approx \lambda_0^{bch} \approx 0.7 \ \AA^{-1}$. This is, however unlikely since λ_0^{bch} represents a strongly confined bonding orbital whereas λ_0^{quin} and λ_0^{bph} represents weakly bonding anion radical states, so that $\lambda_0^{bch} \gg \lambda_0^{quin}, \lambda_0^{bph}$. This condition means that the value $\lambda_0^{quin} \approx 0.7 \ \AA^{-1}$ emerges from recombination, with maximum overlap close to the (bch)$_2$ dimer acceptor. If λ_0^{quin} is the same for the two reactions, the only result for λ_0^{bph} compatible with both data sets is then $\lambda_0^{bph} \approx \lambda_0^{quin} \approx 0.7 \, \mathring{A}^{-1} \ll \lambda_0^{bch}$.

5.2.2 Strong environmental fluctuation effects

Fig.1, and the discussion in section 2[24] show that $\lambda_i(\theta^* \approx 0) \approx \lambda_{io}$ for activationless processes whereas $\lambda_f(\theta^* \approx 0)$ is significantly smaller than λ_{fo}. If we maintain the constrictions that λ_0^{quin} is the same for the two reactions, and $\lambda_0^{bph} \approx \lambda_0^{quin} \ll \lambda_0^{bch}$, then either of the following two outcomes are formally compatible with the data:

(1) If $\lambda_0^{quin} < \lambda^{bch}(\theta^* \approx 0)$ where $\lambda^{bch}(\theta^* \approx 0) < \lambda_0^{bch}$ then $\lambda_0^{quin} \approx 0.7 \, \mathring{A}^{-1}$ from the second reaction. Moreover, since fig.1 shows that $\lambda^{bch}(\theta^* \approx 0) \approx 0.5\lambda_0^{bch}$, then $\lambda_0^{quin} \approx 0.7 \, \mathring{A}^{-1}$ means that $\lambda_0^{bch} > 1.4 \, \mathring{A}^{-1}$. This value is an upper limit for a medium with a dielectric constant of 10 and could be lower, i.e. approaching 1-1.2 Å^{-1} for less polar media. For this to be compatible with the first reaction, λ_0^{bph} must also be about 0.7 Å^{-1}, i.e. $\lambda_0^{bph} \approx \lambda^{bph} \approx \lambda_0^{quin} \approx \lambda^{quin}$ for this reaction. This implies that modulation must be insignificant for the forward reaction, since otherwise $\lambda^{quin}(\theta^* \approx 0) < \lambda_0^{bph}$ and we should have obtained α smaller than 0.7 Å^{-1} for this reaction.

(2) If $\lambda_0^{quin} > \lambda^{bch}(\theta^* \approx 0)$ in the second reaction, then $\lambda_0^{quin} > 0.7 \, \mathring{A}^{-1}$ and $\lambda_0^{bch} \approx 1.4 \, \mathring{A}^{-1}$ emerges from the data. This condition for λ_0^{quin} is only compatible with the first reaction if also $\lambda_0^{quin} > 1.4 \, \mathring{A}^{-1}$ so that $\lambda_0^{bph} < \lambda^{quin}(\theta^* \approx 0)$. Otherwise $\lambda_0^{bph} > \lambda^{quin}(\theta \approx 0)$ and we should have observed a smaller α.

On the other hand, $\lambda_0^{quin} > 1.4 \, \mathring{A}^{-1}$ is a far too large value for a weakly bound anion radical state. The most likely outcome is therefore that environmental fluctuation effects on the electron tunnel factor can only be significant for the very long-range back ET and must be insignificant for the shorter-range forward ET. The most likely equilibrium orbital exponents are $\lambda_0^{bph} \approx \lambda_0^{quin} \approx 0.7 \, \mathring{A}^{-1} \ll \lambda_0^{bch}$. The data do not provide an unambiguous clue as to whether the fluctuation effects are with certainty important. If they are, then λ_0^{bch} must exceed a value of approximately 1-1.4 Å^{-1} depending on the dielectric properties of the protein medium. This value is reasonable for a stable and strongly confined state.

Environmental fluctuation effects should be reflected in free energy relations for the recombination reaction and in particular in temperature shifts of these correlations. Extensive data of this kind based both on quinone substitution[66] and external field effects have been reported[64]. We have recently provided an analysis of these data in the framework of section 2.[35] The conclusion was that the system parameters appropriate to these reactions should lead to noticeably temperature dependent free energy red-shifts. The data do not conflict with this view, but since most of the data points are located in a broad shallow activationless free energy range, they are unfortunately insufficiently diagnostic on this important but subtle point of ET theory.

6. CONCLUDING REMARKS

Investigations of the free energy and temperature dependence of the electron tunnel factor in long-range ET suggest that this effect can be significant but also that its magnitude is rather system specific. It appears little important in the primary superexchange charge separation(17 Å) and forward bph^-/Q_A ET (13 Å) in bacterial photosynthesis. It is possibly important in singlet recombination from bph^- to $(bch)_2^+$ if this process is regarded as an inverted two- or three- level transition, and it is likely to be important in the very long-range (28 Å) charge recombination from Q_A^-. Finally, it is possibly important in the intramolecular ET reactions in the modified metalloprotein systems. The diverse picture which emerges could reflect the additional effects of local charges and spatial dielectric dispersion in the inhomogeneous protein medium, both of which exert a profound influence on the sensitivity of the electronic charge distribution to environmental nuclear configurational fluctuations[24].

ACKNOWLEDGEMENT

We would like to thank the Danish Natural Science Research Council for financial support of this work.

REFERENCES

1. Frauenfelder, H., Petsko, G.A. and Tsernoglou, D. (1979) Nature 280, 558.
2. McGammon, J.A. and Karplus, M. (1983) Ann.Rev.Phys.Chem. 31, 29.
3. Welch, G.R. (ed.) (1986) The Fluctuating Enzyme, Whiley, New York.
4. Sumi, H. and Ulstrup, J. (1988) Biochim.Biophys.Acta 955, 26.
5. For a review, see: Kuznetsov, A.M. and Kharakats, Yu.I. (1987) in Kazarinov, V.E. (ed.) The Interface Structure and Electrochemical Processes at the Boundary between Two Immiscible Liquids, Springer-Verlag, Berlin, p. 11.
6. Kjær, A.M. and Ulstrup, J. (1987) Inorg.Chem. 26, 2052.
7. For a recent review, see: Kuznetsov, A.M. , Ulstrup, J. and Vorotyntsev, M.A. (1988) in Dogonadze, R.R., Kalman, E., Kornyshev, A.A. and Ulstrup, J. (eds.) The Chemical Physics of Solvation. Part C, Elsevier, Amsterdam, p. 163.
8. Sykes, A.G. (1985) Chem.Soc.rev. 14, 283.
9. Nettesheim, D.G., Johnson, W.V. and Feinberg, B.A. (1980) Biochim.Biophys.Acta 593, 371.
10. Armstrong, F.A. (1982) in Sykes, A.G. (ed.) Adv.Bioinorg.Mech. 1, 65.
11. Nocera, D.G., Winkler, J.R., Yocom, K.M., Bordignon, E. and Gray, H.B. (1984) J.Amer.Chem.Soc. 106, 5145.
12. Isied, S.S., Kuehn, C. and Worosila, G. (1984) J.Amer.Chem.Soc. 106, 1722.
13. Osvath, P., Salmon, G.A. and Sykes, A.G. (1988) J.Amer.Chem.Soc. 110, 7114.
14. Kostic, N.M., Margalit, R., Che, C.-M. and Gray, H.B. (1983) J.Amer.Chem.Soc. 105, 7765.
15. Jackman, M.P., Lim, M.-C., Sykes, A.G. and Salmon, G.A. (1988) J.Chem.Soc.Dalton Trans. 2843.
16. Axup, A.W., Albin, M., Mayo, S.L., Crutcley, R.J. and Gray, H.B. (1988) J.Amer.Chem.Soc.110, 435.

17 Peterson-Kennedy, S.E., McGourty, J.L., Kalweit, J.A. and Hofman, B.M. (1986) J.Amer.Chem.Soc. 108, 1739.
18. a. McLendon, G. (1988) Acc.Chem.Res. 21, 160; b. Conklin, K.T. and McLendon, G. (1988) J.Amer.Chem.Soc. 110, 3345.
19. For a review, see: Armstrong, F.A., Hill, H.A.O. and Walton, N.J. (1988) Acc.Chem.Res. 21, 407.
20. Degani, Y. and Heller, A. (1987)J.Phys.Chem. 91, 1285.
21. Kuznetsov, A.M., Bogdanovakaya, V.A., Tarasevich, M.R. and Gavrilova, E.F. (1987) FEBS Lett. 215, 219, and references there.
22. Deisenhofer, J., Epp, O., Miki, K., Huber, R. and Michel, R. (1984) J.Mol.Biol. 180, 385.
23. For recent reviews, see: a. Michel-Beyerle, M.E. (ed.) (1985) Antennas and Reaction Centres of Photosynthetic Bacteria, Springer-Verlag, Berlin; b. Breton, J. and Vermeglio, A. (eds.) (1988) The Photosynthetic Bacterial Reaction Centre, Structure and Dynamics, Plenum, New York.
24. a. Kuznetsov, A.M. (1981) Nouv.J.Chim. 5, 427; b. Kuznetsov, A.M. and Ulstrup, J. (1982) Faraday Disc. Chem.Soc. 74, 31; c. Kuznetsov, A.M. and Ulstrup, J. (1986) in Jortner, J. and Pullman, B. (eds.) Tunnelling, Reidel, Dordrecht, p. 345.
25. Dogonadze, R.R., Kuznetsov, A.M., Vorotyntsev, M.A. and Zakaraya, M.G. (1977) J.Electroanal.Chem. 75, 315.
26. Itskovitch, E.M. and Kuznetsov, A.M. (1980) Elektrokhimiya 16, 755.
27. Jortner, J. (1976) J.Chem.Phys. 64, 4860.
28. Ray, P.S. (1972) Appl.Optics 11, 1836.
29. Fletcher, N.H. (1970) The Chemical Physics of Ice, Cambridge University Press, Cambridge.
30. Go, N. Noguti, T. and Nishikawa, T. (1983) Proc.Nat.Acad.Sci.USA 80, 1683.
31. Frolich, H.A. (1958) Theory of Dielectrics, 2nd ed., Clarendon, Oxford.
32. Ovchinnikov, A.A. and Ovchinnikova, M.Ya. (1969) Zhur.Eksp.Tero.Fiz. 56, 1278.
33. Kuznetsov, A.M. and Ulstrup, J. (1986) Chem.Phys. 107, 381.
34. Mikkelsen, K.V., Ulstrup, J. and Zakaraya, M.G. (1989) J.Amer.Chem.Soc. 111, 1315.
35. Kuznetsov, A.M. and Ulstrup, J. (1989) Bioelectrochem.Bioeng., in press.
36. a. Anderson, P.V. (1950) Phys.Rev. 79, 350; b. Halpern, J. and Orgel, L.E. (1960) Disc.Faraday Soc. 29, 32; c. McConnell, H.M. (1961) J.Chem.Phys. 35, 508.
37. Jortner, J. and Bixon, M. (1987) in Austin, R., Bukhs, E., Chance, B., DeVault, D., Dutton, P.L., Frauenfelder, H. and Goldanskij, V.I. (eds.) Protein Structure, Molecular and Electronic Reactivity, Springer-Verlag, New York, p.277.
38. Zakaraya, M.G. and Ulstrup, J. in preparation.
39. Kornyshev, A.A. (1985) in Dogonadze, R.R., Kalman, E., Kornyshev, A.A. and Ulstrup, J. (eds.) The Chemical Physics of Solvation. Part A, Elsevier, Amsterdam, 1985 p. 77.
40. a. Levich, V.G., Madumarov, A.K. and Kharkats, Yu.I. (1972) Dokl.Akad.Nauk SSSR, Ser.Fiz.Khim. 203, 135; b. Dogonadze, R.R., Kharkats, Yu.I. and Ulstrup, J. (1972) J.Electroanal.Chem. 39, 47; Kharkats, Yu.I., Madumarov, A.K. and Vorotyntsev, M.A. (1974) J.Chem.Soc.Faraday Soc.II 70, 1578; d. Kuznetsov, A.M. and Kharkats, Yu.I. (1976) Elektrokhimiya 12, 1277;e. Kuznetsov, A.M. and Ulstrup, J. (1981) J.Chem.Phys. 75, 2047.
41. Newton, M.D. (1980) Int.J.Quant.Chem.Symp. 14, 363.
42. Larsson, S. (1981) J.Amer.Chem.Soc. 103, 4034.

43. Ohta, K., Closs, G.L. Murukuma, K. and Green, N.J. (1986) J.Amer.Chem.Soc. 108, 1319.
44. Beratan, D.N. and Hopfield, J.J. (1984) J.Amer.Soc. 106, 1584.
45. Mikkelsen, K.V., Dalgaard, E. and Swanstrøm, P. (1987) J.Phys.Chem. 91, 3081.
46. Duke, C.B. (1979) in Chance, B., DeVault, D.C., Frauenfelder, H., Marcus, R.A., Schrieffer, J.R. and Sutin, N. (eds.) Tunnelling in Biological Systems, Academic Press, p. 31.
47. Kuznetsov, A.M. and Ulstrup, J. (1982) Phys.Stat.Sol. 114, 673.
48. Ulstrup, J. (1980) Surf.Sci. 101, 564, and references there.
49. Schmickler, W. and Schultze, J.W. (1986) Mod.Asp.Electrochemistry 17, 347.
50. Dogonadze, R.R., Kuznetsov, A.A. and Ulstrup, J. (1977) J.Theor.Biol. 69, 239.
51. Sumi, H. (1984) J.Phys. C 17, 6071.
52. Zakaraya, M.G. and Ulstrup, J. (1989) Chem.Phys., in press.
53. Ogrodnik, A., Volk, M. and Michel-Beyerle, M.E. (1987) in ref. 23b, p. 177.
54. Fleming, G.R., Martin, J.L. and Breton, J. (1988) Nature 333, 190.
55. Bixon, M., Jortner, J., Michel-Beyerle, M.E., Ogrodnik, A., and Lersch, W. (1987) Chem.Phys.Lett. 140, 626.
56. Bixon, M. and Jortner, J. (1988) J.Phys.Chem. 92, 7148.
57. Michel-Beyerle, M.E.; Bixon, M. and Jortner, J. (1988) Chem.Phys.Lett. 151, 188.
58. Michel-Beyerle, M.E., Plato, M., Deisenhofer, J., Michel, M., Bixon, M. and Jortner, J. (1988) Biochim.Biophys.Acta 932, 52.
59. Schenck, C.C., Blankenship, R.R. and Parson, W.W. (1985) Biochim.Biophys.Acta 680, 44.
60. Goldenstein, R.A., Tahiff, K. and Boxer, S.G. (1988) Biochim.Biophys.Acta 934, 253, and references there.
61. a. Marcus, R.A. (1987) Chem.Phys.Lett. 133, 471; b. (1988) Chem.Phys.Lett. 146, 13.
62. Itskovitch, E.M., Ulstrup, J. and Vototyntsev, M.A. (1986) in Dogonadze, R.R., Kalman, E., Kornyshev, A.A. and Ulstrup, J. (eds.) in The Chemical Physics of Solvation. Part B, Elsevier, Amsterdam, p. 223.
63. Kjær, A.M. and Ulstrup, J. (1987) J.Amer.Chem.Soc. 109, 1934.
64. Feher, G., Okamura, M. and Kleinfeld, D. (1987) in Austin, R., Bukhs, E., Chance, B., DeVault, D.C., Dutton, P.L., Frauenfelder, H. and Goldanskij, V.I. (eds.) Molecular and Electronic Reactivity, Springer-Verlag, New York, p. 399.

SOLVENT POLARITY EFFECTS ON THERMODYNAMICS AND KINETICS OF INTRAMOLECULAR ELECTRON TRANSFER REACTIONS

H. HEITELE, S. WEEREN, F. PÖLLINGER, AND M. E. MICHEL-BEYERLE
Institut für Physikalische und Theoretische Chemie
Technische Universität München
Lichtenbergstraße 4
8046 Garching, FRG

ABSTRACT. We present time-resolved fluorescence measurements on an intramolecular electron transfer compound in solvents of different polarity as a function of temperature. In a few solvents the free energy change of the electron transfer and its dependence on temperature and the dielectric constant of the solvent can be determined experimentally. Electron transfer rates are analyzed in terms of activation energies and preexponential factors. Both are found to depend on the polarity of the solvent. The latter is interpreted as evidence for significant entropies of activation.

1. Introduction

Since the classical papers by Marcus [1] and Hush [2] on the theory of electron transfer reactions in condensed media the solvent has been identified as an essential factor influencing the rates of this kind of processes. In many electron transfer reactions charges are separated or recombine. Solvent stabilization of these charges determines the thermodynamics of the overall reaction [3] to a great extent. Solvent reorganization is a major contribution to the activation barrier [4-8] which must be overcome during the transfer. In very fast reactions or in solvents with long dielectric relaxation times solvent reorientation dynamics [9] can become the rate determining factor of the process. Finally the mediation of the electronic coupling via the solvent by a superexchange mechanism is likely to be responsible for fast electron transfer over large distances between unconnected donors and acceptors [10] and possibly even in some bridged systems [11].

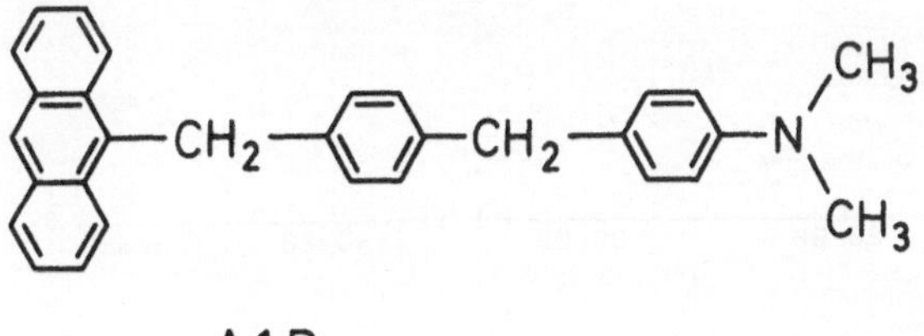

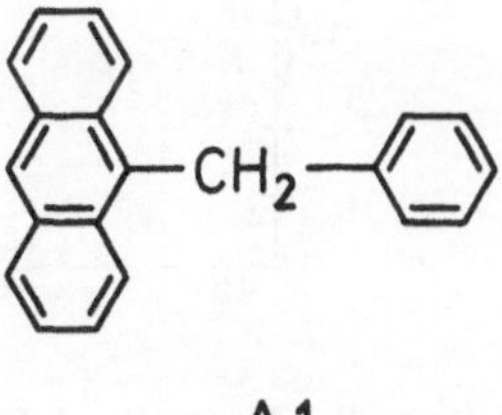

Figure 1. Structure formulas of the donor/acceptor compound A1D and the reference substance A1.

J. Jortner and B. Pullman (eds.), Perspectives in Photosynthesis, 261–271.
© 1990 *Kluwer Academic Publishers.*

In the course of our own investigations on electron transfer reactions [12–16] we have studied a series of molecules in which an electron donor is covalently bound to an electron acceptor via different molecular bridges with variing donor/acceptor distances. One of these molecules (A1D) together with a reference substance (A1) is shown in Figure 1.

In a previous publication [14] we used the temperature dependence of the charge separation rate in these molecules to determine the solvent reorganization energy. To this aim the assumption of temperature independent energies of reaction and solvent reorganization was explicitly made. In this work we present experimental data in other solvents which indicate that the inclusion of entropic terms in these quantities could be necessary to account for the solvent dependence of the electron transfer rate.

2. Measurements and Results

We have measured fluorescence decay curves of A1D and A1 in the following series of solvents with the static dielectric constants ϵ at room temperature given in parentheses [17]: trichloroethylene (3.4), diisopropyl ether (3.9), diethyl ether (4.35), pentyl acetate (4.75), butyl acetate (5.0), ethyl acetate (6.0), methyl acetate (6.7), 1,2-dichloroethane (10.4), pyridine (12.3), acetone (21.6), propionitrile (29.6), methanol (32.6), and dimethylformamide (36.7) in the temperature range from 190 to 350 K.

In the nonpolar solvents hexane [12] and trichloroethylene the fluorescence decay times of A1 and A1D are practically the same. There is no evidence for intramolecular charge transfer. In the polar solvents from ethyl acetate to dimethylformamide the fluorescence decay in A1D is essentially monoexponential. The fluorescence lifetime, however, is much shorter in A1D compared to A1. In the weakly polar solvents from diisopropyl ether to butyl acetate the fluorescence behaviour is more complex. For illustration fluorescence decay curves of A1D in butyl acetate at different temperatures are shown in Figure 2.

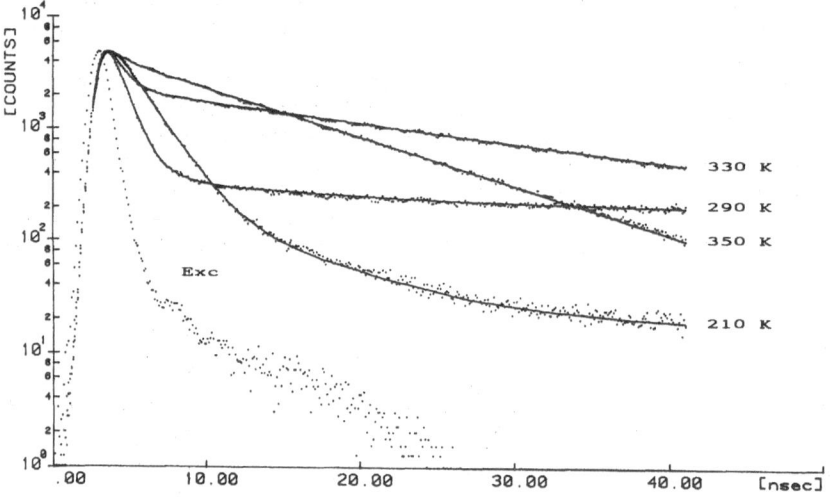

Figure 2. Normalized fluorescence decay curves of A1D in butyl acetate at different temperatures. (exc = excitation pulse)

The fluorescence decay in A1D is clearly biexponential with a short-lived component whose relative amplitude rapidly grows to almost 100 % for lower temperature T and a long-lived component with an increasing lifetime up to about 200 ns but rapidly decreasing amplitude.

As was shown in refs. [12-15] optical excitation of the anthracene moiety of A1D in polar solvents induces an intramolecular electron transfer from the dimethylaniline-residue to the excited anthracene thereby quenching its fluorescence. The pertinent kinetic scheme is shown in Figure 3.

In strongly polar solvents the free energy of charge separation ΔG_{CS} is high enough to neglect the back reaction k_{-CS}. This case is realized in the polar solvents from ethyl acetate to dimethylformamide. In nonpolar solvents such as hexane [12] and trichloroethylene the (free) energy of the radical pair state A^-1D^+ is above the energy of the locally excited singlet state A^*1D, i.e., the charge transfer reaction is thermodynamically unfavourable. In the weakly polar solvents from diisopropyl ether to butyl acetate the energy of the radical ion pair state is close to the locally excited state. In this case the locally excited state can be repopulated by thermal activation from the ion pair state which leads to the observed biexponential fluorescence decay.

In the case of essentially monoexponential fluorescence decay the intramolecular charge separation rate k_{CS} is given by: $k_{CS} = 1/\tau - 1/\tau_0$ where τ is the fluorescence lifetime in A1D and τ_0 in A1.

In the weakly polar solvents with biexponential decay the charge separation rate k_{CS}, the free energy change ΔG_{CS} of the charge separation (or, alternatively, the back reaction rate k_{-CS}) as well as the recombination rate into the ground state k_{Cr} can be determined from an analysis of the two fluorescence components and their relative amplitudes. The free energy change as a function of temperature is shown in Figure 4. ΔG_{CS} decreases with rising ϵ and strongly increases with T.

In polar solvents k_{CS} exhibits a conventional activated temperature dependence as was previously found in propionitrile [14]. There is a good linear correlation between $\ln(\sqrt{T}k_{CS})$ and $1/T$ as suggested by nonadiabatic electron transfer theory [18]. The pertinent preexponential factor A and the activation energy ΔE^* defined by $k_{CS} = A/\sqrt{T} \cdot \exp(-\Delta E^*/k_B T)$ are summarized in Table I.

TABLE I. Experimentally determined activation energies ΔE^* and preexponential factors A in polar solvents.

Solvent	A $[\sqrt{K}s^{-1}]$	ΔE^* [kJ/mol]
ethyl acetate	$1.5 \cdot 10^{11}$	4.2
methyl acetate	2.2 ·	4.2
1,2-dichloroethane	3.3 ·	3.1
pyridine	29 ·	6.7
acetone	5.9 ·	4.2
propionitrile	7.8 ·	4.5
methanol	12 ·	7.7
dimethylformamide	53.6 ·	7.5

The activation energies are small and no simple relation between ϵ and the energy of activation is recognizable whereas the preexponential factor changes by a factor of 40 and seems to increase with the dielectric constant.

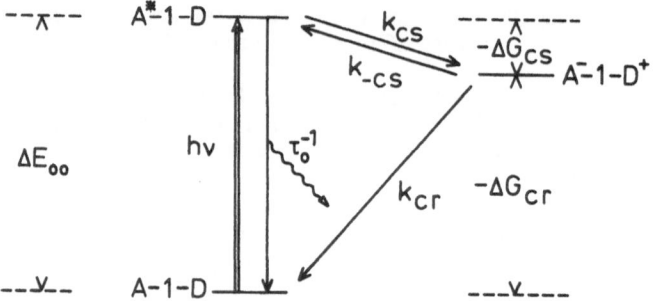

Figure 3. Photoinduced electron transfer cycle in A1D.

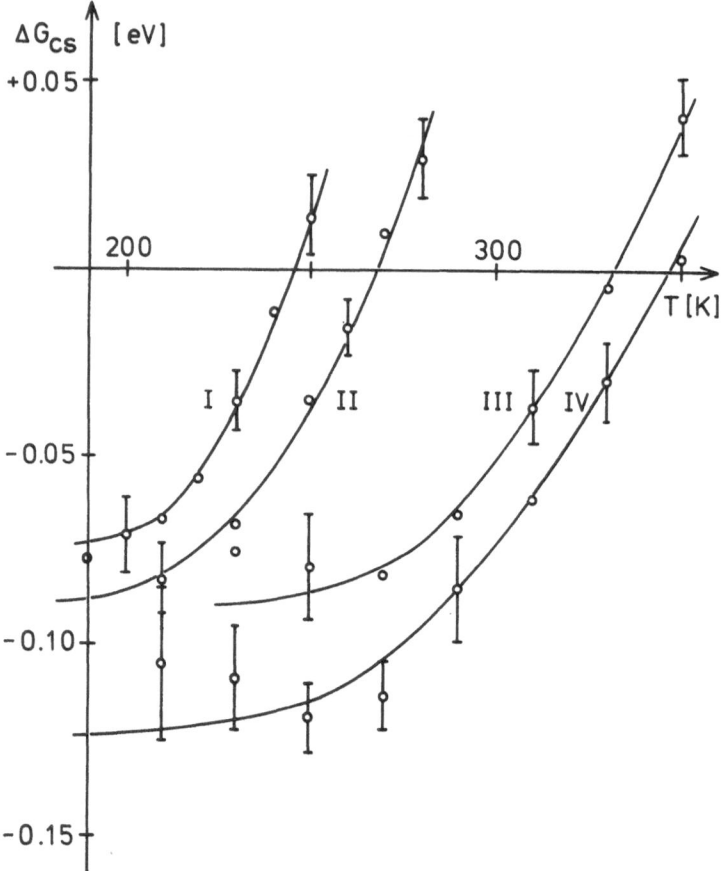

Figure 4. Temperature dependent free energy of charge separation ΔG_{cs} in moderately polar solvents determined experimentally from the biexponential fluorescence decay: I diisopropyl ether, II diethyl ether, III pentyl acetate, IV butyl acetate.

3. Discussion

3.1. THERMODYNAMICS

For the quantitative interpretation of the experiments we begin with the effects of solvent polarity and temperature on the free energy change ΔG_{CS} of the charge separation reaction. ΔG_{CS} is approximately given by:

$$\Delta G_{CS} = E_{ox} - E_{red} - \frac{e^2}{d\epsilon} - \Delta E_{oo} \qquad (1)$$

with the excitation energy ΔE_{oo} of the acceptor, the redox potentials E_{ox} and E_{red} of donor and acceptor and the Coulomb energy $e^2/d\epsilon$ of the ion pair with the center to center distance d (d=9 Å). ΔE_{oo} in A1 or A1D is very weakly dependent on the solvent. The redox potentials are only known in strongly polar solvents, however, and must be corrected for different solvation in less polar ones. This is done using Born's equation for the free energy of solvation ΔG_{solv} of an ion with an effective radius $R_{\pm}$: $\Delta G_{solv} = e^2/2R_{\pm} \cdot (1/\epsilon - 1)$. In the continuum approximation (CA) $R_{\pm}$ is estimated from the effective volumes of the neutral molecules [14,15]. In the mean sphere approximation (MSA) [19,20] the molecular nature of the solvent is incorporated in a dynamic correction $\Delta_{\pm}$ to the radius of the solute $R_{\pm}'=R_{\pm}+\Delta_{\pm}$ which depends on ϵ and the effective radius r_s of a solvent molecule [20]:

$$\Delta_{\pm} \simeq 3r_s \cdot \frac{1}{4.76 \cdot \epsilon^{1/6} - 2} \qquad (2)$$

Using ΔG_{CS}=-0.59 eV [14] in acetonitrile (ϵ=37.5) we get the curves in Figure 5 for the dependence of ΔG_{CS} on ϵ in the CA and the MSA.

The experimental free energy of reaction appears to lie halfway between the CA and MSA predictions - at least for small $|\Delta G_{CS}|$. For a crude estimate of ΔG_{CS} for solvents with dielectric constants between 6 and 37.5 it seems reasonable to interpolate between the CA and the MSA [21]. Numerically this is afforded by the relation:

$$\Delta G_{CS} \simeq 0.67 \cdot \left[\frac{4.4}{\epsilon} - 1 \right] \quad \text{(in eV)} \qquad (3)$$

shown as the dashed line in Figure 5. The enthalpies ΔH_{CS} and entropies ΔS_{CS} of reaction are given by: $\Delta S_{CS} = - \partial \Delta G_{CS}/\partial T$, $\Delta H_{CS} = \Delta G_{CS} - T \cdot \partial \Delta G_{CS}/\partial T$. Experimentally $\Delta S_{CS} = -140 \pm 20$ J/K·mol and $\Delta H_{CS} = -50 \pm 10$ kJ/mol (-0.52 eV) are obtained in butyl acetate at room temperature. On the basis of eq. (3) the temperature dependence of ΔG_{CS} is due to the dependence on ϵ:

$$\Delta S_{CS} = - \frac{\partial \Delta G_{CS}}{\partial \epsilon} \cdot \frac{\partial \epsilon}{\partial T} \simeq 2.85 \cdot 10^5 \cdot \frac{1}{\epsilon^2} \cdot \frac{\partial \epsilon}{\partial T} \quad \text{(in J/K·mol)} \qquad (4)$$

Using tabulated data for $\epsilon(T)$ [17] eq. (4) gives $\Delta S_{CS} \simeq -160$ J/K·mol and $\Delta H_{CS} \simeq -55.7$ kJ/mol (-0.58 eV) in butyl acetate, i.e., the experimental entropies (and enthalpies) of reaction are consistent with the predictions based on these simple energetic estimates. In methanol $\Delta S_{CS} \simeq -54$ J/K·mol and $\Delta H_{CS} \simeq -71.4$ kJ/mol (-0.74 eV) are obtained which implies that except for nonpolar solvents the dependence of ΔG_{CS} on ϵ is essentially an entropic effect.

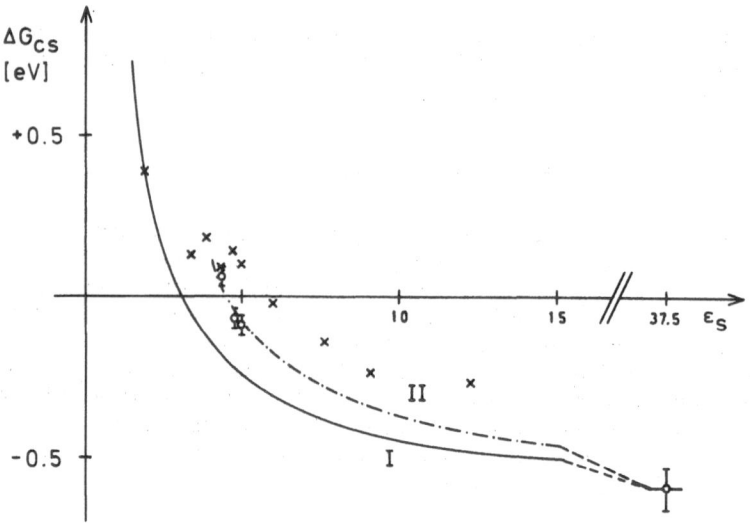

Figure 5. Calculated dependence of the free energy of charge separation ΔG_{CS} on the static dielectric constant ϵ of the solvent; (I) continuum approximation, (x) mean sphere approximation, (o) experiment, (II) interpolation between experiments.

3.2. ACTIVATION PARAMETERS

For the discussion of the rates the following well-known expression for the nonadiabatic electron transfer rate will be used [22,23]:

$$k_{CS} = \frac{2\pi}{\hbar} \frac{V^2 \cdot e^{-S}}{\sqrt{4\pi k_B T \lambda_s}} \cdot \sum_{m=0}^{\infty} \frac{S^m}{m!} \cdot e^{-\frac{(\Delta G_{CS} + \lambda_s + m\hbar\omega)^2}{4k_B T \lambda_s}} \tag{5}$$

($S = \lambda_i / \hbar\omega$, λ_s = solvent reorganization energy, λ_i = intramolecular reorganization energy, ω = frequency of a typical intramolecular vibration mode, V = electronic coupling matrix element). The free energies of reaction ΔG_{CS} are taken from above. The solvent reorganization energies are estimated using [1]:

$$\lambda_s = e^2 \left[\frac{1}{2R_+} + \frac{1}{2R_-} - \frac{1}{d} \right] \cdot \left(\frac{1}{n^2} - \frac{1}{\epsilon} \right) = B \cdot \left(\frac{1}{n^2} - \frac{1}{\epsilon} \right) \tag{6}$$

with B=2.0 eV (R_-=4.1 Å, R_+=3.9 Å, d=9 Å [14]) and the refractive index n. The other reorganization parameters are λ_i=0.44 eV and ω=1500 cm^{-1} [14]. The electronic factor V^2 is chosen so that eq. (5) reproduces the experimental rate in propionitrile at 298 K.

Experimental rates at 298 K and theoretical estimates based on eqs. (5,6) are compared in Figure 6. Experiment and theory roughly agree. In the moderately polar esters the theoretical prediction seems to underestimate the rate which might be due to an overestimate of the solvent reorganization energy. Yet, the rate changes only weakly with the polarity of the solvent. As other authors have pointed out before [6] this is due to a partial compensation of the changes of ΔG_{CS} and λ_S for different solvents.

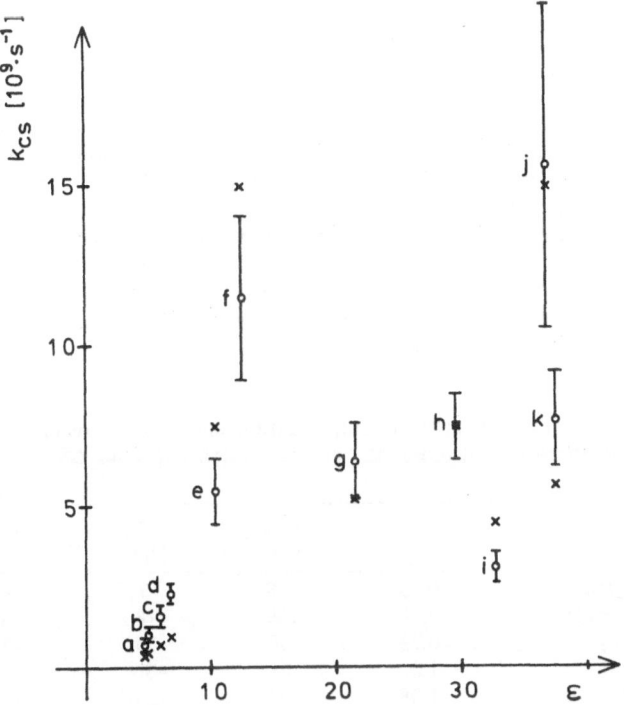

Figure 6. Comparison of experimental rates (o) at 298 K and theoretical estimates (x) depending on the dielectric constant of the solvent; (a) pentyl acetate, (b) butyl acetate, (c) ethyl acetate, (d) methyl acetate, (e) 1,2-dichloroethane, (f) pyridine, (g) acetone, (h) propionitrile, (i) methanol, (j) dimethylformamide, (k) acetonitrile. Theoretical estimates are normalized to the experimental rate in propionitrile.

For a qualitative analysis of the temperature dependence of k_{CS} the following simplified version of eq. (5) is sufficient:

$$k = \frac{2\pi}{\hbar} \frac{V^2 \cdot e^{-S}}{\sqrt{4\pi k_B T \lambda_S}} \cdot e^{-\frac{(\Delta G_{CS} + \lambda_S)^2}{4 k_B T \lambda_S}} = \frac{2\pi}{\hbar} \frac{V^2 \cdot e^{-S}}{\sqrt{4\pi k_B T \lambda_S}} \cdot e^{\frac{\Delta S^*}{k_B}} \cdot e^{-\frac{\Delta H^*}{k_B T}} \qquad (7)$$

where the weak temperature dependence of λ_s in the preexponential factor $\propto 1/\sqrt{\lambda_s}$ is neglected. The activation enthalpy ΔH^* is then equated with the experimental activation energy ΔE^* whereas the solvent dependence of the preexponential factor is incorporated in $1/\sqrt{\lambda_s}\cdot\exp(\Delta S^*/k_B)$. Introducing entropies ΔS_λ and enthalpies ΔH_λ of solvent reorganization as: $\lambda_s = \Delta H_\lambda - T\cdot\Delta S_\lambda$, enthalpy ΔH^* and entropy ΔS^* of activation are given by:

$$\Delta S^* = -\frac{\partial \Delta G^*}{\partial T} = \frac{2(\Delta G_{cs}+\lambda_s)(\Delta S_\lambda+\Delta S_{cs})\cdot\lambda_s-\Delta S_\lambda(\Delta G_{cs}+\lambda_s)^2}{4\lambda_s^2} \tag{8}$$

$$\Delta H^* = \frac{(\Delta G_{cs}+\lambda_s)^2}{4\lambda_s} + T\cdot\Delta S^* \tag{9}$$

ΔG_{cs}, λ_s, and ΔS_{cs} are taken from above; $\Delta S_\lambda = -(\partial\lambda_s)/(\partial T)$ is estimated from eq. (6) using temperature dependent dielectric constants and refractive indices. The results for ΔS_{cs}, λ_s, ΔS_λ and the activation entropy ΔS^* are summarized in Table II. Experimental and theoretical preexponential factors A and $1/\sqrt{\lambda_s}\cdot\exp(\Delta S^*/k_B)$ are compared in Figure 7. The theoretical values are normalized by a common factor so as to optimize the agreement with the experimental ones. Figure 7 shows that the above estimate correctly predicts changes of about 40 in the preexponential factor. There is indeed a general increase of the activation entropy with the dielectric constant of the solvent. Altogether the agreement between theory and experiment is encouraging.

TABLE II. Calculated solvent reorganization energies λ_s, reaction entropies ΔS_{cs}, reorganization entropies ΔS_λ and activation entropies ΔS^*.

Solvent	λ_s [eV]	ΔS_{cs} [J/K·mol]	ΔS_λ [J/K·mol]	ΔS^* [J/K·mol]
ethyl acetate	0.73	-118	7	-42
methyl acetate	0.78	-139	18	-45
pyridine	0.73	-60	-21	-17
acetone	1.0	-61	-36	-21
methanol	1.08	-54	-31	-18
dimethylformamide	0.93	-51	-27	-14

When these activation entropies are inserted into eq. (9) as for example for methanol and ethyl acetate small positive or even negative activation enthalpies of $\Delta H^* = 184$ J/mol and -2.6 kJ/mol result, however, in contrast to the experiments. The reason for this discrepancy is not yet clear. It might be due to the crude approximations made above or to a small steric activation energy to achieve the transition state which could be "hidden" in the experimental activation energies ΔE^*. At present there is no further evidence for the latter argument. Force field calculations indicate that the optimum conformation for electron transfer is also the energetically most favoured one (or very close to it). Nor is there any correlation between the viscosity of the solvents used here and the electron transfer rate.

The possibility of a negative activation enthalpy was discussed before [24]. As the estimates above show this should not be an uncommon situation. Yet, there is scant experimental evidence for such a behaviour. The only reaction we are aware of where a negative enthalpy of activation was measured is the oxidation of $Fe(H_2O)_6^{2+}$ by $Ru(terpy)_2^{3+}$ [25]. In other cases a decrease of the rate with increasing temperature was attributed to an acti-

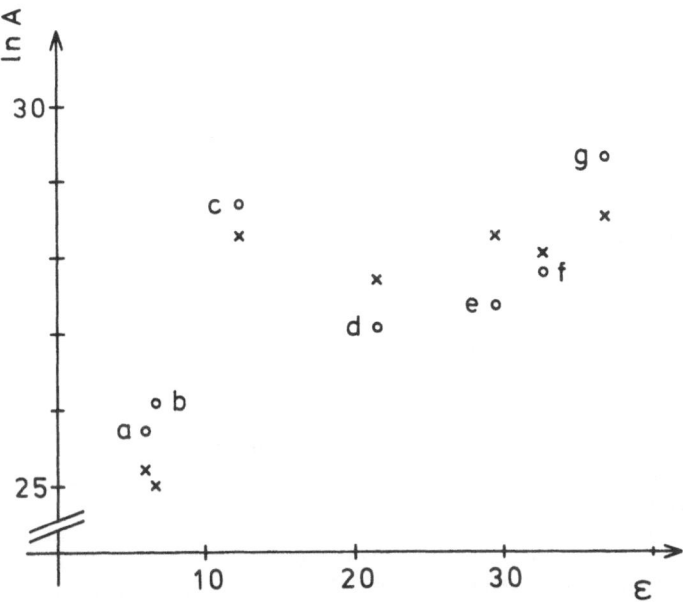

Figure 7. Experimental (o) and calculated (x) preexponential factors as a function of the dielectric constant ϵ of the solvent; (a) ethyl acetate, (b) methyl acetate, (c) pyridine, (d) acetone, (e) propionitrile, (f) methanol, (g) dimethylformamide. Calculated values are normalized by a common factor so as to optimize the agreement with the experiments.

vationless reaction such as in the first electron transfer step between the special pair and the bacteriopheophytin in the photosynthetic reaction center of certain purple bacteria or in reactions of certain transition metal compounds [26] to a more complex mechanism, i.e., the observed negative enthalpy of activation is an effective one and not due to a single 'elementary' electron transfer step.

4. Conclusions

In the preceding discussion we have shown that temperature and polarity changes may affect electron transfer reactions in different ways, even in a seemingly simple system. We have presented experimental data for polarity effects on the free energy change of a charge separation reaction as well as on the preexponential factors and activation energies of the corresponding rates. The strong temperature dependence of the free energy of reaction measured in weakly polar solvents shows that in the system investigated here different entropies of reaction are mainly responsible for the influence of solvent polarity on the driving force of the reaction. Conventional thermodynamic estimates provide a satisfactory description of this effect. A completely consistent quantitative description of the temperature dependence of the rate is not yet possible. Yet, there is strong evidence that the activation entropy associated with the reorganization of the solvent is the reason for the variation of the preexponential factor of the rate in different solvents.

Acknowledgements

We gratefully acknowledge the financial support by the Deutsche Forschungsgemeinschaft.

References

[1] R. A. Marcus, *J. Chem. Phys.* **24**, 988 (1956); *J. Chem. Phys.* **43**, 679 (1965).
[2] N. S. Hush, *Elektrochem.* **61**, 734 (1957); *J. Chem. Phys.* **28**, 962 (1958); *Trans. Faraday Soc.* **57**, 557 (1961).
[3] A. Weller, *Z. Phys. Chem. Neue Folge* **133**, 93 (1982).
[4] M. D. Newton and N. Sutin, *Annu. Rev. Phys. Chem.* **35**, 437 (1984); and references therein.
[5] R. A. Marcus and N. Sutin, *Biochim. Biophys. Acta* **811**, 265 (1985); and references therein.
[6] H. Oevering, M. N. Paddon-Row, M. Heppener, A. M. Oliver, E. Cotsaris, J. W. Verhoeven, and N. S. Hush, *J. Am. Chem. Soc.* **109**, 3258 (1987).
[7] J. A. Schmidt, A. Siemiarczuk, A. C. Weedon, and J. R. Bolton, *J. Am. Chem. Soc.* **107**, 6112 (1985).
[8] M. D. Archer, V. P. Y. Gadzekpo, J. R. Bolton, J. A. Schmidt, and A. C. Weedon, *J. Chem. Soc., Faraday Trans. 2* **82**, 2305 (1986).
[9] L. D. Zusman, *Chem. Phys.* **49**, 295 (1980). A. B. Helman, *Chem. Phys.* **65**, 271 (1982). I. Rips and J. Jortner, *J. Chem. Phys.* **87**, 2090 (1987). H. Sumi and R. A. Marcus, *J. Chem. Phys.* **84**, 4272 (1986); 4894 (1986). A. Garg, J. N. Onuchic, and V. Ambegaokar, *J. Chem. Phys.* **83**, 4491 (1985). D. F. Calef and P. G. Wolynes, *J. Phys. Chem.* **87**, 3387 (1983). D. Huppert, V. Ittah, and E. M. Kosower, *Chem. Phys. Lett.* **144**, 15 (1988).
[10] J. R. Miller and J. V. Beitz, *J. Chem. Phys.* **74**, 6746 (1981). V. Krongauz, R. K. Huddleston and J. R. Miller, to be published.
[11] A. M. Oliver, D. C. Craig, M. N. Paddon-Row, J. Kroon, and J. W. Verhoeven, *Chem. Phys. Lett.* **150**, 366 (1988).
[12] H. Heitele and M. E. Michel-Beyerle, *J. Am. Chem. Soc.* **107**, 8068 (1985).
[13] H. Heitele, M. E. Michel-Beyerle, and P. Finckh, *Chem. Phys. Lett.* **134**, 273 (1987).
[14] P. Finckh, H. Heitele, M. Volk, M. E. Michel-Beyerle, *J. Phys. Chem.* **92**, 6584 (1988).
[15] H. Heitele, S. Weeren, F. Pöllinger, and M. E. Michel-Beyerle, *J. Phys. Chem.*, in press.
[16] H. Heitele, P. Finckh, and M. E. Michel-Beyerle, *Angew. Chem.* **101**, in press.
[17] R. C. Weast, *Handbook of Chemistry and Physics*, CRC Press, Cleveland, 1971. J. A. Riddick and W. B. Bunger, *Organic Solvents*, Wiley-Interscience, New York, 1970.
[18] V. O. Levich, *Adv. Electrochem. Electrochem. Eng.* **4**, 249 (1966). N. R. Kestner, J. Jortner, and J. Logan, *J. Phys. Chem.* **78**, 2148 (1974). R. P. Van Duyne and S. F. Fischer, *Chem. Phys.* **5**, 183 (1974). J. Ulstrup and J. Jortner, *J. Chem. Phys.* **63**, 4358 (1975).
[19] P. G. Wolynes, *J. Chem. Phys.* **86**, 5133 (1987). A. L. Nichols and D. F. Calef, *J. Chem. Phys.* **89**, 3783 (1988).
[20] I. Rips, J. Klafter, and J. Jortner, *J. Chem. Phys.* **88**, 3246 (1988); **89**, 4288 (1988).
[21] This estimate ignores specific solvent-solute interactions. Recently, the existence of a simple relation between ΔG_{cs} and ϵ has been questioned [8].
[22] J. Jortner, *J. Chem. Phys.* **64**, 4860 (1976).
[23] R. A. Marcus, *J. Chem. Phys.* **81**, 4494 (1984).
[24] R. A. Marcus and N. Sutin, *Inorg. Chem.* **14**, 213 (1975). N. Sutin in *Supramolecular*

Photochemistry, ed. V. Balzani, NATO ASI Series C Vol. 214, D. Reidel, Dordrecht, 1987, p. 73.

[25]   J. N. Braddock and T. J. Meyer, *J. Am. Chem. Soc.* **95**, 3158 (1973).

[26]   N. Kitamura, S. Tazuke, and S. Okano, *Chem. Phys. Lett.* **90** (1982) 13. S. Tazuke, N. Kitamura, and Y. Kawanishi, *J. Photochem.* **29** (1985) 123. S. Tazuke, N. Kitamura, and H.-B. Kim in *Supramolecular Photochemistry*, ed. V. Balzani, NATO ASI Series C Vol. 214, D. Reidel, Dordrecht, 1987, p. 87.

SOLVATION DYNAMICS AND ULTRAFAST ELECTRON TRANSFER

Paul F. Barbara, Tai Jong Kang, Włodzimięrz Jarzeba
Department of Chemistry
University of Minnesota
Minneapolis, Minnesota 55455

Teresa Fonseca
Department of Chemistry
Colorado State University
Fort Collins, Colorado 80523

ABSTRACT. Ultrafast fluorescence spectroscopy has been used to study two processes: (i) the transient solvation of electronically excited coumarin probes and (ii) the solvent mediated excited state intramolecular electron (charge) transfer of 9,9'–bianthryl and related compounds. The solvation measurements have been analyzed in terms of contemporary theory to gain insight on the molecular aspects of microscopic motion in polar liquids. The excited state electron transfer (et) examples are well modeled by an electronically adiabatic approach, employing an "outer sphere" generalized Langevin equation description of motion along the et reaction coordinate. The relationship between solvation and electron transfer dynamics is discussed.

1. INTRODUCTION

In the last few years ultrafast emission spectroscopy on polar fluorescent molecules has led to important new information on the microscopic motion involved in solute/solvent interactions. In particular, there has been a great deal of activity in the study of dynamics of solvation of polar aromatic molecules in polar liquids. The experimental results, along with several innovative and important theoretical advances, have led to an unprecedented knowledge of the molecular aspects of solvation. This paper briefly reviews the present state of knowledge on solvation dynamics in section 2 with emphasis on work from Minnesota.

The other major subject of this paper is the ultrafast electron transfer of electronically excited aromatics in solution, see section 3. This type of process has been recently used by several groups as a prototype for fast electron transfer reactions, in general. The great experimental and theoretical activity in this field lately has emphasized the role of solvation dynamics in fast electron transfer reactions. Section 3 describes an analysis of the new femtosecond resolved data on electron transfer from Minnesota in terms of an outer sphere electron transfer model, involving simulations made at Colorado State employing the generalized Langevin equation.

J. Jortner and B. Pullman (eds.), Perspectives in Photosynthesis, 273–292.

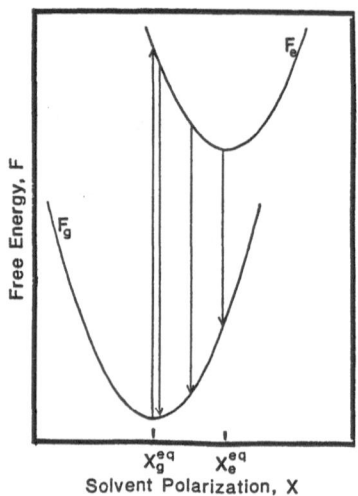

Figure 1.   Nonequilibrium free energy as a function of instantaneous solvent polarization for the ground electronic state S_0 and the excited state S_1 of an ideal probe.   In this example the equilibrium solvent polarization in S_1 (X_e^{eq}) is larger than in S_0 (X_e^{eq}) because the dipole moment is larger in S_1 than S_0.

2. SOLVATION

Time resolved fluorescence spectroscopy of polar fluorescent "probes" that have a dipole moment that depends upon electronic state has recently been used extensively to study microscopic solvation dynamics of a broad range of solvents. The basic concept is outlined in Figure 1, which shows the dependence of the non–equilibrium free energies (F_g and F_e) of solvated ground state and electronically excited probes, respectively, as a function of a generalized solvent coordinate.   Optical excitation (vertical) of an equilibrated ground state probe produces a non–equilibrium configuration of the solvent about the excited state of the probe. Subsequent relaxation is accompanied by a time–dependent fluorescence spectral shift toward lower frequencies, which can be monitored and analyzed to quantify the dynamics of solvation via the solvation dynamics function C(t), which is defined by eq. 1.

$$C(t) = \frac{\nu(t) - \nu(\infty)}{\nu(0) - \nu(\infty)} \tag{1}$$

Here $\nu(0)$, $\nu(t)$, $\nu(\infty)$ represent the frequency of the intensity maximum of the fluorescence spectrum immediately after photon excitation, at some time t after excitation, and at a time sufficiently long to ensure the excited state solvent configuration is at equilibrium.

Until recently, due to the lack of sufficiently short time–resolution, C(t) measurements were limited to slowly relaxing and associated viscous solvents, typically at low temperature[1–14]. In the last two years, the first C(t) measurements of the solvation of ordinary, non–viscous room temperature liquids has been made by subpicosecond and femtosecond fluorescence spectrometers[7,15–18]. Very recently, the first report of a C(t) measurement of the solvation dynamics of water has been published[18]. A few papers have dealt with the potential sources of errors in C(t) measurements, particularly with respect to the properties of the fluorescent probes[7,8,16]. Theoretical activity on solvation dynamics has also blossomed in the recent past[19–35]. Traditionally, solvation dynamics have been described in terms of a simple continuum model[36,37], which treats the solvent as a uniform dielectric medium with exponential dielectric response[37]. The associated dielectric parameters are ϵ_∞, ϵ_0, and τ_D, which are,

respectively, the infinite frequency dielectric constant (approximately n^2), the static dielectric constant, and the dielectric relaxation time.

According to the simple continuum model, the microscopic solvation function C(t) should decay exponentially with a time constant that is approximately given by,

$$\tau_1 \cong \frac{\epsilon_\infty}{\epsilon_0} \tau_D \qquad\qquad (2)$$

where τ_1 is the solvent longitudinal dielectric relaxation time[38–46].

Considerable progress has been made in going beyond the simple Debye continuum model. Non Debye relaxation solvents have been considered. Solvents with non–uniform dielectric properties, and translational diffusion have been analyzed. Furthermore, models which mimic microscopic solute/solvent structure, (such as the linearized mean spherical approximation) but still allow for analytical evaluation have been extensively explored[23,26–28]. Finally, detailed molecular dynamics calculations have been made on the solvation of water[47,48].

Time resolved fluorescence spectra for C(t) determinations have usually been measured by recording several transients at different wavelengths, a technique which is denoted by *spectral reconstruction*. The intensity from each of the transient is adjusted to correct for the wavelength dependent sensitivity of the apparatus by setting the time integral of the transient intensity equal to the intensity from the static fluorescence spectrum. The corrected transients can then be used to reconstruct fluorescence spectra at different times after excitation. Usually, to obtain better time resolution, deconvoluted multiexponential functions are used for reconstruction instead of the experimental transients. Following a procedure of Maroncelli and Fleming [7] the reconstructed time resolved fluorescence spectra are fitted to log–normal line shape function [7,49]. Figure 2 shows examples of time resolved reconstructed spectra and log normal fits for DMACAA in water. The Stokes shift correlation function C(t) can now be calculated using the spectra maxima or their first moments from the log–normal function. The calculated function C(t) for DMACAA in water is shown in figure 3.

2.1 SUMMARY OF C(t) MEASUREMENTS

Transient solvation dynamics of a variety of excited state probe molecules have been studied in the last few years. Table 1 summarizes the measurements that

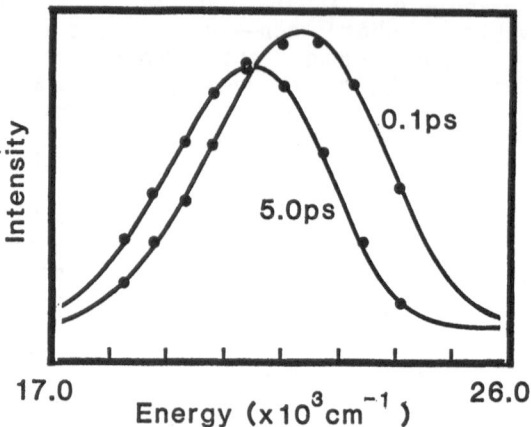

Figure 2. Reconstructed fluorescence spectra of DMACA ion in water at 0.1 and 1 ps after excitation. The solid line represents the best fit of the log–normal distribution function to the data.

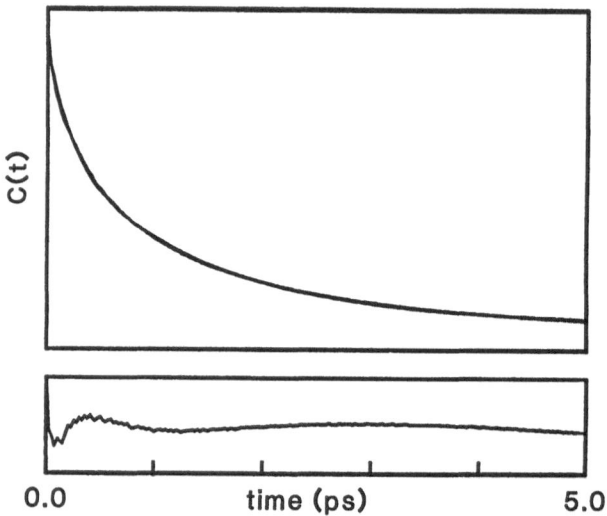

Figure 3. The upper section of the figure is a superposition of an experimentally determined C(t) function for the fluorescing probe DMACA ion in water and a biexponential fit of C(t). The lower section represents the difference between the experimental C(t) and the biexponential fit on a scale that is expanded by a factor of 3 in the Y direction.

have been made at room temperature. For each entry in Table 1, the measured $C(t)$ was fit by either a single exponential function $[\exp(-t/\tau_1)]$ or a biexponential function $[A_1\exp(-t/\tau_1) + A_2\exp(-t/\tau_2)]$. For example, where a single exponential fit was satisfactory, a dash is found in the τ_2 column. On the other hand, when an observed $C(t)$ was better fit by a biexponential form, the best fit times τ_1 and τ_2 and their relative amplitudes (as a percentage) are listed. The average relaxation time $<\tau_s>$ is the zero moment of fitted $C(t)$ (i.e. the amplitude weighted average of the two times). For some measurements, the biexponential fit was only a slight improvement over the single exponential form. In these cases two entries are given for the same measurement.

Some of the measurements in Table 1 were made with $\simeq$ 500 fs resolution. Other results [from 17,50,51] were made with $\simeq$ 50 fs time resolution. The shorter time resolution measurements tend to exhibit short components ($<$ 400 fs) at early times which were not resolved in the subpicosecond measurements. As a result the 50 fs resolved measurements tend to be more obviously biexponential and tend to have shorter average relaxation times. The results for *different probes* in the same solvent are in good agreement, especially if it is taken into account that the 50 fs measurements tend to give shorter times because of better time resolution, as mentioned above.

2.2 AN EVALUATION OF THE DEBYE–ONSAGER MODEL

The simplest treatment for solvation dynamics is the Debye – Onsager model. It assumes that the solvent (i) is well modeled as a uniform dielectric continuum and (ii) has a single relaxation time, τ_D (i.e. the solvent is a "Debye solvent"). The model predicts that $C(t)$ should be a single exponential with a decay time, τ_1. Many of the solvents in Table 1 have been studied by dielectric relaxation methods and the observed $\hat{\epsilon}(\omega)$ values can be well fit by a single relaxation time, Debye time. τ_1 values are therefore available for comparison with experiment. In most cases however, dielectric measurements have not been made at sufficiently high frequency to test whether potentially important, additional relaxation components with short time scales are present for these solvents.

Ignoring the potential limitations of the dielectric data, we can evaluate the Debye–Onsager model for a number of apparently roughly Debye solvents, like propylene carbonate and other solvents [17, 50]. First of all, $C(t)$ is often strongly non–monoexponential, in contradiction to the theoretical prediction. Second, the observed average solvation time $<\tau_s>$ is often much different from τ_1 [17, 50].

2.3. MODERN THEORIES OF SOLVATION

It was suggested by Onsager that the molecular structure of the solvent would cause deviations [19] from the simple uniform dielectric continuum picture, such that the relaxation time for solvent molecules in the vicinity of the probe should be $\simeq \tau_D$, while the relaxation time far from the probe should be $\simeq \tau_1$. Hence relaxation starts far from the probe and proceeds toward the probe (the inverted snowball). In simple terms the liquid far from the probe is like the bulk, hence the solvation time

Table 1. Experimentally observed solvation dynamics at 298 K determined by the correlation function C(t). τ_1, τ_2 relaxation times, $<\tau_s>$ average relaxation time (zero moment of C(t)).

Solvent	Probe	method	τ_1(ps)	τ_2(ps)	$<\tau>$	ref.
MeCN	C102	sw	0.9	—	0.9	16
	C311	sw	0.7	—	0.7	16
	LDS–750	sr	≃0.4	—	≃0.4	7
	C152	sw	0.56	—	0.56	17
EtCN	C102	sw	1.5	—	1.5	16
	C311	sw	1.1	—	1.1	16
	C152	sr	0.31(48%)	1.3(52%)	0.85	17
		sw	0.33(68%)	1.5(32%)	0.7	17
PrCN	C102	sr	1.5	—	1.5	16
		sw	2.1	—	2.1	16
	C311	sw	1.6	—	1.6	16
BuCN	C102	sw	3.6	—	3.6	16
BenzoCN	C152	sw	2.1(39%)	6.1(61%)	4.5	50
		or	4.7[a]	—	4.7	50
MeOAc	C102	sw	1.5	—	1.5	16
	C311	sw	1.8	—	1.8	16
EtOAc	C102	sr	2.7	—	2.7	16
		sw	2.6	—	2.6	16
	C311	sw	2.3	—	2.3	16
PrOAc	C102	sw	4.0	—	4.0	16
BuOAc	C102	sw	6.6	—	6.6	16
DMSO	LDS–750	sr	3.1	—	3.1	7
	C152	sw	0.33(57%)	2.3(43%)	1.2	50
	C153	sw	0.33(44%)	2.2(56%)	1.4	50
DMF	C152	sw	0.40(55%)	1.7(45%)	1.0	50
	C153	sw	0.75(55%)	2.5(45%)	1.5	50
			or 1.4[a]	—	1.4	50
Acetone	C152	sw	0.31(47%)	0.99(53%)	0.67	50
		or	0.70[a]	—	0.70	50
	C153	sw	0.83	—	0.83	50

PC	C102	sw	4.9	—	4.9	16
	C152	sr	0.43(46%)	4.1(54%)	2.4	17
		sw	0.64(58%)	4.8(42%)	2.4	17
	C153	sr	0.48(50%)	6.2(50%)	3.4	50
		sw	0.60(40%)	4.6(60%)	3.0	50
MeOH	LDS–750	sr	3.3	—	3.3	7
	C152	sr	1.2(40%)	9.6(60%)	6.2	17
		sw	1.3(60%)	8.4(40%)	4.1	17
n–PrOH	C153	sr	9.3(32%)[b]	78.8(68%)	57	7
		sr	14(30%)	40(70%)	33	50
		or	33[a]	—	33	50
n–BuOH	LDS–750	sr	61	—	61	7
EG	DMAPS	sr	100	—	100	42
water	DMACA	sr	0.16(33%)	1.2(67%)	0.86	18
	C343	sr	0.25(50%)	0.96(50%)	0.61	51
		sw	0.29(51%)	1.1(49%)	0.54	51

a) for this case the observed solvation dynamics can be represented reasonably well by a single exponential decay.
b) temperature T = 295 K.
c) PC – propylene carbonate, EG – ethylene glycol, DMF – dimethyl formamide, C – coumarin, DMACA – 7–dimethylaminocoumarin–4–acetate–ion, DMAPS – 4,4'–(dimethylamino)phenyl sulfone.
d) sr – spectral reconstruction, sw – single wavelength method

should be τ_1, while the solvent near the probe is more like a single molecule relaxing, hence τ_D is more relevant close to the probe. The first qualitative treatment of Onsager's prediction can be found in the work of Hubbard and Onsager in 1977 who showed that the relaxation of an ion's mobility in polar liquids occurs with a distribution of time scales from τ_1 to τ_D[35]. Calef and Wolynes in 1983 looked in detail at the solvation problem and verified that a distribution of time scales are relevant by numerically solving the Smoluchowski equation for the solvent orientational relaxation[22].

An important advance in the understanding of microscopic solvation and Onsager's snowball picture has recently been made through the introduction of the linearized mean spherical approximation (MSA) model for the solvation dynamics around ionic and dipolar solutes. The first model of this type was introduced by Wolynes who extended the equilibrium linearized microscopic theory of solvation to handle dynamic solvation[23]. Wolynes further demonstrated that approximate solutions to the new dynamic MSA model were in accord with Onsager's predictions. Subsequently, Rips, Klafter, and Jortner published an exact solution for the solvation dynamics within the framework of the MSA[28]. For an ionic solute the exact results from these authors' calculations are in agreement with Onsager's inverted snowball model and the previous numerical calculations of Calef and Wolynes[22]. Recently, the MSA model has been extended by Nichols and Calef and Rips, Klafter, and Jortner [24–28] to solvation of a dipolar solute.

It is interesting to compare the MSA theory with the experimental results. Independently, Maroncelli and Fleming [29], Rips et al [28], and our group [17] have noticed that the MSA model is qualitatively in accord with experiment: both in average solvation time $<\tau_s>$ and the shape of $C(t)$. However, both the MSA ion and MSA dipole theory, tend to predict longer relaxation times than experiment for the long time decay of $C(t)$. The $C(t)$ theoretical predictions for MSA ion and MSA dipole both decay more slowly than experiment at long times. Indeed, the average solvation times $<\tau_s>$ for experiment and theory further demonstrate that the MSA based theories tend to decay more slowly than experiment.

Recent work on the theory of solvation dynamics have attempted to go beyond the linearized MSA model of Wolynes that considers the rotational dynamics of the solvent as the only relaxation mechanism. Certain translational and hydrodynamic–like motions of the solvent are neglected. Bagchi and co–workers [32–35] have explored the role of translational diffusion in the dynamics of solvation by employing a Smoluchowski–Vlasov equation, see also Calef and Wolynes [22] and Nichols and Calef [27]. A significant contribution to polarization relaxation is observed in certain cases. It is found that the Onsager inverted snow ball model is correct only when the rotational diffusion mechanism of solvation dominates the polarization relaxation. The Onsager model significantly breaks down when there is an important translational contribution to the polarization relaxation [32–35]. In fact, translational effects can rapidly accelerate solvation near the probe. In certain cases, the predicted behavior can actually approach the uniform continuum result that $\tau_s = \tau_l$.

2.4 SOLVATION DYNAMICS IN WATER

Water offers a unique opportunity to evaluate theoretical models for solvation. Only for water have extensive molecular dynamics simulations been accomplished. The results in Table 1 clearly indicate that solvation in water is nonexponential. Uniform dielectric continuum estimates give approximately the correct average solvation time.

Recently, several authors have studied solvation dynamics of aqueous solutions using molecular dynamics (MD) computer simulations [21,47,48,52]. The simulations offer a detailed molecular approach to interpreting the experimental results, as they focus particularly on the microscopic, molecular aspects of the solvation process.

Engström et al [52] used molecular dynamics simulations to study quadrapole relaxation mechanism for Li^+, Na^+, and Cl^- ions in dilute aqueous solutions. They found that NMR relaxation rate for these ions was determined by the relaxation of water molecules in the first solvation shell. The simulations show nonexponential solvation dynamics which can be modeled by two relaxation time constants $\tau_1 \lesssim 0.1$ ps and $\tau_2 \cong 1$ ps.

Maroncelli and Fleming [47] studied the time dependence of solvation dynamics of monoatomic ions immersed in large spherical clusters of ST2 water. The simulations for solutes of different size and charge predict nonexponential hydration dynamics with a very short component (10 − 20 fs) due to librational motions of water molecules and a longer nonexponential component with average time constant of a few hundred femtoseconds. Our measurements would not be able to resolve the 10 − 20 fs component. The simulations show that response of the first shell is much faster than that of the second and third shells, which seems to be in contrast to Onsager's inverted snow ball picture. Similar results were reported by Karim et al [48].

3. ULTRAFAST ELECTRON TRANSFER

There is a growing appreciation[53] that fast electron transfer reactions are not well characterized by traditional models (such as Marcus–Hush theory or related non–adiabatic theories) which are based on a quasi–equilibrium approximation for the population of thermally excited states of the solute/solvent systems[54–55]. Contemporary theories treat solvation dynamics in detail. A key result is that the rate of a charge transfer reaction should be a function of the microscopic dynamics of the specific solvent. In fact, in the case of a very small intrinsic charge transfer activation barrier, the rate is predicted to be roughly equal to the rate of solvation (i.e. τ_l^{-1} for a solvent with a single relaxation time, τ_D). This result was first derived over twenty years ago by Mozumder for the neutralization of an isolated ion pair in polar media[56]. The predictions are considerably more complex for solvents with a distribution of relaxation times and barrier energies that are comparable or larger than the available thermal energy.

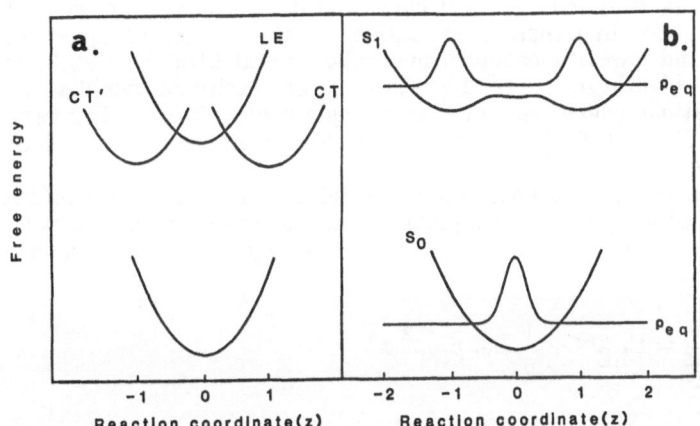

Figure 4.   (a) The zero order energies for the S_0, LE, CT and CT' states as a function of the solvent coordinate z. The parameters have been adjusted in order to bring agreement between predicted and observed spectra for BA in propylene carbonate.   (b) The adiabatic energy for the first excited singlet state (S_1) of BA in propylene carbonate. The equilibrium distribution function is denoted by ρ_{eq}.

The first experimental observation correlating k_{et} with τ_1^{-1} was made by Huppert and Kosower [57]. Many of the theoretical predictions of new models are indeed observed experimentally. Unfortunately, many aspects of the experiments remain poorly defined, so the comparison to theory is often qualitative at best. In other cases, the complexity of the solvation dynamics (non–exponential) and the complexity of the solute solvent interactions make it difficult to make an unambiguous interpretation of the data.

3.1 A PROTOTYPE CHARGE TRANSFER MOLECULE: BIANTHRYL

Excited state charge transfer has been extensively investigated for the molecule BA by a number of research groups [15,58–69]. The fluorescence spectrum of BA is strongly solvent polarity dependent. In polar solvents the emission can be roughly interpreted as being due to the sum of bands from two isomers of S_1; a non–polar LE (locally excited) form and CT (charge transfer) form, as proposed over a decade ago by Nakashima et al [62] and Grabowski et al [70].

In order to explain the polarity dependence of the fluorescence of BA it is necessary to consider the effect of solvent polarity on the energy of the LE and CT states. For simplicity we will consider three elementary electronic configurations resulting from hypothetical excitations of the highest occupied molecular orbital HOMO and lowest unoccupied molecular orbital LUMO on each of the anthracene rings of bianthryl. The LE state is an excitonic doublet of two nonpolar configurations which we model as a single configuration. The two charge transfer forms, CT and CT', have large dipole moments that are in directions opposite to each other.

The effect of solvent on the total free energy of a BA/solvent system is represented in figure 4a for a polar solvent such as propylene carbonate. These zero order (uncoupled) curves for the various configurations correspond to the following equations.

$$F_{LE} = F_{LE}{}^{eq} + \frac{1}{2} B_{or} \mu_{CT}{}^2 z^2 \tag{3}$$

$$F_{CT} = F_{CT}{}^{eq} + \frac{1}{2} B_{or} \mu_{CT}{}^2 (z-1)^2 \tag{4}$$

$$F_{CT'} = F_{CT'}{}^{eq} + \frac{1}{2} B_{or} \mu_{CT'}{}^2 (z+1)^2 \tag{5}$$

Here μ_{CT} and $\mu_{CT'}$ are the dipole moments of the CT and CT' forms. They have the same magnitude but opposite direction. F_x and F_x^{eq} are equilibrium and nonequilibrium free energies, respectively. B_{or} is the orientational component of the force constant of solvation. z is the solvent coordinate which physically corresponds to the dielectric polarization of the solvent that is the result of reorientation of the solvent molecules. The actual values for the parameters employed in constructing figure 4a were determined empirically as described elsewhere.[69]

It is obvious from figure 4a that for a particular configuration of the solvent (particular z value), except for $z=0$, the CT and CT' states are of unequal energy. This effect is an example of local symmetry breaking which was recognized as important for BA several years ago[62].

3.2 AN ADIABATIC POTENTIAL FOR S_1

The various energy curves in figure 4a ignore mixing (configuration interaction) between the various zero order states. In fact, the LE/CT mixing is quite substantial and the adiabatic energy curves of figure 4b are a more accurate representation of S_1 BA than are the diabatic curves of figure 4a. These curves come from mixing the zero order curves of figure 4a assuming two independent matrix elements: $<LE|H'|CT> = <LE|H'|CT'> = 1$ kcal/mol and $<CT|H'|CT'> = 0$. More details are given elsewhere[69]. In terms of the adiabatic curves, the electronic wave function of S_1 BA is represented as follows

$$|S_1> = c_1|LE> + c_2|CT> + c_3|CT'> \qquad (6)$$

The coefficients c_1, c_2, c_3, and the energy of S_1, $E(S_1)$ are *functions of z* that are found by diagonalizing the 3 by 3 Hamiltonian matrix of zero order energies and off diagonal mixing between LE, CT and CT' at each z. For this adiabatic model, the LE to CT interconversion occurs on a single potential energy surface. The methods for describing charge transfers of this type have recently made major progress.

3.3 THE PROBABILITY DISTRIBUTION FUNCTION AND SPECTRA

A key variable for the BA problem is the normalized probability distribution function $\rho(z,t)$. Its equilibrium value $\rho(z,t=\infty)$ can be found from the Boltzman distribution on the S_1 surface, see figure 4b. The physical significance of figure 4b is firmly established by considering how emission spectra are calculated from this model. First it can be assumed that the solvent is stationary on the time scale of emission so that for each specific value of z we can make the Franck–Condon approximation for the emission intensity.

$$I(\omega) \ \alpha \ \beta_{el}^2 \ [<X|X'>]^2 \qquad (7)$$

The electronic part β_{el} is given by

$$\beta_{el} = <\psi_{s1}|\mu|\psi_{s0}> = <c_1 LE + c_2 CT + c_3 CT'|\mu|\psi_{s0}>$$

$$= c_1 \mu_{LE,S_0} + c_2 \mu_{CT',S_0} + c_3 \mu_{CT',S_0} \qquad (8)$$

Here μ_{LE,S_0}, μ_{CT,S_0} and μ_{CT',S_0} are the z independent transition moment matrix elements in terms of zero order states, but $|\mu_{CT,S_0}| = |\mu_{CT',S_0}|$. The Franck – Condon factors in eq. 7 are assumed to be z independent since LE and CT have similar vibrational spectra. It follows simply that each z value contributes the following element to the spectrum for an arbitrary distribution $\rho(z,t)$,

$$dI(\omega,t) \propto g[\omega_0(z),\omega - \omega_0(z)] \, \beta_{el}^2(z) \, \rho(z,t) \, dz \qquad (9)$$

where ω_0 is the origin of the electronic state, i.e. $\omega_0 = [E_{S_1}(z) - E_{S_0}(z)]/\hbar$ and $g(\omega_0, \omega - \omega_0)$ is the normalized shape of the spectrum from the Franck – Condon factors shifted to the appropriate $\omega_0(z)$ value.
The total spectrum is the integral

$$I(\omega,t) \propto \int_{-\infty}^{\infty} g(\omega_0, \omega - \omega_0) \, \rho(z, t) \, \beta_{el}^2(z) \, dz \qquad (10)$$

We have been able to reproduce the equilibrated emission spectra of BA remarkably well by using eq. 10, the potential and $\rho(z,t)$ in figure 4b. The emission shape $g(\omega_0, \omega - \omega_0)$ function was determined empirically by assuming that it is simply equal to the emission spectra of BA in hexane where the S_1 of BA is dominated by LE in the relevant region of the potential and ω_0 is not a function of z because LE and S_0 are both nonpolar. The band positions and band widths agree qualitatively well with experiment, see figure 5. The theoretical results also correctly predict that there is not a strong solvent effect on the absorption spectra of BA [69].

In summary, an adiabatic model for the first excited state of BA is able to account well for the equilibrium absorption and emission properties of BA. *It seems reasonable to assume that many of the dual fluorescent molecules would be well described by an adiabatic model of this type.*

3.4   SIMULATION OF THE CHARGE TRANSFER DYNAMICS OF BIANTHRYLS

Recently, we developed a theoretical model for charge transfer in BA and related molecules employing an approach that allowed for each of the special characters of these reactions, i.e. strong adiabaticity, non–mono–exponential solvation dynamics, and multiple well charge transfer[69]. Our goal was to calculate $\rho(z,t)$, the time dependent distribution function for the solvent coordinate, in order to model the extensive new data that are available on the time and wavelength resolved fluorescence of excited state charge transfer molecules.

The starting point is the well known generalized Langevin Equation, GLE, as adopted for stochastic motions involving coupling to a solvent coordinate.

$$m\ddot{z} = -\frac{\partial \, V(z)}{\partial z} - m \int \eta(t-\tau) \, \dot{z}(\tau) \, d\tau \, + F(t) \qquad (11)$$

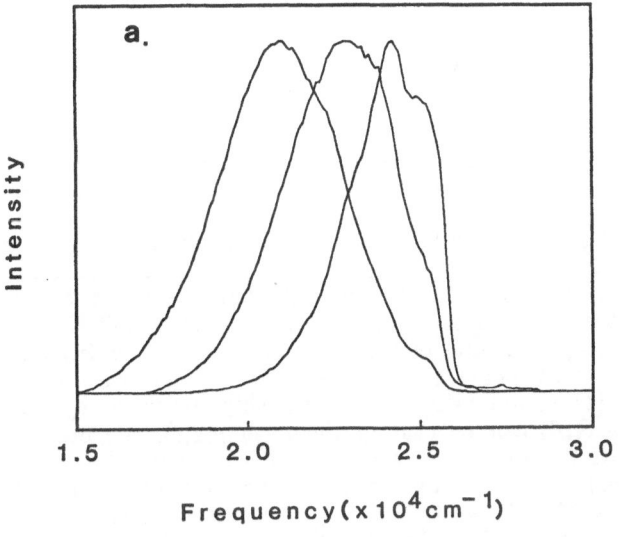

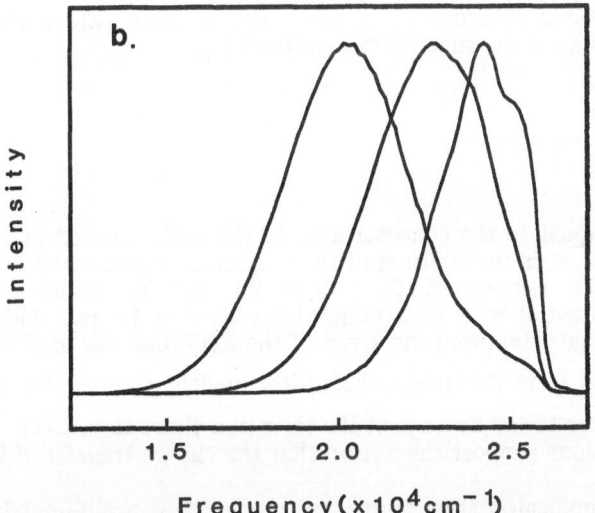

Figure 5. (a) Steady state fluorescence spectra of BA in hexane (right), glycerol triacetate (middle) and propylene carbonate (left). (b) Simulated emission spectra for BA in a nonpolar, moderately polar and very polar solvent. The potential energy parameters are optimized to represent the solvents indicated in figure 5a. In both figures, the peak intensities are normalized to the spectrum in hexane.

The GLE is a stochastic equation of motion for the coordinate z, see figure 4b. The l.h.s. of eq. 11 is the inertial force along z in terms of the effective polarization mass m of the solvent and the acceleration $\ddot{z}$. The term $-\dfrac{\partial V[z(t)]}{\partial z}$ is associated with the force due to the potential $V(z) = E_{s_1}(z)$. The term $F(t)$ is a random force due to fluctuations of the solvent. The remaining term on the r.h.s. of eq. 41 accounts for the friction (or retarding motion) along the reaction coordinate due to the lag of the solvent motion.

The quantity $\eta(t)$ is the time—dependent friction kernel. It characterizes the dissipation effects of the solvent motion along the reaction coordinate. The dynamic solute—solvent interactions in the case of charge transfer are analogous to the transient solvation effects manifested in $C(t)$, see SOLVATION section. We assume that the underlying dynamics of the dielectric function for BA and other molecules are similar to the dynamics for the coumarins. Thus we quantify $\eta(t)$ from the experimental $C(t)$ values using the relationship discussed by one of us[71]. The solution to the GLE can be cast in the form $\rho(z,t)$, the probability distribution function. Figure 6 shows $\rho(z,t)$ calculated for BA with the appropriate static and dynamic parameters for BA in the polar aprotic solvent, propylene carbonate. The results show how the charge transfer in S_1 occurs. At early times $\rho(z,t)$ is highly peaked near $z=0$, where the LE probability is high. As time progresses, the charge transfer occurs and the system approaches the equilibrium configuration, see figure 7. It is interesting at this point however to consider how the dynamics of $\rho(z,t)$ is related to the usual measure of kinetics, the reaction rate constant. The key quantity to consider is the survival population $S(t)$,

$$S(t) = \int_{-z_c}^{z_c} \rho(z,t) \, dz \tag{12}$$

which is proportional to the concentration in the well centered at $z=0$, i.e. the LE well in figure 4b. z_c is the position of the maximum of the small barrier separating the LE and CT regions of S_1. $S(t)$ for BA in propylene carbonate is non—mono—exponential with an average decay time of 1.7 ps. This is in fact very close to the average solvation time $<\tau_s>$ of the $C(t)$ that was used to calculate $\eta(t)$. Thus, as expected from the simple theoretical models, the reaction time ($\int_0^\infty S(t)dt$) is indeed very close to the average of the solvation times from $C(t)$. In other words, the GLE simulations support the notion that the charge transfer of BA is controlled by solvation!

For other molecules, simulations and theory show a different behavior. If the barrier is comparable or greater than k_BT the rate is of course partly controlled by thermal activation. On the other hand, if the barrier is zero and the reaction is very exoergic, then the average relaxation time can be much shorter than the average solvation time [71], as it is the case for the molecule ADMA[69].

3.5 EXPERIMENTS ON 9,9'—BIANTHRYL

The time and wavelength resolved fluorescence dynamics of bianthryl has been

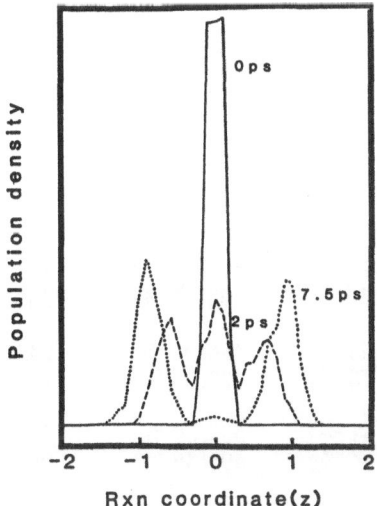

Figure 6.   The time dependent probability distribution function $\rho(z,t)$ for the excited state charge transfer of BA from a GLE simulation.   The S_1 potential employed in the simulation is shown in figure 4b.

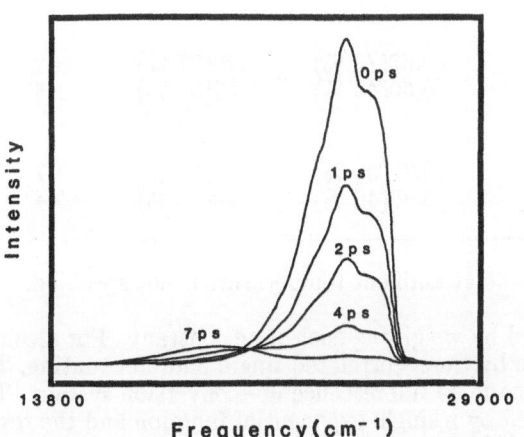

Figure 7.   Simulated time resolved fluorescence spectra for BA in propylene carbonate.

Table 2. A comparison of the average electron transfer times $<\tau_{et}>$ of 9,9'–bianthryl(BA) and 9–(9–anthryl)carbazole(CA) in various polar solvents to the average solvation times $<\tau_s>$ of coumarin probes.

solvent[a]	compound	τ_1, ps	τ_2, ps	$<\tau_{et}>$[b], ps	$<\tau_s>$[c], ps
propanol	BA	91(62 %)	276(38 %)	161	138±21
(273 K)	CA	82(87 %)	223(13 %)	100	
butanol	BA	115(64 %)	486(36 %)	249	
(273 K)	CA	116(75 %)	420(25 %)	192	
pentanol	BA	183(41 %)	635(59 %)	450	
(273 K)	CA	109(54 %)	495(46 %)	287	
pentanol	BA	138(75 %)	366(25 %)	195	142±2
	CA	105(90 %)	433(10 %)	138	
acetone	BA	0.86			0.83
	CA	0.85			
benzonitrile	BA	1.48(46 %)	8.75(54 %)	5.4	4.5–5.9
	CA	2.93(49 %)	13.1(51 %)	8.1	
dimethyl sulfoxide	BA	0.67(73 %)	2.70(27 %)	1.2	1.2–1.4
	CA	0.50(49 %)	2.39(51 %)	1.5	
propylene carbonate	BA	0.72(66 %)	3.59(34 %)	1.7	2.4–3.4
	CA	0.78(48 %)	3.98(52 %)	2.4	

a. Measurements were made at ambient temperature if not specified.

b. Average times obtained by weighting each time constant. For alcohol solvents, measurements were made by time–correlated single photon counting. The rest of the measurements were made by fluorescence up–conversion system. The transient data of acetone were fitted by a single exponential function and the rest of the transient data were modeled with a bi–exponential function.

c. Average solvation times were obtained with several coumarin probes (see ref. 11 for details). The average solvation time of propanol at 273 K is taken from ref. 12b.

investigated by several groups [15,58,62,63,69]. In addition, this molecule has been studied by picosecond absorption spectroscopy [62], electric field induced fluorescence anisotropy measurements [64] and optically induced dielectric absorption (microwave) measurements [65,66]. The results are generally in accord with the theoretical model just discussed. One of the challenges of studying the photodynamics of BA is that the LE and CT interconversion is so rapid (i.e. on the time scale of solvation) that it is necessary to employ UV subpicosecond and even femtosecond fluorescence spectroscopy which has only recently been available [15,58,69].

We have studied the time evolution of the fluorescence spectrum of BA in acetone[69]. The results are in qualitative agreement with the simulated time dependent spectra using the simulated $\rho(z,t)$ and eq. 11. This strongly supports the validity of the adiabatic, GLE model for the charge transfer of S_1 BA. Many of our measurements on BA have been confined to single emission wavelength near the short wavelength edge of the BA fluorescence. The emission at these wavelengths monitors population in the LE region of figure 4b. The results show that population in this region evolves with non–mono–exponential charge transfer kinetics due to solvent kinetic control, and in turn, the underlying non–mono–exponential solvation dynamics. Fitting parameters for the emission dynamics are compared in Table 2 with the fitting parameters of C(t) measured with coumarin probes in the same solvent. It is remarkable how closely the population dynamics of LE parallel dynamics of C(t). Thus experimental and theoretical data strongly indicate that the excited state LE/CT interconversion of BA is controlled by solvation dynamics and that the barrier for the reaction is small compared to $k_B T$.

The relationship between the intensity dynamics near the peak wavelength of LE ($\approx$420 nm) and the conventional description of charge transfer rate constants is clearly established with GLE simulations. We observe that the simulated *intensity* dynamics for a given solvent are nearly identical to the simulated *survival* dynamics, which supports the notion that the intensity dynamics on the short wavelength edge of the spectrum does indeed give an accurate measure of the underlying charge transfer dynamics.

Acknowledgements

Acknowledgement is made to the National Science Foundation (Grant CHE–8251158). TF was supported by NIH GM27945.

REFERENCES

1. W. R. Ware, S. K. Lee, G. J. Brandt and P. P. Chow, *J. Chem. Phys.* **54**, 4729 (1971).
2. R. P. DeToma and L. Brand, *Chem. Phys. Lett.* **47**, 231 (1977).
3. Y. T. Mazurenko and V. S. Udaltsov, *Opt. Spectrosc.* **45**, 765 (1978) [*Opt. Spektrosk.* **45**, 909 (1978)].
4. L. A. Hallidy and M. R. Topp, *Chem. Phys. Lett.* **48**, 40 (1977).
5. S. L. Shapiro and K. R. Winn, *Chem. Phys. Lett.* **71**, 440 (1980).
6. T. Okamura, M. Sumitani and K. Yoshihara, *Chem. Phys. Lett.* **94**, 339

(1983).

7. M. Maroncelli and G. R. Fleming, *J. Chem. Phys.* **86**, 6221 (1987); E. W. Castner, Jr., M. Maroncelli, and G. R. Fleming, *J. Chem. Phys.* **86**, 1090 (1987).

8. V. Nagarajan, A. M. Brearley, T. J. Kang and P. F. Barbara, *J. Chem. Phys.* **86**, 3183 (1987).

9. A. Declémy and C. Rullière, *Chem. Phys. Lett.* **145**, 262 (1988); *ibid.* **146**, 1 (1988); A. Declémy, C. Rullière and Ph. Kottis, *ibid.* **133**, 448 (1987); C. Rullière, Z. R. Grabowski and J. Dobkowski, *ibid.* **137**, 408 (1987).

10. S.–G. Su and J. D. Simon *J. Phys. Chem.* **91**, 2693 (1987).

11. S. Kinoshita, N. Nishi and T. Kushida, *Chem. Phys. Lett.* **134**, 605 (1987).

12. H. E. Lessing and M. Reichert, *Chem. Phys. Lett.* **46**, 111 (1976).

13. A. M. Brearley, A. J. G. Strandjord, S. R. Flom and P. F. Barbara, *Chem. Phys. Lett.* **113**, 43 (1985).

14. S. R. Meech, D. V. O'Connor, D. Philips and A. G. Lee, *J. Chem. Soc., Faraday Trans 2*, **79**, 1563, (1983).

15. M. A. Kahlow, T. J. Kang and P. F. Barbara, *J. Phys. Chem.* **91**, 6452 (1987).

16. M. A. Kahlow, T. J. Kang and P. F. Barbara, *J. Chem. Phys.* **88**, 2372 (1988).

17. M. A. Kahlow, W. Jarzęba, T. J. Kang and P. F. Barbara, *J. Chem. Phys.* **90**, 151, (1989).

18. W. Jarzęba, G. C. Walker, A. E. Johnson, M. A. Kahlow and P. F. Barbara, *J. Phys. Chem.* **92**, 7039 (1988).

19. L. Onsager, *Can. J. Chem.* **55**, 1819 (1977).

20. J. Hubbard and L. Onsager, *J. Chem. Phys.* **67**, 4850 (1977).

21. M. Rao and B. J. Berne, *J. Phys. Chem.* **85**, 1498 (1981).

22. D. Calef and P. G. Wolynes, *J. Chem. Phys.* **78**, 4145 (1983).

23. P. G. Wolynes, *J. Chem. Phys.* **86**, 5133 (1987).

24. V. Friedrich and D. Kivelson, *J. Chem. Phys.* **86**, 6425 (1987).

25. R. F. Loring and S. Mukamel, *J. Chem. Phys.* **87**, 1272 (1987); R. F. Loring, Y. J. Yan, and S. Mukamel, *J. Phys. Chem.* **91**, 1302 (1987).

26. I. Rips, J. Klafter and J. Jortner, *J. Chem. Phys.* **88**, 3246 (1988).

27. A. L. Nichols III and D. F. Calef, *J. Chem. Phys.* **89**, 3783, (1988).

28. I. Rips, J. Klafter and J. Jortner, *J. Chem. Phys.* **89**, 4288 (1988).

29. M. Maroncelli and G. R. Fleming, *J. Chem. Phys.* **89**, 875 (1988).

30. E. W. Castner, Jr, G. R. Fleming, B. Bagchi and M. Maroncelli, *J. Chem. Phys.* **89**, 3519 (1988).

31. E. W. Castner, Jr., G. R. Fleming and B. Bagchi, *Chem. Phys. Lett.* **143**, 270 (1988); E. W. Castner, Jr, B. Bagchi and G. R. Fleming, *Chem. Phys. Lett.* **148**, 269 (1988).

32. A. Chandra and B. Bagchi, *J. Chem. Phys.* in press.

33. A. Chandra and B. Bagchi, *Chem. Phys. Lett.* **151**, 47 (1988).

34. B. Bagchi and A. Chandra, *Chem. Phys. Lett.* in press.

35. A. Chandra and B. Bagchi, *J. Chem. Phys.* in press.

36. R. A. Marcus, *J. Chem. Phys.* **38**, 1858 (1963); R. A. Marcus, *J. Chem. Phys.* **43**, 1261 (1965).

37. C. J. F. Böttcher and P. Bordewijk, *Theory of Electronic Polarization*, Elsevier, Amsterdam, 1978, Vol. 2.

38. N. G. Bakshiev, *Opt. Spectrosc.* **16**, 446 (1964) [*Opt. Spektrosk.* **16**, 821 (1964)] and earlier papers in this series.

39. Y. T. Mazurenko and N. G. Bakshiev, *Opt. Spectrosc.* **28**, 490 (1970) [*Opt. Spektrosk.* **28**, 905 (1970)].

40. Y. T. Mazurenko and V. S. Udaltsov, *Opt. Spectrosc.* **44**, 417 (1978) [*Opt. Spektrosk.* **44**, 714 (1978)].

41. Y. T. Mazurenko, *Opt. Spectrosc.* **48**, 388 (1980) [*Opt. Spektrosk.* **48**, 704 (1980)].

42. J. D. Simon, *Acc. Chem. Res.* **21**, 128 (1988).

43. B. Bagchi, D. W. Oxtoby and G. R. Fleming, *Chem. Phys.* **86**, 257 (1984).

44. G. van der Zwan and J. T. Hynes, *J. Phys. Chem.* **89**, 4181 (1985).

45. H. L. Friedman, *J. Chem. Soc., Faraday Trans. 2*, **79**, 1465 (1983).

46. H. Sumi and R. A. Marcus, *J. Chem. Phys.* **84**, 4272 (1986).

47. M. Maroncelli and G. R. Fleming, *J. Chem. Phys.* **89**, 5044 (1988); M. Maroncelli, E. W. Caster, Jr., B. Bachi, and G. R. Fleming, *Faraday Discuss. Chem. Soc.*, **85**, 1 (1988).

48. O. A. Karim, A. D. J. Haymet, M. J. Banet and J. D. Simon, *J. Phys. Chem.* **92**, 3391 (1988).

49. D. B. Siano and D. E. Metzler, *J. Chem. Phys.* **51**, 1856 (1969).

50. W. Jarzęba, G. C. Walker, A. E. Johnson and P. F. Barbara, in preparation (C 153).

51. G. C. Walker, W. Jarzęba, A. E. Johnson, and P. F. Barbara, in preparation (C 343).

52. S. Engström, B. Jönsson and R. W. Impey, *J. Chem. Phys.* **80**, 5481 (1984).

53. L. D. Zusman, *Chem. Phys.* **49**, 295 (1980); ibid **80**, 29 (1983) ; ibid **119**, 51 (1988); G. Van der Zwan and J. T. Hynes, *J. Chem. Phys.* **76**, 2993 (1982); J. T. Hynes, *J. Phys. Chem.* **90**, 3701 (1986); D. F. Calef and P. G. Wolynes, *J. Phys. Chem.* **87**, 3387 (1983); H. Sumi and R. A. Marcus, *J. Chem. Phys.* **84**, 4894 (1986); W. Nadler and R. A. Marcus, *J. Chem. Phys.* **86**, 3906 (1987); I. Rips and J. Jortner, *J. Chem. Phys.* **87**, 2090 (1987); ibid **87**, 6513 (1987); **88**, 818 (1988); *Chem. Phys. Lett.* **133**, 411 (1987); M. Sparpaglione and S. Mukamel, *J. Chem. Phys.* **88**, 3266 (1988); ibid **88**, 1465 (1988); M. D. Newton and H. L. Friedman, *J. Chem. Phys.* **88**, 4460 (1988); I. Rips and J. Jortner, *Chem. Phys. Lett.* **133**, 411 (1987); G. E. McManis, A. K. Mishra, and M. J. Weaver, *J. Chem. Phys.* **86**, 5550 (1987); A. Warshel and J–K. Hwang, *J. Chem. Phys.* **84**, 4938 (1986); J–K. Hwang, S. Creighton, G. King, D. Whitney, and A. Warshel, *J. Chem. Phys.*, in press; M. Morillo and R. I. Cukier, *J. Chem. Phys.*, in press; J. Jortner and M. Bixon, *J. Chem. Phys.* **88**, 167 (1988); I. Rips, K. Klafter, and J. Jortner, Proc. IPS–7, Chicago, 1988, Elsevier Science Publishers, Amsterdam.

54. R. A. Marcus, *J. Chem. Phys.* **24**, 966 (1956); ibid **24**, 979 (1956); *Ann. Rev. Phys. Chem.* **15**, 155 (1964); V. G. Levich and R. R. Dogonadze, *Dokl. Akad. Nauk. SSSR* **124**, 123 (1959).

55. J. Ulstrup, "Charge Transfer Processes in Condensed Media", Springer Verlag, Berlin, 1079.

56. A. Mozumder, *J. Chem. Phys.* **50**, 3153, 3162 (1969).

57. E. M. Kosower and D. Huppert, *Ann. Rev. Phys. Chem.* **37**, 127 (1986).

58. T. J. Kang, M. A. Kahlow, D. Giser, S. Swallen, V. Nagarajan, W. Jarzęba and P. F. Barbara, *J. Phys. Chem.* **92**, 6800 (1989).

59. D. W. Anthon and J. H. Clark, *J. Phys. Chem.* **91**, 3530 (1987).

60. F. Schneider and E. Lippert, *Ber. Bunsenges. Physik. Chem.* **72**, 1155 (1968); ibid **74**, 624 (1970).

61. E. M. Kosower, K. Tanizawa, *Chem. Phys. Lett.* **16**, 419 (1972).

62. N. Nakashima, M. Murakawa, and N. Mataga, *Bull. Chem. Soc. Japan*, **49**, 854 (1976).

63. W. Rettig and M. Zander, *Ber. Bunsenges. Phys. Chem.* **87**, 1143 (1983); M. Zander and W. Rettig, *Chem. Phys. Lett.* **110**, 610 (1984); N. Mataga, H. Yao, T. Okada, and W. Rettig, *J. Phys. Chem.* **93**, 3383 (1989).
64. W. Baumann, E. Spohr, A. Bischhof, W. Liptay, *J. Luminescence*, **37**, 227 (1987)
65. R. J. Visser, P. C. M. Weisenborn, P. J. M. van Kan, B. H. Hvizer, C. A. G. O. Varma, J. M. Warman, and M. P. de Haas, *J. Chem. Soc. Faraday Trans. 2*, **81**, 689 (1985).
66. D. B. Toublanc and R. W. Fessenden and A. Hitachi, *J. Phys. Chem.* **93**, 2893, 1989.
67. K. Yamasaki, K. Arita, O. Kasimoto, and K. Hara, *Chem. Phys. Lett.* **123**, 277 (1986).
68. R. L. Khundkar, A. H. Zewail, *J. Chem. Phys.* **86**, 1302 (1986).
69. T. J. Kang, W. Jarzeba, T. Fonseca, P. F. Barbara, to be submitted.
70. Z. R. Grabowski, K. Rotkiewicz, A. Siemiarczuk, D. J. Cowley, and W. Baumann, *Nouv. J. Chimie* **3**, 443 (1979).
71. T. Fonseca, *J. Chem. Phys.* in press.

FREQUENCY FACTOR FOR THE OUTER-SPHERE ELECTRON TRANSFER. BRIDGING BETWEEN THE DIABATIC, ADIABATIC AND SOLVENT-CONTROLLED LIMITS.

ILYA RIPS AND JOSHUA JORTNER
School of Chemistry,
Tel-Aviv University
69978, Tel-Aviv, Israel

ABSTRACT. A general expression for the thermally activated electron transfer (ET) rate is derived. The frequency factor reduces to the transition state theory (TST) result, to the Fermi Golden Rule and to the solvent-controlled expression in the appropriate limits. The crossover between the uniform and the diffusive motion along the reaction coordinate is described in terms of a single parameter, the ratio of the mean-free path and the root-mean-square displacement of the reaction coordinate.

1. Introduction.

Considerable progress has been made in theoretical description of the electron transfer (ET) kinetics [1-3] in polar solvents during the last few years [4-17]. This progress has been due mainly to the realization of the delicate interplay between the strength of the electronic coupling of the two states and the dynamics of motion along the reaction coordinate. When the electronic coupling matrix element, V, is small and/or the time that the system spends in the vicinity of the crossing point of the two diabatic terms is sufficiently short (cf. below), the transition at the crossing point constitutes the rate determining step. The rate in this limit is given by the well-known Fermi Golden Rule expression [2,3]:

$$k_{NA} = A_{NA} \exp(-E_a/k_B T) = \frac{2\pi}{\hbar} \frac{V^2}{(4\pi E_r k_B T)^{1/2}} \exp(-E_a/k_B T) \qquad (1)$$

where E_a is the activation energy. For the conventional model of the two crossing parabolic diabatic terms the latter is given by [1]

$$E_a = (E_r - \Delta E)^2/4E_r \qquad (2)$$

where E_r is the solvent reorganization energy and ΔE is the (free) energy gap between the two states. The most important physical result implicit in Eq.(1) is that the solvent relaxation dynamics drops out from the expression for the frequency factor, which is not surprising taking into account that the electronic step is the rate-determining one. Therefore it is irrelevant how the system actually gets to the crossing point.

The second physical limit is realized when the dynamics of the reaction coordinate is uniform so that the thermal equilibrium prevails up to the crossing point. In this case the

293

J. Jortner and B. Pullman (eds.), Perspectives in Photosynthesis, 293–299.
© 1990 *Kluwer Academic Publishers.*

frequency factor in the rate expression is given by the transition state theory (TST) result [18]

$$A_{TST} = \frac{\Omega}{2\pi}$$

(3)

where Ω is the well frequency of the parabolic term and is closely related to the mean-square thermal rotational frequency of the free solvent molecules [10,11a]

$$\Omega^2 = \frac{(2\epsilon_s + \epsilon_\infty)}{3\epsilon_s \kappa} \frac{2k_B T}{I}$$

(4)

In this expression ϵ_s and ϵ_∞ are the static and the optical dielectric constants of the solvent, respectively; κ is the Kirkwood factor [19] and I is the moment of inertia of the solvent molecule.

Finally, if the motion along the reaction coordinate is strongly overdamped the dynamics of reaching the crossing point may constitute the rate-determining step. This situation is realized in solvents, where the polarization (dielectric) relaxation is slow (e.g., due to hydrogen bonding), provided that the electronic coupling is large enough. The latter condition is satisfied in the **intramolecular** ET, when the donor and the acceptor are the functional groups of the same molecule [15]. In this case the reaction coordinate dynamics enters explicitly the frequency factor and the reaction is solvent-controlled (more precisely, solvation controlled) [4-14]. In the particular case of the Debye solvents with the dielectric susceptibility function

$$\epsilon(\omega) = \epsilon_\infty + \frac{\epsilon_s - \epsilon_\infty}{1 - i\omega\tau_D}$$

(5)

and for the so-called resonant ET ($\Delta E = 0$) the frequency factor is given in the extremely overdamped limit by [4,5]

$$A_{sc} = \frac{1}{\tau_L} \left[\frac{E_r}{16\pi k_B T} \right]^{1/2}$$

(6)

where τ_L is the longitudinal dielectric relaxation time [20] related to the Debye relaxation time τ_D by $\tau_L = (\epsilon_\infty/\epsilon_s)\tau_D$. Elsewhere we have shown that this result is almost insensitive to the details of the crossing of the two terms [11b]. The case of the Debye relaxation is particularly simple as the reaction coordinate dynamics is the classical diffusion. A more complicated situation, in which the relaxation kinetics is non-exponential and is characterized by a static [12] and dynamic [9] distribution of relaxation times, has also been studied.

The limiting physical situations being well understood, the natural question arises of bridging the results. The problem is to derive a general expression for the frequency factor, which would reduce to the above results in the appropriate limits (the activation energy is independent of the mechanism). There have been a number of papers devoted to this subject.

1.   Crossover from the non-adiabatic limit to the TST.   This crossover is the most simple one and historically the first one that has been described [2].   Since the motion of the reaction coordinate in the vicinity of the crossing point is uniform, one can use the Landau-Zener (LZ) [21,22] expression for the transmission coefficient of the barrier to bridge the two limits.

2.   Crossover from the uniform adiabatic (TST) to the solvent-controlled limit.   This crossover is associated with the change in character of the motion in the vicinity of the crossing point with the increase of damping (decrease of the mean-free path of the reaction coordinate).   For the   particular case of Markoffian dissipation and smooth potential barrier this crossover has been described by Kramers [18].   His result has been applied to the ET problem by Calef and Wolynes [6a].   It has been further extended to the case of non-Markoffian dissipation (non-Debye dielectric relaxation) by Hynes [9].   This situation is more delicate for the case of the cusped barrier.

3.   Crossover from the non-adiabatic to the solvent-controlled limit.   This has been described by Zusman [4] using the approach based on the stochastic Liouville equation. Note that his expression is actually of the mean-first passage time character and is therefore a physically plausible (and useful) interpolation between the limits rather than a real bridging.   There were a few attempts to improve this result [8], but they were based on the higher order Pade approximants [13] to the rate expression and therefore also of an interpolative nature.

The purpose of this paper is to provide a unified description of the ET kinetics as a function of the relevant physical parameters (electronic coupling, mean-free path, and relaxation dynamics).   This amounts to derivation of the rate expression (frequency factor) which continuously bridges between the physical limits mentioned above.

2.   Results.

For simplicity we shall consider below the case of resonant ET ($\Delta E = 0$).   It should be pointed out, however, that the results can readily be extended to the general case.   We shall also assume the applicability of the continuum dielectric model for the solvent and the Debye dielectric relaxation.   To bridge the results to the non-adiabatic limit, we shall require the solution for the case of the cusped barrier.   In the extremely overdamped limit the problem has been solved by Kramers [18].   His result has been generalized to the intermediate friction domain by Helman [7].   The particular case of the symmetric double well with Markoffian (Debye) relaxation has been treated by Dekker [23].   The rate expression is given by

$$k_{ET} = A_{ad} \exp(-E_r/4k_B T) \tag{7}$$

where A_{ad} is the adiabatic frequency factor

$$A_{ad} = (\pi^{1/2} x) \exp(x^2) \, \text{erfc}(x) \, A_{TST} \tag{8}$$

A_{TST} is the TST frequency factor, Eq.(3) and the parameter x is given in the case of Debye relaxation by

$$x \equiv \frac{(E_r/\tau_L)}{\Omega \, (4 E_r k_B T)^{1/2}}$$

(9)

This parameter determines the crossover from the uniform to the diffusive dynamics along the reaction coordinate. Small values of the longitudinal relaxation time τ_L correspond to the large values of the x parameter ($x \gg 1$). In this limit the motion is uniform and the frequency factor reduces to the TST result, Eq. (3). In the opposite limit, $x \ll 1$, which corresponds to large values of the longitudinal dielectric relaxation time, the frequency factor reduces to that appropriate in the solvent-controlled limit [cf. Eq. (6)], i.e.,

$$A_{sc} = \pi^{1/2} x \, \frac{\Omega}{2\pi} = \frac{1}{\tau_L} \left[\frac{E_r}{16\pi k_B T} \right]^{1/2}$$

(10)

The x parameter has a simple physical meaning. It follows from the classical fluctuation-dissipation theorem [24] that the root-mean-square displacement of the reaction coordinate ℓ is given by

$$\ell = (2 E_r k_B T)^{1/2}$$

(11)

On the other hand, a simple analysis shows [11a] that the mean-free path (coherence length), ℓ_{mfp}, on which the reaction coordinate motion is uniform is

$$\ell_{mfp} \simeq E_r/(\Omega \tau_L)$$

(12)

Thus the x parameter is given by the ratio of the coherence length of the reaction coordinate to its root-mean-square displacement under the equilibrium condition. Alternatively, one can use the root-mean-square velocity $\nu \equiv \Omega \ell$ together with an averaged velocity $\nu_{mfp} \equiv \Omega \ell_{mfp}$ and write the parameter as

$$x \simeq \nu_{mfp}/\nu$$

(13)

Thus far the non-adiabaticity of the barrier has not been taken into account. To bridge the above result with the non-adiabatic limit it has to be multiplied by the transmission coefficient of the barrier,

$$A = T(g) \, A_{ad}$$

(14)

The Landau-Zener expression [21,22] is not the appropriate one because it assumes that the reaction coordinate dynamics in the vicinity of the crossing point is uniform. We require a result, which is valid both for the uniform and for the diffusive motion along the reaction coordinate. For the transmission coefficient it is reasonable to use the Holstein [2] expression

$$T(g) = T_H(g) = \frac{2 T_{LZ}(g)}{1 + T_{LZ}(g)}$$

(15)

where

$$T_{LZ}(g) = 1 - \exp(- 2\pi g) \tag{16}$$

is the Landau–Zener expression for the transmission coefficient and g is the adiabaticity parameter, which depends upon the character of the motion along the reaction coordinate. The form of the adiabaticity parameter in the uniform adiabatic and in the solvent-controlled limit can be obtained from the condition that in the non–adiabatic limit the frequency factor should be given by the Fermi Golden Rule expression, Eq.(1). As a result we obtain for the adiabaticity parameters

$$g_{unif} = \frac{\pi^{1/2} V^2}{\hbar\Omega \, (4E_r k_B T)^{1/2}} \tag{17}$$

and

$$g_{dif} = \frac{V^2 \tau_L}{\hbar E_r} \tag{18}$$

It should be noted that the ratio of the adiabaticity parameters in the two limits is simply related to the x parameter, which was introduced previously to describe the crossover from the TST to the solvent controlled limit

$$g_{unif}/g_{dif} = \pi^{1/2} x \tag{19}$$

We now assume that generally the adiabaticity parameter has the form

$$g \equiv g(x) = g_{unif} \, F(x) \tag{20}$$

Here F(x) is the crossover function, which depends only on the details of reaction coordinate dynamics via the x parameter introduced above. From expressions (17) and (18) one can determine the limiting physical behaviour of the crossover function

$$F(x) \sim \begin{cases} 1 & (x \gg 1) \\ 1/\pi^{1/2} x & (x \ll 1) \end{cases}$$

The simplest choice of the crossover function is

$$F(x) = 1 + 1/\pi^{1/2} x$$

which corresponds to the linear interpolation between the uniform and diffusive limits.

The crossover function can be determined uniquely from the condition that, in the limit of extremely weak coupling, $V \rightarrow 0$, the frequency factor is given by the Fermi Golden Rule expression irrespective of the dynamics of the reaction coordinate. In other words, one can always choose the electronic coupling in such a way that the Landau–Zener length is shorter than the mean-free path of the reaction coordinate. From Eqs.(14-16,20) we obtain

$$4\pi g_{unif} \, F(x) \, A_{ad} = \frac{2\pi}{\hbar} \frac{V^2}{(4\pi E_r k_B T)^{1/2}} \tag{21}$$

which leads to the following expression for the crossover function $F(x)$

$$F(x) = \{ \pi^{1/2} x \, \exp(x^2) \, \mathrm{erfc}(x) \}^{-1} \tag{22}$$

3. Discussion.

The expression for the adiabaticity parameter, Eqs.(20) and (22) is the main result of this work. Together with Eqs.(8,14-16) it gives the frequency factor for the whole range of the physical parameters of the problem. It also reduces to the well-known results in the non-adiabatic, the uniform adiabatic (TST) and in the solvent-controlled limits. Extension for an arbitrary type of crossing of the diabatic terms and for the case of non-Debye dielectric relaxation presents no difficulty.

Eq.(20) for the adiabaticity parameter assumes implicitly that the reaction coordinate dynamics effect upon adiabaticity can be described in terms of a single parameter. At present the only justification of this simplifying assumption is that it allows one to solve the crossover problem in a closed form. Whether it is correct, requires further study.

References

[1]   Marcus, R.A. (1956) J. Chem. Phys. 24, 966, 979.
[2]   Holstein, T. (1959) Ann. Phys. (N-Y) 8, 325, 343. (cf. Appendix III).
[3]   Levich, V.G. and Dogonadze, R.R. (1959) Dokl. Akad. Nauk SSSR 124, 123.
        [(1959) Proc. Acad. Sci. USSR, Phys. Chem. Sect. 124, 9].
        Levich, V.G. and Dogonadze, R.R. (1960) Dokl. Akad. Nauk SSSR 133, 158.
        [(1960) Proc. Acad. Sci. USSR, Phys. Chem. Sect. 133, 591].
[4]   Zusman, L.D. (1980) Chem. Phys. 49, 295.
[5]   Alexandrov, I.V. (1980) Chem. Phys. 51, 499.
[6]   (a) Calef, D.F. and Wolynes, P.G. (1983) J. Phys. Chem. 87, 3387;
        (b) Calef, D.F. and Wolynes, P.G. (1983) J. Chem. Phys. 78, 470.
[7]   Helman, A.B. (1983) Chem. Phys. 79, 235.
[8]   Frauenfelder, H. and Wolynes, P. (1985) Science 229, 337.
[9]   Hynes, J.T. (1986) J. Phys. Chem. 90, 3701.
[10]   Sumi, H. and Marcus, R.A. (1986) J. Chem. Phys. 84, 4894.
[11]   (a) Rips, I. and Jortner, J. (1987) J. Chem. Phys. 87, 2090;
        (b) Rips, I. and Jortner, J. (1987) J. Chem. Phys. 87, 6513.
[12]   Rips, I. and Jortner, J. (1987) Chem. Phys. Lett. 133, 411 (1987);
        (1987) Int. Journ. Quant. Chem. Symp. 21, 313.
[13]   Sparpaglione, M. and Mukamel, S. (1988) J. Chem. Phys. 88, 3263.
[14]   Onuchic, J.N. and Wolynes, P.G. (1988) J. Phys. Chem. 92, 6495.
[15]   Kosower, E.M and Huppert, D. (1986) Annu. Rev. Phys. Chem. 37, 127 and
        references therein.
[16]   Lippert, E., Rettig, W., Bonacic-Koutecky, V., Heisel, F., and Miehe, J.A. (1987)
        Adv. Chem. Phys. 68, 1.
[17]   Barbara, P.F. and Jarzeba, W. (1989) 'Ultrafast Photochemical Intramolecular
        Charge Transfer and Excited State Solvation', in press.

[18]   Kramers, H. (1940) Physica 8, 284.
[19]   Bottcher, C.J.F. and Bordewijk, P. (1978) Theory of Electric Polarization, Elsevier, Amsterdam,. Vol. 2, Ch. IX.
[20]   Fröhlich, H. (1958) Theory of Dielectrics, Clarendon, Oxford, Sect.10.
[21]   Landau, L.D. and Lifshitz, E.M. (1977) Quantum Mechanics. Non-relativistic Theory, Pergamon, Oxford, Sect.90.
[22]   Nikitin, E.E. and Umanskii, S.Ya. (1984) Theory of Slow Atomic Collisions, Springer Verlag, Berlin, Ch. 8.
[23]   Dekker, H. (1985) Phys. Lett. 113A, 193.
[24]   Landau, L.D. and Lifshitz, E.M. (1980) Statistical Physics I. Pergamon, Oxford, Sect. 124.

STATIC AND DYNAMIC ELECTROLYTE EFFECTS ON EXCITED STATE
INTRAMOLECULAR ELECTRON TRANSFER AND EXCITED STATE SOLVATION

Dan Huppert and Varda Ittah
School of Chemistry, Sackler Faculty of Exact Sciences,
Tel-Aviv University, Tel- Aviv 69978
Israel

ABSTRACT. In a solution of an electrolyte in a liquid of low
permittivity after excitation of molecules which exhibit either a
large Stokes shift (coumarins) or undergo an intramolecular electron
transfer (DMABN) process, the fast solvent longitudinal dielectric
relaxation, $\tau_1 = (\epsilon_\infty/\epsilon_S)\tau_D$ (few picosecond in non associative liquids)
is succeeded by nanosecond time scale translational relaxation of the
"ionic atmosphere".
    In such solutions one may assume that single ions, ion pairs,
triple ions and larger aggregates are present. At an electrolyte
concentration range of 0.005M-0.5M dissolved in neat liquid with
permittivities of 4.5-7 most of the salt is present as ion-pairs. The
dissolved electrolyte ionic relaxation time measured by time resolved
fluorescence was found to be inversely dependent on the electrolyte
concentration and linearly dependent on the ion size. Thus new ways of
probing medium effects on excited state behavior are opened.

1. INTRODUCTION

Two classes of molecules that exhibit a large band shift between
absorption and fluorescence, are those in which for the emitting
excited state dipole moment is markedly different from that of the
ground state (e.g., coumarins), or those in which the emitting excited
state results from intramolecular charge transfer (e.g., 4-dimethyl-
aminobenzonitrile (DMABN),
    The role of solvent dynamics on excited state solvation of these
polar molecules has attracted intense interest over the past years,
reflecting the influence of new experimental and theoretical results.
[1-13] The expression for the electron transfer rate in the adiabatic
limit where the rate is limited by the solvent motion is given by [6]

$$k^{AD} = (1/\tau_L)(E_r/16\pi \ kT)^{1/2} \ exp\text{-}E_a/kT \tag{1}$$

where E_r is the solvent reorganization energy, E_a is the activation

J. Jortner and B. Pullman (eds.), Perspectives in Photosynthesis, 301–316.

energy   of the process and τ_L is the solvent   longitudinal   relaxation
time

$$\tau_L = (\epsilon_\infty/\epsilon_S)\tau_D \qquad\qquad\qquad\qquad\qquad (2)$$

where τ_D is the solvent dielectric relaxation time, ϵ_S is the   static
dielectric constant and ϵ_∞ is the high frequency dielectric constant.

The analysis of the results has been mainly based on a dielectric
continuum model for the solvent. Within the model, microscopic relaxation can be related to bulk dielectric susceptibility function for the
pure solvent. For Debye solvents (exponential solvent relaxation), the
rates of excited state intramolecular electron transfer processes were
found   to   be   activationless and hence are directly   related   to   the
longitudinal   relaxation   time, τ_L. However, experimental   measurements
have shown that there are deviations from the simple model.

Nonexponential   decay kinetics are found for the fluorescence   of
the   initially   formed excited state of several molecules,   among   them
DMABN   in several alcoholic (monools and diols) and   nitrile   solvents
[2,3],   N-arylaminonaphthalenesulfon-N,N-dimethylamides   (TNSDMA)   in
monools   and   pentanediols [1h,i], bianthryl   (BA)   [9],   and   bis(4-
dimethyl aminophenyl) sulfone (DAMPS) [10].

Certain classes of molecules exhibit time dependent   fluorescence
shifts   in   polar solvents. As the consequence   of   the   instantaneous
change   in the dipole moment of a probe molecule   on   excitation,   the
distribution   of solvent molecules around the probe molecule   is   nonequilibrium   per se   or with respect to a rapidly formed product.   The
relaxation   time of a correlation function, C(t) is given by

$$C(t) = (\nu(t) - \nu(\infty))/(\nu(0) - \nu(\infty)) \qquad\qquad\qquad (3)$$

in which $\nu(0)$, $\nu(t)$ and $\nu(\infty)$ are the mean fluorescence frequencies   of
the spectrum at time zero, t and at long times respectively. The   time
dependence   of   the   correlation function for   several   coumarins   and
similar molecules was found to be similar to the activationless intramolecular electron transfer process found for the molecules   mentioned
above.

In addition to the   stabilization energy of an excited dipole   or
ion   resulting from solvent dipole reorientation, a   relatively   large
stabilization   energy   can   result   from   electrolytes   [1j].   This
phenomenon   can   be simply measured by observing the   time   integrated
(steady-state) fluorescence spectral band-shift of the probe   molecule
(coumarins)   to   lower   energies   (red-shift) as   a   function   of   the
electrolyte concentration.

Debye-Hückel theory [14a] and later, Onsager [15b],   provided   the
basis for our current understanding of the concentration-dependence of
the   conductivity   of dilute aqueous   ($\epsilon_s=81$)   electrolyte   solutions.
Debye-Hückel theory suggests that an ionic atmosphere is formed around
a   central ion with a distribution determined   by the balance   between
Coulomb forces and thermal motion.

Fuoss and Kraus [16a,b] studied the conductance of several electrolytes in solutions with solvent dielectric constants between ϵ_s=2.38 and ϵ_s=38.

The conductance of electrolytes in solvents of low dielectric constant varies with concentration in a way characterized by regions of behavior dependent upon the structure of the salt and the properties of the solvent.

In order to explain the conductance dependence on salt concentration, Fuoss and Kraus used extended the Bjerrum ion pair model [16a,b,17]. The Debye-Hückel model assumed the ions to be in almost random positions. However, ion pairs a species with a net charge of zero, may be formed as suggested by Bjerrum [17]. Fuoss and Kraus interpreted the observed conductances in terms of ion pair and higher aggregates formation.

The important feature emerging from detailed analysis of the ionic species distribution in low dielectric constant solvents is that at low and medium salt concentrations (10^{-3}M-0.1M), the fraction of ion-pairs is much larger than that of both free ions and triple ions. The fact that the fraction of ion-pairs is close to unity over the salt concentration range used in our experiments is also manifested in the monotonic increase of the solution dielectric constant as a function of the electrolyte concentration.

The theory of frequency dependent conductivity was first formulated by Debye and Falkenhagen[14b,c]. In the case of aqueous electrolyte solutions where the ions are loosely associated the behavior of the ionic cloud around a central ion in the presence of an external alternating electric field is dependent on the relaxation time necessary for the distribution of the ionic atmosphere density to appear or disappear. The theory predicts that the relaxation time, τ will be proportional to $\epsilon_s/\Lambda C$ where ϵ_s is the dielectric constant, Λ is the equivalent conductivity at infinite dilution and C is the ion concentration. The Debye-Falkenhagen concept that the ionic atmosphere relaxation time has an influence on the motion of charges can in principle be extended to the current studied case of ion pair (loosely bound) relaxation around excited state molecular large dipoles.

The excited state molecule would be surrounded by a nonequilibrium distribution of solvent molecules, ions, ion pairs and triple ions. The ions and ion pairs are expected to move from the equilibrium position determined by the ground state to form a new distribution that would accommodate the large dipole created in the excited state. The translational relaxation time needed to form the new ionic distribution around the excited dipole can be measured by time resolved emission techniques in the way which has been used for the study of dipolar solvent orientational relaxation times around such molecules in the excited state.

Steady state fluorescence spectral shift measurements can be used to determine the reorganization energy increase due to the addition of a term involving the electrolyte atmosphere to the value for solvent stabilization of the emitting state.

2. EXPERIMENTAL

Two experimental systems were used for following the kinetic behavior of excited states:

a)   Picosecond range

A neodymium high power low repetion rate YAG oscillator was passively mode-locked with a saturable dye to yield a 1.06μm pulse train (half-width 25ps, separation 7ns) [18]. A single pulse was extracted with a Pockels cell between crossed Glan polarizers, and amplified with a neodymium YAG amplifier to 10 mJ/pulse, partially converted to 532 nm with a KDP crystal in 10% efficiency. A small portion of the 532 nm light was collimated onto a vacuum diode to generate a streak camera trigger (Hamamatsu model C939, resolution 10-12 ps). The streak camera output was imaged onto a vidicon connected to an optical multi-channel analyzer (PAR 1205D). The records were averaged and analyzed by a microcomputer system. Samples were irradiated in 5 mm quartz cells. The fluorescence was collected from the front surface.

b) Time resolved emission high repetition low power

When high sensitivity, large dynamic range and low intensity illumination are important in fluorescence measurements, time correlated single photon counting is in preference to using system (a) which has a better time resolution but a poor dynamic range. The use of picosecond tunable dye lasers as the excitation source and the development of microchannel plate (MCP) photomultipliers improved the time resolution of single photon counting. A CW mode locked Nd/YAG pumped dye laser (Coherent) providing high repetition rate short pulses (1ps) suitable as the sample excitation source for the photon counting system. The detection system is based on a Hamamatsu 1654U-01 or 1564U-05 (MCP) photomultipliers and an IBM personal computer.

c) Reagents

Solvents of spectroscopic quality were obtained from Merck, Fluka or BDH. Coumarin 153 (Exciton), DMABN (Aldrich) were used without further purification.

d) Steady state fluorescence spectra were measured using a Shimadzu spectro-fluorimeter, RF-540, absorption spectra with a Varian Cary 219 spectrophotometer.

3. RESULTS AND DISCUSSION

This section is divided into three subsections: the conductance of $LiClO_4$ solutions in ethyl acetate, steady-state fluorescence of coumarin 153 or DMABN in low permittivity liquids - $LiClO_4$ solutions, and time-resolved fluorescence measurements of these fluorofors in electrolyte solutions.

a. Conductance of LiClO$_4$ solutions in ethyl acetate

Fuoss and Kraus observed four distinctive regions in plots of conductance versus concentration for electrolytes in solvents of low dielectric constant. At very low concentrations, the conductance decreases in a manner proportional to the square root of concentration, in accordance with the Onsager equation [15]. In the second region, the conductance falls more rapidly with increasing concentration. The third and fourth regions are observed only in solvents of sufficiently low dielectric constant, such as ethyl acetate (ϵ_s=6), the solvent used in the conductance concentration dependence measurements. In the third region there is a minimum in the conductance-concentration plots. In the fourth region, the conductance increases very rapidly as the concentration increases.

Fuoss and Kraus interpreted the behavior of the conductance in terms of ion pair and triple ion formation. The dissociation constant of an ion-pair is strongly dependent on the dielectric constant of the solvent and can be evaluated numerically if the "ion size" parameter, a, and the solution dielectric constant are known.

The parameter, a, depends on both solute and solvent properties, and is roughly an additive function for the constituent ions. The ion pair dissociation equilibrium constant, K, is given approximately by

$$K^{-1} = (4\pi \, N_0 a^3/1000)(a\epsilon_s kT/e^2)\exp[e^2/(\epsilon_s akT)] \qquad (4)$$

where N_0 is the Avogadro number, a, the "ion size" parameter and ϵ_s the dielectric constant.

An expression for the dissociation constants of triple ions was derived by Fuoss and Kraus using an electrostatic model similar to that used in the derivation of the ion-pair dissociation constant eq. (4).

We have found that the conductance of LiClO$_4$ in ethyl acetate varies with concentration in the way (four regions) described above [1j]. Our conductance data is analyzed with the Fuoss & Kraus equations to determine both the ion-pair and triple ion equilibrium association constants. We found only a qualitative agreement between the experimental and the calculated curves in the concentration range 10^{-4}M-1M. The important feature emerging from the detailed analysis of ionic species is that, at low and medium salt concentrations, the ion-pair fraction, is much larger than that of both free ions and triple ions. The monotonic increase of the solution dielectric constant as a function of the electrolyte concentration indicates that the ion pair concentration also increases [19-21].

b. Steady-state fluoresecence

Time integrated (steady-state) fluorescence spectra of coumarin 153 in tetrahydrofuran (THF) and diethyl ether (DEE) solutions containing various concentrations of LiClO$_4$ are shown in Fig.1.

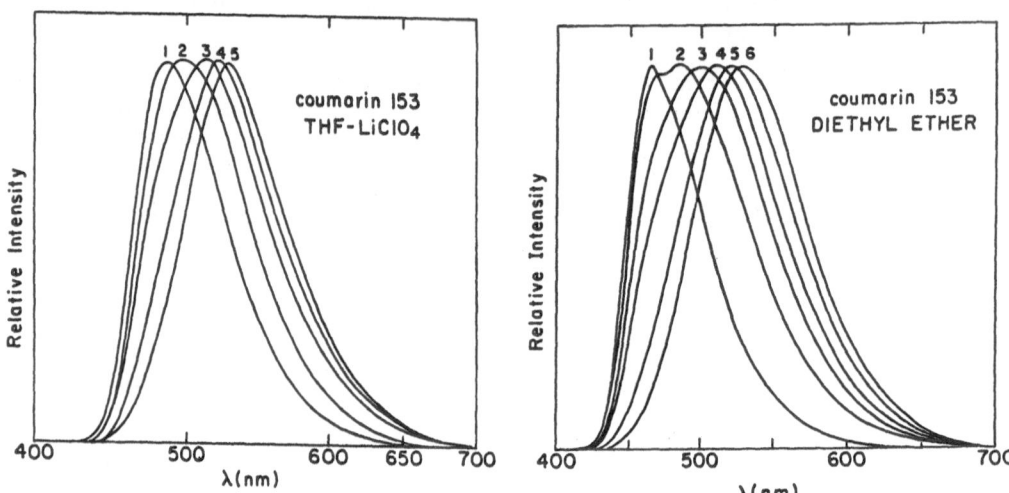

Figure 1. Steady state emission spectra of coumarin 153 in $LiClO_4$ THF and diethyl
ether solutions. From left to right; pure solvent 0.03M; 0.09M; 0.12M; 0.18M.

The position of the band maximum is strongly dependent on the
electrolyte at concentrations between 7.5×10^{-3} - 0.1M; at higher
concentrations, the maximum is less sensitive to electrolyte concentration and levels off, the maximum change in emission energy being
4.2 kcal/mole, 4.5 kcal/mole and 7 kcal/mole for THF ($\epsilon_s=7$), ethyl
acetate ($\epsilon_s=6$) and DEE ($\epsilon_s=4.2$) solutions respectively. The band
maximum at high electrolyte concentration (527nm) is similar to all
solvents used and comparable to the position of the maximum in a pure
polar solvent like methanol (530nm) ($\epsilon_s=32$).

The Stokes shift derived from the steady state fluorescence is
usually directly correlated with the solvent reorganization energy or
with one of the solvent polarity parameters. The correlation is justified only in the case for which the rate of the process which forms
the relaxed emitting state ("ES formation rate") is much larger than
the radiative rate of the initial excited state [9a]. As will be shown
in the next section, the "ES formation rate" is linearly dependent on
the salt concentration. At low salt concentrations, the relaxation
time for the ions is longer than the radiative lifetime and hence the
energy shift of the steady state fluorescence band maximum does not
reach its limit. In this case, the shift in the band maximum is
smaller than expected. In order to correlate the reorganization energy
with the electrolyte concentration, we estimated the band maximum
location for the virtual case where the "ES formation rate" is larger
than the radiative rate.

The solvent reorganization energy E_r is given by

$$E_r = 1/(8\pi C_p)\int \Delta D^2 dr \qquad C_p = 1/\epsilon_\infty - 1/\epsilon_s \qquad (5)$$

where ϵ_∞ and ϵ_s are the high frequency and the static dielectric constants respectively, ΔD is the change in the electric displacement of the excited molecule. In the solution of electrolyte, most of the ions form electrically neutral ion-pairs, species with a large dipole (~10D for tetrabutylammonium chloride in ethyl acetate [19c]), an association accompanied by a large increase in the static dielectric constant with increasing $LiClO_4$ concentration. The increase in the static dielectric constant, about 10 units at the highest electrolyte concentration, in all solvents used increases the solvent reorganization energy through the large change in the Pekar factor C_p between pure solvent and solvent containing a high concentration of $LiClO_4$. The plot of Δ_{max} (the emission energy shift) for coumarin 153 in ethyl acetate as a function of the electrolyte contribution to the Pekar factor, ΔC_p, is shown in Fig. 2.

$$\Delta C_p = 1/\epsilon_s^S - 1/\epsilon_s^T \qquad (6)$$

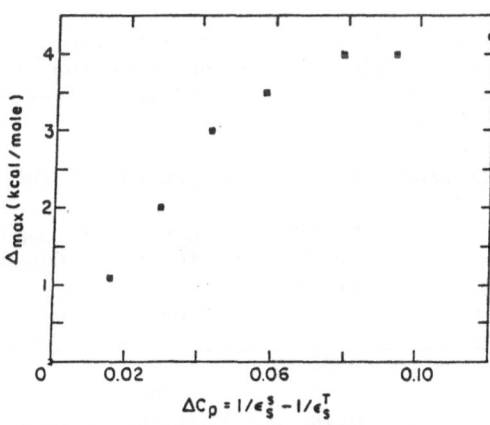

Figure 2. The corrected emission energy for the maximum (kcal/mole) of coumarin 153 as a function of the change of the Pekar factor (ΔC_p) due to the presence of $LiClO_4$ in ethyl acetate.

where ϵ_s^S is the solvent contribution to the static dielectric constant and ϵ_s^T is the total static dielectric constant of solutions of $LiClO_4$. The slope of the plot of Δ_{max} versus ΔC_p is linear in dilute solutions (Fig.2) but falls to a much lower value at high electrolyte concentrations. The emission energy of coumarin 153 and Z-value changes with electrolyte concentration are parallel [22]. The lack of change in the

emission energy at high electrolyte concentration may be explained as
a consequence of screening of the dipolar interaction by the ions and
triple ions present in ionic atmosphere around the probe molecule. The
triple ion concentration increases from 0.004M at 0.1M to 0.2M at 1M.
The average distance between the triple ions in a salt solution of
0.1M is ~74Å while at 1M solution the average distance is reduced to
~17Å. The number of ion-pairs and solvent molecules contributing to
the "solvent" reorganization energy decreases due to ionic screening
at high electrolyte concentrations; the emission energy change which
gauges the total reorganization energy does not thus follow the
increase in ΔC_p at high electrolyte concentrations.

c. Time resolved measurements

To gain more insight into the time scales for and the mechanism of
electrolyte effects on an excited probe molecule, we measured the time
resolved fluorescence Stokes shifts of coumarin 153 in various
solvents of low permittivity as a function of electrolyte concentration. Another probe molecule used is DMABN which exhibits two emitting
states. For coumarin 153, the relaxation of polar solvent around the
newly formed excited state dipole leads to a red shift of the fluorescence spectrum represented by the correlation function, C(t), in
eq.(3). Barbara et al [9a] developed time-saving procedure for evaluating C(t) with a single emission transient. We found that the
transient emission from coumarin 153, measured at 460 mm, provides the
C(t) function. For several electrolyte concentrations the decay times
decrease rapidly with increasing electrolyte concentration (Fig. 3).
The experimental relaxation times for coumarin 153 in various solvents
used are given in Tables 1, 2 and 3.

TABLE 1. Ionic relaxation times of coumarin 153 in THF

$LiClO_4$ conc(M)	$\epsilon_s{}^a$	Experimental τ(ps) Long component	short component amplitude	Longitudinal Dielectric relaxation τ_L(ps)^a	specific conductivity σ^a(ohm^{-1}m^{-1})
0.03	7.75	2580	0.08		
0.05				78	0.002
0.06	8.2	2140	0.18		
0.09	8.7	1760	0.2		
0.1				88	0.005
0.12	9.4	1250	0.2		
0.24	11.8	880	0.26	76	0.039
0.3	12.7	814	0.27		
0.42	14.3	530	0.2		
0.6	16.3	465	0.4	62	0.171

a. from ref [19]

TABLE 2. Ionic relaxation times of coumarin 153 in diethyl ether

LiClO$_4$ conc(M)	Experimental τ(ps)	short component amplitude
0.06	1060	0.08
0.09	940	0.1
0.12	830	0.1
0.18	700	0.1
0.3	660	0.1
0.42	640	0.2
0.6	590	0.3

TABLE 3. Ionic relaxation times of coumarin 153 in ethyl acetate

LiClO$_4$ conc(M)	$\epsilon_s{}^a$	Experimental τ(ps)		Longitudinal Dielectric relaxation time τ_L(ps)^a	specific conductivity σ^a(ohm^{-1}m^{-1})
		long component τ(ps)	short component amplitude		
0.0075	6.1	11,600	0.04		
0.015	6.1	3,300	0.03		
0.03	6.5	2,600	0.13		
0.09	7.4	1,400	0.1		
0.1	7.6			107	0.0027
0.12	8	1,170	0.15		
0.18	9	950	0.1		
0.21	9.4			113	0.0131
0.24	10	680	0.1		
0.30	10.8	750	0.28		
0.42	12.6	680	0.3	97	0.0695
0.6	14.4			89	0.139
0.8	15.7			79	0.246
1	17.0			77.5	0.359

a. from ref [19]

The orientational relaxation time for pure ethyl acetate around the related coumarin 311 was found by Barbara et al [9b] to be 2.3 ps, close to the value of the longitudinal relaxation time τ_L = 1.7 ps. This relaxation time is much shorter than the electrolyte relaxation time found in the present study. The solvent relaxation measured by coumarin 153 fluorescence at 460 nm in the pure solvents is hardly observable with our streak camera (10ps resolution).

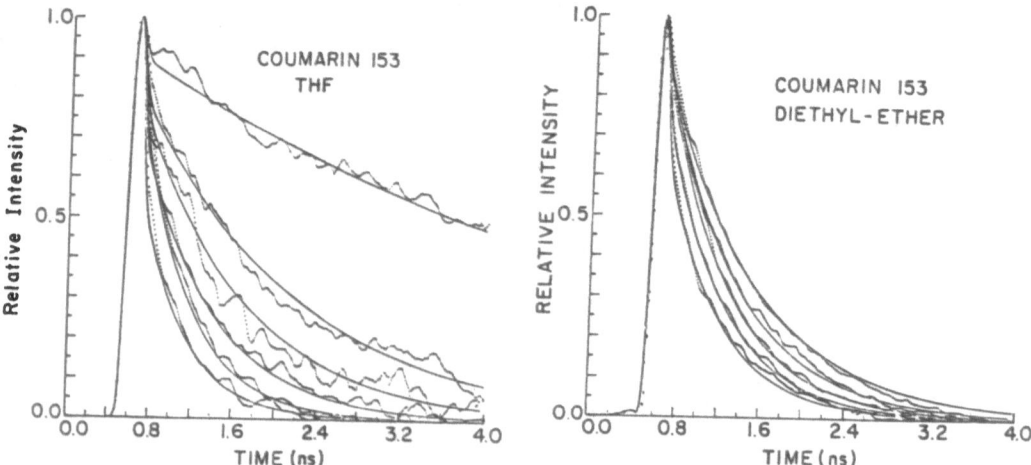

Figure 3. Time resolved luminescence of coumarin 153 in THF and diethyl ether containing various concentrations of electrolyte measured at 460nm. The solid line through the data points is generated by computer fit using a pulse convolution with two exponential decay time constants.

The fluorescence decay curves were fitted by a two exponential computer program. The analysis indicates that the relative amplitude of the short-lived component (due to the solvent reorientation) increases with the electrolyte concentration. In neat solvent the relative amplitude of the solvent component is ~0.05 while at 0.42M LiClO$_4$ it is 0.3. The reason for the large increase in the amplitude of the solvent component is two fold. The fluorescence band maximum is shifting to the red with an increase of the electrolyte concentration and hence the band overlap between the initial unrelaxed emission and the relaxed one decrease upon the addition of electrolyte to the solution. The amplitude of the short-lived fluorescence component follows the steady state fluorescence relative amplitude at 460nm, (see Fig.1). The second reason is that the solvent reorientation time in the non associative liquids used increases by about a factor of two at 0.4M of LiClO$_4$. The above statement was deduced from viscosity measurements of electrolyte solutions. The short-lived and the long-lived components must thus represent the relaxation of the solvent and the electrolyte respectively. The identification of the two components is strongly supported by time resolved emission measurements taken at low temperatures. The short-lived time component of coumarin 153 emission in amyl acetate solution ($\tau_L \simeq$ 3ps at room temperature) at temperatures close to the freezing point ~-70°C is easily time resolved (~70ps). The lifetime of the solvent component is almost identical for both neat liquid and 0.4M electrolyte solution therefore its origin is electrolyte independent and we attributed it to the solvent reorientation. The electrolyte component is also strongly temperature dependent and at ~-20°C it is much longer than the observed radiative lifetime of the relaxed emission (~5ns). Thus at lower temperatures the time resolved fluorescence decay curves of coumarin 153 in neat liquids or in electrolyte solutions are almost identical.

Electrolytes dissolved in low permittivity liquids affect not only the fluorescence of molecules like coumarins (single fluorescent band with large Stokes shift in polar liquids) but also molecules that exhibit dual fluorescence bands like DMABN. The relative intensity ratio between the initial "locally excited" state, LE (350nm) and the charge transfer state, CT (450nm) depends on the solvent polarity and in the current study it was found to be a function of the salt concentration. Time integrated (steady state) fluorescence measurements show that the CT fluorescence band intensity increases at high electrolyte concentration and it is accompanied by a decrease of the LE band intensity. The dynamics of the electron transfer is observed by the fluorescence decay of the LE band of DMABN in $LiClO_4$ - ethyl acetate solutions measured at 380nm (see Fig.4). The time correlated single photon counting (TCSPC) technique was used to monitor over three decades, the fluorescence intensity versus time. The decay curves comprise of three distinctive time components; the short component is due to the solvent reorientation. The instrumental time resolution limits the actual determination of the short component lifetime. The solvent longitudinal relaxation time is ~2ps thus we expect a very small amplitude for the solvent component since the system time resolution is relatively slow 100ps. The amplitude, ~0.75 in neat liquid is much larger than we found for coumarin 153 (~0.05) in the same solvent. The reason for that might be the absence of overlap between the LE and the CT bands the small rate of back electron transfer, and that the activation barrier for the intramolecular electron transfer rate is quite large and therefore the rate is of the order of few tenth of picoseconds. The intermediate fluorescence component is due to the salt relaxtion and its lifetime is electrolyte concentration dependent as was found for coumarin 153. The third component lifetime is 2.2 ± 0.1ns, the same lifetime is also measured for the CT band at 450nm and its value is almost salt concentration independent as expected in a case where the salt does not quench the CT state due to collisions. The salt effect data analysis for DMABN is given in Table 4.

TABLE 4. Ionic relaxation times of DMABN in ethyl acetate

$LiClO_4$ conc(M)	$\epsilon_s{}^a$	Experimental τ(ps)	short component amplitude
0.03	6.5	6000	0.86
0.06	7	4500	0.82
0.12	8	2000	0.78
0.18	9	1290	0.75
0.60	14.4	900	0.75

a. from ref [19]

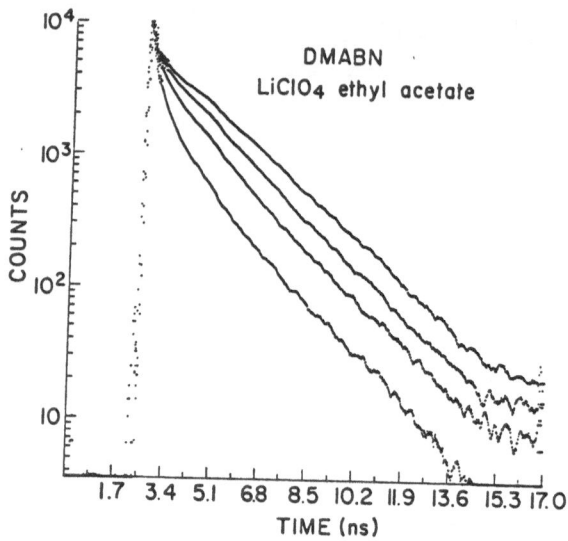

Figure 4. Time resolved emission of DMABN in ethyl acetate containing electrolyte at 380nm. (Note a semilog plot).

The electrolyte, LiClO$_4$ in the low dielectric solvent, used in the experiment is presented in solution as ions, ion-pairs and triple ions and larger aggregates. If an ion were created in the excited state of the probe molecule (coumarin, DMABN) rather than a dipole, and only ions were present in the solution, then we would expect a relaxation of the ionic species to form an atmosphere (ions and triple ions) with a relaxation time which follows Debye-Falkenhagen theory. In the present case of excited aromatic probes, a large dipole is created, implying that the interaction energy with the surrounding ions will decrease more rapidly with distance than $1/r$. However, according to the MSA theory, the solvent relaxation time constant is similar for an excited ion or a large dipole [6b]. The major difference between the physical system for which the Debye Falkenhagen theory is valid and our system seems to be the fact that most of the electrolyte is in the form of electrically neutral ion-pairs, for which the theory is inadequate. Ion pair relaxation should be similar that of a polar solvent dissolved in a non-polar solvent. We would then expect two relaxation modes with two characteristic relaxation times. The faster mode is the ion pair reorientation for which a reorientation time can be estimated from the Stokes-Einstein equation $\tau = 4\pi a^3 \eta / kT$, where a is the radius of the rotating spherical dipole, and η the solvent viscosity (for ethyl acetate, 0.45cp). The ion pair size can range from a short distance such as the contact radius ~3Å up to $e^2/2\epsilon kT$ [17]; which for $\epsilon=6$ is ~45Å. The corresponding reorientation times can vary from a few picoseconds to a nanosecond. Thus the ion pair reorientation relaxation should be non exponential (due to a distribution of relaxation times) and almost independent on the electrolyte concentration. This is in contradiction to the experimental findings which show that the relaxation time exhibit a linear concentration dependence. Furthermore, the substitution of the Li^+ ions (small size positive ions) with large ions like tetrabutylammonium decreases the observed ionic relaxation rate by less than a factor of two. This finding is in contrast to the cube dependence expected from the Stokes-Eisntein relation of the reorientation time on the ion size. The second relaxation mode will be the ion pair drift toward the excited dipole to form a "dipolar atmosphere" analogous to the way that ions create an ionic atmosphere. This translational relaxation should depend linearly on the ion size in a similar way as the ionic conductivity. The similar ionic relaxation rate of tetrabutylammonium and Li^+ ions favor the translational relaxation mechanism.

Although we consider that the Debye-Falkenhagen theory is inadequate for description of the relaxation of ion pairs around the excited dipole, there are parallels between the experimental data and the theory. In addition, at high electrolyte concentrations, about 10% of the electrolyte is in the form of triple ions for which the interaction with the excited dipole is similar to ion-ion interaction which govern the Debye Falkenhagen relaxation.

The Debye-Falkenhagen theory for the frequency dependence of electrolyte solution conductivity [14b,c] yields an estimate of the relaxation time for the density distribution of the ionic atmosphere to appear or disappear. For a 1:1 electrolyte ($LiClO_4$) the relaxation rate, $1/\tau$ is given by

$$1/\tau = \Lambda k T \chi^2 / 15.3 \times 10^{-8} \qquad (7)$$

$1/\chi$ is the ionic atmosphere radius and Λ is the equivalent conductance of the electrolyte.

$$\chi^2 = [4\pi/(\epsilon_s k T)](e^2 N_o C/1000) \qquad (8)$$

where N_0 is the Avogadro number, e is the electronic charge, C is the ion concentration (in liquids with low dielectric constant the total ion concentration is much lower than the electrolyte concentration). Eliminating χ^2 in eq.(9), the relaxation rate is given by

$$1/\tau = 1.13 \cdot 10^{10} C \Lambda / \epsilon_s \quad (sec^{-1}) \qquad (9)$$

The relaxation rate defined in eq.9 is linearly dependent on both the concentration of ions and their equivalent conductance but inversely dependent on dielectric constant (which, in turn, is strongly dependent on the electrolyte concentration). A plot of the log of the observed electrolyte relaxation rate for coumarin 153 in THF versus $log(C/\epsilon_s)$ is shown in Fig. 5. The linear correlation of the observed data with C/ϵ_s for $LiClO_4$ in both ethyl acetate, and THF solutions has a slope of ~0.9. The computed relaxation rate taking into account only the actual concentration of the ions and triple ions is smaller than the observed rate by almost two orders of magnitude at low electrolyte concentration and by a factor of 4 at the highest concentration. This analysis indicate that the ion-pairs contribution to the salt effect is larger than that of the ions and the application of the Debye Falkenhagen theory is not adequate or it might be more complicated to apply it to the electrolyte relaxation.

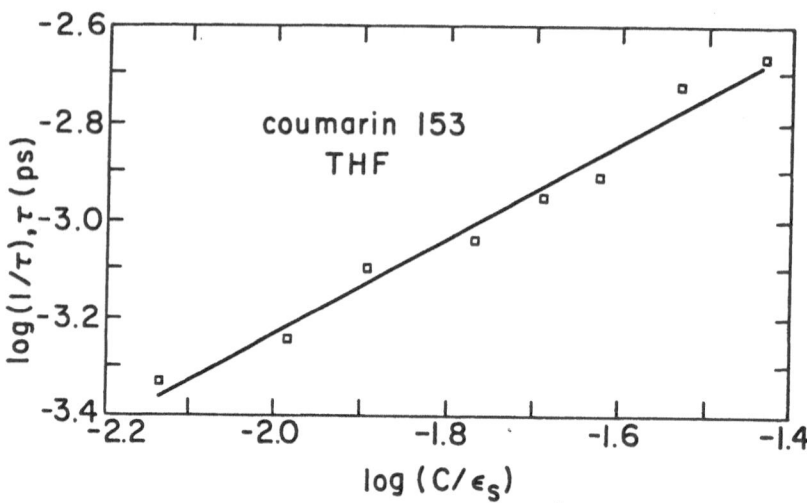

Figure. 5. Plot of the log of the observed electrolyte relaxation rate versus log C/ϵ_s

4. CONCLUSIONS

Steady state fluorescence emission energies of coumarin 153 are decreased by LiCLO$_4$ in liquids of low permittivity. In DMABN, the ratio between the charge transfer fluorescence band (450nm) and the locally excited band (350nm) is increased with the electrolyte concentration. Electrolyte creates a new, observable relaxation rate for the emitting state. This relaxation rate is relatively slow (ns) compared to solvent orientational relaxation rates (ps). The electrolyte in low dielectric constant solvents is mainly in the form of ion-pairs. The "slow" relaxation rate is attributed to the ion pair translational relaxation rate around the excited probe molecule. It was found to exhibit almost a linear dependence on the ion-pair concentration. New ways of probing medium effects on excited state behavior are thus opened.

Acknowledgement: Helpful discussion with Prof. E.M. Kosower, Dr. Ilya Ripps and Dr. E.Pines contributed much to the progress of the research.

5. REFERENCES

[1]   a. D.Huppert, H.Kanety & E.M.Kosower, Chem. Phys. Lett. 84 (1981)
         48-53.
      b. D. Huppert, H. Kanety & E.M. Kosower, Disc. Faraday Soc. 74
         (1982) 161-175 (disc. 194-203).
      c. E.M. Kosower & D. Huppert, Chem. Phys. Lett. 96 (1983) 433-435.
      d. E.M. Kosower, H. Kanety, H. Dodiuk, G. Striker, T. Jovin,
         H. Boni & D. Huppert, J. Phys. Chem. 87 (1983) 2479-2484.
      e. R. Giniger, D. Huppert & E.M. Kosower, Chem. Phys. Lett. 118
         (1985) 240-245.
      f. E.M. Kosower, J. Am. Chem. Soc. 107 (1985) 1114-1118.
      g. E.M. Kosower & D. Huppert, Ann. Rev. Phys. Chem. 37 (1986)
         127-156.
      h. D. Huppert, V. Ittah & E.M. Kosower, Chem.Phys. Lett. 144
         (1988) 15-23.
      i. D. Huppert, V. Ittah, A. Masad & E.M. Kosower, Chem. Phys.
         Lett. 150 (1988) 349-356.
      j. D. Huppert, V. Ittah, & E.M. Kosower, Chem. Phys. Lett., in press.
      k. D. Huppert, S.D. Rand, P.M. Rentzepis, P.F. Barbara, W.S. Struve
         & Z.R. Grabowski, J. Chem. Phys. 75 (1981) 5714-5719.
[2]   a. G. van der Zwan and J.T. Hynes, J. Phys. Chem. 89 (1985)
         4181-4188.
      b. J.T. Hynes, J. Phys. Chem. 90 (1986) 3701-3706.
      c. E.A. Carter and J.T. Hynes 93 (1989) 2184-2187.
[3]   a. F. Heisel and J.A. Miehe, Chem. Phys. 98 (1985) 233-241.
      b. F. Heisel, J.A. Miehe & J.M.G.Martinho, Chem. Phys. 98 (1985)
         243-249.
      c. F. Heisel & J.A. Miehe, Chem. Phys. Lett. 128 (1986) 323-329.
[4]   M. Morillo and R.I. Cukier, J. Chem. Phys. 89 (1988) 6736-6743.
[5]   P.G. Wolynes, J. Chem. Phys. 86 (1987) 5133-5136.
[6]   a. I. Rips & J. Jortner, J. Chem. Phys. 88 (1988) 818-822.
      b. I. Rips, J. Klafter & J. Jortner, J. Chem. Phys. 88 (1988)
         3246-3252.
[7]   a. M. Maroncelli and G.R. Fleming, J. Chem. Phys. 86 (1987)
         6221-6239.
      b. M. Maroncelli & G.R. Fleming, J. Chem. Phys. 89 (1988) 875.
[8]   M. Maroncelli, E.W. Castner Jr., B. Bagchi & G.R. Fleming,
      Faraday Disc. 85 (1988) 199-210.
[9]   a. V. Nagarajan, A.M. Brearley, T.J. Kang & P.F. Barbara,
         J. Chem. Phys. 86 (1987) 3183-3196.
      b. M.A. Kahlow, T.J. Kang & P.F. Barbara, J. Phys. Chem. 91
         (1987) 6452-6455.
      c. P.F. Barbara & W.Jazerba, Acc. Chem. Res. 21 (1988) 195-199.
[10] a. J.D. Simon & S.-G.Su, J. Chem. Phys. 87 (1987) 7016-7023.
      b. J.D. Simon, Acc. Chem. Res. 21 (1988) 128-134.
      c. J.D. Simon & S.-G.Su, J. Phys. Chem. 92 (1988) 2395-2397.
[11] H. Sumi & R.A. Marcus, J. Chem. Phys. 84 (1986) 4894-4914.
[12] M. Sparpaglione & S. Mukamel, J. Phys. Chem. 91 (1987) 3938-3943.

[13] M.D. Newton & H.L. Friedman, J. Chem. Phys. **88** (1988) 4460-4468.
[14] a. P. Debye & E. Hückel, Phys. Zeit. **24** (1923) 185.
     b. P. Debye & H. Falkenhagen, Phys. Zeit. **29** (1928) 121.
     c. P. Debye & H. Falkenhagen, Phys. Zeit. **29** (1928) 401.
[15] L. Onsager, Phys. Zeit. **28** (1927) 277.

[16] a. R.M. Fuoss & C.A. Kraus, J. Am. Chem. Soc. **55** (1933) 1019.
     b. R.M. Fuoss & C.A. Kraus, J. Am. Chem. Soc. **55** (1933) 2387.
[17] N. Bjerrum, Kgl. Danske-Vidensk-Selskab. Skr. **7** (1926) 9.
[18] D. Huppert & E. Kolodney, Chem. Phys. **63** (1981) 401-410.
[19] a. J.P. Badiali, H. Chacet & J.C. Lestrade, Ber. Bunsen Ges. **75**
        (1971) 297.
     b. J.P. Badiali, H. Chacet & J.C. Lestrade, Electrochim. Acta **16**
        (1971) 731.
     c. J.C. Lestrade, J.P. Badiali & H. Chachet, *'Dielectric and
        Related Molecular Processes*, 2 (1975) Chap. 3, Chem. Soc.,
        London.
[20] T. Sigvartsen, B. Gestblom, E. Noreland & J. Songstad, Acta
     Chimica Scandinavica 43 (1989) 103-115.
[21] J. Padova in *'Water and Aqueous solutions'* R.A. Horne, editor
     p.109-174, Wiley-Interscience New York 1972.
[22] a. M. Mohammad & E.M. Kosower, J. Phys. Chem. **74** (1970) 1153.
     b. E.M. Kosower, J. Am. Chem. Soc. **80** (1958) 3253-3261.

HIERARCHICAL RELAXATION : ULTRAMETRIC SPACES

J. KLAFTER* , A. BLUMEN# and G. ZUMOFEN$

* School of Chemistry, Tel-Aviv University, Tel-Aviv,
  69978 Israel

Physics Institute and BIMF , University of Bayreuth,
  D-8580 Bayreuth , West Germany

$ Laboratorium fur physikalische Chemie , ETH- Zentrum,
  CH-8092 Zurich , Switzerland

ABSTRACT. Relaxation processes are studied in ultrametric spaces which model energetic complexity through hierarchically distributed energy barriers. The relaxation processes are described as diffusion-limited reactions and are shown to follow stretched-exponential and power-law patterns in time.

1. Introduction

Much effort has been recently devoted in order to understand the physics which underlies non-exponential relaxation patterns in complex systems such as glasses [1-4], polymers [5-7] and also proteins [8,9]. The decay forms which are frequently observed experimentally are:

(1) The stretched-exponential, Kohlrausch-Williams-Watts [1-7] law

$$\phi(t) = \exp[-(t/\tau)^{\alpha}] \qquad (1)$$

This form has been found in measurements on defect annihilation by diffusing of hydrogen in amorphous Si:H [10]. The same form is also present when monitoring dielectric relaxation in glasses [1,2,11] and in polymers [5-7]. Recent works on the time evolution of solvation in polar liquids suggest that Eq(1) fits well the dynamics of solvation [12,13].

(2) The exponential- logarithmic relaxation, or the enhanced power law,

317

J. Jortner and B. Pullman (eds.), Perspectives in Photosynthesis, 317–324.

$$\phi(t) = \exp[-Cln^{\beta}(t/\tau)] \qquad\qquad (2)$$

which is often used in describing electron scavenging and electron-hole recombination, and which also appears in the analysis of relaxation phenomena related to hole burning in glasses [14,15].

(3) Algebraic decays,

$$\phi(t) \propto (t/\tau)^{-\gamma} \qquad\qquad (3)$$

as for instance reported for the relaxation processes of photogenerated carriers, which occur after electron-hole-pair creation in amorphous semiconductors [16]. It is also observed in measurements of rebinding of carbon monoxide to myoglobin [8,9].

In order to be able to apply these decay patterns to relaxation in real systems one needs to understand the possible microscopic origins of such temporal behaviors. Otherwise they only serve as convenient fitting forms. The study of relaxation in *model systems* that capture certain aspects of disorder may provide some of the required insight. Models of various degrees of sophistication have been introduced to mimic different physical realizations. The most elementary ones employ geometrical or energetic hierarchical structures [3,4,17-23] which have been shown to naturally lead to stretched-exponential and power law shapes In some of them a common mathematical framework underlies the non-exponential relaxation [3-5,24].

In this contribution we focus on one type of hierarchical models, the ultrametric spaces (UMS). The UMS are hierarchical in their energy barriers (*vide infra*) and they model systems where transitions take place through activated jumps over hierarchies of energy barriers [19-23].

In realistic situations sites in a disordered material , such as a glass , are separated by energy barriers, whose height is, in general, random. Taking for simplicity a nonquantal picture and for all sites identical ground states, a "walker" (as for example a charge carrier or a localized excitation) needs thermal energy to surmount these barriers. A given activation energy allows a walker to visit only a subset of sites around the starting point, the subset being separated from the other sites by barriers higher than the prescribed activation energy. One may then classify the sites through the energy required to reach them [25].To such a classification corresponds an ultrametric space [26-28]. It has been recently proposed that proteins are similar in some ways to glasses, as exemplified by heat capacity and hole burning experiments [29,30],and that a UMS description might apply to proteins too [9].

Here we would not discuss the applicability of UMS to any real system. We have chosen to introduce the UMS concept and to demonstrate how its hierarchical nature relates to non-exponential relaxation. In what follows we briefly summarize our works on UMS which appear in Refs.[4,21,22,25].

2. Ultrametric Spaces

The UMS we consider here consists of the set of *tips* of a finite Bethe-lattice. Figure 1 shows the UMS Z_3. Note that only the points on the baseline of the figure belong to the space, and the structure above the baseline defines connections. A distance between two sites may now be defined as being the minimal number of branches one has to walk on the tree in order to get from one tip to the other.Such a distance has a physical meaning as being proportional to the energy required to overcome the intersite barrier. It is straightforward to verify that the so-defined distance $d(x,y)$ satisfies the strong triangle inequality [26-28]:

$$d(x,y) \leq \max(d(x,y), d(y,z)) \qquad (4)$$

for all sites x,y,z, of the UMS. Furthermore, specifying a value E for the activation energy leads to the partition of the UMS into a set of disjoint clusters,where any two points of energy lower that E, and any two points belonging to different clusters by barriers higher than E. Here we take for simplicity the barrier heights to be hierarchically distributed, so that all consecutive energy levels differ by Δ, and assume that the branching ratio z is constant over the whole Bethe lattice. (In Fig.1, z equals 3.) The UMS considered here are thus *homogeneous*.

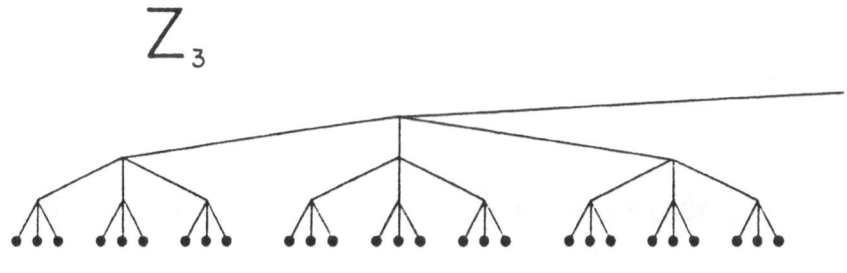

Figure 1. An ultrametric structure (UMS) with z=3

For more detailed description of the topological properties of UMS the reader is referred to Refs.[25-28]. We now investigate the dynamics over UMS of the motion of a single particle which, while temporarily localized, is still able (due to the thermal activation, $\beta=1/\kappa T$) to hope over energy barriers.

As mentioned, a simple qualitative argument describes the basic role of the hierarchical nature of UMS. The idea is that after the activation by the energy E the particle may reach only sites which belong to a sphere centered on the original site, the sphere radius depending, of course, on E. Using Fig. 1, and equidistant energy barriers, a particle activated by E with $\Delta \leq E < 2\Delta$ may reach three sites (including the original one) and generally for $j\Delta \leq E < (j+1)\Delta$ may reach 3^j sites. Here the number three appears because of the specific geometry of Fig. 1, which depicts the UMS Z_3. To a general space, in which the clusters are nested hierarchically in groups of z objects, corresponds the UMS Z_z , and an activation energy of $j\Delta \leq E < (j+1)\Delta$ allows to reach z^j sites.

Consider random walks on the UMS. Basic quantities for such random walks are the distribution $R(t)$ of distinct sites visited in time t, its mean $S(t) = <R(t)>$ over all realizations, and the probability $P_0(t)$ of being at the origin [19,20]. Jumps over the energy barrier E_m display the typical activated rates $\omega \cdot e^{-m\beta\Delta}$. Thus, during the time interval from time t=0 to time t_m, where $e^{m\beta\Delta} \le \omega \cdot t_m < e^{(m+1)\beta\Delta}$, the particle may have access to the z^m points of the sphere, and one has

$$z^m \propto z^{(1/\beta\Delta)\ln(\omega \cdot t_m)} = e^{\delta\ln(\omega \cdot t_m)} = (\omega \cdot t_m)^\delta, \qquad (5)$$

where we set

$$\delta = (1/\beta\Delta) \ln z \qquad (6)$$

The temperature and the characteristic energy barrier Δ enter the dynamical quantities through the power δ. Similar behavior is well known in dispersive transport studies in amorphous semiconductors [16,17,31].

One may now distinguish between two cases, depending on δ. For $\delta < 1$, i.e., $\beta\Delta > 1$, z^m increases more slowly than $\omega \cdot t_m$, and the walker explores practically all points(compact exploration). Therefore

$$S(t_m) \propto z^m \propto (\omega \cdot t_m)^\delta \qquad (7)$$

i.e.,

$$S(t) \propto (\omega \cdot t)^\delta \qquad (8)$$

Furthermore, assuming an equipartition of probabilities

$$P_0(t) \propto z^{-m} \propto (\omega \cdot t_m)^{-\delta} \qquad (9)$$

so that $P_0(t) \propto (\omega \cdot t)^{-\delta} \propto [S(t)]^{-1}$. On the other hand, for $\delta > 1$ (i.e., $\beta\Delta < 1$), z^m increases more rapidly than $\omega \cdot t_m$, and the mean number of distinct sites visited stays proportional to t_m,

$$S(t) \propto \omega \cdot t \qquad (10)$$

Here the exploration is noncompact [4,25]. $P_0(t)$, on the other hand, is still proportional to $(\omega \cdot t)^{-\delta}$ [19,20,28].

In general we note that the first moments of the distribution $R(t)$, the mean $S(t)=<R(t)>$ and the variance $\sigma^2(t)=<R^2(t)>-<R(t)>^2$, are analytically difficult to obtain even for walks on regular lattices (an exact expression for $R(t)$ is known only in one dimension). The knowledge of $R(t)$ is fundamental , however, for various reactions and one can obtain this information by simulations. For a detailed analysis of the distribution $R(t)$ and its peculiar temperature dependence see Ref [22].

Those familiar with random walks on regular lattices and on fractals will recognize that Eqs. (8) and (10) are similar to the lattice and fractal expressions for the mean number of distinct sites visited [4,17,21]. here δ plays the role of a temperature dependent "dimension". $\delta<1$ corresponds to low dimensions while $\delta>1$ corresponds to higher dimensions. According to Eqs. (6), (8) and (10) changes in temperature allow to shift the "dimension" parameter δ smoothly through all positive values, and to study the transition in behavior around and below $\delta=1$. Note that unlike regular lattices and fractals, UMS do not have any specific underlying geometry.

3. Relaxation Patterns

We now turn our attention to those relaxation mechanisms on UMS which can be modelled by diffusion-limited reaction schemes in the framework of random walk theories. In the previous section we derived random walk quantities which are essential for the description of diffusion-limited reactions: the distribution R(t) and its mean and variance S(t) and σ^2(t) respectively. We focus only on reactions whose decay is given in terms of S(t).Here again the reader is referred to Ref.[25] for a more elaborate discussion of the reaction scheme approach to relaxation and for more examples of reactions on UMS.

Let us start with the *target* problem, in which a static A target is annihilated by diffusing B particles. This is a pseudounimolecular reaction of the type $A+B \rightarrow B$. When applied to geometrically disordered systems the target picture provides a generalized formulation of the Glarum defect mechanism for dielectric relaxation in polymers and glasses [5-7]. On UMS the target model results in a stretched exponential decay at low temperatures ($\delta<1$) which changes into a simple exponential at high temperatures ($\delta>1$). This is an example of the transition in behavior described in the previous section and which is the fingerprint of hierarchies in energy barriers.

All sites of our UMS (fixed z and equidistant barriers) are equivalent. We denote the site on which the target sits as the origin, r=0 and assign integer numbers by counting to the other sites. $F_m(r)$ is the probability that a random walker starting from r reaches the origin 0 for the first time in the m-th step. Because of the symmetry of the walk, $F_m(r)$ is also the first-passage time from 0 to r. The probability $H_n(r)$ that a first passage from r to 0 occurred in the first n steps is

$$H_n(r) = \sum_{m=1}^{n} F_m(r)$$

$$(11)$$

The probability therefore that a walker from r did *not* reach 0 in the first n steps is thus

$$\phi_n(r) = 1-H_n(r)$$

$$(12)$$

The decay law for the target follows by weighting Eq(12) with the initial particle distribution g(j) and by multiplying over all UMS sites. The survival probability of the target at the origin is now

$$\phi_n = \prod_r \left[\sum_j g(j) \left[\phi_n(r) \right]^j \right]$$

(13)

In the case of a Poisson distribution one has

$$g(j) = \exp[-p]\ p^j/j!$$

(14)

where $g(j)$ is the normalized probability of having j B particles at one site, with the mean particle density being p. With this Poisson distribution we obtain,

$$\phi_n = \prod_r \exp\left[-p + p\phi_n(r) \right]$$

(15)

Eq(15) can be expressed in terms of $H_n(r)$,

$$\phi_n = \exp\left[-p \sum_r H_n(r) \right]$$

(16)

Eq(16) may be further simplified by noting that according to Eq(11)

$$\sum_r H_n(r) = \sum_{m=1}^{n} \sum_r F_m(r)$$

(17)

Furthermore, the sum over $F_m(r)$ gives the increase in the total number of visited sites in the m-th step, and thus equals $S_m - S_{m-1}$. Consequently, we obtain as the decay law of the A particles due to their being annihilated by diffusing B species [4,25]

$$\phi_n = \exp\left[-p\ (S_n - 1) \right]$$

(18)

When translating number of steps n into time t Eq(18) can be rewritten as [25]

$$\phi(t) \propto \exp\left[-pS(t) \right]$$

(19)

where $S(t)$ is given by Eqs. (8) and (10). Eq(19) represents a decay form which is a stretched exponential at low temperatures ($\delta < 1$) and a single exponential at high temperatures ($\delta > 1$). The temperature dependence enters in this model through the parameter δ. As suggested in some works on glassy relaxation p may also be temperature dependent.

Another reaction scheme which describes relaxation in terms of the quantity S(t) is the *bimolecular reaction* A+A →0.This reaction has been also widely studied on regular lattices and on fractals and has been shown to result in power law decays [4,25,32]. On UMS, as in the target case, the parameter δ replaces the role of dimension in the geometrical models. Here we obtain temperature dependent patterns which decay algebraically in time [25],

$$\phi^{AA}(t) \propto t^{-1}$$

$$(20)$$

for $\delta > 1$ (high temperatures), and

$$\phi^{AA}(t) \propto t^{-\delta}$$

$$(21)$$

for $\delta < 1$ (low temperatures).

The target model and the bimolecular scheme are two examples of relaxation in a complex system where the complexity is introduced through hierarchies in energy barriers, the UMS. Typical to this model of complexity are interesting temperature dependencies of the relaxation patterns. The frameworks within which we described relaxation naturally lead to stretched exponential or power law decays.

Acknowledgments

The Deutsche Forschungsgemeinscaft , the Fonds der Chemischen Industrie and the Funds for Basic Research administrated by the Israel Academy of Science and Humanities are gratefully acknowledged for the support of this research.

References

1. G. Williams and D.C. Watts, Trans. Faraday Soc. 66, 80 (1970).

2. G. Williams, Adv. Polymer Sci. 33, 59 (1979).

3. R.G. Palmer, D. Stein, E.S. Abrahams and P.W. Anderson, Phys. Rev. Lett. 53, 958 (1984).

4. A. Blumen, J. Klafter and G. Zumofen in 'Optical Spectroscopy of Glasses', Ed. I.Zschokke, (Reidel, Dordrecht 1986), p. 199.

5. M.F. Shlesinger and E.W. Montroll, Proc. Natl. Acad. Sci. USA 81, 1280 (1984).

6. A.A. Jones, in 'Molecular Dynamics in Restricted Geometries', Ed. J. Klafter and J.M. Drake (John Wiley, New York, 1989).

7. J.T. Blender, D.G. Le Grand and W.V. Olszewski, in 'Transport and Relaxation in Random Materials', Ed. J. Klafter, R.J. Rubin and M.F. Shlesinger (World Scientific, Singapore, 1986).

8. R. Elber and M. Karplus, Science 235, 318 (1987).

9. H. Frauenfelder, F. Parak and R.D. Young, Ann. Rev. Bioph. Chem. 17, 451 (1988).

10. J. Kakalios, R.A. Street and W.B. Jackson, Phys. Rev. Lett. 59, 1037 (1987).

11. W.B. Jackson and Kakalios, in 'Advances in Amorphous Semiconductor I. Amorphous Silicon Related Materials',Ed. H. Fritzche (World Scientific, Singapore, 1986).

12. D. Huppert, V. Ittah, A.Masad and E.M. Kosower, Chem. Phy. Lett. 150, 349 (1988).

13. D. Huppert, in 'Dynamical Processes in Condensed Molecular Systems', Ed. J. Klafter, J. Jortner and A. Blumen, (World Scientific, Singapore, 1989).

14. J. Friedrich and D, Haarer, Angew. Chem. Int. Ed. Eng. 23, 113 (1984).

15. J, Freidrich and A. Blumen, Phys. Rev. B32, 1434 (1985)

16. J. Tauc, Semicond. Semimet. 21B, 299 (1984).

17. A. Blumen, J. Klafter, and G. Zumofen in 'Fractals and Physics' Ed. L. Pietronero and E. Tossati, (North Holland, Amsterdam 1986), p.399; for applications of the Weierstrass function see: D. Haarer and A. Blumen, Agnew. Chem. 100, 1252 (1988); H. Schnorer, D. Haarer and A. Blumen, Phys. Rev. B38, 8097 (1988).

18. S. Grosmann, F. Wegner, and K.H. Hoffmann, J. Physique Lett. 46, L575 (1985).

19. A.T. Ogielski and D.L. Stein, Phys. Rev. Lett. 57, 1634 (1985).

20. B.A.Huberman and M. Kerszberg, J. Phys. A18, 4565 91985).

21. A. Blumen , J. Klafter and G. Zumofen, J. Phys A19, L77 (1986).

22. A. Blumen, G. Zumofen and J. Klafter, J. Phys. A19, L861 (1986).

23. G.H. Kohler, E.W. Knapp, and A. Blumen, Phys. Rev B38, 6774 (1988).

24. J. Klafter and S. Shlesinger, Proc. Natl. Acad. Sci. USA 83, 848 (1986).

25. G. Zumofen, A. Blumen and J. Klafter, J. Chem. Phys. 84, 6679 (1986).

26. A.D. Gordon, 'Classification', (Chapman and Hall, London 1981).

27. W.H. Schikhof, 'Ultrametric Calculus' (Cambridge Univ. Press 1984).

28. R. Rammal, G. Toulouse and M.A. Virasoro, Rev. Modern Phys. 58, 765 (1986).

29. V.I. Goldanski, Yu. F. Krupyanskii and V.N. Fleurov, Dok. Akad. Nauk. SSSR 272, 978 (1983).

30. V. Fleurov, in 'Dynamical Processes in Condensed Molecular Systems', Ed. J. Klafter, J. Jortner and A. Blumen,(World Scientific, Singapore, 1989).

31. H. Scher and E.W. Montroll, Phys. Rev. B12, 2455 (1975).

32. G. Zumofen, A. Blumen and J. Klafter, J. Chem. Phys .82, 3198 (1985).

THE SUPEREXCHANGE MODEL FOR THE PRIMARY CHARGE SEPARATION IN BACTERIAL PHOTOSYNTHESIS

M. BIXON and JOSHUA JORTNER
School of Chemistry
Raymond and Beverly Sackler Faculty of Exact Sciences
Tel Aviv University, 69978 Tel-Aviv, Israel
and
M.E. MICHEL-BEYERLE
Institut für Physikalische und Theoretische Chemie
Technische Universität München, Lichtenbergstrasse 4
D-8049 Garching, Federal Republic of Germany

ABSTRACT. A number of physical phenomena and observables can be accounted for and are consistent with the superexchange mechanism for the primary electron transfer in the reaction centre of *Rb.sphaeroides*. These include electric field effects on the quantum yield and polarization of the prompt fluorescence, the unidirectionality of the charge separation and the magnetic properties and recombination dynamics of the primary radical pair. A rationalization for the prevalence of the superexchange mechanism in the primary charge separation is provided on the basis of a kinetic optimization criterion in conjunction with energy constraints on the medium reorganization energy.

I. Introduction

The conversion of solar energy into photochemical energy in photosynthetic reaction centres (RC) of purple bacteria proceeds via a sequence of well organized, highly efficient, directional and specific electron transfer (ET) steps across the membrane. The primary process in the RC is

$$^1P^*BH \rightarrow P^+BH^- \qquad (I.1)$$

which involves ET from the excited singlet state of the bacteriochlorophyll dimer (P) to the bacteriopheophytin (H) across the L branch of the RC. This primary process is ultrafast, occurring on the time scale of τ_{ET} = 2.8 ± 0.2 psec at T = 300K for both *Rb.sphaeroides* [1] and *Rps.viridis* [2]. No chemical intermediate involving the accessory bacteriochlorophyll (B) was recorded on a time scale of $\geq$ 100 fsec [3]. This ultrafast ET time scale precludes any energy waste by backtransfer to the antenna. In the theoretical modelling of the primary charge separation process in the RC we assert that it is possible to construct a consistent theory, which will provide an interpretation of the wealth of experimental observables. As the RC constitutes a very complex object, the theoretical framework has to identify and incorporate only the basic and most important physical phenomena, which are relevant to the charge separation process.

II. Non-Adiabatic Electron Transfer

The primary electron transfer can be modelled in the framework of the non-adiabatic

325

J. Jortner and B. Pullman (eds.), Perspectives in Photosynthesis, 325–336.
© 1990 *Kluwer Academic Publishers*.

electron transfer theory.   The validity of this approximation rests on two central assumptions.

(1) The applicability of the non-adiabatic limit. This issue pertains to the weak Landau-Zenner coupling, which is determined by the parameter

$$\gamma_{LZ} = \frac{2\pi V^2}{\hbar\omega(2\lambda k_B T)^{1/2}} \ll 1 \tag{II.1}$$

where $V = 25$ cm^{-1} [4] is the electronic coupling, $\lambda \simeq 2000$ cm^{-1} [4] is the medium reorganization energy and $\hbar\omega \simeq 100$ cm^{-1} [5] is the medium characteristic frequency. Accordingly, $\gamma_{LZ} \simeq 0.04$ and the non-adiabatic limit prevails.

(2)   The separation of time scales for electronic process and medium damping. The experimental electron transfer rate is slow as compared to the characteristic relaxation time of the protein medium.   This relaxation time is proportional to the inverse effective average medium frequency which is $\simeq 100$ cm^{-1} and results in a relaxation time $\tau \simeq 100$ fs. The rough estimate of $\tau = 100$ fsec is consistent with recent molecular dynamics simulations of the protein structure in the RC accompanying electron transfer [6]. Accordingly, $\tau_{ET}^{-1} \ll \tau^{-1}$, ensures the validity of the conventional theory.

The basic equation for the electron transfer rate is then given by

$$k = \frac{2\pi}{\hbar} V^2 F \tag{II.2}$$

where V is the electronic coupling and F is the thermally averaged nuclear Franck-Condon factor.

Consider first the nuclear contribution to k.   The major contribution to F involves medium vibrational modes.   Under these circumstances F can be expressed within the effective single-mode approximation.   The full quantum mechanical expression for F assumes the form [7]

$$F = (\hbar\omega)^{-1} \exp[-S(2v+1)]I_p\{2(v(v+1))^{1/2}\}\{(v+1)/v\}^{p/2} \tag{II.3}$$

where

$$p = \Delta G/\hbar\omega \tag{II.4}$$

and

$$S = \lambda/\omega \quad . \tag{II.5}$$

ΔG is the free energy of the reaction, λ is the medium reorganization energy and ω is the (mean) vibrational frequency of the medium.   The mean thermal vibrational excitation is

$$v = \{\exp(\hbar\omega/k_B T-1)\}^{-1} \tag{II.6}$$

Finally, $I_p(\cdot)$ is the modified Bessel function of order p.

Provided that the frequencies of the nuclear modes, $\hbar\omega$, are sufficiently low relative to the thermal energy $k_B T$, the classical high temperature limit of F is applicable, being given by the Marcus relation

$$F = (4\pi\lambda k_B T)^{-1/2} \exp(-E_a/k_B T) \qquad (II.7)$$

where the activation energy E_a is given by

$$E_a = (\Delta G + \lambda)^2/4\lambda \quad . \qquad (II.8)$$

III. Temperature Dependence of the ET Rate

The rate, k, of the primary ET exhibits a very weak temperature dependence from liquid He temperature up to room temperature. Recent subpicosecond studies by Fleming, Martin and Breton [3] have established that the ET rate slightly decreases with increasing temperature. In the temperature range of 8K-295K, k decreases by a numerical factor of 2 for *Rb.sphaeroides* and by a numerical factor of 4 for *Rps.viridis*.

The weak temperature dependence of the primary charge separation rate, which exhibits a slight negative activation energy, is ubiquitous for ET processes in the RC. Such a behaviour was also recorded for a number of later ET reactions, i.e., ET from H^- to the quinone (Q), the backrecombination reaction $P^+Q^- \rightarrow PQ$ and the triplet recombination reaction $P^+H^- \rightarrow {}^3P^*H$. These latter reactions were attributed [5] to activationless ET, which involves the crossing of the nuclear potential surfaces at the minimum of the initial donor-acceptor state. Reaction (1) also falls into the category of activationless ET.

In analysing the temperature dependence of the ET rate, one notes that the energetic and nuclear parameters ΔE and λ may be temperature dependent, while the electronic coupling V depends on the donor-acceptor distance and orientation, which are affected by complicated thermal expansion processes of the protein. Under the assumption of temperature independence of ΔE, λ and V activationless behaviour of k prevails for $-\Delta E = \lambda$.

The notion of an activationless ET clearly constitutes a limiting situation. One may expect that a weak temperature dependence of the rate will be exhibited not only for the crossing of the potential surfaces at the exact minimum of the initial nuclear surface, but for a range of the relevant energetic and nuclear parameters. A weak temperature dependence can be exhibited when one of the following conditions is satisfied:
(A) The ET is strictly activationless, i.e., p = S. In this case (for S >> 1) the ET reduces to [5]

$$k = \frac{2\pi V^2}{\hbar^2\omega(2p)^{1/2}} \left[\frac{\exp(\hbar\omega/k_B T)-1}{\exp(\hbar\omega/k_B T)+1} \right]^{1/2} \qquad (III.1)$$

(B) The ET rate is pseudo-activationless. We expect that when the crossing of the nuclear potential surfaces occurs within the energy range between the minimum of the initial state potential surface and its ground vibrational level $\hbar\omega/2$, a weak temperature dependence will be exhibited.

Conditions (A) and (B) were combined into the following relation for a weak temperature dependence of the ET rate [8]

$$1-(2/S)^{1/2} \leq p/S \leq 1 + (2/S)^{1/2} \qquad (III.2)$$

This relation defines the relevant range of the energetic parameters spanning the activationless and pseudoactivationless regimes.

Utilization of the thermodynamic $\Delta E = -2000$ cm^{-1} [4] and vibrational $\hbar\omega = 100$ cm^{-1} [5] data results in p = 20 for the primary process. We then infer from Eq. (III.2) that the weak temperature dependence of k will be exhibited for $0.75 \leq p/S \leq 1.25$, as is evident from the data of Fig. 1, while Fig. 2 provides a quantitative determination of the activationless and pseudo-activationless domain.

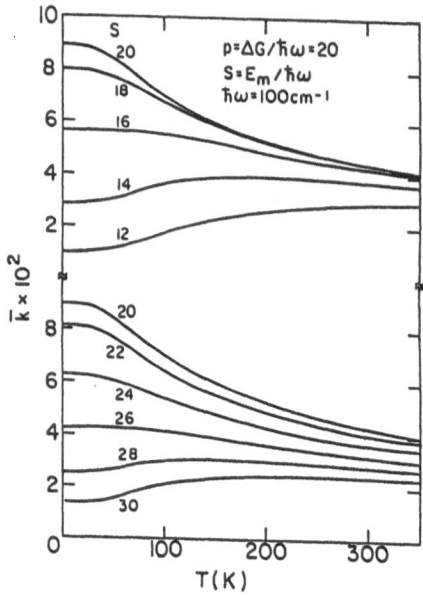

Figure 1. The temperature dependence of the relative ET rate $\bar{k} = k/(2\pi V^2/\hbar^2\omega)$ for $\Delta E = 2000$ cm^{-1} (p = 20) over a broad range of the medium reorganization energy parameter $E_m = 1200$-3000 cm^{-1} (S = 12-30). The characteristic frequency of the protein medium is $\hbar\omega = 100$ cm^{-1}. The case of activationless ET corresponds to p = S.

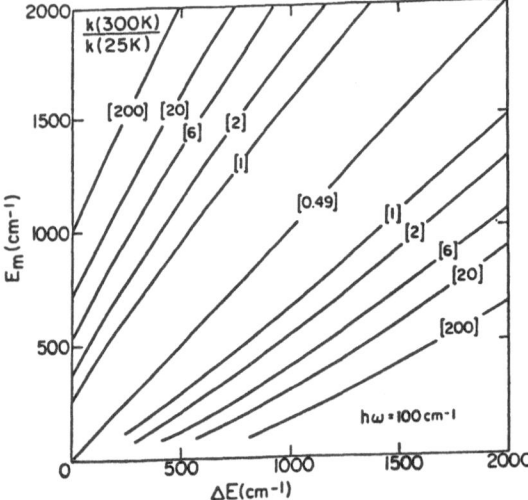

Figure 2. The dependence of the ratio k(T=300K)/(k(T=25K) of the ET rates (marked as numbers in square brackets [ ]) on $E_m(\lambda)$ and ΔG.

This analysis considered only the coupling with the medium vibrations, which were handled within the framework of quantum mechanical treatment for mthe single frequency approximation. Small but finite coupling with high frequency intramolecular vibrational modes of the prosthetic groups will result in the increase of the k vs ΔE curve at the inverted region ($-\Delta E > \lambda$) [9]. The shape of the (free) energy relationships in the RC was not yet fully explored.

The gross features of the temperature dependence of k [3] for both *Rb.sphaeroides* (fig. 3) and *Rps.viridis* (fig. 4) are consistent with activationless ET with a characteristic frequency $\hbar\omega = 80\text{-}100$ cm^{-1} for both RCs [8].

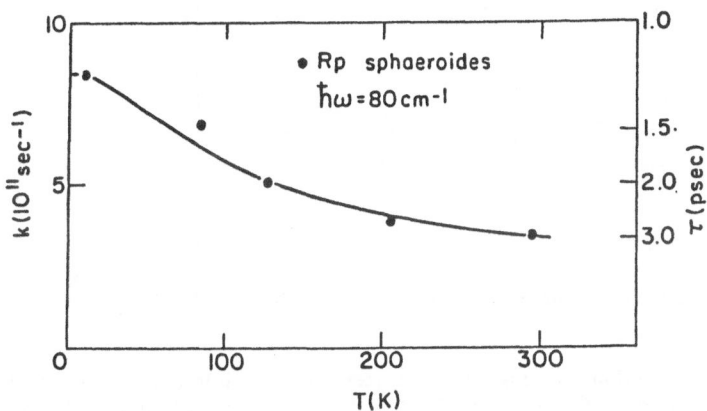

Figure 3. The temperature dependence of the primary ET rate in the RC of *Rb.sphaeroides*. Experimental data (•) from reference 3. The solid curve corresponds to the activationless limit with $\Delta\omega = 80$ cm^{-1}.

This characteristic frequency is very close to the values of $\hbar\omega = 120\text{-}80$ cm^{-1} inferred for later ET processes in the RC [5]. The analysis of the activationless nature of k establishes some universal features of the primary charge separation in the RC, and inspires confidence in the validity of non-adiabatic ET theory in providing a conceptual framework for the description of the primary process in photosynthesis.

IV. The Electronic Coupling

The mechanism responsible for the electronic coupling, V, in the primary process rests on two experimental attributes:
(1) The non-observability of a chemical intermediate involving B^+ or B^- (on a time scale of > 100 fsec) seems to rule out sequential ET, favouring unistep primary charge separation [3,4]. The experimental ET rate (k = 3.6x10^{11} sec^{-1} at 300K) in conjunction with the analysis of the F factor (section III) imply that the electronic coupling is V = 25 cm^{-1}.
(2) The (17Å) center to centre P-H distance is too large to account for the large V = 25 cm^{-1} coupling.

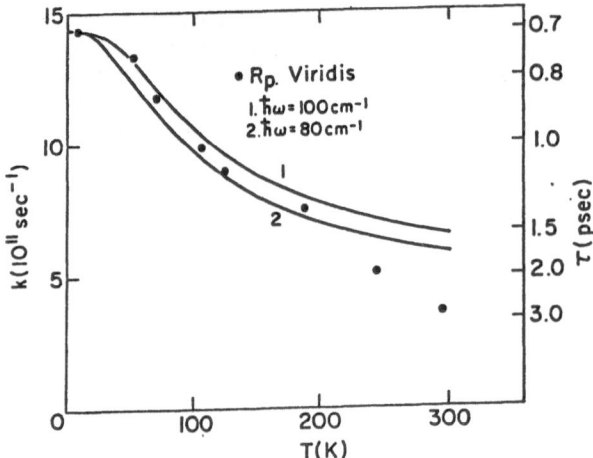

Figure 4.   The temperature dependence of the primary ET rate in the RC of *R.viridis*.
Experimental data (•) from reference 3. The solid curves correspond to the activationless
limit with $\hbar\omega = 100$ cm^{-1} (curve 1) and $\hbar\omega = 80$ cm^{-1} (curve 2).

A possible resolution of this dilemma rests on the introduction of superexchange
electronic interaction mediated via the P$^+$B$^-$H state (Fig. 5). The superexchange electronic
coupling is

$$V_{super} = V_{PB}V_{BH}/\delta E \qquad \text{(IV.1)}$$

where V_{PB} is the electronic coupling between ^{1}P*BH and P$^+$B$^-$H, while V_{BH} is the
electronic coupling between P$^+$B$^-$H and P$^+$BH$^-$. δE is the vertical energy difference
between the potential surfaces for P*BH and P$^+$B$^-$H at the intersection point of the
potential surfaces of ^{1}P*BH and P$^+$B$^-$H (Fig. 5)

An inevitable consequence of the superexchange mechanism involves the occurrence of
a parallel activated ET channel (k_1), which occurs at the intersection of the potential
surfaces of P*BH and P$^+$B$^-$H (Fig. 5) [10,11].   The relevant nuclear and electronic
parameters are the vertical gap δE, the reorganization energies λ (< 2500 cm^{-1}) and λ_1
(≥ 800 cm^{-1}) for k and k_1, respectively and the electronic couplings V_{BH} and V_{PB}, which
are related by $V_{BH}/V_{PB} = 6$ [10]. Utilizing the experimental constraints [P$^+$B$^-$]/[^{1}P*] ≤ 0.1
at 300K and [P$^+$B$^-$]/[^{1}P*] < 0.02 at 10K results in the following acceptable parameters
[10,11]

$$\begin{aligned}
\delta E &\geq 1100 \text{ cm}^{-1} \\
\Delta E &\geq 300 \text{ cm}^{-1} \\
V_{PB} &\geq 60 \text{ cm}^{-1} \\
V_{BH} &\geq 360 \text{ cm}^{-1}
\end{aligned} \qquad \text{(IV.2)}$$

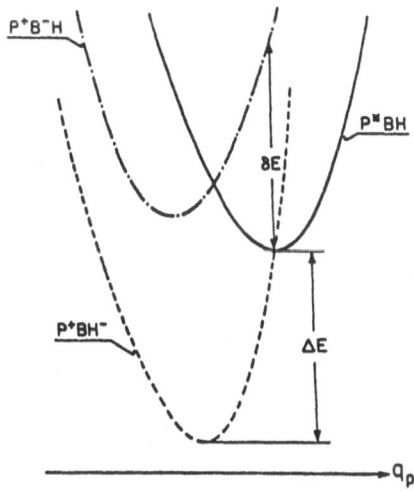

Figure 5. Nuclear potential energy surfaces for the superexchange electronic interaction mechanism in the RC.

V. Observables

A critical scrutiny of the superexchange model rests on several experimental sources:

(A) Electric field effects on the primary charge separation.
The primary electron transfer rate is affected by an external electric field, $\vec{\epsilon}$, through the shifts in the energies of the ion pair states [12-15]. The electric field dependence of the fluorescence quantum yields $Y_f(\epsilon)$ from isotropic samples is

$$y(\epsilon) = \frac{Y_f(\epsilon)}{Y_f(0)} = k(0) < k(\vec{\epsilon})^{-1} > \qquad (V.1)$$

where $\vec{\epsilon}$ denotes orientational average. The electric field dependence of $y(\epsilon)$ (fig. 6) exhibits an increase of the relative fluorescence quantum yield with increasing ϵ.

In confronting theory and experiment the interrelationship between the externally applied field, ϵ_E, and the internal field, ϵ, is of crucial importance. Local field corrections, which result in $\epsilon > \epsilon_E$, surface polarization effects and electrode contact effects, and which may result in $\epsilon < \epsilon_E$, among others, have to be elucidated. The experimental result of Lockhart and Boxer [14], $y(\epsilon_E = 9 \text{ mV}/\text{Å}) = 1.25$ at 75K, is close to the calculated value $y(\epsilon=5\text{mV}/\text{Å}) = 1.39$ at this temperature [15]. The agreement is not conclusive as it implies that $\epsilon_E > \epsilon$, and further work is required to calibrate the internal field.
    Fluorescence polarization measurements from isotropic samples at constant electric field provide a measurement of the direction of the electric dipole of the ion pair produced in the primary charge separation process [16]. The total fluorescence yield $F(\chi,\epsilon)$ for radiation polarized in the angle χ relative to the electric field $\vec{\epsilon}$, is used to evaluate the relative change of the polarized fluorescence

$$\theta(\chi) = \frac{F(\chi,\epsilon)-F(0)}{F(\chi=90^\circ,\epsilon)-F(0)} \qquad (V.2)$$

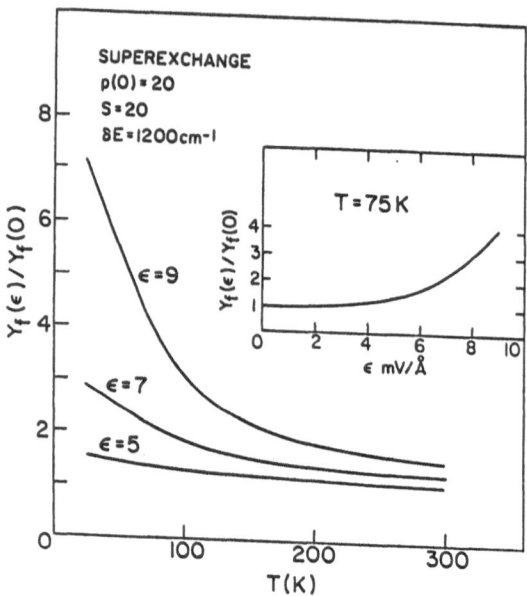

Figure 6. The electric field and temperature dependence of the fluorescence quantum yields from isotropic samples over the temperature range 20-300K, as calculated for the superexchange model. The electric field (ϵ) is given in units of mV/Å. The insert shows the electric field dependence at 75K.

which is independent of ϵ for sufficiently low ϵ. The geometric input information involves the angle $\phi = 50^0$ between the dimer transition moment $\vec{\mu}_t$ and $\vec{\mu}(P^+B^-)$, and the angle ψ between $\vec{\mu}_t$ and $\vec{\mu}(P^+H^-)$, which according to theoretical data is $\psi = 55^0$-61^0. Good agreement between theory and experiment was obtained for $\psi = 61^0$ (Fig. 7), establishing the consistency of the superexchange mechanism [17].

(B) Unidirectionality of charge separation.
The remarkable effect of unidirectionality of the charge separation across the A branch of the RC was attributed to structural symmetry breaking [4,10]. The total asymmetry of the ET transfer rates k(A) and k(B) across the A and B branches of the RC, respectively, is

$$k(A)/k(B) = \psi \qquad\qquad\qquad\qquad (V.3)$$

where ψ is the ratio of the electronic superexchange couplings and r is the ratio of the nuclear Franck-Condon factors.

Regarding the r ratio, it was previously shown [4] that the energetics of the charge separation across the A and B branches is different, due to the extra stabilization of H_A^- by the polar GLU104 residue. This ratio is r = 1.35 (+0.3,-0.45) at T = 300K and r = 2.5(+3.1,-1.8) at 80K. Accordingly, the nuclear contribution to k(A)/k(B) is small.

The major contribution to the asymmetry of the rates originates from the electronic

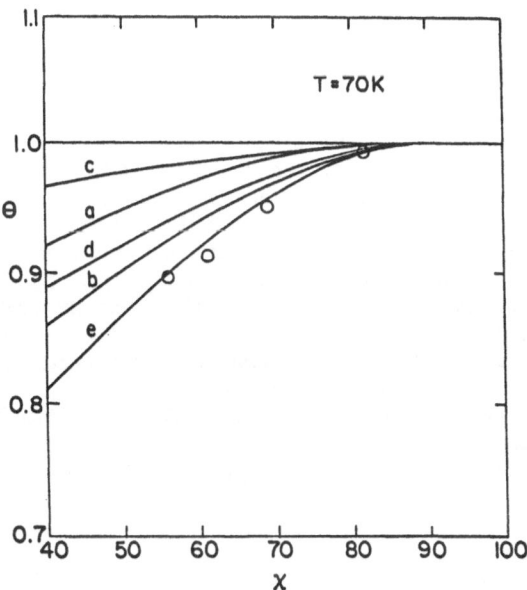

Figure 7. The dependence of the relative polarized fluorescence $\theta(\chi)$ on the angle χ between the polarization direction of the fluorescence and the electric field. Calculations were performed at 70K for the superexchange model:

a. $p = 20$; $S = 20$; $\psi = 58°$;
b. $p = 20$; $S = 16$; $\psi = 58°$;
c. $p = 20$; $S = 24$; $\psi = 58°$;
d. $p = 20$; $S = 20$; $\psi = 61°$;
e. $p = 20$; $S = 16$; $\psi = 61°$;
Open circles (O) represent the experimental data of Lockhart, Goldstein and Boxer (reference 14).

contribution ψ, which for the superexchange mechanism is given by

$$\psi = |V_{PB}(A)/V_{PB}(B)|^2 \, |V_{BH}(A)/V_{BH}(B)|^2 \quad . \tag{V.4}$$

The utilization of the intermolecular overlap approximation leads to the ratios [10]

$$|V_{PB}(A)/V_{PB}(B)|^2 = 7.7 \pm 3.8 \tag{V.5}$$

$$|V_{BH}(A)/V_{BH}(B)|^2 = 4.3 \pm 2.0 \tag{V.6}$$

resulting in the asymmetry factor for the electronic coupling

$$\psi = 33 \pm 16 \quad . \tag{V.7}$$

Eqs. (V.3) to (V.7) result in $k(A)/k(B) = 45$ (+38,-30) at $t = 300K$ and $k(A)/k(B) = 82$ (+190,-70) at $T = 80K$. This last result is in agreement with the best experimental value

$k(A)/k(B) \geq 25$ at 80K [17,18]. It should be noted that the superexchange mechanism is exclusively consistent with the large experimental electronic asymmetry, as it incorporates both contribution (V.5) and (V.6) into ψ.

(C) Magnetic interactions.

The interrelationship between the primary electron-transfer kinetics in the RC and the properties of the radical pair P^+BH^-, i.e., the singlet-triplet splitting, J, and the triplet recombination rate, k_T, constitutes a long-standing problem. We have recently examined [19] the magnetic interactions of the primary P^+H^- radical pair, establishing the interrelationship between the singlet energy shift and the triplet energy shift with the primary ET rate, k, and the triplet recombination rate k_T whereupon the singlet-triplet splitting of P^+H^- may be approximated by $J = \alpha k - \beta k_T$, where the coefficients α and β depend on energetic parameters and Franck-Condon factors. The estimate of J within the superexchange mechanism rests on the incorporation of configurational relaxation and essential cancellation effects. Therefore no correlation between J and the rate constants k and k_T is expected. Finally, the rate k_T can be accounted for by invoking differences in the charge distribution of $^1P^*$ and $^3P^*$.

The details of this analysis are provided elsewhere [11,17,19].

VI. Concluding Remarks

We have demonstrated that a number of physical phenomena and observables can be accounted for and are consistent with the superexchange mechanism for the primary electron transfer in the reaction center of *Rb.sphaeroides*. Such a consistency check provides, of course, a necessary condition for the validity of the superexchange model. A conceptual question we are facing pertains to the issue: why the primary process proceeds via superexchange? The elements for the resolution of this issue are

(1) A kinetic optimization criterion.

The primary process must be sufficiently fast, occurring on the psec time scale, to compete effectively with backtransfer of the excitation energy to the antenna. The most efficient process will involve activationless electron transfer, optimizing the medium FC factor by establishing that ΔE equals approximately the protein reorganization energy.

(2) Energetic constraints on the reorganization energy.

The medium reorganization energy for the primary electron transfer is determined by the protein environment for the donor and acceptor sites. In the protein medium the reorganization energy is a local property, being essentially determined by the interaction of the co-factors and their ions with amino acid residues in their vicinity.

One can advance a plausible argument that the medium reorganization energy, λ, for the primary process $^1P^* \rightarrow P^+H^-$ has to be about equal to the corresponding medium reorganization energy $\lambda_T \simeq 1800$ cm^{-1} for the triplet recombination $P^+H^- \rightarrow {}^3P^*$ as both λ and λ_T should include a substantial contribution from the polar amino acid residues located in the vicinity of P and H. If in fact $\lambda = \lambda_T \simeq 1800$ cm^{-1}, this energetic constraint would exclude the possibility of primary electron transfer via two sequential activationless processes with energy gaps $|\Delta E_1|$ and $|\Delta E_2|$ satisfying $|\Delta E_1| + |\Delta E_2| = |\Delta E| = 2000$ cm^{-1}. In this case $\lambda > |\Delta E_1|$ and/or $|\Delta E_2|$, and the kinetic optimization criterion (1) would be violated. Accordingly, the superexchange mechanism constitutes the only way to satisfy the kinetic optimization, speeding up the rate by selectively increasing the electronic coupling without affecting the Franck-Condon factor, so that the activationless nature of the primary step is maintained.

ACKNOWLEDGMENT. This research was supported by the Deutsche Forschungs-gemeinschaft (SFB 143) and by the Z. Weinberg Fund for Research in Chemical Physics at Tel Aviv University.

References

1. Martin, J.L., Breton, J., Hoff, A.J., Migus, A., and Antonetti, A. (1986) 'Femtosecond spectroscopy of electron transfer in the reaction center of the photosynthetic bacterium *Rb.sphaeroides* R-26: Direct electron transfer from the dimeric bacteriochlorophyll primary donor to the bacteriopheophytin acceptor with a time constant of 2.8±0.2 psec', Proc. Natl. Acad. Sci. USA 83, 957-961.

2. Breton, J., Martin, J.L., Migus, A., Antonetti, A., and Orszag, A. (1986) 'Femtosecond spectroscopy of excitation energy transfer and initial charge separation in the reaction center of the photosynthetic bacterium *R.viridis*', Proc. Natl. Acad. Sci. USA 83, 5121-5125.

3. Fleming, G.R., Martin, J.L., and Breton, J. (1988) 'Rates of primary electron transfer in photosynthetic reaction centers and their mechanistic implications', Nature 333, 190-192.

4. Michel-Beyerle, M.E., Plato, M., Deisenhofer, J., Michel, H., Bixon, M., and Jortner, J. (1988) 'Unidirectionality of charge separation in reaction centers of photosynthetic bacteria', Biochim. Biophys. Acta 932, 52-70.

5. Bixon, M. and Jortner, J. (1986) 'Coupling of protein modes to electron transfer bacterial photosynthesis', J. Phys. Chem. 90, 3795-3800.

6. Treutlein, H., Schulten, K., Deisenhofer, J., Michel, H., Branger, A., and Karplus, M. (1988) 'Molecular dynamics simulation of the primary processes in the photosynthetic reaction center of *R.viridis*', in J. Breton and A. Vermeglio (eds.), The Photosynthetic Reaction Center, NATO ASI Series A: Life Sciences, Plenum, New York, pp. 369-377.

7. Jortner, J. (1976) 'Temperature dependent activation energy for electron transfer between biological molecules', J. Chem. Phys. 64, 4860-4867.

8. Bixon, M. and Jortner, J. (in press) 'Activationless and pseudoactivationless primary electron transfer in photosynthetic bacterial reaction centers', Chem. Phys. Lett.

9. Efrima, S. and Bixon, M. (1976) 'Vibrational effects in outer-sphere electron-transfer reactions in polar media', Chem. Phys. 13, 447-460.

10. Plato, M., Möbius, K., Michel-Beyerle, M.E., Bixon, M., and Jortner, J. (1988) 'Intermolecular electronic interactions in the primary charge separation in bacterial photosynthesis', J. Am. Chem. Soc. 110, 9279-9285.

11. Bixon, M., Michel-Beyerle, M.E., and Jortner, J. (1988) 'Formation dynamics, decay kinetics and singlet-triplet splitting of the (bacteriochlorophyll)$^+$ (bacteriopheophytin)$^-$ radical pair in bacterial photosynthesis', Isr. J. Chem. 28, 155-168.

12. Popovic, Z.D., Kovacs, G.J., Vincett, P.S., Alegria, G., and Dutton, P.L. (1986) 'Electric field dependence of recombination kinetics in reaction centers of photosynthetic bacteria', Chem. Phys. 110, 227-237.

13. Gopher, A., Blatt, Y., Schönfeld, M., Okamura, M.Y., Feher, G., and Montal, M. (1985) 'The effect of an applied electric field on the charge recombination kinetics in reaction centers reconstructed in planar lipid bilayers', Biophys. J. 48, 311-320.

14. Lockhart, D.J. and Boxer, S.G. (1988) 'Electric field modulation of the fluorescence from *Rb.sphaeroides* reaction centers', Chem. Phys. Lett. 144, 243-249.

15. Bixon, M. and Jortner, J. (1988) 'Electric field effects on the primary charge

separation in bacterial photosynthesis', J. Phys. Chem. 92, 7148-7156.

16. Lockhart, D.J., Goldstein, R.F., and Boxer, S.G. (1988) 'Structure-based analysis of the initial electron transfer step in bacterial photosynthesis: Electric field induced fluorescence anisotropy', J. Chem. Phys. 89, 1408-1415.

17. Bixon, M., Jortner, J., Michel-Beyerle, M.E., and Ogrodnik, A. (in press) 'A superexchange mechanism for the primary charge separation in photosynthetic reaction centers', Biochim. Biophys. Acta.

18. Aumaier, W., Uberl, U., Ogrodnik, A., and Michel-Beyerle, M.E. (to be published).

19. Michel-Beyerle, M.E., Bixon, M., and Jortner, J. (1988) 'Interrelationship between primary electron transfer dynamics and magnetic interactions in photosynthetic reaction centers', Chem. Phys. Lett. 151, 188-194.

ROLE OF CHARGE TRANSFER STATES OF THE SPECIAL PAIR IN PRIMARY PHOTOSYNTHETIC CHARGE SEPARATION

Richard A. Friesner and Youngdo Won
Department of Chemistry
University of Texas at Austin
Austin, Texas 78712 U. S. A.

ABSTRACT. We propose a model for primary charge separation in the reaction centers of photosynthetic bacteria based upon an internal charge transfer state of the special pair dimer. In our model, this state serves as a trigger for primary electron transfer, linking the excited state of the special pair (with which it is nearly in resonance) to a superexchange pathway coupled to the initial bacteriopheophytin acceptor. The recombination reaction is not similarly facilitated because the charge transfer state is not close in energy to the ground state; this could be a key factor in explaining the high quantum yield of the primary electron transfer event. The model provides a qualitative explanation for three important experimental measurements (Stark effect, photochemical holeburning, singlet-triplet splitting of the charge separated species) on the reaction center. In addition, its predictions are in excellent accord with recent studies of a genetically modified reaction center in which one bacteriochlorophyll molecule of the special pair is changed to a pheophytin. For this system, one can directly observe the proposed charge transfer state participating in the electron transfer kinetics, located at an energy compatible with our model calculations.

1. Introduction

Despite the availability of highly resolved crystal structures for the bacterial photosynthetic reaction center (RC) [1–3], a microscopic understanding of the dynamics of primary charge separation has not yet been rigorously achieved. The unusual features of this reaction – near unit quantum efficiency, extreme rapidity at both room and cryogenic temperatures – have been quite difficult to reproduce in synthetic model systems, or to easily explain theoretically in the natural system. While a wide variety of theoretical approaches have been adopted in attempting the latter task, the complexity of the RC precludes the use of reliable, first principles calculations and requires the construction and evaluation of approximate models, often with numerous adjustable parameters. The large set of experiments that have been carried out on the RC provide the possibility for selecting the correct mechanism among these alternative models; at present, however, all proposals have some discrepancies with experimental data, and important controversies remain to be resolved.

We have recently reviewed the current experimental and theoretical situation with regard to interpretation of spectroscopic experiments on the RC, and their consequences for charge separation models [4]. In this article, we focus on one particular physical

J. Jortner and B. Pullman (eds.), Perspectives in Photosynthesis, 337–346.

hypothesis, involving the participation of a charge transfer (CT) state of the the initial donor complex (henceforth referred to as P_{CT}) in primary charge separation. It is argued that this hypothesis can simultaneously explain unusual results from three different experiments on the RC. While alternate explanations exist for each of these experiments, no direct evidence for the required ad hoc assumptions have been found. In contrast, our proposal has recently been afforded strong confirmation by experiments on a genetically modified photosynthetic bacterium [5], in which the above CT state (proposed on the basis of indirect evidence) can be directly observed to participate in the charge separation process, and is located at an energy consistent with our model.

The paper is organized into six sections. In Section 2, important (for our purposes) features of the ET kinetics (as measured by transient absorption spectroscopy) of the RC are briefly reviewed. Section 3 discusses three experiments (Stark, holeburning, and magnetic field measurements), analysis of which led to the formulation of our theoretical model. Section 4 presents our model and indicates how the experimental results of Section 3 can be rationalized (as well as what problems with our interpretations remain). Section 5 considers the genetically modified RC experiments. Section 6, the conclusion, examines prospects for testing the theoretical models experimentally, and suggests how one might utilize insights into RC dynamics to construct more effective photosynthetic model compounds or artificial charge separation systems.

2. Electron Transfer Dynamics of the Reaction Center

Charge separation is initiated from the lowest excited state P* of the special pair P of bacteriochlorophyll (BChl) molecules in the RC . These molecules have a small center to center separation, and are strongly coupled electronically. The allowed character of the P* derives from excitonic mixing of the Q_y states of the two BChl monomers.

In roughly 3 picoseconds, one can observe directly in transient absorption experiments the formation of the cation radical of P and the anion of the bacteriopheophytin (BPh) H_L (L designates the active L branch of the approximately C_2 symmetric chromophore distribution of the RC; henceforth, we suppress the L subscript in referring to both H and to the "intermediate" bacteriochlorophyll B_L), indicating that charge separation has taken place [6]. However, attempts to detect participation of B have been unsuccessful, despite increasingly precise and better resolved experiments [7,8]. The apparent nonparticipation of B, which is difficult to reconcile with the rapid ET over the long (17Å) distance from P to H , is one central fact which must be explained by theoretical models.

An attractive explanation as to why B is not observed as an explicit intermediate is that it facilitates rapid ET via a superexchange mechanism involving the configuration P^+B^-, which is thus assumed to lie at higher energy than P* [9]. However, as will be discussed later, there are difficulties with this mechanism, in that it requires quite large electronic coupling matrix elements between configurations, in apparent contradiction with an important experiment. Thus, all proposed simple models have significant problems explaining the observed primary kinetics.

From H, the electron proceeds under physiological conditions to a quinone acceptor in 200 picoseconds [6]. The high quantum yield of the charge separation process is a consequence of the fact that this reaction successfully competes with the recombination reaction of the charge separated state P^+H^- to the ground state, which occurs on a nanosecond timescale. There is thus a three to four order of magnitude difference in rates between the primary forward and recombination reactions. A discrepancy of this magnitude has yet to be produced in rigid model systems, where it is found that a rapid charge separation is almost always accompanied by equally rapid recombination. In view of the fact that there is little change in the kinetics at cryogenic temperatures in the RC, the key to photosynthetic efficiency is likely to be found, to a great extent, in the magnitude of the electronic coupling of the charge separated state to P* as opposed to its coupling to the ground state of the special pair dimer.

The excitonic model of the P* state described above can quantitatively account for the absorption, circular dichroism, and polarized absorption spectra of the RC [10]. The absorption spectrum displays a number of unusual properties: a substantial red shift compared to solution spectra of BChl, an anomalously large width, and a strong temperature dependence of the width and the absorption maximum. The latter two observations can be explained by postulating that a ~100 cm^{-1} intermolecular mode of the dimer is strongly coupled to the excited state, and that the equilibrium position of this mode is temperature dependent (more evidence for the presence of this mode will be discussed in Section 3), while the former probably arises from a combination of dispersion forces and chromophore-protein interactions [11]. However, there is nothing in this picture which suggests why charge separation from P* should be so efficient. To obtain insight into this question, we must examine other, more complex experiments. These will, at the same time, suggest a way of resolving the difficulties inherent in the kinetic mechanisms described above.

3. Three Anomalous Experiments

3.1. STARK SPECTROSCOPY

The Stark spectrum of the RC has been measured by several groups [12,13]. The important observation is that the P* band has an extraordinarily large Stark amplitude as compared to BChl monomers in solution or as compared to the BChl monomer in the RC, indicating a greater change in dipole moment between the ground and excited states. Additionally, the angle that the difference dipole makes with the polarization of the P* transition is inconsistent with what one would obtain from an exciton analysis of the dimer in terms of monomer properties.

This experiment provides initial evidence that the P* state possesses significant CT character. This character is unlikely to arise from coupling to P$^+$B$^-$, as in this case one would expect to see an initial, instantaneous bleaching of B in the kinetics, which is not observed experimentally. The only reasonable remaining candidate for an admixed CT configuration is an internal CT state of P (P$_{CT}$), i.e. one in which charge is separated between the two monomers of the special pair.

3.2. PHOTOCHEMICAL HOLEBURNING EXPERIMENTS

Initial studies by Boxer and coworkers [14,15] and Hoff, Wiersma and coworkers [16] indicated that the P* band displays a broad, featureless hole spectrum, in contrast to the sharp zero-phonon line (ZPL) which can be observed for a porphyrin in a glass or even for chlorophyll in myoglobin [17]. More recent experiments by Small and coworkers, utilizing samples prepared in a glycerol/water glass (as opposed to the polyvinyl alcohol films employed in the earlier experiments) have revealed a very weak, 10 cm^{-1} hole at selected burn wavelengths, as well as a series of broad, featureless holes spaced by ~120 cm^{-1}[18]. Assuming that the differences in the experiments are due to the smaller inhomogeneous broadening induced by the glycerol/water environment (and perhaps other aspects of sample preparation), we focus on the latter experiments as those that are most highly resolved. While our previous published articles have analyzed the initial set of data (in which the ZPL was completely destroyed), the results of Small and coworkers can be understood in a similar framework.

The broad hole sequence is readily understood as a Franck-Condon progression in the intermolecular dimer mode of P mentioned above. A Huang-Rhys S factor between 1 and 2 is extracted from the relative intensities; this value is consistent with other optical properties and with the temperature dependence of the P* absorption as it is modeled in refs. 10 and 11.

The question then arises as to why the ZPL is so weak, and why sharp side holes are not observed. While the latter are often not visible if the Franck-Condon factors of the modes in question are small, the present strong progression in the 120 cm^{-1} mode should, under normal conditions, yield such side holes, at least for the 0–1 transition. One can, of course, hypothesize that there is exceptionally strong dephasing by medium phonons, which increases as one goes to more quanta in the dimer mode progression. This is, in fact the explanation offered by Small and coworkers [19]. While it is a plausible possibility, it does represent an ad hoc hypothesis unsupported by other experimental data.

An alternate mechanism for broadening is that proposed by Won and Friesner [20,21]. In their model, coupling of the P* band to a CT state is shown, for a reasonable range of parameters, to destroy sharp line structure in the homogeneous lineshape. If the coupling is sufficiently strong, all structure, including the ZPL, is eliminated via chaotic vibronic mixing. On the other hand, a smaller coupling constant leaves the ZPL intact (although weakened) but eliminates any further sharp structure due to the dimer mode (note that such behavior is apparent in the results of ref. 20, despite the fact that the experiments of Small and coworkers had not yet been carried out when this paper was published). This picture, of course, also requires a specific hypothesis. It is, however, a hypothesis suggested by the Stark experiments, and hence is not entirely ad hoc.

3.3. SINGLET-TRIPLET SPLITTING OF THE CHARGE SEPARATED STATE

The singlet-triplet splitting ΔE_{ST} of the radical pair state P$^+$H$^-$ has been measured experimentally and is found to have a very small value, 1.9×10^{-7} eV [22]. Because the unpaired electrons in this state are separated by a large distance, it is not expected that dipolar interaction in this state is the major contributor to ΔE_{ST}. Rather, the splitting derives from interactions with other states, presumably P* or P$^+$B$^-$. In fact, these interactions are identical to those involved in mediating electron transfer in the superexchange mechanism, except that they must be evaluated at the equilibrium geometry of the radical pair, as opposed to the crossing point of the electron transfer reaction.

Several years ago, Marcus pointed out that the small value of ΔE_{ST} leads to an apparent inconsistency in the superexchange mechanism for the primary reaction [23]. If one neglects the dependence of the superexchange matrix elements on nuclear geometry (Condon approximation), the magnitude of these quantities can be estimated from ΔE_{ST}, provided one makes some assumptions concerning the energy of the superexchange intermediate. If these matrix elements are then used to calculate the effective electronic coupling for electron transfer, and the overall ET rate is evaluated via nonadiabatic multiphonon theory, the result is roughly three orders of magnitude less than the experimental rate. While numerous guesses of parameter values had to be made to obtain a final number for the rate, the size of the discrepancy is still such that it requires explanation.

The straightforward way to rationalize this conflict is to readjust parameter values by making different model assumptions. Once one discards the Condon approximation, the matrix elements relevant to the calculation of ΔE_{ST} can in principle differ substantially from those used to calculate the ET rate, as can in any case the diagonal energy of the intermediate state. A possible explanation of large alterations in matrix element values is a conformational change of the radical pair after charge separation. This approach has been extensively pursued by Bixon, Jortner, and coworkers, and finds some support in molecular dynamics simulations [24]. Again, however, there is no firm experimental basis for these assumptions, and the unaltered functioning of the RC at cryogenic temperatures does not suggest that such a conformational change is relevant to charge separation.

A very different explanation has been proposed by Won and Friesner recently. If one simply includes the P_{CT} state in the superexchange calculations, and assumes that most of the coupling to the superexchange pathway occurs through this state (which is then quasiresonantly coupled to P*), a small ΔE_{ST} is obtained independent of the effective ET rate constant. The numerical demonstration of this can be found in ref. 25; here, we give a physical picture of how such a result comes to pass. In the usual model without P_{CT}, the P* state has a singlet-triplet splitting of 0.4 eV [26]; this splitting is transmitted to the radical pair via the superexchange coupling. However, the singlet-triplet splitting of P_{CT} is expected to be small, because this state also has substantial physical separation of the unpaired electrons (one residing on each BChl monomer). Consequently, the splitting to be transmitted via superexchange is much smaller, and in fact leads to consistency with experimental results under a reasonable set of assumptions.

4. Energetics and Coupling of the Internal Dimer Charge Transfer State

The most likely interpretation of the Stark experiments is that the P* state possesses substantial charge separated character. While it is possible that this character derives from fundamental alterations of the monomeric molecular orbitals by the protein environment, this picture is not suggested by any other spectroscopic measurements. Admixture of the Q_y states with an internal CT configuration is thus, by default, a highly probable explanation.

Indeed, most semiempirical electronic structure calculations have included such a state in their configuration interaction matrix, and have invoked it to explain the Stark results. The critical issues are then the diagonal energy and magnitude of the coupling of the state to the Q_y states. In what follows, we will assume that P_{CT} has no intrinsic oscillator strength; this assumption, made in all theoretical models to date, is consistent with experiments on the RC and with observations on analogous model compounds.

We do not believe (see ref. 4) that existing electronic structure methods are capable of computing the energy or coupling strength of P_{CT} reliably. Consequently, one must consider the consistency of any proposal with respect to all of the experiments discussed above. For example, the calculations of Fischer and coworkers [27] and of Parson and Warshel [28] place the CT state at high energy (in the Q_x region) and assign a very strong coupling of it to the dimer Q_y states. While this explains the CT character of P*, it also predicts that the CT state at high energy will acquires substantial oscillator strength, and should thus be visible in the absorption spectrum of the RC. Furthermore, this new state would be expected to display a substantial Stark effect. No experimental evidence for such a state exists, despite a considerable experimental effort to look for it [12]. While one can again suggest ad hoc reasons for this negative result (e.g., the state is so broad as to be undetectable), such hypotheses run the danger of rendering a model immune from confrontation with experiment. However, as all models display some quantitative difficulties with regard to experimental data at this point, one cannot dismiss this as a possibility.

If one instead attempts to infer the position and coupling strength of P_{CT} from experiment, as opposed to relying on semiempirical calculations, a qualitatively consistent picture can be developed. In this picture, P_{CT} is close in energy to the low energy exciton Q_y state, and has an intermediate (~100 cm^{-1}) coupling strength with that state. This is the type of parameter set that we have used to explain the holeburning and singlet-triplet splitting results. Furthermore, no discrepancy in the absorption spectrum along the lines described in the previous paragraph arises. There are potential difficulties in quantitatively reproducing the Stark lineshape; this is a subject of current investigation.

The picture described here is at this point a qualitative one; additional quantitative calculations are needed to specify more precisely an acceptable range of parameter values. For example, what are the values of the position and coupling strength of P_{CT} which do not produce inconsistency with some set of experiments? It is difficult to answer this question, because there are a large number of additional parameters (e.g., those describing the P_{CT} vibrational potential surface) which can also affect the results. Nevertheless, it should be possible to considerably improve present models, particularly given the new, more highly resolved holeburning data of Small and coworkers [18].

5. Experiments on Genetically Modified Reaction Centers

Site directed mutagenesis has been carried out on RCs of the bacterium *Rps. capsulatus* by Youvan and coworkers [29]. While the crystal structure of this RC has not been determined, the amino acid sequence, spectroscopy, and electron transfer dynamics are very similar to those of *Rps. viridis* and *Rb. sphaeroides*. It is therefore reasonable to use the known structures of the latter RCs to draw conclusions concerning experiments on the former.

The most interesting mutation to date is one in which His^{M200} is changed to Leu. There is now a considerable amount of experimental data which indicates that the principal effect of this mutation is to remove a magnesium atom from the M BChl (the histidine served as a ligand of the magnesium, whereas the replacement amino acid is incapable of this function) of P, thus converting this molecule to a pheophytin. The resulting special pair has been designated a "heterodimer".

While the mutant strain will not grow photosynthetically, it does carry out primary charge separation, with a reduced (50%) quantum yield [29]. Spectroscopic measurements (e.g. linear dichroism) indicate that there is little change in any of the pigment orientations in the mutant as compared to the wild type [30]. An interpretation of experimental results based upon the assumption that pheophytinization of P_M is the only modification of importance for primary charge separation is thus reasonable (although not definitively demonstrated at this point in time).

Femtosecond transient absorption measurements on the modified RC reveal the existence of a new intermediate, which is formed almost immediately after excitation of P [5]. The spectrum of this intermediate contains features characteristic of a BPh anion. However, this species cannot be the anion of the primary acceptor H, because the absorption band of that molecule is not bleached on such a short timescale. Assuming that the spectrum has been properly identified, the only reasonable possibility is that one is directly observing the internal CT state of the heterodimer, which has now been shifted in energy below the Q_y special pair states. The absorption and Stark spectra of the mutant are consistent with this interpretation as well. One sees in absorption a long, weak tail in the far red region, which could arise from the upper vibronic levels of the CT state borrowing intensity from the remaining BChl Q_y state (the BPh molecule has its absorption shifted to higher energy). Two second derivative features are seen in the low energy region of the Stark spectrum, one at around the energy of the BChl state and the other in the tail region [31]. The latter is particularly intense compared to its weak absorption amplitude, which is expected if it has a strong CT component.

The ET kinetics of the heterodimer are easily understood in this picture. The internal CT state does not fluoresce efficiently (as it has no direct oscillator strength), but it can relax to the ground state via radiationless decay. The lowered energy of the internal CT state implies that it will be populated significantly in the quasiequilibrium established with P^+H^-, thus decreasing the quantum yield of charge separation. The modification of the kinetics of charge separation can similarly be attributed to altering the free energy gap between reactants and products and/or to changing the energy separation with a virtual intermediate in a superexchange pathway.

From this analysis, we can extrapolate the energy of the P_{CT} state in the wild type bacterium. The redox potential of a BPh in solution is approximately 0.2 eV lower than that of a BChl [32]. Assuming that this energy difference is roughly preserved in the RC, the P_{CT} state is clearly quite close to the P^* absorption band in the natural system. The precise location is difficult to pinpoint, because the magnitude of the shift in the protein probably differs somewhat from that obtained in solution, and because the position of the state in the heterodimer is not known accurately (recall that the tail in absorption probably represents upper vibronic components of P_{CT} , so that the zero of energy is likely to lie below this). However, two conclusions are clear. First, there are two distinct states, and the P_{CT} state is intimately involved in charge separation. Secondly, the position of the state in the natural system is compatible with the model put forth by us, whereas it is in serious disagreement with the electronic structure calculations described above.

6. Conclusion

The above analysis provides a simple physical picture with which to understand the extraordinary efficiency of charge separation in the RC. The P_{CT} state is coupled to the superexchange pathway (i.e., to P^+B^-, which is then coupled to P^+H^-) and is quasiresonant with P^*, and thus serves as a trigger for forward charge separation. On the other hand, P_{CT} is energetically distant from the ground state of P, and hence is not effective in enhancing recombination. If the coupling to P_{CT} were very strong (e.g., an order of magnitude larger than it is), the lack of resonance would be unimportant, and recombination to the ground state would also be expected to be very rapid. The argument is thus that a high quantum yield is obtained by specifically tuning the energy and coupling strength of P_{CT}. Presumably, these parameters have been evolutionarily optimized in the RC by modification of the amino acid environment. This seems quite plausible, as a non-optimized system would still be capable of separating charge, albeit less efficiently (cf. the heterodimer).

A great deal of additional experimental and theoretical work will be required before this picture can be accepted as definitively correct. For example, experiments measuring the effect of strong electric fields on the ET kinetics and fluorescence yield must be explained; more experiments along these lines need to be carried out as well. We have proposed a resonance Raman experiment with direct excitation into the P^* band as a possible probe for the existence of a component of P_{CT} within P^*. Site-directed mutagenesis experiments should provide further results of interest, particularly if the mutant crystal structures can be analyzed.

A different approach to testing these ideas is to construct synthetic model systems which can mimic the proposed electronic coupling pathways of the RC. Recently, advances in this direction have been made; a model system consisting of two porphyrins and a quinone appears to exhibit rapid ET mediated via superexchange through one porphyrin [33], and a synthetic complex in which a CT state has been tuned into resonance with the lowest excited state has been prepared [34]. However, assembling all of the features of the RC in a single molecular model still remains as a challenging task. Hopefully, the concepts presented here will facilitate the realization of this objective.

Acknowledgments

This work was supported by a grant from the NIH. RAF is a Camille and Henry Dreyfus Teacher-Scholar and the recipient of a Research Career Development Award from the NIH, Institute of General Medical Sciences.

References

1. Deisenhofer, J., Epp, O., Miki, K., Huber, R. and Michel, H. (1985) 'Structure of the protein subunits in the photosynthetic reaction center of *Rhodopseudomonas viridis* at 3 Å resolution', Nature, 318, 618–624.
2. Allen, J. P., Feher, G., Yeates, T. O., Komiya, H. and Rees, D. C. (1988) 'Structure of the reaction center from *Rhodobacter sphaeroides* R-26: Protein-cofactor (quinones and Fe^{2+}) interactions', Proc. Natl. Acad. Sci. USA, 85, 8487–8491.
3. Chang, C- H., Tiede, D. M., Tang, J., Smith, U., Norris, J. R. and Schiffer, M. (1986) 'Structure of *Rhodopseudomonas sphaeroides* R-26 reaction center', FEBS Lett., 205, 82–86.
4. Friesner, R. A. and Won, Y. (1989) 'Spectroscopy and electron transfer dynamics of the bacterial photosynthetic reaction center', Biochim. Biophys. Acta, submitted.
5. Kirmaier, C., Holten, D., Bylina, E. J. and Youvan, D. C. (1988) 'Electron transfer in a genetically modified bacterial reaction center containing a heterodimer', Proc. Natl. Acad. Sci. USA, 85, 7562–7566.
6. Kirmaier, C. and Holten, D. (1987) 'Primary photochemistry of reaction centers from the photosynthetic bacteria', Photosynthesis Res., 13, 225–260.
7. Breton, J., Martin, J- L., Migus, A., Antonetti, A. and Orszag, A. (1986) 'Femtosecond spectroscopy of excitation energy transfer and initial charge separation in the reaction center of the photosynthetic bacterium *Rhodopseudomonas viridis*', Proc. Natl. Acad. Sci. USA, 83, 5121–5125.
8. Martin, J- L., Breton, J., Hoff, A. J., Migus, A. and Antonetti, A. (1986) 'Femtosecond spectroscopy of electron transfer in the reaction center of photosynthetic bacterium *Rhodopseudomonas sphaeroides* R-26: Direct electron transfer from the dimeric bacteriochlorophyll primary donor to the bacteriopheophytin acceptor with a time constant of 2.8 ± 0.2 psec.', Proc. Natl. Acad. Sci. USA, 83, 957–961.
9. Bixon, M., Jortner, J., Michel–Beyerle, M. E., Orgodnik, A. and Lersch, W. (1987) 'The role of the accessory bacteriochlorophyll in the reaction centers of photosynthetic bacteria: Intermediate acceptor in the primary electron transfer?' Chem. Phys. Lett., 140, 626–630.

10. Won, Y. and Friesner, R. A. (1988) 'Simulation of optical spectra from the reaction center of *Rhodopseudomonas viridis*', J. Phys. Chem., 92, 2208–2214.
11. Won, Y. and Friesner, R. A. (1988) 'A thermal expansion model for the special pair of the bacterial reaction center', Israel J. Chem, 28, 67–72.
12. Lockhart, D. J. and Boxer, S. G. (1988) 'Stark effect spectroscopy of *Rhodobacter sphaeroides* and *Rhodopseudomonas viridis* reaction centers', Proc. Natl. Acad. Sci. USA, 85, 107–111.
13. Lösche, M., Feher, G. and Okamura, M. Y. (1987) 'The stark effect in reaction centers from *Rhodobacter sphaeroides* R-26 and *Rhodopseudomonas viridis*', Proc. Natl. Acad. Sci. USA, 84, 7537–7541.
14. Boxer, S.G., Lockhard, D.J. and Middendorf, T.R. (1986) 'Photochemical holeburning in photosynthetic reaction centers', Chem. Phys. Lett., 123, 476–482.
15. Boxer, S.G., Midendorf, T.R. and Lockhart, D.J. (1986) 'Reversible photochemical holeburning in *Rhodopseudomonas viridis* reaction centers', FEBS Lett., 200, 237–241.
16. Meech, S.R., Hoff, A.J. and Wiersma, D.A. (1985) 'Evidence for a very early intermediate in bacterial photosynthesis. A photon-echo and hole-burning study of the primary donor band in *Rhodopseudomonas sphaeroides*', Chem. Phys. Lett., 121, 287–292.
17. Boxer, S. J., Gottfried, D. S., Lockhart, D. J. and Middendorf, T. R. (1987) 'Nonphotochemical holeburning in a protein matrix: Chlorophyllide in apomyoglobin', J. Chem. Phys. 86, 2439–2441.
18. Tang, D., Jankowiak, R., Small, G. and Tiede, D. M. (1989) 'Structured hole burned spectra of the primary donor state absorption region of *Rhodopseudomonas viridis*', Chem. Phys., 131, 99–113.
19. Hayes, J. M., Gillie, J. K., Tang, D. and Small, G. (1988) 'Theory for spectral hole burning of the primary electron donor state of photosynthetic reaction centers', Biochim. Biophys. Acta, 932, 287–305.
20. Won, Y. and Friesner, R. A. (1987) 'Simulation of photochemical hole-burning experiments on photosynthetic reaction centers', Proc. Natl. Acad. Sci. USA, 84, 5511–5515.
21. Won, Y. and Friesner, R. A. (1988) 'Theoretical studies of photochemical hole burning in photosynthetic bacterial reaction centers', J. Phys. Chem., 92, 2214–2219.
22. Ogrodnik, A., Lersch, W., Michel-Beyerle, M. E., Deisenhofer, J. and Michel, H. (1985) 'Spin dipolar interaction of radical pairs in photosynthetic reaction centers', in Michel-Beyerle, M. E. (ed.), Antennas and Reaction Centers of Photosynthetic Bacteria - Structure, Interactions, and Dynamics, Springer-Verlag, Berlin, pp. 198–206.
23. Marcus, R. A. (1987) 'Superexchange versus an intermediate BChl⁻ mechanism in reaction centers of photosynthetic bacteria', Chem. Phys. Lett., 133, 471–477.
24. Treutlein, H., Schulten, K., Niedermeier, C., Deisenhofer, J., Michel, H. and DeVault, D. (1988) 'Electrostatic control of electron transfer in the photosynthetic reaction center of *Rhodopseudomonas viridis*', in Breton, J. and Vermèglio, A. (eds.) Photosynthetic Bacterial Reaction Center: Structure and Dynamics, vol. 149, Plenum, New York, pp. 369–377.
25. Won, Y. and Friesner, R. A. (1988) 'On the viability of the superexchange mechanism in the primary charge separation step of bacterial photosynthesis', Biochim. Biophys. Acta, 935, 9–18.

26. Shuvalov, V. A. and Parson, W. W. (1981) 'Energetics and kinetics of radical pairs involving bacteriochlorophyll and bacteriopheophytin in bacterial reaction centers', Proc. Natl. Acad. Sci. USA, 78, 957–961.
27. Fischer, S. F. and Scherer, P. O. J. (1987) 'On the early charge separation and recombination processes in bacterial reaction centers', Chemical Physics, 115, 151–158.
28. Parson, W. W., and Warshel, A. (1987) 'The spectroscopic properties of photosynthetic reaction centers II. Application of the theory to *Rhodopseudomonas viridis*', J. Am. Chem. Soc., 109, 6152–6163.
29. Bylina, E. J. and Youvan, D. C. (1988) 'Directed mutations affecting spectroscopic and electron transfer properties of the primary donor in the photosynthetic reaction center', Proc. Natl. Acad. Sci. USA, 85, 7226–7230.
30. Breton, J., Bylina, E. J. and Youvan, D. C., this volume.
31. Norris, J. R., Bylina, E. J. and Youvan, D. C., this volume.
32. Felton, R. H. (1978) 'Primary redox reactions of metalloporphyrins', in Dolphin, D. (ed.), The Porphyrins, Academic Press, New York, pp. 53–125.
33. Kirmaier, C., Holten, D. and Sessler, J. L., personal communication.
34. Wasielewski, M. R., this volume.

PHOTOSYNTHETIC MODEL SYSTEMS THAT ADDRESS THE ROLE OF SUPEREXCHANGE IN ELECTRON TRANSFER REACTIONS

Michael R. Wasielewski, Mark P. Niemczyk, Douglas G. Johnson, Walter A. Svec, and David W. Minsek
Chemistry Division, Argonne National Laboratory, Argonne, IL 60439, USA

ABSTRACT. Four fixed-distance porphyrin-quinone molecules, 1-syn, 1-anti, 2-syn, and 2-anti, were synthesized. These molecules possess a zinc 5-phenyl-10,15,20-tripentylporphyrin electron donor attached to a naphthoquinone via a rigid pentiptycene spacer. The central benzene ring of the spacer is unsubstituted in 1 and possesses p-dimethoxy substituents in 2. The naphthoquinone is oriented either syn or anti to the porphyrin across the spacer. These molecules provide information concerning the orientation dependence of electron transfer between the porphyrin and the quinone, and the dependence of this transfer on low-lying ionic states of the spacer. The rate constants for the oxidation of the porphyrin lowest excited singlet state by the naphthoquinone are 1-syn: 8.2×10^9 s^{-1}; 1-anti: 1.7×10^{10} s^{-1}; 2-syn: 8.5×10^9 s^{-1}; 2-anti: 1.9×10^{10} s^{-1}. The corresponding rate constants for the porphyrin cation - naphthoquinone anion recombination reaction are 1-syn: 1.4×10^{10} s^{-1}; 1-anti: 2.5×10^{10} s^{-1}; 2-syn: 5.0×10^{10} s^{-1}; 2-anti: 8.2×10^{10} s^{-1}. The rate constants for the syn isomers are uniformly a factor of about 2 slower than those of the anti isomers. The charge separation reaction rates for 1 and 2 are similar, while the ion pair recombination reactions are about 3-4 x faster in 2 than in 1. The conformational effect is attributed to better overlap of the spacer wave functions in the anti vs the syn conformation, while the increase in recombination rate for 2 over 1 is attributed to a superexchange interaction involving an electronic configuration of the spacer in which the dimethoxybenzene cation contributes.

1. INTRODUCTION

The chlorophyll and quinone electron donors and acceptors in photosynthetic reaction centers are positioned at precise distances and orientations to promote efficient charge separation and to impede charge recombination.[1] Moreover, the nature of the medium that lies between the donor and acceptor is thought to have a large influence on the observed rates of electron transfer.[2-4] Covalently-linked porphyrin-quinone molecules have been studied extensively as models for the light-initiated charge separation in photosynthesis.[5] Studies performed to date have been concerned primarily with the dependence of the electron transfer reactions on free energy, distance, and solvent.[6-9] In general, our approach to this problem is to synthesize molecules in which both the porphyrin-quinone distance and mutual orientation are restricted. This entails the use of rigid hydrocarbon spacer molecules as part of the overall molecular structure.

The role of the intervening medium that lies between a donor and an acceptor is beginning to be studied both theoretically and experimentally. In the bacterial photosynthetic reaction center a bacteriochlorophyll molecule lies between the dimeric bacteriochlorophyll donor and the bacteriopheophytin acceptor. It is thought that mixing low-lying ionic states of the intermediate bacteriochlorophyll with those of the donor and acceptor lead to a greatly

347

J. Jortner and B. Pullman (eds.), Perspectives in Photosynthesis, 347–360.
Kluwer Academic Publishers.

increased rate of electron transfer.[10-12] This concept, known as superexchange, has its origins in the work of McConnell, which treats electron transfer between aromatic molecules across a hydrocarbon spacer.[13] This idea has been elaborated in a number of papers regarding the dependence of electron transfer on the energies and spatial characteristics of both hydrocarbon [14-17] and protein spacer orbitals.[2,3,18-20] In general, molecules possessing low-lying π molecular orbitals can contribute strongly to a superexchange mechanism for electron transfer. It is possible that aromatic amino acids, such as tyrosine, phenylalanine, and tryptophan, which lie between an electron donor and an acceptor in proteins, can facilitate electron transfer reactions. However, there are few experimental tests of superexchange in donor-acceptor molecules, especially those involving excited state electron transfers.[21,22] The problem lies in producing a series of rigid donor-acceptor molecules in which the effects of changing the orbital energies of the intervening spacer molecule are not convolved with changes in conformation. Molecules 1-syn, 1-anti, 2-syn, and 2-anti, described in this paper, are designed to address this problem. These molecules utilize a polycyclic hydrocarbon spacer to maintain a fixed distance and restricted orientation between the porphyrin donor and the

1-anti: R = H and 2-anti: R = OCH$_3$

1-syn: R = H and 2-syn: R = OCH$_3$

quinone acceptor. This spacer belongs to a general class of hydrocarbons that have been named "iptycenes" by Hart.[23] In particular, the hydrocarbon spacer used in 1 and 2 is a "pentiptycene" because it contains 5 aromatic rings. There are several advantages to using this ring system as a spacer. First, it is closely related to triptycene, a spacer that we have studied extensively.[5] Second, the presence of two positional isomers, in which the porphyrin and quinone are either **syn** or **anti** relative to one another, affords us the opportunity to study the orientation dependence of photoinduced electron transfer reactions at a fixed distance. Third, the presence of a central benzene ring, rigidly fixed between the two triptycene moieties, allows us to use substituents on the remaining free positions of this ring to alter the energy of the HOMO and LUMO of this intervening spacer fragment. This alters the relative contribution of ionic states of the spacer to a superexchange description of electron transfer.

   In addition to pentiptycene containing molecules, 1 and 2, we have prepared the corresponding triptycene containing reference compound, 3, and the appropriate porphyrin reference compound 5-phenyl-10,15,20-tripentylporphyrin, 4.

3

4

2. ABSORPTION AND FLUORESCENCE DATA

The ground state optical absorption spectrum of 1-anti is shown in Figure 1. The spectrum is identical to that of porphyrin 4, which indicates that the attachment of both the spacer and the naphthoquinone to the porphyrin do not strongly perturb its electronic structure. Figure 2 shows the fluorescence emission spectra of 1-anti and 4. The emission spectra of the two molecules are similar. The biggest difference between the emission spectra of 1-anti and 4 are their relative intensities. The fluorescence spectra of 1-3 all possess lineshapes and maxima similar to those of 4, but display varying intensities. The fluorescence quantum yields of compounds 1-3 are all strongly quenched relative to that of 4, Table 1. The data in Table 1 also show that the fluorescence lifetimes of 1-3 decrease in parallel with their respective fluorescence quantum yields. This suggests that the radiative rate constants for the zinc porphyrins in compounds 1-4 remain fairly constant across this series of molecules. This further implies that the fluorescence quenching in 1-3 is due to a fast nonradiative process that depletes the lowest excited singlet state population of the porphyrin in compounds 1-3. It is interesting to note that the fluorescence lifetimes of 1 and 2 exhibit a dependence on the orientation of the donor relative to the acceptor, i.e. syn or anti. Moreover, the fluorescence lifetimes of 1-syn and 1-anti are similar to those of 2-syn and 2-anti, respectively, and therefore, are not influenced by the addition of the methoxy groups in 2.

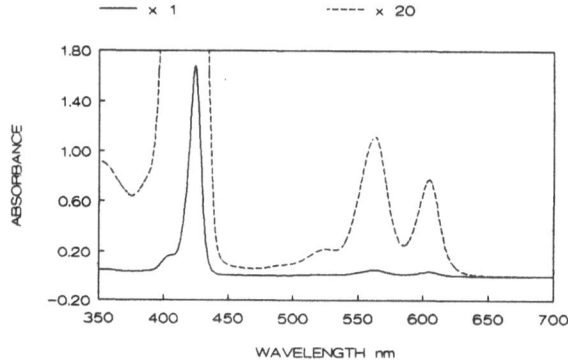

Figure 1. Ground state absorption spectrum of 1-anti in PrCN.

TABLE 1. Fluorescence data in PrCN

Compound	ϕ_F	τ_F (ps)
1-anti	0.0012	57 ± 2
1-syn	0.0026	115 ± 2
2-anti	0.0014	52 ± 2
2-syn	0.0034	111 ± 2
3	0.00013	<10
4	0.036	1850 ± 15

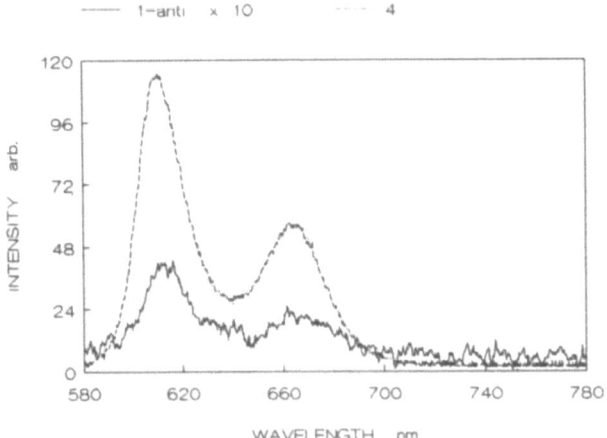

Figure 2. Fluorescence spectra of **1-anti** and **4** in PrCN, excitation at 560 nm.

3. ENERGETICS

An energy level diagram for molecules **1** and **2** is shown in Figure **3**. The excited singlet state energy was obtained from the maximum of the (0,0) band in the fluorescence emission spectra of **1** and **2**. This maximum was the same for **1** and **2**. The triplet state energy shown is that reported earlier for **4**.[24] The approximate energy of $P^+ - S - Q^-$ was obtained by summing the half-wave potentials for one-electron oxidation of the zinc porphyrin donor and one-electron reduction of the naphthoquinone acceptor in **1** and **2**. The redox potentials of these molecules were measured in butyronitrile containing 0.1M tetra-n-butylammonium perchlorate electrolyte and are the same for **1** and **2**. The redox potentials for the tetra-alkylbenzene central rings in the pentiptycene spacers were estimated from the literature values for the analogous benzene [25,26] and p-dimethoxybenzene [27] derivatives. The energetics for **3** are the same as those for **1** and **2** excluding the availability of the ionic spacer states.

Figure **3** shows that the driving force for oxidation of the porphyrin excited singlet state by the quinone is approximately -0.8 eV. Our earlier work on the dependence of charge separation rates in porphyrin-quinone molecules on free energy suggests that the energetics of **1-3** should result in near maximum rates for both charge separation and and recombination in these molecules.[6] States involving anions of the spacers will lie > 4 eV above the ground state, while states involving cations of the spacers lie at 2.4 eV for the benzene spacer and 1.8 eV for the p-dimethoxybenzene spacer. Moreover, the state $P - S_O^+ - Q^-$ is 0.6 eV lower in energy than $P - S_H^+ - Q^-$. Thus, $P - S_O^+ - Q^-$ may contribute more significantly than $P - S_H^+ - Q^-$ to a superexchange mechanism for ion pair recombination.

4. TRANSIENT ABSORPTION DATA

Figure **4** shows the transient absorption spectra obtained for **1-anti** at 10 ps and 60 ps following excitation. The transient absorption spectra for **1-syn**, **2-syn**, and **2-anti** are similar to those in Figure **4**. The spectra for compound **3** are also similar to those in Figure **4**, except

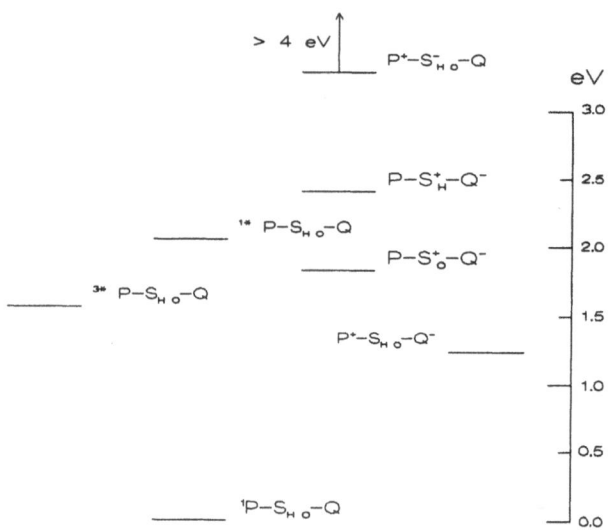

Figure 3. Energy level diagram for 1 and 2. P = porphyrin; S_H = spacer in 1; S_O = spacer in 2; Q = quinone.

that the analogous spectra for 3 occur at 1 ps and at 5 ps, respectively. The transient absorption spectrum of porphyrin 4 is shown in Figure 5. Since porphyrin 4 has no electron acceptor attached to it, the spectrum in Figure 5 is the difference between the ground state spectrum and that of the lowest excited singlet state of the porphyrin. This spectrum serves as a reference for the spectra in Figure 4. The spectra in Figure 4 are a superposition of the spectra of both P^+ - S - Q^- and $^{1*}P$. The spectra in Figures 4 and 5 were obtained at similar concentrations and excitation intensities. A comparison of Figures 4 and 5 shows that the absorption feature near 460 nm of 1-anti is due to both $^{1*}P$ and P^+ - S - Q^-, while the absorbance in the near-infrared is dominated by P^+ - S - Q^-.[28] Although the spectra of 1-anti show that the mechanism of excited singlet state quenching is electron transfer from the porphyrin to the quinone, even 60 ps after excitation the spectrum of 1-anti contains significant contributions from the transient absorption of $^{1*}P$. This is supported by the presence of absorption troughs near 650 nm in both Figures 4 and 5 that are due to stimulated emission from $^{1*}P$. These spectra strongly suggest that the rate of charge separation is on same order of magnitude as that of the ion pair recombination. The rate constants for charge separation, k_{cs}, and ion pair recombination, k_{cr}, are obtained from the following analysis. Constants k_{cs} in 1-3 are obtained from their respective fluorescence lifetimes given in Table 1. Assuming that the enhanced nonradiative decay route observed in 1-3 is due entirely to the charge separation process,

$$k_{cs} = 1/\tau_f - 1/\tau_4$$

where τ_f is the fluorescence lifetime of the porphyrin-quinone and τ_4 is the fluorescence lifetime of reference porphyrin 4. The relative contributions of $^{1*}P$ - S - Q and P^+ - S - Q^- to the transient absorption spectra of 1-3 as a function of time are obtained by comparing the transient absorption spectra of 1-3, e.g. Figure 4, with that of 4, Figure 5. Using these two pieces of information and assuming a series A -> B -> C mechanism, where A is

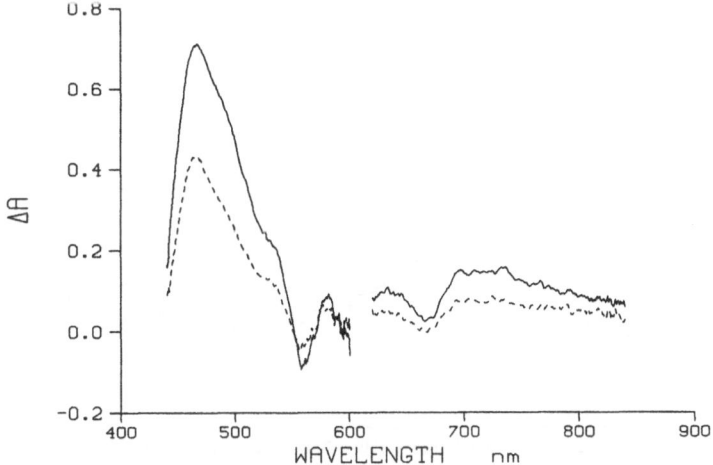

Figure 4. Transient absorption spectra of **1-anti** in PrCN at 10 ps —— and 60 ps - - - following a 1 ps laser flash at 610 nm.

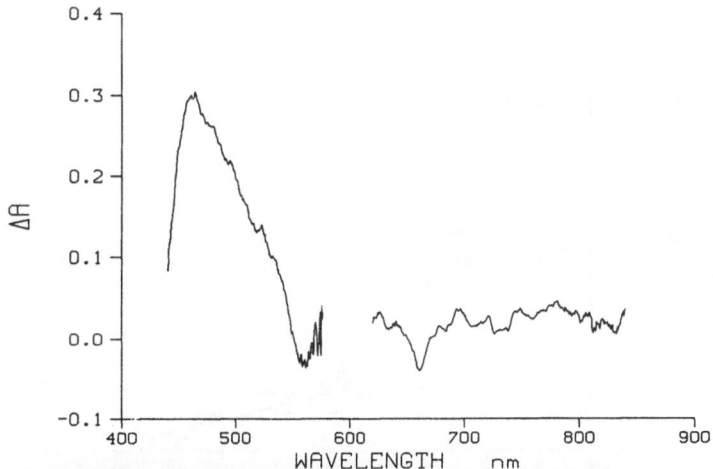

Figure 5. Transient absorption spectrum of **4** in PrCN at 60 ps following a 1 ps laser flash at 610 nm.

^{1*}P - S - Q, B is P$^+$ - S - Q$^-$, and C is P - S - Q, the kinetics for the decay of the transient absorption features for **1-3** can be fit to determine k_{cr}. Data obtained at 460 nm and at 700 nm for **1-anti** are shown in Figures 6 and 7, respectively, along with the corresponding fits. The values of k_{cs} and k_{cr} obtained in this fashion are listed in Table 2.

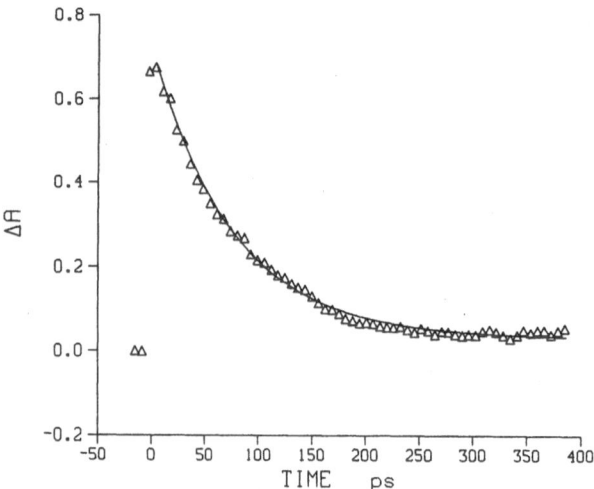

Figure 6. Transient absorption kinetics at 460 nm for
1-anti in PrCN following a 1 ps, 610 nm laser flash.

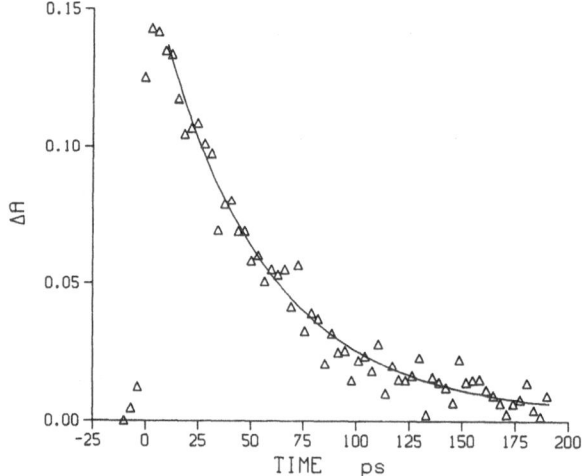

Figure 7. Transient absorption kinetics at 700 nm for
1-anti in PrCN following a 1 ps, 610 nm laser flash.

The rate constants for the ion-pair recombination reactions are all somewhat faster than those of the charge separation reactions. In addition, ion pair recombination occurs about a factor of 2 faster in the **anti** isomers than the **syn** isomers. This is similar to the observed difference between the rates of charge separation for these isomers. The recombination rates for the dimethoxy substituted compounds **2** are about 3-4 x faster than are those of the unsubstituted compounds, **1**. This substituent effect is observed for both the **syn** and **anti** isomers. Both the charge separation and recombination rates for **3** are very fast as predicted by our earlier work.[6] The rate constants for **3** are about an order of magnitude faster than those of **1** and **2**.

TABLE 2. Electron Transfer Rate Constants for **1-3**

Compound	k_{cs} (s^{-1})	$\bar{k}_{cr}$ (s^{-1})
1-anti	1.7×10^{10}	2.5×10^{10}
1-syn	8.2×10^{9}	1.4×10^{10}
2-anti	1.9×10^{10}	8.2×10^{10}
2-syn	8.5×10^{9}	5.0×10^{10}
3	1.0×10^{12}	3.0×10^{11}

5. DISCUSSION

Introduction of the pentiptycene spacer between the porphyrin and quinone in **1** and **2** results in a decrease in the rate constants for both charge separation, k_{cs}, and charge recombination, k_{cr}, by about a factor of 10 relative to those for **3**. This diminution in rate is approximately the same as we observed earlier in comparing a porphyrin - triptycene -quinone molecule to a porphyrin - trans-1,2-diphenylcyclopentane - quinone molecule.[7] The triptycene spacer possesses one saturated carbon atom between the π systems of the donor and acceptor, whereas the cyclopentane spacer possesses two saturated carbon atoms between them. At that time we determined that the difference in rate for these two molecules reflected the difference in edge-to-edge donor-acceptor distance. The rate diminished by a factor of 10 when the donor-acceptor edge-to-edge distance was increased from 2.4 Å using the triptycene spacer to 3.7 Å using the cyclopentane spacer.[7] The pentiptycene spacer in **1** and **2** possesses a central benzene ring along with two saturated carbon atoms between the donor and acceptor. The edge-to-edge distance between the π system of the porphyrin phenyl group and the naphthoquinone in **1** and **2** is 6 Å, and is similar for the syn and anti isomers, whereas the respective porphyrin-quinone distance in **3** is only 2.4 Å.[29] Even though the donor-acceptor distance has increased by 3.6 Å in going from **3** to **1** and **2**, the presence of the benzene ring in the pentiptycene spacer does not appreciably slow the rates of electron transfer in **1** and **2** more than the factor of about 10 that is expected for a change from a spacer with one saturated carbon atom to one possessing two such atoms.

The data in Table 2 show that both k_{cs} and k_{cr} for **1** and **2** are slower for the syn isomer than for the anti isomer. The difference in rate constant between the two isomers is about a factor of 2. It is well-established that rate constants of electron transfer reactions most often decrease exponentially with distance.[5] The center-to-center distances between the porphyrin donor and quinone acceptor in **1** and **2** are isomer dependent: 11 Å for the syn isomers and 16 Å for the anti isomers. Even though the center-to-center porphyrin-quinone distance in the syn isomer is about 5 Å shorter than that of the anti isomer, k_{cs} and k_{cr} are actually faster for the anti isomers. This strongly suggests that the principal pathway for the electron transfer from the porphyrin to the quinone is through the bonds of the hydrocarbon spacer. Oliver et al.[30] recently observed similar behavior in molecules possessing 1,4-dimethoxynaphthalene donors attached via rigid hydrocarbon spacers to 1,1-dicyanoethylene acceptors. Their hydrocarbon spacers have either an all-trans arrangement of single bonds or a single s-cis kink in the chain. In each case the all-trans isomer gives the faster rate. The fact that k_{cs} and k_{cr} for the anti isomers of **1** and **2** are faster than those for the syn isomers is consistent with considerations of maximum orbital interactions in hydrocarbons which have a zig-zag or all trans configuration relative to those which have cisoid segments in the chain.[14] Ohta et al. [15] recently calculated the matrix element for electron transfer between two methylene groups across a cyclohexane ring as a function of stereochemistry. They found that the equatorial-equatorial conformation of the ring, in which an all <u>trans</u> arrangement of the C-C bonds occurs, results in the larger matrix element, and therefore, the faster electron transfer rate. Examination of the structures of 1 and 2 show that the **anti** isomer maintains an effective all-<u>trans</u> arrangement of the C-C bonds. This phenomenon has also been demonstrated for electron transfer in radical anions of 1,4-diarylcyclohexanes.[31]

Since it is likely that electron transfer between the porphyrin and the quinone occurs through the bonds in 1 and 2, it is appropriate to ask whether the rate of electron transfer is dependent on the energies of the orbitals comprising the spacer. Placing substituents on the π system of the central benzene ring of the pentiptycene spacer provides a direct avenue for modifying the energy of the HOMO and LUMO of the spacer. These energies will determine the degree to which the states P^+ - S - Q and P - S^+ - Q^- will mix with the states $^{1*}P$ - S - Q and P^+ - S - Q^-, respectively, to enhance the electronic coupling matrix element for charge separation and ion pair recombination, respectively.

The electronic matrix element, V_S for the superexchange interaction between states A, B, and C, where B is the virtual state is given by [32]

$$V_S = V_{AB} \, V_{BC} \, / \, \delta E_{AB}$$

where V_{AB} and V_{BC} are the respective electronic coupling terms between states A and B, and B and C, and δE_{AB} is the energy difference between states A and B. For the charge separation reaction in 1 and 2 A = $^{1*}P$ - S - Q, B = P^+ - S^- - Q, and C = P^+ - S - Q^-, while for the ion pair recombination in 1 and 2 A = P^+ - S - Q^-, B = P - S^+ - Q^-, and C = P - S - Q. Since the rate of the electron transfer reaction depends on $V_S{}^2$, the reaction rate will be a function of $(1/\delta E_{AB})^2$. Since the structures of 1 and 2 are esssentially the same, the terms V_{AB} and V_{BC} should remain similar for compounds 1 and 2.

Benzene and dimethoxybenzene can be reduced electrochemically at about -3.4 V and -3.6 V vs SCE, respectively. Since the zinc porphyrin in 1 and 2 oxidizes at 0.61 V vs SCE, we estimate that the energy of P^+ - $S_H{}^-$ - Q is about 4.0 eV, whereas that of P^+ - $S_O{}^-$ - Q is about 4.2 eV. Since participation of the state P^+ - S^- - Q in a superexchange mechanism for charge separation depends on the square of the energy difference between that state and $^{1*}P$ - S - Q, we can estimate the ratio of the charge separation rate constants $(k_{cs})_1/(k_{cs})_2$. This ratio should be $[(1/\delta E_{AB})^2]_1/[(1/\delta E_{AB})^2]_2$, or $[1/(2.0)^2]/[1/(2.2)^2] = 1.2$. This estimate predicts that k_{cs} for both 1 and 2 should be similar. This prediction is borne out by the data for k_{cs} in Table 2. We find that substitution of two methoxy substituents on the benzene ring of the spacer does not significantly change k_{cs}.

There are two ways in which ion pair recombination in 1 and 2 can involve ionic states of the spacer. One way is via states such as P^+ - S^- - Q. However, these states are sufficiently high in energy relative to P^+ - S - Q^- as to mix only weakly. A second route back to ground state is via states such as P - S^+ - Q^-. Figure 3 shows that P - $S_H{}^+$ - Q^- is relatively low-lying for 1 and P - $S_O{}^+$ - Q^- is especially so for 2. This suggests two things. First, a superexchange mechanism for ion pair recombination involving low-lying cationic spacer states should enhance k_{cr} for both 1 and 2. Second, since the energy of P - $S_O{}^+$ - Q^- in 2 is 0.6 eV lower than that of P - $S_H{}^+$ - Q^- in 1, if superexchange is important, ion-pair recombination should be faster in 2 than in 1. The data in Table 2 show that k_{cr} is faster than k_{cs} for each isomer of 1 and 2. However, it is difficult to attribute this result solely to superexchange because the free energies of the charge separation and ion pair recombination reactions are not the same. In addition, the total reorganization energies for these two reactions are probably not equal either.

On the other hand, a much clearer case for superexchange emerges when one compares k_{cr} for 1 and 2. In this case the free energies of the ion pair recombination reactions in 1 and 2 are equal, and most likely the respective total reorganization energies for the ion pair recombination reactions in 1 and 2 are also very similar. We observe that k_{cr} is indeed 3-4 x faster for **2-syn** and **2-anti** than for **1-syn** and **1-anti**, respectively. Using the definition given above for states A and B in the ion pair recombination reaction, the data in Figure 3 show that δE_{AB} is 1.2 eV for 1 and only 0.6 eV for 2. Thus, if the superexchange mechanism is dominant for ion pair recombination in 1 and 2, the relative rates of ion pair recombination should be $[(1/\delta E_{AB})^2]_1/[(1/\delta E_{AB})^2]_2$, or $[1/(1.2)^2]/[1/(0.6)^2] = 4$. Since, our data show that the ion pair recombination rate for 2 is 3-4 x faster than that for 1, it is likely that the superexchange mechanism is responsible for the faster recombination rate in 2 relative to 1.

6. CONCLUSIONS

Excited state electron transfer reactions and ion pair recombination reactions in porphyrin-quinone molecules both show significant orientation dependencies. The nature of these dependencies strongly suggests that electron transfer occurs through the bonds of rigid hydrocarbon spacer molecules. Modulating the orbital energies of the spacer molecules, while maintaining conformational integrity, allows one to test the superexchange mechanism of electron transfer. The availability of a low-lying ionic spacer state that mixes with the ion pair state results in enhanced rates of electron transfer through the spacer. Since this effect occurs for simple substituted benzene spacer fragments, it is likely that similar enhancements of electron transfer reaction rates can occur when aromatic amino acids are positioned in strategic orientations between electron donors and acceptors within proteins.

7. EXPERIMENTAL

Butyronitrile (PrCN) was refluxed over $KMnO_4$ and Na_2CO_3, then twice distilled retaining the middle portion each time. The PrCN was then dired and stored over Linde 3 Å molecular sieves.

UV-visible absorption spectra were taken on a Shimadzu UV-160. The fluorescence spectra were obtained using a Perkin-Elmer MPF-2A fluorimeter interfaced to a PDP 11/34 computer. All samples for fluorescence were purified by prep-TLC on Merck silica gel plates. Samples for fluorescence measurements were 10^{-6} M in 1 cm cuvettes. The emission was measured 90° to the excitation beam. Fluorescence quantum yields were determined by integrating the digitized emission spectra from 580 to 780 nm and referencing the integral to that for Zn meso-tetraphenylporphyrin in benzene.[33]

The transient absorption spectra were obtained using a Rh-6G dye laser synchronously-pumped by a mode-locked, frequency-doubled CW Nd-YAG laser. The 1.0 psec pulses of 610 nm light were amplified by a 4-stage dye amplifier (Rh-640) pumped by the frequency-doubled output of a Nd-YAG laser possessing a 10 Hz repetition rate. Saturable absorber dye jets between stages 2 and 3, and between stages 3 and 4 of the amplifier chain minimized the amplified stimulated emission generated in the amplifier. The amplification produced a 1.5 mJ/pulse at a 10 Hz repetition rate. This pulse was sent through a 60/40 beam splitter. The smaller portion was focused down to a 2 mm diameter and used as the excitation pulse. The larger portion was tightly focused into a 2 cm path length cell containing either 2/1 $CCl_4/CHCl_3$ or 1/1 H_2O/D_2O. This generated a white-light continuum probe pulse, which was used as the probe light. The arrival at the sample of the probe pulse was delayed relative to the excitation pulse by an optical delay. The probe pulse was divided into reference and sampling pulses by a 50/50 beam splitter. Both probe pulses passed through the sample. The reference pulse passed through an area that was not illuminated by the excitation pulse, while the sampling pulse passed through the same portion of the sample through which the excitation pulse passed. Both pulses were then focused onto the slit of a monochromator. The monochromator dispersed the pulses onto the face of an intensified SIT detector, which is part of an optical multichannel analyzer (PAR OMA II). Solutions of 1-4 with an absorbance of about 0.6 at 610 nm (2 mm pathlength cells) were used.

Fluorescence lifetime measurements used 1.0 psec, 1 mm diameter, 100 μJ pulses from the same source as described for the transient absorbance experiments. The samples of 1-4 were placed in 1 cm cells (optical density ca. 0.1 at 610 nm) and emission 90° to the excitation was collected and focused onto the slit of a Hamamatsu C979 streak camera. The temporally dispersed image was recorded by the intensified SIT vidicon of the PAR OMA II. The geometry of the experimental set up results in a 10 ps instrument response function.

Measurements of one-electron redox potentials vs SCE were carried out at a Pt disc electrode in butyronitrile containing 0.1 M tetra-n-butylammonium perchlorate using AC voltammetry as described previously.[34]

8. ACKNOWLEDGEMENT

This work was supported by the Division of Chemical Sciences, Office of Basic Energy Sciences, Department of Energy under contract W-31-109-Eng-38. DWM was supported by the National Science Foundation

9. REFERENCES

1. Deisenhofer, J., Epp, O., Miki, K., Huber, R., and Michel, H. (1984) 'X-ray structure analysis of a membrane protein complex. Electron density map at 3 Å resolution and a model of the chromophores of the photosynthetic reaction center from Rhodopseudomonas viridis', J. Mol. Biol., 180, 385-398.
2. Larsson, S. (1981) 'Electron transfer in chemical and biological systems. Orbital rules for nonadiabatic transfer', J. Am. Chem. Soc., 103, 4034-4040.
3. Larsson, S. (1983) 'Electron transfer in proteins', J. Chem. Soc., Faraday Trans. 2 , 79, 1375-1388.
4. Miller, J. R., Beitz, J. V., and Huddleston, R. K. (1984) 'Effect of free energy on rates of electron transfer between molecules', J. Am. Chem. Soc., 106, 5057-5068.
5. Wasielewski, M. R. (1988) 'Distance dependencies of electron transfer reactions', in M. A. Fox and M. Chanon (eds.), Photoinduced Electron Transfer, Part A, Elsevier, Amsterdam, pp. 161-206.
6. Wasielewski, M. R., Niemczyk, M. P., Svec, W. A., and Pewitt, E. B. (1985) 'Dependence of rate constants for photoinduced charge separation and dark charge recombination on the free energy of reaction in restricted distance porphyrin-quinone molecules', J. Am. Chem. Soc., 107, 1080-1082.
7. Wasielewski, M. R. and Niemczyk, M. P. (1986) 'Distance-dependent rates of photoinduced charge separation and dark charge recombination in fixed-distance porphyrin-quinone molecules' in M. Gouterman, P. M. Rentzepis and K. D. Straub (eds.), Porphyrins - Excited States and Dynamics, ACS Symposium Series No. 321, American Chemical Society, Washington, D.C., pp. 154-165.
8. Joran, A. R., Leland, B. A., Geller, G. G., Hopfield, J. J., and Dervan, P. B. (1984) 'Models for photochemical electron transfer at fixed distances. Porphyrin-bicyclo(2.2.2.)octane-quinone and porphyrin-bibicyclo(2.2.2.)octane-quinone', J. Am. Chem. Soc., 106, 6090-6092.
9. Schmidt, J. A., Siemiarczuk, A., Weedon, A. C., and Bolton, J. R. (1985) 'Intramolecular photochemical electron transfer 3. Solvent dependence of fluorescence quenching and electron transfer rates in a porphyrin-amide-quinone molecule', J. Am. Chem. Soc., 107, 6112-6114.

10. Marcus, R. A. (1988) 'Superexchange versus an intermediate BChl⁻ mechanism in reaction centers of photosynthetic bacteria', Chem. Phys. Lett., 133, 471-477.
11. Won, Y. and Friesner, R. A. (1988) 'On the viability of the superexchange mechanism in the primary charge separation of bacterial photosynthesis', Biochim. Biophys. Acta, 935, 9-18.
12. Bixon, M., Jortner, J., Plato, M., and Michel-Beyerle, M. E. (1988) 'Mechanism of the primary charge separation in bacterial photosynthetic reaction centers', in The Bacterial Reaction Center, Structure and Dynamics, J. Breton and A. Vermeglio, (eds.) Plenum, New York, pp. 399-419.
13. McConnell, H. M. (1961) 'Intramolecular charge transfer in aromatic free radicals', J. Chem. Phys., 35, 508-515.
14. Paddon-Row, M. N. (1982) 'Some aspects of orbital interactions through bonds: Physical and chemical consequences', Acc. Chem. Res., 15, 245-251.
15. Ohta, K., Closs, G. L., Morokuma, K., and Green, N. J. (1986) 'Stereoelectronic effects in intramolecular long-distance electron transfer in radical anions as predicted by ab initio MO calculations', J. Am. Chem. Soc., 108, 1319-1320.
16. Beratan, D. N. and Hopfield, J. J. (1984) 'Calculation of electron tunneling matrix elements in rigid systems. Mixed-valence dithiaspirocyclobutane molecules', J. Am. Chem. Soc., 106, 1584-1594.
17. Larsson, S. and Volosov, A. J. (1986) 'Distance dependence in photo-induced intramolecular electron transfer', J. Chem. Phys., 85, 2548-2554.
18. Redi, M and Hopfield, J. J. (1988) 'Theory of thermal and photoassisted electron tunneling', J. Chem. Phys., 72, 6651-6660.
19. Marcus, R. A. (1988) 'An internal consistency test and its implications for the initial steps in bacterial photosynthesis', Chem. Phys. Lett., 146, 13-22.
20. Joachim, C. (1987) 'Ligand-length dependence of the intramolecular electron transfer through-bond coupling parameter', Chem. Phys., 116, 339-349.
21. Heitele, H. and Michel-Beyerle, M. E. (1985) 'Electron transfer through aromatic spacers in bridged electron donor-acceptor molecules', J. Am. Chem. Soc., 107, 8286-8288.
22. Heitele, H. and Michel-Beyerle, M. E. (1985) 'Electron transfer through aromatic spacers in bridged electron donor-acceptor molecules', in M. E. Michel-Beyerle, (ed.), 'Antennas and Reaction Centers of Photosynthetic Bacteria', Springer, Berlin, pp. 250-255.
23. Hart, H., Bashir-Hashemi, A., Luo, J., and Meador, M. A. (1986) 'Iptycenes. Extended triptycenes', Tetrahedron, 42, 1641-1654.
24. Wasielewski, M. R., Johnson, D. G., Svec, W. A., Kersey, K. M., and Minsek, D. W. (1988) 'Achieving high quantum yield charge separation in porphyrin-containing donor-acceptor molecules at 10 K', J. Am. Chem. Soc., 110, 7219-7221.
25. Howell, J. O., Goncalves, J. M., Amatore, C., Klasinc, L., Wightman, R. M., and Kochi, J. A. (1984) 'Electron transfer from aromatic hydrocarbons and their π complexes with metals. Comparison of the standard oxidation potentials and vertical ionization potentials', J. Am. Chem. Soc., 106, 3968-3976.
26. Mortensen, J. and Heinze, J. (1984) 'The electrochemical reduction of benzene. First direct determination of the reduction potential', Angew. Chem. Int. Ed. Engl., 23, 84-85.
27. Henton, D. R., McCreery, R. L., and Swenton, J. S. (1980) 'Anodic oxidation of 1,4-dimethoxy aromatic compounds. A facile route to functionalized quinone bisketals', J. Org. Chem., 45, 369-378.
28. Fajer, J., Borg, D. C., Forman, A., Dolphin, D., and Felton, R. H. (1970) 'π-Cation radicals and dications of metalloporphyrins', J. Am. Chem. Soc., 92, 3451-3460.
29. Distances were determined from Corey-Pauling-Koltun molecular models.
30. Oliver, A. M., Craig, D. C., Paddon-Row, M. N., Kroon, J., and Verhoeven, J. W. (1988) 'Strong effects of the bridge configuration on photoinduced charge separation in rigidly-linked donor-acceptor systems', Chem. Phys. Lett., 150, 366-373.
31. Closs, G. L., Calcaterra, L. T., Green, N. J., Penfield, K. W., and Miller, J. R. (1986) 'Distance, stereoelectronic effects, and the Marcus inverted region in intramolecular

electron transfer in organic radical anions', J. Phys. Chem., 90, 3673-3683.
32. Plato, M., Mobius, K., Michel-Beyerle, M. E., Bixon, M., and Jortner, J. (1988) 'Intermolecular electronic interactions in the primary charge separation in bacterial photosynthesis', J. Am. Chem. Soc., 110, 7279-7285.
33. Seybold, P. G. and Gouterman, M. (1969) 'Porphyrins XIII: Fluorescence spectra and quantum yields', J. Mol. Spectrosc., 31, 1-13.
34. Wasielewski, M. R., Smith, R. L., and Kostka, A. G. (1980) 'Electrochemical production of chlorophyll a and pheophytin a excited states', J. Am. Chem. Soc., 102, 6923-6928.

Electronic Excitations and Electron Transfer Coupling within the Bacterial Reaction Center Based on an INDO/S–CI Supermolecule Approach including 615 Atoms

P.O.J. Scherer and Sighart F. Fischer

Department of Physics
Technical University Munich
D–8046 Garching, Germany

Abstract. A representative part of the reaction center of Rhodopseudomonas viridis consisting of four bacteriochlorophylls, two pheophytins, one menaquinon, seven protein residues and two water molecules is treated as a supermolecule by means of an INDO–S/CI program. The electronic coupling matrix elements for the initial charge separation are determined. Optical absorption, linear dichroism and electrochromicity spectra are evaluated on the basis of an extended exciton model which accounts for charge transfer states within the special pair dimer.

I. Introduction

The most direct information about the initial charge separation comes from the electronic spectra, either as stationary or time resolved absorption emission, polarized or difference spectroscopy /1/. To draw from these observations conclusions about the mechanism of the electronic charge separation after excitation is not easy, since the charge separation is linked to charge transfer (CT–) states, which change the spectra due to bleaching of transitions from the ground state due to addition of transitions from the CT–states and due to electrochromic shifts and electronic state mixing. Quantum calculations can help to unravel the complexity and to determine directly the intermolecular electronic couplings /2–5/. For such large systems like the complex of the prosthetic groups of a reaction center only semiempirical methods can be applied and a close interplay between experimental observations and theoretical predictions is needed. In particular one needs to learn about the limitations of the theoretical approximation and the possible effects of the protein surroundings. Fortunately a large multitude of electronic spectra are available such as the absorption /6/, the linear dichroism /7/ fluorescence /8/, triplet minus singlet difference spectra /9/, electrochromicity spectra /10–11/, hole burning / 6/ and time resolved spectra /12/. Taking all these observations together clearly reduces the freedom in the choice in which parameters like energy positions can be adjusted.

In this note we apply INDO/S–CI calculations. In chapter I we present results on individual pigments based on structural data and modified structural data in order to learn about the sensitivity of geometry changes. In chapter II the electronic intermolecular couplings between different pigments and selected residues are represented. In chapter III we provide a set of adjusted parameters and simulate absorption, linear dichroism and electrochromicity spectra. This way we are able to

J. Jortner and B. Pullman (eds.), Perspectives in Photosynthesis, 361–370.

draw conclusions concerning the location of the CT—states. In particular it is shown, that the lowest CT—states between the two pigments of the dimer are very asymmetric, such that only one mixes appreciably with the lowest excited state P^*.

II. The Isolated Pigments

The methods of the INDO/S—CI calculations have been described elsewhere /2/. Here we use the refined structure data of Deisenhofer et al /13/ and alternatively allow the molecules to relax by means of an energy minimization program /14/. The results are shown in table 1. As expected from earlier studies we find for all cases that the excitation energies come out too high and the transition dipoles become too large. These deficiencies can be reduced if multiple excitations are included in the CI procedure /15/. Keeping the necessary corrections in mind we note that the calculations give a larger excitation energy to the pheophytine on the M—branch H_M as compared to H_L (for the nomenclature see Fig. 1). In comparing the unrelaxed with the relaxed structure a reduction of the very strong change of the permanent dipole due to excitation is found. Smaller values as resulting for the relaxed structural data are more consistent with experiments. For the simulation of the electrochromicity spectrum we used a value of 2 Debyes for the isolated pigments..

	Qy transition frequency intensity		Angle with the NI/NIII NII/NIV		change of permanent	
	(cm^{-1})	Debyes^2	axis (degrees)		dipole	(Debyes)
B_M	13182	95	8.1	80.8	13.4	
B_L	13410	96	8.8	80.2	13.1	
P_M	13217	99	6.5	83.5	12.1	
P_L	12802	118	4.6	84.1	7.6	
BC	13124	125	11.6	78.2	1.5	
H_M	13151	67	10.6	76.7	9.1	
H_L	12553	74	10.4	74.8	8.0	
BP	13019	80	14.0	74.8	1.0	

Table 1: INDO/SCI results calculated using the x—ray structure of rps.viridis for the isolated molecules (bacterio—chlorophylls B_M, B_L, P_M, P_L and bacteriopheophytins H_L, H_M as well as for relaxed structures BC and BP)

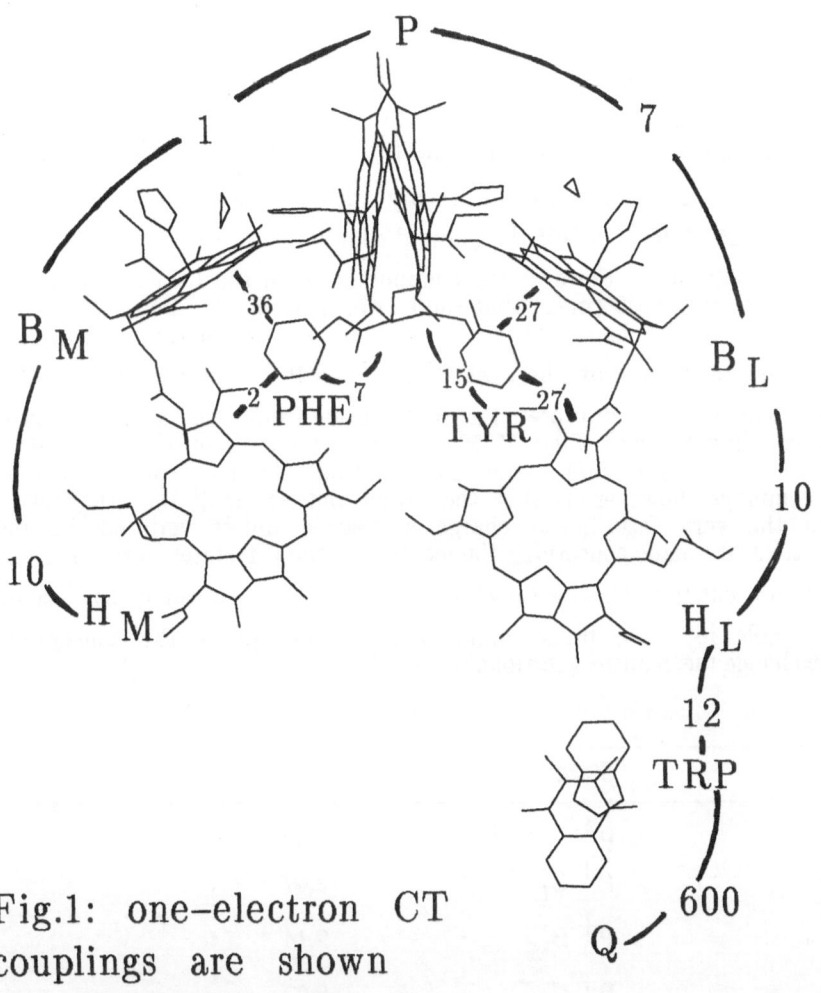

x–ray structure from Deisenhofer et al refined 1987

Fig.1: one–electron CT couplings are shown as calculated from the INDO wavefunctions in units of cm-1

III. Electronic Couplings of the Reaction Center Complex

Fig. 1 shows the pigments and the intermolecular one electron coupling matrix elements between the LUMO's of the pigments. Within the frame of a non adiabatic electron transfer theory they control the rate of the initial charge separation. It should be noted that the coupling between the dimer P and the accessory monomer B_L on the L—branch is considerably larger than the corresponding coupling on the M—branch. In addition there is a coupling route from P to H_L via a Tyrosin (TYR M208), which is essentially lacking on the M—branch. It should be further noted that the coupling between the bacteriopheophytin H_L and the Quinone is mediated by the Tryptophan (TRP M250). In low order perturbation theory with an energy difference between $P^+H_L^-$ and P^+ TRP$^-$ of 3eV the induced coupling between $P^+H_L^-$ and P^+Q^- amounts to 0.2 cm-1. However the coupling between the LUMO's of TRP 250 and Q is so large (600 cm-1) that these two molecules are better treated as a super molecule. We have done the corresponding calculation and found an enhancement of the coupling between $P^+H_L^-$ and P^+Q^- by a factor of two. The enhancement might be interpreted as contribution from a hole transfer, since the strong coupling between TRP M250 and the Quinone admixes the HOMO of TRP M250 into the LUMO of the supermolecule. The main conclusion from all these couplings, however, is that they come out so small that they can not explain the very fast initial charge separation unless very special model assumptions are made concerning the location of the CT states during the charge separation event. In particular the state $P^+B_L^-$ has to be assumed so close to P^* that it falls into the broad dimer band which makes the concept of a superexchange mechanism questionable.

State	Energy (eV)
P^*	1.50
$P^+ B_L^-$	2.07
$P^+ B_M^-$	2.14
$P^+ H_L^-$	2.03
$P^+ H_M^-$	2.36
$P^+ Q_A^-$	3.37
P^+ TRP$^-$	5.06

Table 2 : Calculated energies of the charge separated (CT) states

In Table 2 we present the energies of CT states as they come out of the calculations. They correspond to the vertical transition of the ground state configuration for the whole reaction center complex. They are too high since polarization of the medium is not included.

IV. Simulation of the spectra

In order to interpret electronic spectra some adjustments of the parameters which come out of the quantum calculations must be made. In particular the energy positions can not come out precisely since we have limited CI interactions and have not incorporated the full protein surroundings. In Table 3 a set of parameters is presented which can reproduce the electronic absorption, the linear dichroism and the Stark–spectrum (Fig. 2) simultaneously.

	P_L^*	P_M^*	$P_M^+ P_L^-$	$P_L^+ P_M^-$	H_M^*	B_M^*	B_L^*	H_L^*
P_L^*	−402	−450	750	1072	−16	−100	−20	4
P_M^*		−534	1072	750	−5	3	106	18
$P_M^+ P_L^-$			816	0	0	0	0	0
$P_L^+ P_M^-$				584	0	0	0	0
H_M^*					880	−93	5	2
B_M^*						239	18	5
B_L^*							140	−106
H_L^*								549

Table 3 : Excitonic couplings for the simulation of the spectra of rps.viridis in cm^{-1}. The diagonal energies are relative to 11824 cm^{-1} which is the center of the experimental absorption below 15000 cm^{-1}.

The main assumption we had to make is that the charge transfer state $P_L^+ P_M^-$ is much lower than $P_M^+ P_L^-$. The calculations give here a relatively small difference (in the right direction). The larger spread is essential to reproduce the electrochromicity and the absorption simultaneously. The origin of this shift might be partially due to the water molecules bridging the histidines which are linked to the Mg atoms of the special pair to the keto group of ring V from the accessory monomers. The location of the molecules is sufficiently asymmetric to produce different hydrogen bonding conditions on the two branches. Only on the L–branch a hydrogen bonding to the oxygen keto group of B_L is conceivable. This induces also an asymmetry in the location of the CT–states with $P^+B_L^-$ being below $P^+B_M^-$, a feature which again favors the electron transfer on the L–branch. Since the location of the CT states is strongly dependent on the orientation of the surrounding dipoles one expects a distribution for the energy location of these states. This inhomogeneous distribution induces also an increase in width of the state P^*. We were able to simulate the spectra using the same line profile for all transitions and a Gaussian distribution for the CT states. The width of this distribution can be seen in the 750 nm regime of Fig. 3. It should be noted also that the relatively high contributions from the first derivative of the absorption profile to the electrochromicity spectrum for a specific conformation is strongly suppressed in favor of the second derivative if the average over the distributions is performed (Fig. 2). In Table 4 we summarize the decomposition of the eigenstates in terms of their localized components. The strong Stark amplitude of P^* is clearly due to the interaction with the internal CT–state $P_L^+P_M^-$. This state, however, does not fall into the dimer band P^* itself. Such a location would cause a much too strong first derivative component in the Stark spectrum.

λ nm	I D^2	Δp D	P_L^*	P_M^*	$P_M^+P_L^-$	$P_L^+P_M^-$	H_M^*	B_M^*	B_L^*	H_L^*
1065	161	7.2	.61	−.58	−.11	−.47	.00	−.03	−.03	−.01
929	10	2.8	−.19	−.28	−.03	−.03	.00	.03	−.05	−.01
917	5	2.8	.02	−.29	−.01	−.01	−.00	−.00	−.06	−.01
894	62	2.8	−.58	−.59	−.03	.16	.02	.17	−.31	−.06
877	118	2.3	−.15	−.20	−.01	−.01	.05	.35	.87	.22
868	54	2.4	.17	.19	.01	−.08	.13	.91	−.28	−.07
844	63	3.0	.01	.01	.00	−.00	.00	−.00	−.24	−.97
822	63	3.0	−.01	−.00	.00	−.01	.99	−.14	−.00	−.00
813	17	12	.28	−.22	−.03	.58	.02	.03	.02	.01
804	6	2.9	.02	−.00	−.01	.02	.00	.00	−.00	.00
782	14	13	.33	−.08	−.08	.63	.01	.03	.00	.00
720	0.3	6.2	.08	.07	−.00	.05	.00	.00	−.00	−.00
719	15	6.4	−.09	.04	.04	−.13	−.00	−.00	−.00	−.00
509	3.9	32	.08	−.11	.99	.02	.00	.00	.00	.00

Table 4: Decomposition of the lowest exciton states
The lowest eigenstates of the exciton model are decomposed into contributions of local excitations of the pigments and intradimer charge transfer states. Transition wavelengths λ, intensities I and permanent dipole moment change Δp are also shown.

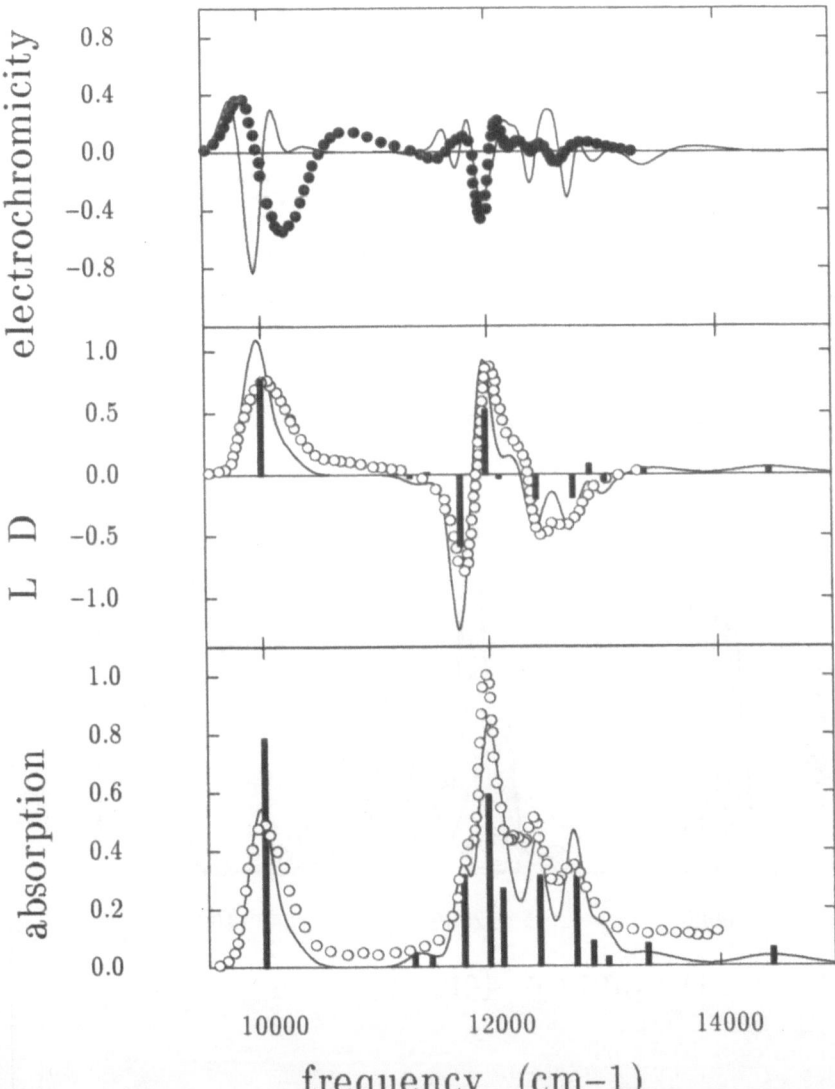

Fig.2 Model calculations for the optical absorption (bottom), linear dichroism (middle) and electrochromicity (top) spectra of rps.viridis are shown as solid curves. Experimental data (absorption and LD at 4K from /6/ , electrochromicity at 77K from /10/) are redrawn as circles for comparison. The bars show positions and intensities of the exciton states.

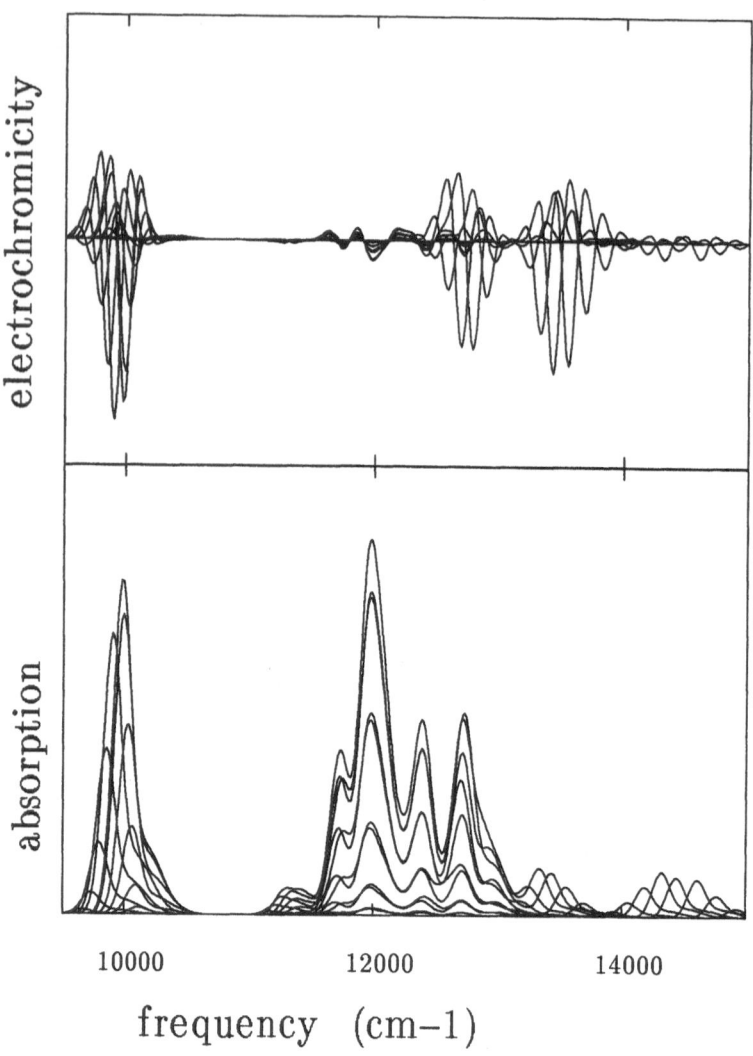

Fig.3   Absorption (bottom) and electrochromicity (top) spectra are calculated for
several positions of the intradimer charge transfer state. The spectra shown
in Fig.2 are averages over a Gaussian distribution of CT energies. The
exciton transitions are dressed with an asymmetric profile which mimics the
typical absorption profile of the Bacteriochlorophyll Q_y–transition. The
dimer band width increases due to its strong coupling to the CT states.

In order to understand further some of the fine peaks in the Stark spectrum around 840 nm and the long wavelength tail of the LD in the same regime we had to incorporate also vibronic coupling to the C–C and C–N stretching modes with a frequency of 1300 cm^{-1} for the special pair pigments P_L and P_M, resulting in 8 additional states (two excitonic and two CT–states of the dimer with either P_L or P_M vibronically excited).

V. Conclusions

In summary we like to point out that our simulations suggest that the internal CT state $P_L{}^+ P_M{}^-$ of the special pair is in Rps. viridis located above the excited states H_L and H_M. The rather broad linewidth of the dimer band is partially due to inhomogeneous distributions which result from the sensitive coupling of CT states to the electrostatic potential of the surrounding protein. For a better understanding of the charge separation one needs to identify the relaxation processes within the excited state P^*. The electronic couplings suggest that the possible virtual intermediate $P^+ B_L{}^-$ must get very close to P^* as the relaxation within P^* develops. In this connection the bridging water on the L–branch (Fig. 1) may play a role.

Acknowledgement

We like to thank Prof. Deisenhofer for providing us with the refined structure data prior to publication

References

/1/   The Photosynthetic Bacterial Reaction Center, Structure and Dynamics ed. J.Breton and A.Vermeglio, NATO ASI Series A Vol.149, Plenum Press N.Y. and London
/2/   S..F. Fischer and P.O.J. Scherer, Chem. Phys.,115, 151(1987)
/3/   P.O.J. Scherer and S.F.Fischer, Chem.Phys. 131, 115 (1989)
/4/   M.Plato, K.Moebius, M.E. Michel–Beyerle, M.Bixon and J.Jortner, J.Am.Chem.Soc. 110,7279 (1988)
/5/   W.W. Parson and A. Warshel, J. Am. Chem. Soc., 109, 6152 (1987), A. Warshel and W. W. Parson, J. Am. Chem. Soc.,109, 6143 (1987)
/6/   G.J.Small see article in this book
/7/   J.Breton, Biochim.Biophys.Acta, 810, 235 (1985)
/8/   S.G.Boxer, R.A.Goldstein, D.J.Lockhart, T.R. Middendorf and L. Takiff in The Photosynthetic Reaction Center ..., pp165
/9/   E.J.Lous   and   A.J.   Hoff,   Proc.Natl.Acad.Sci.   84,6147   (1987)

/10/   S.G. Boxer, D. J. Lockhart, and T.R. Middendorf, Chem.Phys.Lett.,
         123, 476 (1986)
/11/   P.O.J. Scherer, S.F. Fischer, Chem. Phys. Lett.,141,179 (1987)
/12/   J.L. Martin, J. Breton, A. J. Hoff, A. Migus, and A.Antonetti, Proc.
         Natl. Acad. Sci US,83, 957 (1986)
         J.Breton, J. L. Martin, A. Migus, A. Antonetti, and A.Orsay, Proc.
         Natl. Acad. Sci USA, 83, 5121 (1986)
         G.R. Fleming,J.L.Martin and J.Breton, Nature 333,190(1988)
/13/   J. Deisenhofer, O. Epp, K. Miki, R. Huber, and H. Michel, J. Mol.
         Biol., 180, 385 (1984)
         H. Michel, O. Epp, and J. Deisenhofer, EMDO J., 5, 2445,(1986)
/14/   Alchemy II, Tripos Associates, St. Louis, Missouri

SELF ASSEMBLY OF BACTERIOCHLOROPHYLL a AND BACTERIOPHEOPHYTIN a IN MICELLAR AND NON-MICELLAR AQUEOUS SOLUTIONS; APPLICATION TO THE PIGMENT-PROTEIN ORGANIZATION IN LIGHT-HARVESTING COMPLEXES AND REACTION CENTERS

A. SCHERZ [†], V. ROSENBACH-BELKIN AND J. R. E. FISHER [¥]
Department of Biochemistry
Weizmann Institute of Science
Rehovot, 76100
Israel

ABSTRACT. While investigating the self assembly of photosynthetic pigments in aqueous solutions, it was found that very large aggregates maintained an equilibrium with monomers. This aggregation is cooperative and can be characterized by two equilibrium constants. The first equilibrium constant (K_a) describes dimer formation and the second one (K_b) describes the formation of larger oligomers. K_a and K_b for bacteriochlorophyll a (Bchla) are 1.5×10^3 M^{-1} and 2.2×10^6 M^{-1}, respectively. K_b for bacteriopheophytine a (Bphea) is $\sim 10^9$ M^{-1}. The difference between the spectra of Bchl dimers and large oligomers is insignificant, indicating that the large oligomer consists of repeating dimers which are separated by ~ 15Å from each other.

Self assembly of Bphea and Bchla has also been studied in aqueous solutions containing Triton X-100 (TX-100). To calculate the aggregation number and the corresponding equilibrium constant, we have developed a formalism which considers the pigment distribution in micelles of variable size. Application of the theory to Bchla and Bphea shows dimer formation in the micellar domain with $K_d = 2.2 \times 10^3$ M^{-1} for Bchla in formamide/water (FW), $K_d = 3.9 \times 10^5$ M^{-1} for Bphea in FW and $K_d = 7.5 \times 10^4$ M^{-1} for Bphea in water. The photosynthetic pigments seem to have a major effect on the micelle size. In the pigment-free solutions the micelles contain ~ 150 molecules of TX-100, while the addition of pigments alters this to 4000-40000 molecules, depending on the system.

Comparison with *in vivo* pigments suggests that the Bchla self assembly tunes the photosynthetic pigments to the prevailing light conditions and at the same time affects the organization of the polypeptide units within the intercytoplasmic membrane. This mechanism may explain the relationship between pigmentation and polypeptide assembly in light-harvesting complexes and reaction centers in photosynthesis.

1. Introduction

The reaction centers (RCs) and light harvesting complexes (LHCs) that function in the photosynthetic apparatus to convert solar energy into electrochemical potential (1) are membrane-bound pigment-protein complexes (2). Several LHCs form an energy funnel by means of multiple absorptions and appropriate

---

[†] Incumbent of the Recanati Career Development chair
[¥] In partial fulfillment of the M. Sc. Thesis

J. Jortner and B. Pullman (eds.), Perspectives in Photosynthesis, 371–387.
© 1990 *Kluwer Academic Publishers.*

packing around the RCs (2). The entire structure provides the setting for a unidirectional energy flow towards the primary electron carriers within the lifetime of the singlet excited states. The mechanism for protein-chromophore complex assembly, whether into LHCs or RCs, remains unsolved (2).

Interestingly, the multiple absorption of the pigmented proteins originate in common chromophores: chlorophylls (Chls) in oxygen evolving organisms and bacteriochlorophylls (Bchls) in non-oxygenic bacteria. All Chls and Bchls *in vivo* show similar features of their electronic spectra. In particular, all forms have their lowest energy transition (Q_y) bathochromically shifted, intensified and optically active with respect to the Q_y transition of the *in vitro* monomers (3-6). Since there is no indication of a chemical modification, the spectral diversity of the *in vivo* chromophores must be related to their *in vivo* setting.

Most *in vivo* Chls and Bchls are known to form clusters and are non-covalently attached to membrane proteins (2, 7, 8). In these pigment-protein complexes, there is approximately one Chl (Bchl) molecule for every 4-8 KDa of protein. The separation between individual pigments can be as small as 3.4 Å (face to face) (7-11). At such distances one should expect interactions among both the ground and photo-excited states of the molecules involved (12). For the primary electron donors in purple bacteria (P-860 and P-960), there is a general agreement that interactions among the coupled Bchls' excited states depend upon the chromophores' geometry and lead to part of the bathochromic shift of their Q_y transition when compared with the isolated pigments *in vitro* (13-15). However, the elements involved in determining the chromophore geometry have not yet been elucidated. In addition, there is a controversy over how much interference of the protein (i.e. in the form of charge-transfer state stabilization) (15) is needed to bring about the full extent of the observed shift in particular within LHCs and RCs of the non-oxygenic bacteria (13, 15-17).

Chla aggregates (most likely dimers) have also been observed in concentrated solutions of saturated hydrocarbons by Lavorel (18) and by Weber and Teale (19). The aggregation was suggested to result from donor-acceptor interactions among paired molecules where the C-9 carbonyl of ring V of one molecule interacts with the central Mg of the second. Later on it was shown (20-26) that, in protic organic solvent containing traces of water, Chls oligomerize in such a manner that water molecules bridge the Mg of one molecule and the C-9 keto of a neighbouring molecule. The size of these aggregates is a function of the pigment concentration. When the concentration is low (~10^{-5} M for Bchl) the pigments exist as dimers and trimers and as the pigment concentration increases the pigments exist in the form of large aggregates (23).

The key to this *in vitro* aggregation of the photosynthetic pigments appears to be their donor (the lone-pair electrons of the C=O group) and acceptor (unsaturated coordination of the central Mg) properties. The critical factor in this type of dimerization is the extent to which extraneous nucleophiles (e.g. H_2O) competed for coordination with the Mg. However, raman spectroscopy (27) and recent X-ray crystallography (7-11) studies of bacterial RCs have indicated that most of the *in vivo* Chls and Bchls are hydrogen bonded at their carbonyl functions and strongly ligated at their central Mg to the protein network. Therefore, when Chls and Bchls aggregate *in vivo* they do not rely upon interactions among these sites but rather upon $\pi - \pi$ interactions. These type of interactions have been observed for different non-hydroporphyrins (28), Chls (29) and, recently, for different Bchls, Bphes and Chls in aqueous solutions of formamide (FA) (30-32).

A solution of 3:1 (vol/vol) formamide:water (FW) was found to induce the self organization of different Chls and Bchls into large oligomers with similar spectral properties to those observed *in vivo*. Even though the FW solution is hydrophilic, it imitates the *in vivo* pigments' environment in two ways; the FA chains resemble the polypeptides (33) and provide groups for hydrogen bonding with Bchl molecules (the η transition around 340 nm is evidence that the isocyclic carbonyl oxygen of the Bchl is hydrogen bonded to the solvent) (34, 35)

In order to create a more realistic assay of the photosynthetic pigment organization, we explored the possibility of having the Chls (Bchls), with the attached FA and water molecules, aggregate into small oligomers in a more hydrophobic medium. Micellar solutions seemed to be most suitable since each micelle forms a hydrophobic micro-environment in which a limited number of chromophores can be incorporated.

In a previous article we calculated the apparent equilibrium constant and free energy change of the Bchl's self assembly in FW containing micelles of Triton X-100 (TX-100) (12). In the following manuscript we will focus on this micro-environment in order to find the equilibrium constant and free energy change of chromophore assembly *in situ*. First, we shall present a formalism that was developed in order to determine the chromophore aggregation in FW. Using this formalism we will calculate the minimum oligomer size (the seed) that is responsible for the changes in the optical properties of the assembled chromophores, the equilibrium constant for the formation of oligomers smaller than the seed (K_a) and the equilibrium constant for the formation of the seed and larger oligomers (K_b). Second, the details of the chromophores' distribution within the TX-100 micelles will be discussed and used in calculating the *in situ* dimerization constant and free energy change of Bchla and Bphea. Third, a mechanism for the pigment-protein organization within the photosynthetic membrane shall be put forth based on the interactions among the Chls (Bchls), FW and the amphiphilic molecules.

2. The Formation of Large Oligomers in Aqueous - Formamide Solution

Two spectral forms are observed when Bchla is introduced to FW (12, 32). The forms are named according to their maximum Q_y absorbance peak: Bchl-860 and Bchl-780 (figure 1). Bchl-860 corresponds to the molecules of Bchla in oligomeric form and Bchl-780 corresponds to the molecules of Bchla in monomeric form.

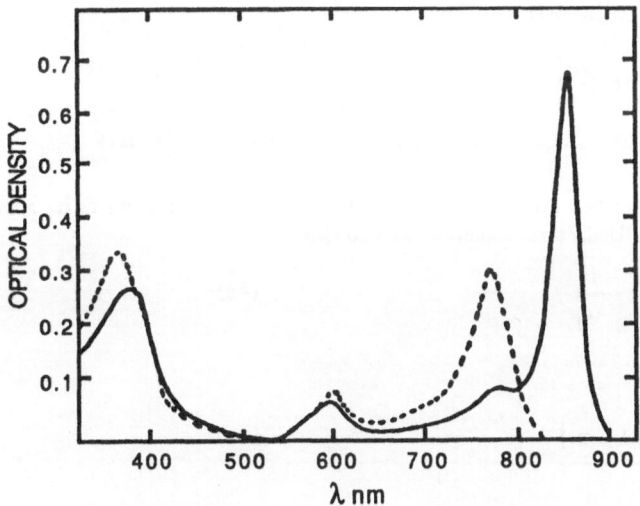

FIGURE 1. Absorption of Bchl-860 (solid line) and Bchl-780 (dashed line), in FW containing no TX-100 and FW containing 0.125% TX-100, respectively.

By calculating the concentration of the two forms at different total Bchla concentrations, Scherz and Rosenbach-Belkin have shown that above a certain total concentration, Bchl-860 increases linearly with the total concentration and Bchl-780 remains constant (figure 2A in ref. 32). In light of this observation, they suggested that the conversion of monomers into oligomers follows a cooperative mechanism. Our purpose is to develop a formalism, making use of the concept of cooperativity, that will enable us to achieve a more quantitative understanding of this system. The formulae we derive are based on Oosawa's model of cooperative polymerization (36).

2.1. THEORY

In Oosawa's model there are two equilibrium constants governing association and dissociation of single monomers from one end of the polymer. When the polymer chain is less than a unique seed size, termed s, the addition of monomeric units is governed by K_a. Whereas the addition of monomeric units to polymers of length greater than or equal to s is governed by K_b. In a cooperative system K_b is much larger than K_a. At equilibrium this phenomenon can be represented by

$$A_n = K_a^{n-1} A_1^n \qquad n < s \qquad (1a)$$

$$A_n = K_a^{s-1} K_b^{n-s} A_1^n \qquad n \geq s \qquad (1b)$$

where A_n is the concentration of polymers of length n, A_1 is the monomer concentration, and s is the seed size. In a recent paper Goldstein and Stryer (37) transformed these concentrations into dimensionless quantities by incorporating the definition $\alpha_n = K_b A_n$ (where K_b was found by Oosawa to be the inverse of the asymptotic value of A_1). In doing so, they were able to isolate the components characterizing the system, namely s and σ, where σ was defined as K_a/K_b. In addition, they introduced a new term, ω which was defined as σ^{s-1}. Applying these definitions to equations 1a and 1b gives

$$\alpha_n = \sigma^{n-1} \alpha_1^n \qquad n < s \qquad (2a)$$

$$\alpha_n = \omega \, \alpha_1^n \qquad n \geq s \qquad (2b)$$

The total concentration of molecules that participate in the process of cooperative polymerization (in units of K_b^{-1}) can be expressed by

$$\alpha_T = \frac{\alpha_1 \left[1 - s \left(\sigma\alpha_1\right)^{s-1}\right]}{1 - \sigma\alpha_1} + \frac{\alpha_1\left[\sigma\alpha_1 - \left(\sigma\alpha_1\right)^s\right]}{\left[1 - \sigma\alpha_1\right]^2} + \frac{s\sigma^{s-1}\alpha_1^s}{1 - \alpha_1} + \frac{\sigma^{s-1}\alpha_1^{s+1}}{\left[1 - \alpha_1\right]^2} \qquad (3)$$

Instead of solving equation 3 precisely, one can focus on the behaviour of α_T as α_1 approaches it's limits, namely 0 and 1. Under these conditions we find (38)

$$\sigma^* \Big|_{\substack{\alpha_1 \to 0}} = \frac{1}{\alpha_1} - \frac{1}{\alpha_T} \qquad (4a)$$

$$\omega^* \Big|_{\substack{\alpha_1 \to 1}} = \alpha_T \left[1 - \alpha_1\right]^2 \qquad (4b)$$

where $\omega = \omega^*$ and $\sigma = \sigma^*$ when $\alpha_1 = 0$ and $\alpha_1 = 1$, respectively.

2.2. APPLICATION

The concentration dependence of Bchl-780 and Bchl-860 on the total concentration of Bchla is shown in figure 2. Each concentration is expressed in units of moles and in units of K_b^{-1} where K_b is given by the inverse of the asymptotic value of Bchl-780 ($K_b = 2.2 \times 10^6$ M^{-1}).

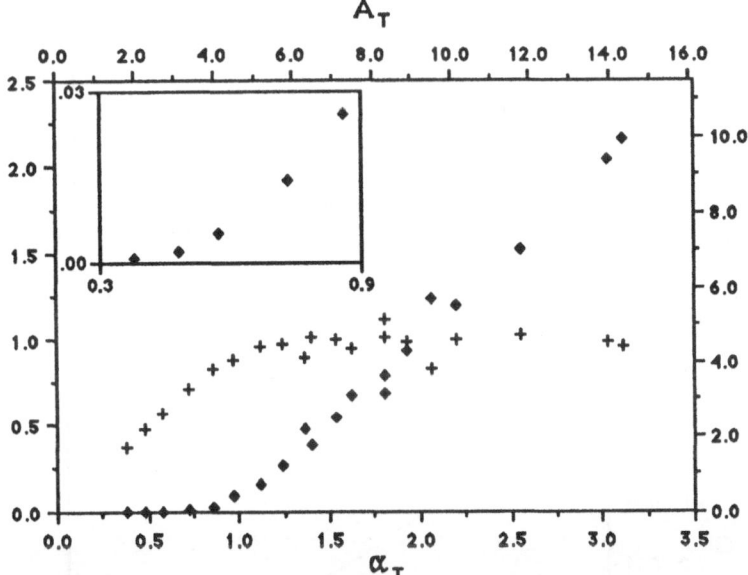

FIGURE 2. α_1 plotted against α_T (+) and $\alpha_T - \alpha_1$ plotted against α_T ($\blacklozenge$) for Bchla in FW. The top and right axes represent the concentration in units of 10^{-7} M.

To find K_a we have plotted σ^* against α_1, where σ^* is given by equation 4a but for all values of α_1 (figure 3). Extrapolating to zero yielded $\sigma = 6.91 \times 10^{-4}$. In a similar manner, ω^* was calculated from equation 4b and plotted against α_1. From figure 4 , ω was found to be 2.5×10^{-5} when α_1 equalled 1. Substituting the values derived for K_b and σ into the equation for σ, K_a was calculated to be 1.5×10^3 M^{-1}. Using the definition of ω, s was calculated to be 2.46. Since the seed size is a whole number it must be either 2 or 3.

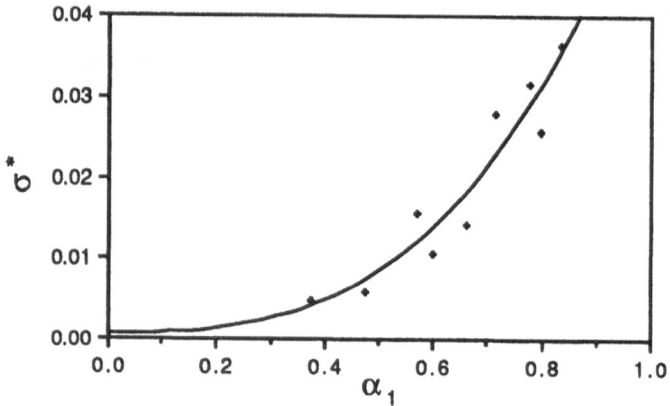

FIGURE 3. σ^* is plotted against α_1, the Bchl-780 concentration given in units of K_b^{-1}. As $\alpha_1 \to 0$, $\sigma^* \to \sigma$.

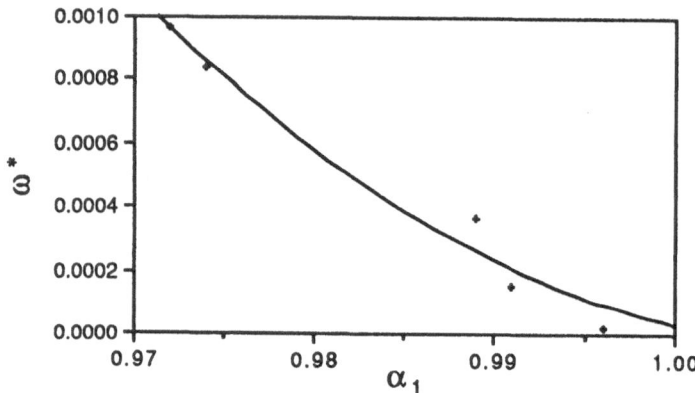

FIGURE 4. ω^* is plotted against α_1, the Bchl-780 concentration given in units of K_b^{-1}. As $\alpha_1 \to 1$, $\omega^* \to \omega$.

Another molecule that was found to aggregate in a cooperative manner when placed in the FW system was Bacteriopheophytin a (Bphea). In this case, we observed that the onset of large oligomer formation occured at approximately 10^{-9} [Bphea$_T$] M or in other words, $K_b \approx 10^9$ M^{-1} (figure 5). Our experimental techniques, to date, did not allow us to acquire significant data in the range of $a_1 \to 0$; therefore, we were unable to find σ and ω for the Bphea system.

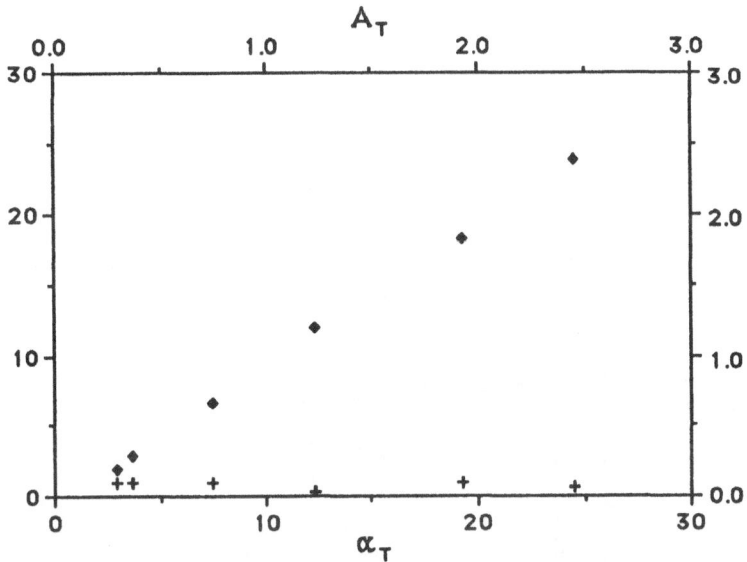

FIGURE 5. α_1 plotted against α_T (+) and $\alpha_T - \alpha_1$ plotted against α_T ($\blacklozenge$) for Bphea in FW. The top and right axes represent the concentration in units of 10^{-8} M.

The theory of cooperative oligomerization predicts that the average aggregation number (the average number of chromophores per oligomer), $\langle i \rangle$, increases abruptly once $A_1 > K_b^{-1}$. Recent theoretical studies have shown that the spectral properties of Bchla's and Bphea's large aggregates are very sensitve to the aggregation number, when it is in the range of approximately 1-50 molecules, and the proximity of the individual molecules in the aggregate is <15Å (center to center) (39). Here, the following equation has been used to calculate $\langle i \rangle$

$$\langle i \rangle = \frac{\sum\limits_{n=2}^{\infty} n\alpha_n}{\sum\limits_{n=2}^{\infty} \alpha_n} \qquad (5)$$

when s=2

$$\langle i \rangle \cong \frac{(\alpha_T - \alpha_1) \times (1 - \alpha_1)}{\sigma\alpha_1^2} \qquad (6)$$

and when s=3

$$\langle i \rangle \cong \frac{(\alpha_T - \alpha_1)}{\sigma\alpha_1^2} \qquad (7)$$

The values of $\langle i \rangle$ are calculated for s=2 and s=3 and plotted against their corresponding α_T values in figs.

6a and b, respectively. Figure 6c illustrates the optical absorption spectra of Bchl-860 for small α_1 values.

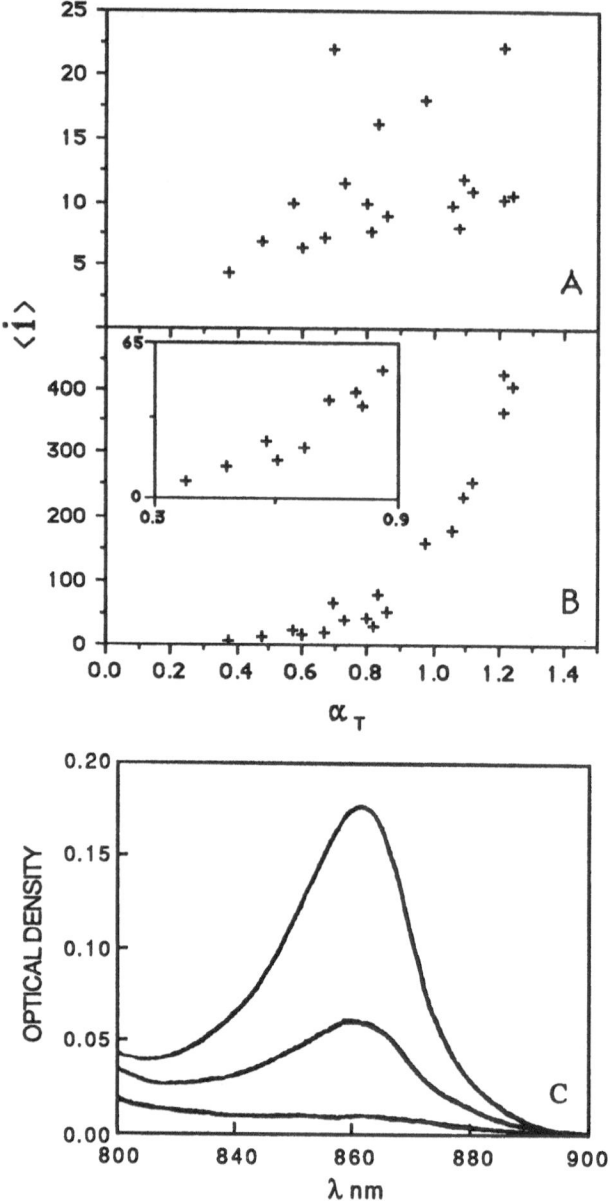

FIGURE 6. $\langle i \rangle$'s dependence on α_T is shown for: (A) s=2 and (B) s=3. (C) shows the optical absorption spectra of Bchl-860 for 3 different total concentrations. The α_T values equal 0.38, 0.80 and 0.95 for curves with increasing intensity.

In the spectra, there is an increase in the absorption intensity but little to no shift in the wavelength for maximum absorption and no change in the bandwidth at half maximum. In this range of concentration, $\langle i \rangle$ increases from approximately 4 to 25 when $s=2$ and from 8 to 500 when $s=3$. One can conclude from this that the bathochromic shift of the Q_y transition and its hyperchromism are primarily a result of interactions within dimers. Moreover, the separation between such dimers must be larger than ~15-20 Å for the large aggregate spectrum to be similar to that of the dimer.

3. The Formation of Small Oligomers in Aqueous Solutions containing Micelles

To observe the Bchla and other pigments' assembly into small clusters we introduced micelles of the detergent TX-100 into the FW solution. As shown by Scherz and Rosenbach-Belkin (12), the pigments' spectral forms were almost invariant, whether they were in the aqueous domain or in the hydrophobic domain of the TX-100 micelle. Optical measurements (Absorbance and CD) indicated that, at equilibrium, all pigments inside the micelles are found in either monomeric or dimeric form (12). Our objective here is to characterize this equilibrium by determining the *in situ* dimerization constant.

3.1. THEORY

Although the spectra indicate the presence of monomers and dimers, their arrangement in the solution is not homogeneous. In order to calculate the true equilibrium constant we must take this arrangement into account. One can conveniently divide the system into three domains: micelles containing two or more pigment molecules, micelles containing only one pigment molecule and the aqueous solution outside the micelles. In the latter domain the pigment behaves in a cooperative manner as described in the previous section. Namely, as long as the concentration of pigment is less than K_b^{-1} this domain will be populated solely by monomers. When this criterion is met, equilibrium occurs exclusively between dimers and monomers in the first domain. Since all the dimers in solution are found in this domain, their concentration can be calculated directly from the spectra. The monomers, however, can be divided into three categories corresponding to the three domains: the monomers in the first domain, $[M]^*$, the monomers which occupy the micelles as single occupants, $[M]_{singles}$, and the monomers in the aqueous solution outside the micelles, $[M]_{out}$. Since there is a constant ratio between $[M]_{out}$ and the monomers inside the micelle (given by the solubility constant K_s), $[M]_{out}$'s contribution to the total monomer concentration is always known (12). The other two types of monomers can be expressed as functions of the total pigment concentration inside the micelle, $[B]_{in}$. As shown below, these expressions provide a method for analyzing the experimental results and estimating the dimerization constant, K_d.

In order to determine the concentration of monomers in the singly occupied micelles we have applied the formula for Poisson distribution[1]. The multiplication of the Poisson probability by the micelle concentration provides an equation for the concentration of such monomers (40).

$$[M]_{single} = [MT] \left\{ \frac{[B]_{in}}{[MT]} \times \exp\left(- \frac{[B]_{in}}{[MT]}\right) \right\} \qquad (8)$$

where $[M]_{single}$ is the concentration of these isolated monomers and $[MT]$ is the concentration of micelles.

---

[1] In using this formula we have assumed that the pigment molecules populate the micelles in a random manner and independent of other existing occupants.

To express the monomer in the first domain, $[M]^*$, in terms of $[B]_{in}$ it is necessary to derive the equilibrium equation; for dimerization, this equation should resemble

$$K_d = \frac{[D]}{[M]^{*2}} \qquad (9)$$

where $[D]$ is the dimer concentration. $[D]$ and $[M]^*$ are also related to $[B]_{in}$ and $[M]_{single}$ in the expression below

$$[B]_{in} = 2[D] + [M]^* + [M]_{single} \qquad (10)$$

Combining equations 6 and 7 results in a quadratic equation with respect to $[M]^*$ whose solution yields an expression for $[M]^*$ (40).

The concentrations referred to here are with respect to the entire volume of the solution (V_T) but, as was mentioned above, under these conditions dimerization occurs in only part of this volume, namely the volume occupied by the micelles of the first domain (V_1). Therefore, in order to calculate the true dimerization constant it is necessary to transform these concentrations from moles/V_T into moles/V_1.

The volume of each micelle depends on the number of TX-100 molecules that it contains and their dimensions. In addition, the micelle size is needed to calculate the micelle concentration appearing in equation 5. While investigating, it was found that the micelle size and, subsequently, their concentration seem to depend on the presence of Bchls or Bphes. Evidence supporting this behaviour is given by fluorescence quenching studies which showed that micelles containing Bchla consisted of 4000-5000 TX-100 molecules (12) whereas empty micelles consisted of 50-150 molecules (39). Considering this behaviour, Poisson's formula was used to derive the equation for [MT], namely

$$[MT] = \frac{[TX]^*}{S_0} \exp\left(-\frac{[B]_{in}}{C}\right) + \frac{[TX]^*}{S}\left\{1 - \exp\left(-\frac{[B]_{in}}{C}\right)\right\} \qquad (11)$$

where $[TX]^*$ is the TX-100 concentration less the critical micelle concentration (C.M.C.), S_0 is the size of the empty micelles, S is the size of the occupied micelles and C regulates the ratio of large/small micelles. Assuming that once a micelle is populated by a pigment molecule its size is S, this corrected [MT] value is only relevant at low pigment concentrations (when there is a significant concentration of empty micelles). Therefore, equation 11 is only incorporated into the expression for $[M]_{single}$ and the micelle concentration, [MT] elsewhere is simply given by $[TX]^*/S$. Finally, the total monomer concentration inside the micellar domain ($[M]_{in}$) can be calculated from the summation of $[M]_{single}$ and $[M]^*$.

The calculated monomer concentration should fit the experimental values, as will be shown below. The complete equation for the monomer concentration inside the micellar domain is

$$
[M]_{in} = \frac{[TX]^* \times \left[1 - \left(1 + \frac{[B]_{in}}{[MT]}\right) \exp\left(-\frac{[B]_{in}}{[MT]}\right)\right]}{2 \times K_d \times A} \left\{ -1 + \sqrt{1 + \frac{4 \times K_d \times A \times [B]_{in}\left[1 - \exp\left(-\frac{[B]_{in}}{[MT]}\right)\right]}{[TX]^* \times \left[1 - \left(1 + \frac{[B]_{in}}{[MT]}\right)\exp\left(-\frac{[B]_{in}}{[MT]}\right)\right]}} \right\}
$$

$$
+ \quad [B]_{in} \exp\left\{ - \frac{[B]_{in}}{\frac{[TX]^*}{S_0}\exp\left(-\frac{[B]_{in}}{C}\right) + \frac{[TX]^*}{S}\left(1 - \exp\left(-\frac{[B]_{in}}{C}\right)\right)} \right\} \tag{12}
$$

where A is a constant given by $\dfrac{2 \times 10^3}{R \times 6.02}$.

Assuming that the large micelles are cylinders with radius R, the volume ratio V_1/V_T is given by (40)

$$
\frac{V_1}{V_T} = \frac{[TX]^* \times R \times 6.02 \times \left\{ [MT] - \left([B]_{in} + [MT]\right)\exp\left(-\frac{[B]_{in}}{[MT]}\right) \right\}}{[MT] \times 10^3} \tag{13}
$$

and K_d is given by

$$
K_d = \frac{[D]}{[M]^{*2}} \times \frac{[TX]^* \times R \times 6.02 \times \left\{ [MT] - \left([B]_{in} + [MT]\right)\exp\left(-\frac{[B]_{in}}{[MT]}\right) \right\}}{[MT] \times 10^3} \tag{14}
$$

3.2. APPLICATION

The formalism was applied to three systems; Bchla in FW, Bphea in FW and Bphea in water, all containing micelles of TX-100. The plots in figure 7 show the experimental points of monomer and dimer concentrations versus the total concentration of pigment inside the micelle (in the case of Bphea, most of the pigment was found inside the micelle so this value was approximately [Bphea]$_T$).

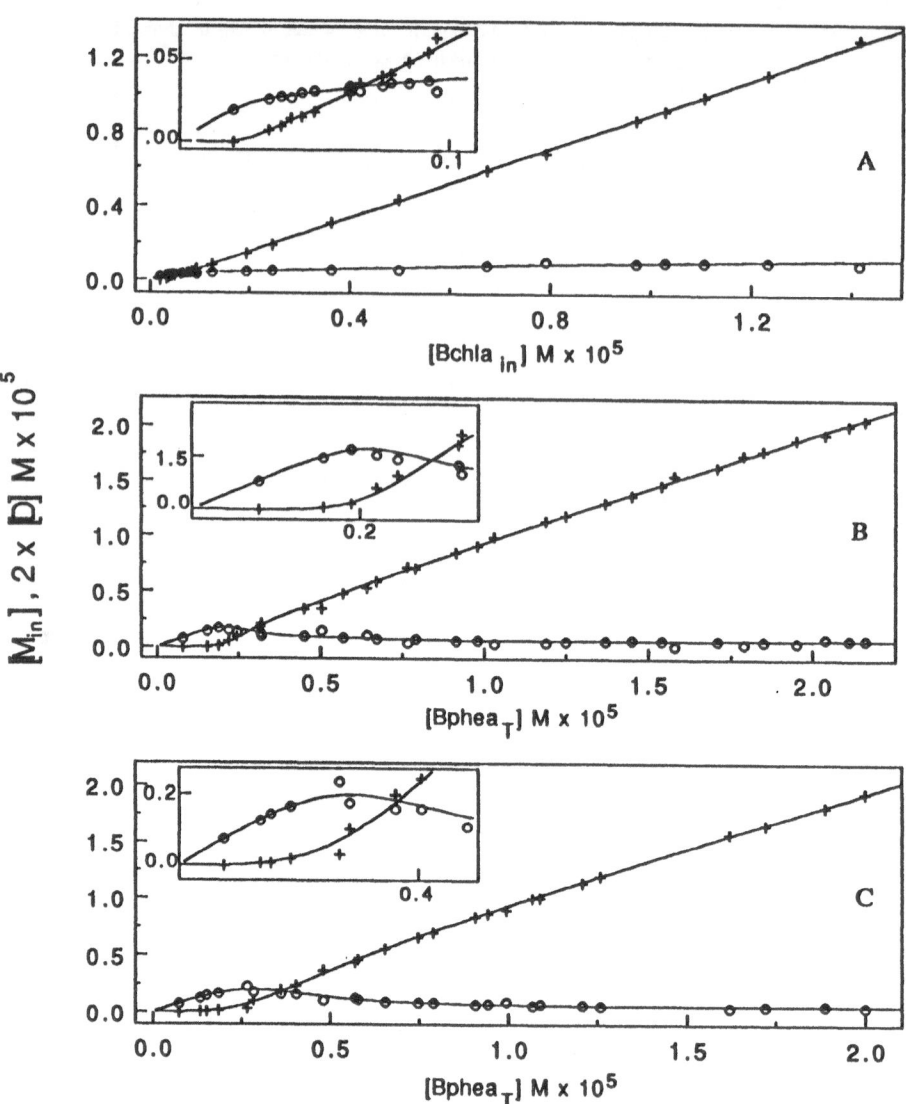

FIGURE 7. Experimental concentrations of monomer (O) and 2 x dimer (+) plotted against the total pigment concentration and their respective calculate curves (solid line) for: (A) Bchla in FW, in the calculated curve $[TX]^{\ast} = 2.45 \times 10^{-3}$ M, $K_d = 2.2 \times 10^3$ M^{-1}, S = 4100 and C = 8 x 10^{-8}, (B) Bphea in FW, in the calculated curve $[TX]^{\ast} = 9.83 \times 10^{-2}$ M, $K_d = 3.9 \times 10^5$ M^{-1}, S = 40000 and C = 4.5 x 10^{-7}, (C) Bphea in water M, in the calculated $[TX]^{\ast} = 1.17 \times 10^{-2}$ curve $K_d = 7.5 \times 10^4$ M^{-1}, S = 4000 and C = 1 x 10^{-6}.[2]

---

[2] The C.M.C. for TX-100 was taken as 3.66×10^{-3} M in FW (12)and 5.0×10^{-4} M in water (41).

The solid lines shown in these plots are the calculated monomer and dimer concentrations that were found to fit the experimental points[3]. In doing these simulations we relied upon the following observations concerning the theoretical curves: a change in K_d affected the high Bchla concentration domain, a change in S affected the intermediate domain and a change in C affected the low concentration domain (40). From the calculated curves, K_d was found to be 2.2×10^3 M^{-1} for the Bchla system, 3.9×10^5 M^{-1} for Bphea in FW and 7.5×10^4 M^{-1} for Bphea in water. The corresponding free energy change (ΔG) for each system is: -4.5 Kcal/mole for Bchla, -7.6 Kcal/mole for Bphea in FW and -6.6 Kcal/mole for Bphea in water. The micelles which incorporated these pigments were found to be of different sizes in the different systems; 4100 TX-100 molecules per micelle occupied by Bchla, 40000 molecules per micelle occupied by Bphea in FW and 4000 molecules per micelle occupied by Bphea in water.

The number of TX-100 molecules per micelle in water has been studied extensively by several groups and found to range from 46 to 150 molecules (41, 42). Yet, the micelle size never approached the values reported above. In addition electron microscopy measurements seem to confirm the existence of large micelles containing pigments. Therefore, we can conclude that the presence of Bchl or Bphe changes the organization of the amphiphilic molecules. The simulation of the experimetnal results suggest that the change in the micelle size occurs when micelles start to be occupied by molecules of Bchla or Bphea. Apparently, the small spherical micelles of TX-100 cannot accomodate the pigments but since the pigment's affinity for the micelle is relatively high, their presence induces cylindrical micelle formation. The cylindrical micelle can grow, eventually, to very large bodies or lyposomes. Further discussion is given elsewhere (40). The micelle size should also depend on the concentration of TX-100 which is different in each case (as reported in figure 7). The significance of the independent parameter, C, is not yet known.

4.   Hypothesis for the Cooperative Assembly of the LHCs and RCs

In previous studies we have shown that the spectral properties of self-assembled dimers and higher aggregates of Bchla and its derivatives in solutions of FW or FW-TX-100 resemble those of Bchla in LHCs and RCs (5, 12, 16, 30, 32). The free energy change that drives the Bchl assembly is relatively high (~-4 to -5 Kcal/mole) and this value remained constant in both the FW and FW-TX-100 systems. In the present study we have shown that the Bchls self assembly influences the organization of the lipid environment (TX-100) in which they are placed.

In applying these observations to the formation of LHCs and RCs it is necessary to consider the characteristics of the *in vivo* complexes. For example, it has been known for some time that in several non-oxygenic bacteria the RCs' polypeptide (PP) assembly within the intercytoplasmic membrane can only occur when phytolated Bchls are present. Similar observations have been reported for the LHCs of photosynthetic bacteria and for the a/b LHCs of higher organisms (for recent review see ref. 43). Two other relevant characteristics of the bacterial RCs and LHCs are: (a) the non-covalent attachment of the Bchls to the protein network through hydrogen bonding and ligation to the central Mg and (b) the relatively high Bchl concentration in the pigment-protein complex (i.e. in *Rs.viridis* the volume of a RC is $\sim 3 \times 10^5$ $\mathring{A}^3$ (7) and, therefore, the Bchl concentration is ~0.02 M).

Combining the *in vivo* and *in vitro* observations, we propose the following model for Chl (Bchl) - protein assembly *in vivo*. Bchls and Chls are attached to single PPs via their central Mg or peripherial carbonyls. Interactions among the PPs lead to the formation of loosely attached dimers which are incorporated into the intercytoplasmic membrane (similar to the Bchl-FW's insertion into the TX-100

---

[3] To calculate V_1 we assumed that $R = 35\mathring{A}$ and the surface area of a TX-100 molecule is $\approx 4\mathring{A}^2$ (41).

micelles). In these Bchl-PP dimers the concentration of Bchl is about 0.01-0.02 M. Assuming that $\Delta G \cong -5$ Kcal/mole, there is a 90% probability of having the Bchls as dimers. Furthermore, the self assembly of two Bchl drives the formation or, at least, the stabilization of a PP-Bchl dimer. This dimer may require further stabilization via a docking protein (i.e. the H subunit in bacterial RCs) (43). The PP-Bchl pair may then act as a seed for further oligomerization of PP-Bchl units.

Our hypothesis seems to provide a clue to the *in vivo* pigment-protein assembly's dependence on the phytolization of bacteriopheophorbide (43); since, only single PPs that are attached to Bchl with an esterfied alcohol can be incorporated into the lipid membrane.

As we have shown, the Bchl dimerization provides the major mechanism for the bathochromic shift of the Q_y transition which may be considered as the coarse tuning of the apparatus to the prevailing light conditions. Additional shift (fine tuning) is determined by the final configuration of the paired molecules within the protein network. This configuration should depend on the interactions between the carbonyl group of the Bchl and the hydrogen bonding amino acid residues (16).

ACKNOWLEDGEMENT We are grateful to Prof. I. Rubenstein from the Department of Applied Mathematics in the Weizmann Institute of Science for helpful discussions on the cooperative oligomerization formulae.

REFERENCES

1. Okamura, M. Y., Feher, G. and Nelson, N. (1982) 'Reaction Center', in Govindjee (ed.), Photosynthesis, Energy Conversion by Plants and Bacteria, Academic Press, New York, pp. 221-227.

2. Zuber, H. (1985) 'Structure and Function of Light-Harvesting-Complexes and their Polypeptides', Photochem. Photobiol. 42, 821-844.

3. Cogdell, R. J. and Thornber, J. P. (1980) 'Light-Harvesting Pigment-Protein Complexes of Purple Photosynthetic Bacteria', FEBS Lett. 122, 1-8.

4. Sauer, K. and Austin, L. A. (1978) 'Bacteriochlorophyll-Protein Complexes from the Light-Harvesting Antenna of Photosynthetic Bacteria', Biochem. 17, 2011-2019.

5. Rosenbach-Belkin, V., Braun, P., Kovatch, P. and Scherz, A. (1988) 'Optical Absorption and Circular Dichroism of Bacteriochlorophyll Oligomers in Triton X-100 and in the Light-Harvesting Complex B850; A Comparative Study', in H. Scheer and S. Schneider (eds.), Photosynthesis Light-Harvesting Systems Organization and Function, Walter de Gruyter, Berlin, New York, pp. 323-337.

6. Wittmershaus, B. P. (1987) 'Measurements and Kinetic Modeling of Picosecond Time-Resolved Fluorescence from Photosystem I and Chloroplasts', in J. Biggins (ed.), Progress in Photosynthesis Research; Proceeding of the VIIth International Congress on Photosynthesis, Martinus Nijhoff Publishers, Dodrecht, Vol. I, pp.75-82.

7. Deisenhofer, J., Epp, O., Miki, K., Huber, R. and Michel, H. (1985) 'Structure of the Protein Subunits in the Photosynthetic Reaction Centre of *Rhodopseudomonas viridis* at 3Å Resolution', Nature 318, 618-624.

8. Michel, H., Weyer, K. A., Gruenberg, H., Dunger, I., Oesterhelt, D. and Lohspeichf (1986) 'The "Light" and "Medium" Subunits of the Photosynthetic Reaction Centre from *Rhodopseudomonas viridis*: Isolation of the Genes, Nucleotide and Amino Acid Sequence', EMBO J. **5**, 1149-1158.

9. Chang, H., Tiede, D., Tang, J., Smith, U., Norris, J. R. and Schiffer, M. (1986) 'Structure of *Rhodopseudomonas sphaeroides* R-26 Reaction Center', FEBS Lett. **205**, 82.

10. Allen, J. P., Feher, G., Yeates, T.O., Komiya, H. and Rees, D. C. (1987) 'Structure of the Reaction Center form *Rhodobacter sphaeroides* R-26: The Cofactors', Proc. Natl. Acad. Sci. USA, **84**, 5730-5734.

11. Allen, J. P., Feher, G., Yeates, T.O., Komiya, H. and Rees, D. C. (1987) 'Structure of the Reaction Center from *Rhodobacter sphaeroides* R-26: The Protein Subunits', Proc. Natl. Acad. Sci. USA, **84**, 6162-6166.

12. Scherz, A and Rosenbach-Belkin, V. (1989) 'Comparative Study of Optical Absorption and Circular Dichroism of Bacteriochlorophyll Oligomers in Triton X-100, the Antenna Pigment B850, and the Primary Donor P-860 of Photosynthetic Bacteria Indicates that all are similar Dimers of Bacteriochlorophyll a', Proc. Natl. Acad. Sci. USA, **86**, 1505-1509.

13. Parson, W. W., Warshel, A. and Scherz, A. (1985) 'Calculation of the Spectroscopic Properties of Bacterial Reaction Centers', in M. E. Michele-Beyerle (ed.), Antennas and Reaction Centers of Photosynthetic Bacteria, Structure, Interaction and Dynamics, Springer Series in Chemical Physic, Springer-Verlag, Berlin, Vol. 42, pp. 122-130.

14. Knapp, E. W., Scherer, P. O. J. and Fischer, S. F. (1986) 'Model Studies of Low-Temperature Optical Transitions of Photosynthetic Reaction Centers A-, LD-, CD-, ADMR- and LD-ADMR-spectra for *Rhodopseudomonas viridis*', Biochem. Biophys. Acta **852**, 295-305.

15. Parson, W. W. and Warshel, A. (1987) 'Spectoscopic Properties of Photosynthetic Reaction Centers. 2. Application of the Theory to *Rhodopseudomonas viridis*', J. Am. Chem Soc. **109**, 6152-6163.

16. Scherz, A. and Rosenbach-Belkin, V. (1989) 'The effect of Non-Excitonic Interactions Among the Paired Bacteriochlorophylls on the Q_y transition of the Primary Donors in Bacterial Reaction Centers; Model Studies with Epimers', J. Am. Chem. Soc., in press.

17. Eccles, J. and Honig, B. (1983) 'Charged Amino Acids as Spectroscopic Determinants for Chlorophyll *in vivo*', Proc. Natl. Acad. Sci. USA, **80**, 4959-4962.

18. Lavorel, J. (1957) 'Infuence of Concentration on the Absorption Spectrum and the Action Spectrum of Fluorescence of Dye Solutions', J. Phys. Chem., **61**, 1600-1605.

19. Weber, G. and Teale, F. W. J. (1958) 'Fluorescence Excitation Spectrum of Organic Compounds in Solution, Part 1.- Systems with Quantum Yield Independent of the Exciting Wavelength', Trans. Faraday Soc., **54**, 640-648.

20. Ballschmiter, K., Truesdell, K. and Katz, J. J. (1969) 'Aggregation of Chlorophyll in Nonpolar Solvents from Molecular Weight Measurements', Biochimica et Biophysica Acta, 184, 604-613.

21. Katz, J. J. and Hindman, J. C. (1982) 'Photoprocesses in Chlorophyll Model Systems', in R. R. Alfano (ed.), Biological Events Probed by Ultrafast Laser Spectroscopy, Academic Press, New York, pp. 119-157.

22. Ballschmiter, K. and Katz, J. J. (1972) 'Chlorophyll-Chlorophyll and Chlorophyll-Water Interactions in the Solid State', Biochimica et Biophysica Acta, 256, 307-327.

23. Katz, J. J., Shipman, L. L., Cotton, T. M. and Janson, T. R. (1978) 'Chlorophyll Aggregation: Coordination Interactions in Chlorophyll Monomers, Dimers and Oligomers', in D. Dolphin (ed.), The Porphyrins, McGraw-Hill, New York, Vol. V, pp. 401-456.

24. Cotton, T. M., Loach, P. A., Katz, J. J. and Ballschmiter, K. (1978) 'Studies of Chlorophyll-Chlorophyll and Chloropyll-Ligand Interactions by Visible Absorption and Infrared Spectroscopy at Low Temperatures', Photochem. Photobiol., 27, 735-749.

25. Katz, J. J. and Ballschmiter, K. (1968) 'Wechselwirkungen Zwischen Chlorophyll und Wasser', Angew. Chem., 80, 283-284.

26. Katz, J. J., Oettmeier, W. and Norris, J. R. (1976) 'Organization of Antenna and Photo-Reaction Centre Chlorophylls on the Molecular Level', Phil. Trans. R. Soc. Lond. B. 273, 227-253.

27. Lutz, M., Robert, B., Zhow, Q., Newmann, J. M., Szponarski, W. and Berger, G. (1988) 'Protein-Prosthetic Group Interactions in Bacterial Reaction Centers', in J. Breton and A. Vermeglio (eds.), The Photosynthetic Bacterial Reaction Centers, Structure and Dynamics, NATO ASI Series A: Life Sciences, Plenum Press, New York, Vol. 149, pp. 41-50.

28. Brown, S. B. and Shillcock, M. (1976) 'Equilibrium and Kinetic Studies of the Aggregation of Porphyrins in Aqueous Solution', Biochem. J. 153, 279-285.

29. Abraham, R. J., Goff, D. A. and Smith, K. M. (1988) 'N.M.R. Spectra of Porphyrins. Part 35. An Examination of the Proposed Models of the Chlorophyll Dimer', J. Chem. Soc. Perkins Trans. I, 2443-2451.

30. Scherz, A. and Parson, W. W. (1984) 'Oligomers of Bacteriochlorophyll and Bacteriopheophytin with Spectroscopic Properties Resembling those found in Photosynthetic Bacteria', Biochimica et Biophysica Acta, 766, 653-665.

31. Scherz, A., Rosenbach, V. and Malkin, S. (1985) 'Small Oligomers of Bacteriochlorophyll as *in vitro* Models for the Primary Electron Donors and Light-Harvesting Pigments in Purple Photosynthetic Bacteria', in M. E. Michel-Beyerle (ed.), Antennas and Reaction Centers of Photosynthetic Bacteria, Springer Series in Chemical Physics, Springer-Verlag, Berlin, Vol. 42, pp. 314-323.

32. Scherz, A. and Rosenbach-Belkin, V. (1988) 'The Spectral Properties of Chlorophyll and Bacteriochlorophyll Dimers; A Comparative Study', in J. Breton and A. Vermeglio (eds.), The Photosynthetic Bacterial Reaction Centers, Structure and Dynamics, NATO ASI Series A: Life Sciences, Plenum Press, New York, Vol. 149, pp. 295-308.

33. Hinton, J. F. and Harpool, R. D. (1977) 'An ab Initio Investigation of (Formamide)$_n$ and Formamide-(H$_2$O)$_n$ Systems. Tentative Models for the Liquid State and Dilute Aqueous Solution', J. Am. Chem. Soc. 99, 349-353.

34. Renge, I. and Avarmaa, R. (1985) 'Specific Solvation of Chlorophyll a: Solvent Nucleophility, Hydrogen Bonding and Steric Effects on Absorption Spectra', Photochem. and Photobiol., 42, 253-260.

35. Rosenbach-Belkin, V. (1988) 'The Primary Reactants in Bacterial Photosynthesis Modeling by in vitro Preparation', Ph. D. Thesis.

36. Oosawa, F. and Kasai, M. (1962) 'A Theory of Linear and Helical Aggregations of Macromolecules', J. Mol. Biol. 4, 10-21.

37. Goldstein, R. F. and Stryer, L. (1986) 'Cooperative Polymerization Reactions - Analytical Approximations, Numerical Examples, and Experimental Strategy', Biophys. J. 50, 583-599.

38. Fisher, J. R. E., Rosenbach-Belkin, V. and Scherz, A. (1989) 'Cooperative Polymerization of Photosynthetic Pigments in Aqueous Solutions: Theory, Application and Relevance to the Photosynthetic Reaction Centers and Light-Harvesting Complexes', submitted.

39. Lalonde, D. E., Petke, J. D. and Maggiora, G. M. (1989) 'Evaluation of Approximations in Molecular Exciton Theory. 2. Application to Oligomeric Systems of Interest in Photosynthesis', J. Phys. Chem. 93, 608-614.

40. Scherz, A., Rosenbach-Belkin, V. and Fisher, J. R. E. (1989) 'Distribution and Self-Assembly of Photosynthetic Pigments in Micelles: Theory and Application to the Pigment-Protein Organization into Light-Harvesting Complexes and Reaction Centers', submitted.

41. Kushner, L. M., Hubbard, W. D. (1954) 'Viscometric and Turbidimetric Measurements on Dilute Aqueous Solutions of a Non-Ionic Detergent', J. Phys. Chem. 58, 1163-1167.

42. Paradies, H. H. (1980) 'Shape and Size of a Non-Ionic Surfactant Micelle. Triton X-100 in Aqueous Solution', J. Phys. Chem. 84, 599-607.

43. Kiley, P. J. and Kaplan, S. (1988) 'Molecular Genetics of Photosynthetic Membrane Biosynthesis in Rhodobacter sphaeroides', Microbiological Rev. 52, 50-69.

Protein dynamics in the photosynthetic reaction center:
the electron transfer from cytochrome c to the special pair.

E. W. Knapp
Physik Department
Technische Universität München
D-8046 Garching, FRG

L. Nilsson
Department of Medical Biophysics
Karolinska Institutet
S-10401 Stockholm, Sweden

Abstract: A protein dynamics simulation of the photosynthetic reaction center of Rps. viridis is performed by using the CHARMM program. The relevant part of the protein contains the special pair and the proximate cytochrome c. In the center is a tyrosine residue (L162) which is believed to support the electron transfer from the cytochrome c to the special pair by mixing the initial and final state wave functions. Preliminary calculations have shown that the 31 water molecules which are in the relevant part of the x-ray structure are much to mobile. Hence 8 extra water molecules have been added to improve the set-up for the dynamics simulation. The dynamics of the water molecules play an important role for the electron transfer due to their large mobility and the formation of hydrogen bridges with the OH-group of the tyrosine residue L162. The anisotropy, asymmetry and anharmonicity of the potential envelope governing the motion of a water molecule can be analysed by evaluating the second order moments $\langle (x_k - \langle x_k \rangle) * (x_l - \langle x_l \rangle) \rangle$ of the cartesian components x_k, $k,l = 1,2,3$ of the oxygen atom trajectory from the water molecule. The fluctuations of the nonbonded interactions of the tyrosine residue with its protein environment are as large as 4 kcal/mol, a value which is of the same order of magnitude as the difference of initial and final state energies of an electron transfer system. The statistical properties of these energy fluctuations must enter a model description of the electron transfer process.

Abbreviations: ET: electron transfer; RC: reaction center; Rps.: Rhodopseudomonas; SP: special pair; BC: bacterochlorophyll b; HE3: proximate cytochrome c; MAS: Mössbauer absoption spectroscopy; rms: root mean square displacement; NS: neutron scattering; RP: redox potential; CYT(h), CYT(l): high and low potential cytochrome; Ch: Chromatium; estat.: electrostatic; VdW: Van der Waals; TYR: tyrosine residue.

J. Jortner and B. Pullman (eds.), Perspectives in Photosynthesis, 389–412.
© 1990 Kluwer Academic Publishers.

1. Introduction

There is now general agreement that protein structure and dynamics are related to the function. This has been clearly demonstrated for reactions of enzymes with substrates [1]. Strong intermolecular interactions of protein subunits in the enzyme-substrate complex give rise to specific conformational changes. Another issue is the role of protein dynamics for reactions which involve mainly a change of the electronic state. This is what we like to study in the present article. Typical reactions of this type are electron transfer (**ET**) processes in proteins. They are important steps of energy conversion in biological systems [2-4]. In a protein of known structure the geometrical arrangement of the donor acceptor system for electron transfer is well defined [5]. One of the most challenging systems for such a study is the photosynthetic reaction center (**RC**) of Rhodopseudomonas (**Rps.**) viridis [6-9]. To convert light energy into chemical energy the RC performs several ET processes. The primary charge separation is very fast. In 2.8ps an electron is transferred from the special pair (**SP**) consisting of two bacteriochlorophyll b molecules (**BC**) to the bacteriopheophytine of the L branch [10]. On the other hand the electron transfer from the quinone A back to the SP is a rather slow process with a rate of about $10^2 s^{-1}$ at 80K [11].

Proteins exhibit underdamped nearly harmonic intramolecular vibrations which are also typical for local modes and acustic phonons in crystals of small molecular units. This can for instance be deduced from Mössbauer absorption spectra (**MAS**) at low temperatures (T < 200K) which show a linear temperature dependence of the mean square displacement (rms: $\sqrt{\langle x^2 \rangle}$) [12]. At higher temperatures the $\sqrt{\langle x^2 \rangle}$ values increase dramatically with temperature (figure 1). At the same time quasielastic lines appear in the MAS. These lines are due to overdamped vibrational modes where the atoms are diffusing in a potential as opposed to an inertial type of motion for underdamped vibrations. A 75% glycerol 25% water mixture with a glass transition temperature of 175K exhibits the same features in the MAS as iron containing proteins like myoglobin [13]. Hence the occurence of overdamped vibrational modes seems to be typical for the dynamics of complex systems. The width of the quasielastic lines of MAS corresponds to a time regime longer than nanoseconds. Neutron scattering (**NS**) experiments on myoglobin [14] and molecular glasses [15] provide quasielastic lines whose width belongs to the picosecond time regime. Thus the specific dynamics of complex systems like proteins extends over a large time regime. Therefore, it can in principle influence the very fast primary charge separation as well as the subsequent slower, electron transfer processes.

One way to decide whether a particular ET process is influenced by protein dynamics is to look at its temperature dependence. A similar temperature dependence of the ET rate and the mean square displacement obtained by MAS and NS experiments can be considered as a good hint for the involvement of protein dynamics. A correlation of the temperature dependence from MAS experiments and protein function in photosynthetic RC's has been demonstrated by Parak et al. [16].

The charge separation process from the SP to the bacteriopheophytine shows only a weak temperature dependence which is inverted [17] i. e. the value of the transfer rate increases with decreasing temperature. Therefore, the primary charge separation process will not be considered here even though it is one of the most chall3nging problems to clarify the role of the accessory BC in this process [17-21].

Another electron transfer process which has not been mentioned so far is the ET from the cytochrome c to the special pair. It neutralizes the positive charge which is left at the SP and would block the next full cycle of charge separation. At room temperature the transfer rate of this process is in the microsecond time regime. At low temperatures the value of the transfer rate depends strongly on the medium redox potential (RP). For the Rps. viridis RC at moderate RP the transfer rate is 4 orders of magnitude smaller at 125K than at room temperature [22]. On the other hand in a low medium RP the transfer rate decreases by only 2 orders of magnitude when the temperature is lowered to 77K [23]. The sensitivity of the ET process on the medium RP is related to two types of cytochromes CYT(l) and CYT(h) which have low and high redox potential respectively [24]. Unfortunately for Rps. viridis a detailed temperature dependence of this ET process is not known.

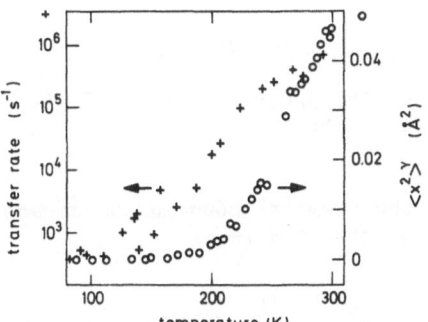

Fig. 1: Comparison of the temperature dependence of the mean square displacements from Mössbauer spectroscopy [25] (open circles) with the ET rate from CYT(h) to SP in photosynthetic RC's of Chromatium vinosum [26] (crosses).

RC's of Chromatium (Ch) vinosum have the same pigment composition and have also a cytochrome c firmly attached. For these reaction centers the full temperature dependence of the transfer rate is available [26]. The rate is constant at temperatures below 130K and rises by more than 3 orders of magnitude to its room temperature value (figure 1). Several theoretical investigations [27,28] have considered this famous experiment.

In a recent paper [29] it was proposed that the temperature dependence of the transfer rate is due to protein specific dynamics. It occurs above a characteristic temperature and populates protein conformations which favor the ET from the cytochrome to the SP. The measurements on Ch vinosum have been made with whole cells where the low RP cytochromes are oxidized [30,31] and the electron can only be transferred from the proximate CYT(h). In Rps. viridis RC's a tyrosine residue (L162) is just half way between the special pair and the proximate CYT(h). It has been speculated [29] that the tyrosine residue can enhance the electronic coupling by mediating the mixing between donor and acceptor states. There is also recent experimental evidence of enhanced ET rates in the presence of aromatic groups between the donor and acceptor system [32]. This has motivated us to do a protein dynamics simulation of Rps. viridis in the neighbourhood of the tyrosine L162.

The scope of the article is as follows: In the next section 2 the method of dynamics simulation of proteins is described in short terms. Furthermore special precautions taken for the simulation of the RC are given. In section 3 and 4 the fluctuations of atomic positions of the protein and the water molecules are described. Section 5 and the appendix yield a moment analysis of the atomic trajectories of the water molecules. The fluctuations of nonbonded interactions of the tyrosine resudue L162 are given in section 6. In the final section 7 the implications of the results for the ET process and future projects are discussed.

2. Molecular dynamics simulation of the reaction center

The method of computer simulation is based on the solution of Newton's equations of motion for all N atoms (i) of the protein

$$m_i \frac{d^2}{dt^2} r_i = -grad_i E(r_1, r_2, \ldots r_N) , \qquad i=1,2,\ldots N. \tag{1}$$

The calculations were performed with the program CHARMM [33,34]. The energy function is split into 3 different contributions

$$E = E_{bond} + E_{nbond} + E_c \tag{2}$$

The internal energy E_{bond} accounts for the chemical bonding and geometry of the different molecular units. It consists of 4 terms

$$E_{bond} = \sum_{[b]} k_b (r_b - r_{0b})^2 + \sum_{[\theta]} k_\theta (\theta - \theta_0)^2 + \sum_{[\phi]} k_{\bar\phi} [1 - \cos(n\bar\phi)] + \sum_{[\omega]} k_\omega (\omega - \omega_0)^2 \tag{3}$$

referring to bond stretching, bond angles, dihedral angles and improper torsions, respectively. The interactions of nonbonded atoms E_{nbond} accounts for electrostatic and Van der Waals interactions

$$E_{nbond} = \sum_{i,j}' \frac{q_i q_j}{\varepsilon r_{ij}} + \sum_{i,j}' \left(\frac{A_{ij}}{r_{ij}^{12}} - \frac{B_{ij}}{r_{ij}^6} \right) \tag{4}$$

where q_i is a partial charge at atom i and ε an effective dielectric constant. ε has been set to 1.0 in the present application. The positive charge at the SP has been distributed evenly among the 2 pigments based on an INDO calculation of the BC monomer [20].

Hydrogen bonds and the interaction of nonbonded atoms are both of electrostatic origin. The parameters of the nonbonded energy terms (4) are
defined such that the effect of hydrogen bonding is included. An explicit evaluation of hydrogen bond interactions would require more computer
time. The third contribution to the energy function (2) is used to
stabilize certain geometries. It can for instance fix an atom coordinate
r_l to its x-ray structure position r_{lx} by using harmonic constraints
$E_c = k_c (r_l - r_{lx})^2$.

The propagation of the equations of motion (1) in time is made with the
Verlet algorithm [35]

$$r_i(t + \Delta t) = 2 r_i(t) - r_i(t - \Delta t) - \frac{(\Delta t)^2}{m_i} \, grad_i \, E(r) \qquad (5)$$

which requires the knowledge of 2 subsequent sets of atomic coordinates.
The propagation is started by using one coordinate set from the x-ray
structure analysis. The second coordinate set is calculated from initial
velocities which are taken at random from a Maxwell distribution of velocities for a given temperature.

Before one can use the coordinates of the x-ray structure hydrogen atoms
have to be added. This is done for the polar hydrogens only. To save
computer time the nonpolar atoms are taken into account by using suitably defined extended atoms as for instance the methyl group. The high
frequency motions can be frozen by constraints. This method is applied
for the bond stretching of the H atoms by using the SHAKE algorithm [35].
It allows to save computer time by increasing the time increment Δt. In
the present application the time increment is 2 fs.

The added hydrogens and minor inconsistencies between the x-ray structure and the energy function (1) lead to stress in the initial structure. The stress is relaxed by minimizing the potential energy. The potential energy of the resulting protein structure does not yet correspond to a particular temperature. This is accomplished by a dynamics
simulation over several ps (in our case 40 ps) where the velocities are
adjusted several times. The subsequent portion of computer simulation
can then be taken for the analysis.

The Rps. viridis RC has almost 10000 nonhydrogen atoms. This system is
too large to do a computer simulation of the whole protein. We are mainly interested in the relevant part of the RC which participates in the
ET process from the proximate CYT(h) to the SP. Therefore, we need only
the protein environment around the tyrosine L162 which is situated between the CYT(h) and the SP. For that purpose only the residues which
have at least one atom in the sphere of radius 16Å around the C_y atom of
residue L162 are considered in the dynamics simulation. The CYT(h) and
the bacteriochlorophylls of the SP except their phytol chains are essentially contained in this sphere (see figure 2). The phytol chains are
not contained in the computer simulation. The atoms in the sphere of radius 14Å around the C_y atom of L162 can move freely according to the
energy function (2). The atoms in the buffer zone outside of the sphere

of 14 Å radius are constrained to the positions obtained after energy minimization of the x-ray structure. The constraining harmonic potentials correspond to the average value of the rms from the x-ray structure of Rps. viridis (force constant k = 0.267 kcal/(mol Å²) in the present application). There are no immobile atoms in the buffer zone since they would reduce the rms values too much [36]. None of the water molecules is constrained to a fixed position. Instead the water oxygen atoms are subject to a repulsive force keeping them inside the 16Å sphere.

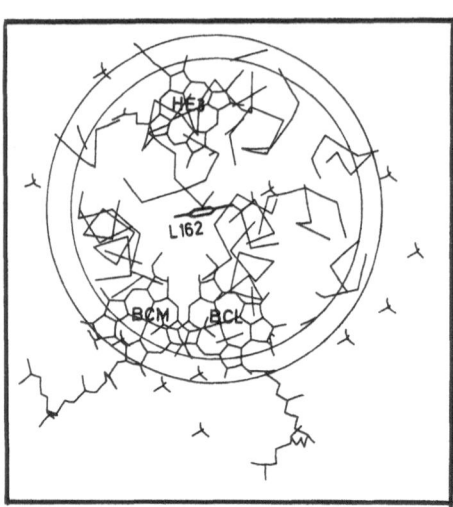

Fig. 2: Relevant part of the x-ray structure of the Rps. viridis reaction center [9] used for the computer simulation of protein dynamics. The C_α atoms f all 134 residues which are within the sphere of radius 16Å around the C_γ atom of the tyrosine residue L162 are depicted. In addition the proximate cytochrome (HE3), the 2 bacteriochlorophylls of the special pair (BCL and BCM) and the 31 water molecules from the x-ray structure (4-leg symbol) are displayed. The BC of the functioning L-branch is at the right side.

The atoms (i) in the buffer zone are also subject to a stochastic force $f_i(t)$ and a friction with rate constant β_i ($\beta_i = 50ps^{-1}$ in the present application). The dynamics of these atoms is given by a set of Langevin equations [36]

$$\ddot{r}_i + \beta_i \dot{r}_i = -\frac{1}{m_i} grad_i E + f_i(t). \qquad (5)$$

The stochastic forces f_i are governed by a Gaussian-Markov process as follows

$$\langle f_i(t) \rangle = 0 \quad \text{and} \quad \langle f_i(t) f_i(t') \rangle = \langle f_i^2(0) \rangle \delta(t - t'). \qquad (6)$$

The strength of the stochastic force ist determined by the friction constant via the fluctuation dissipation theorem [37].

Coordinates of 201 water molecules are given by the x-ray structure of Rps. viridis [9]. 31 water molecules are in the relevant part of the reaction center whose dynamics is simulated. Preliminary computer simulations have indicated that some of the water molecules are so mobile that their motion is similar to a diffusion process in a porous material [38]. This is in conflict with the x-ray structure where mobile water molecules can not show up.

The dynamics of the water molecules with their strong dipole moment may play a significant role for the electron transfer process considered here. Hence we have tried to improve this situation by adding extra water molecules to the x-ray structure. This is done by placing the relevant part of the reaction center in a box of 18.46 Å length containing 216 bulk water molecules. The water model used is TIP3P [39]. The configuration of the waters has been obtained from a Monte Carlo simulation with periodic boundary conditions. All water molecules where the oxygen atoms are closer than 2.8 Å to a nonhydrogen atom of the x-ray structure or outside of the relevant part of the RC are eliminted. Even after addition of the hydrogen atoms to the x-ray structure the 8 remaining water molecules have no significant overlap with atoms of the x-ray structure and are therefore added. Like for the other water molecules the dynamics of the extra water is not constrained.

3. Fluctuations of atomic positions in the protein molecule.

With 2 exceptions, the oxygen atoms of ALA C250 and of GLY M192, the time averaged positions of all atoms of the protein backbone in the sphere of 14Å radius differ by less than 1.5Å from the x-ray structure. The mean difference in atomic positions (except water) between the averaged molecular dynamics structure and the x-ray structure is 0.95Å, similar as for other simulations of the RC [40]. Thermal fluctuations of an individual atom i can be characterized by the root mean square displacement from the time averaged position $\langle r_i \rangle$

$$(rms)_i = (\langle x_i^2 \rangle)^{1/2} = [\langle [r_i - \langle r_i \rangle]^2 \rangle /3]^{1/2} . \tag{7}$$

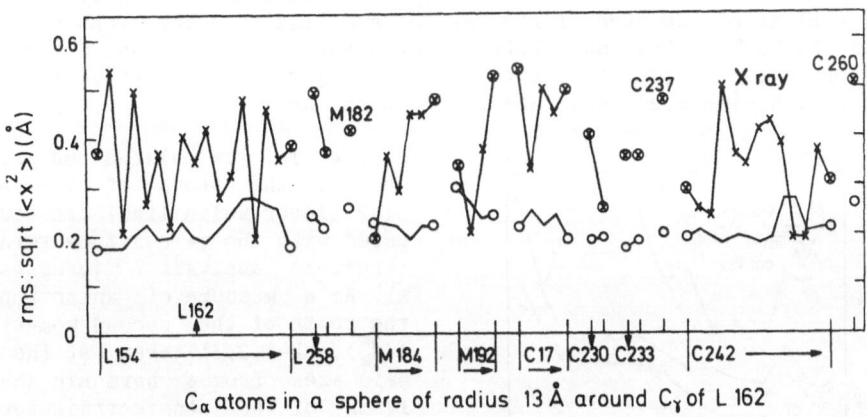

Fig. 3: The calculated (solid line) rms values of all C_α atoms in a sphere of 13Å radius around the C_γ atom of L162 are compared with data from the x-ray structure analysis (solid line with crosses: x) [9].

This quantity can be compared with the experimental B_i factors [9]

$$B_i = 8 \pi^2 \langle x_i^2 \rangle . \tag{8}$$

In figure 3 the rms values of the 54 protein backbone C_α atoms in a sphere of radius 13 Å around the C_γ of L162 are compared with x-ray data. The rms values from the computer simulation are consistently about a factor of two smaller than the experimental values. Such deviations have been found also in computer simulations of lysozyme [41] or of BPTI with crystal water [42]. Note that the rms values from x-ray structure analysis involve not only a time average over several hours but also an ensemble average over roughly 10^{18} molecules of the protein crystal [43]. The latter average is absent in a computer simulation where only one trajectory is generated.

If the dynamics of the relevant part of the protein is ergodic all discrepancies in the rms values may vanish at longer times. The calculated rms values do not vary significantly if only half of the 100 ps time interval is used. Hence the simulation time required for a more complete evaluation of the rms values must be much larger than 100 ps.

The low values of the rms are probably due to a densely packed protein structure in the relevant part of the RC and the constraining forces in the buffer zone which fix the atoms at the x-ray structure positions. Thus large motions of the protein are prevented that may derange the calculated rms values. The rms values of the C_α atoms do not change significantly if the 8 extra water molecules are not included [38].

In figure 4 the rms values of all 14 atoms of the tyrosine residue L162 are displayed (data from simulation: lowest solid line). Also here the calculated rms values are generally a factor of 2 smaller than the values from the x-ray structure analysis. In agreement with the x-ray data the atoms of the phenol ring are more mobile than the atoms of the protein backbone. The qualitative agreement is quite good. The only exception is the oxygen atom at the phenol ring which is relatively more mobile in the simulated data than in the x-ray data.

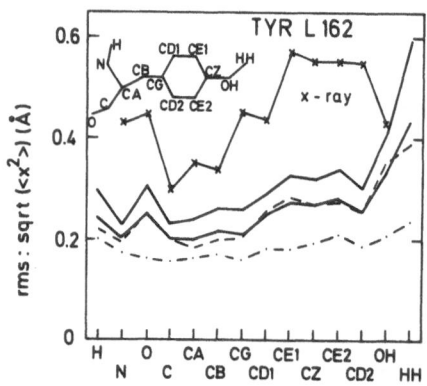

Fig. 4: The calculated rms values of the atoms of tyrosine L162 (lower solid line) are compared with the values from x-ray structure analysis [9] (crosses: x). As a measure of anisotropy the roots of the second moments ($\langle \hat{x}_k^2 \rangle$, $k=1,2,3$) taken at the 3 main axes from a harmonic analysis of the atomic trajectory are displayed (upper solid line, dashed and dashed-dotted line). Details are given in the text and appendix.

4. Fluctuations of water molecules in the protein.

The rms values of the 31 water molecules given by the x-ray structure and of the 8 extra water molecules are depicted in figure 5. In contrast to the rms values of atoms which belong to the protein molecule the rms values of the water molecules have on the average the same magnitude as the one from the x-ray data. Several water molecules have changed their position by more than 2 Å after the 140 ps of simulation. A direct comparison with x-ray data is therefore not justified. The rms values of all other water molecules are marked by open circles. Some water molecules (see Fig. 6 and caption to Fig. 5) are replaced by other water molecules during the initial 40 ps relaxation. In this case the relevant rms values for the comparison with the x-ray data have to be taken from the new water molecules. The corresponding rms values are also marked by open circles. By considering only the calculated rms values given by open circles major discrepancies with the x-ray data are eliminated. The agreement of rms values from x-ray data and computer simulations has been worse [38] without the 8 extra water molecules. Adding more water molecules may improve the situation furthermore and prevent water molecules like W014, W019, W024, W032, W034, W045, W048, W065, W084, W104, W106, W107, W304 to move away from the initial position in the x-ray structure.

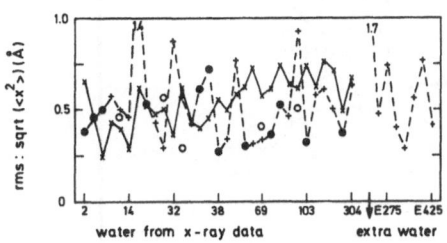

water from x-ray data   extra water

Fig. 5: The rms values from the oxygen atoms of the 38 water molecules are displayed. 31 water molecules taken from the x-ray structure have the residue numbers W002, W003, W011, W012, W013, W014, W019, W024, W026, W029, W032, W033, W034, W035, W037, W038, W045, W048, W051, W065, W069, W072, W082, W084, W088, W103, W104, W105, W106, W107, W302, W304. The 8 extra water molecules are labeled with the numbers from the Monte Carlo simulation as follows: E070, E208, E275, E333, E366, E388, E393, E425. The rms values from the x-ray data [9] (crosses: x) are compared with the rms values obtained from the computer simulation (crosses: +). The rms values of water molecules which after the 140 ps time of simulation have moved away by less than 2 Å are also marked by open circles. Some water molecules have changed their identity during the initial 40 ps of structural relaxation. These are the molecules W069, W014, W029, W033, W088 which are replaced by E333, W013, W033, W029, W107, respectively. The position of the water molecules W029 and W033 are interchanged. The open circles provide the relevant rms values to compare with. The numbers in the figure yield the rms values larger than 1 Å.

An overview of the motion of the water molecules is depicted in figure
6. The x-ray positions of the oxygen atoms of the water molecules are
labeled by 4-leg symbols and the final position of the water molecules
after 140 ps computer simulation is given by triangles.

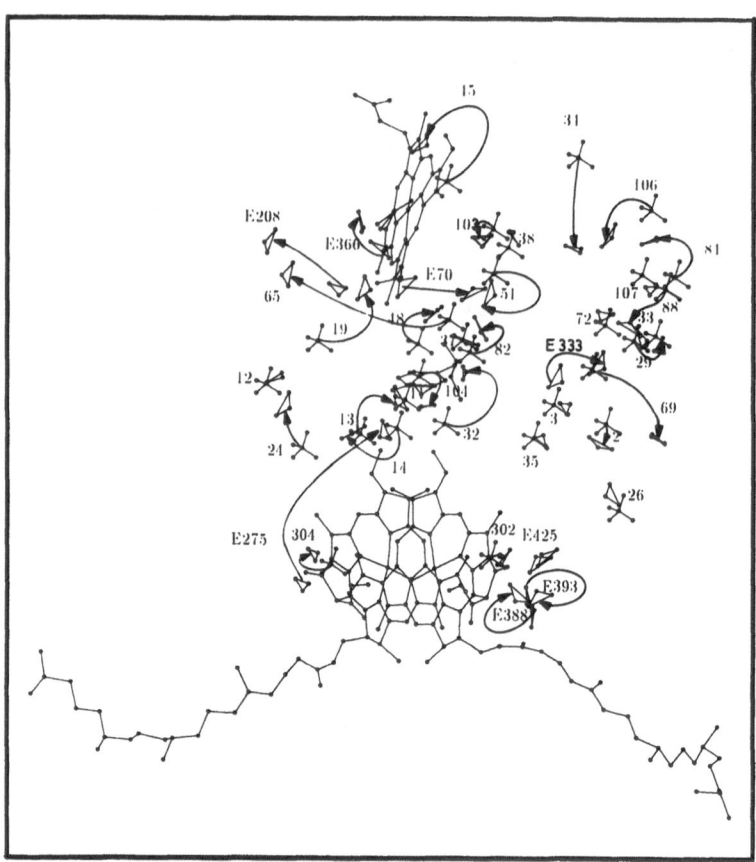

Fig. 6: The oxygen atoms of the water molecules (4-leg symbols), the
tyrosine residue L162 (center of the structure) and the pigments (HE3
and SP) of the x-ray structure of Rps. viridis are displayed. The BC
from the functioning L-branch of the SP is at the right side. The final
position of the water molecules (triangles) after 140 ps computer simulation is also shown. The 8 extra water molecules represented by triangles are also in the initial configuration. The arrows connect initial
and final positions of the water molecules but do not represent the atomic trajectories. Note that the water molecules W069, W014, W088 are replaced by E333, W013, W107 respectively, and that the molecules W029 and
W033 interchange.

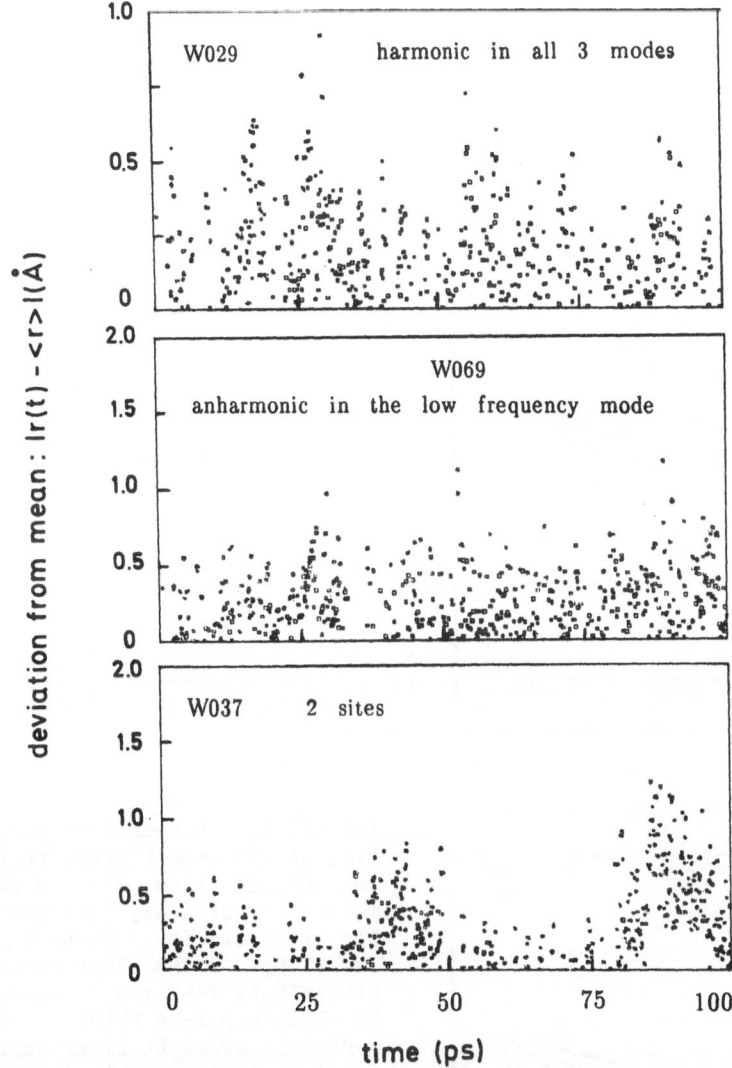

Fig.7: Time series of the deviation from the average position $|r(t)-\langle r\rangle|$ for the last 100 ps of the computer simulation are displayed for the oxygen atoms of 3 selected water molecules. 501 equidistant data points are used corresponding to a time increment of 0.2 ps.

5. Moment analysis of the water trajectories.

The water trajectories can exhibit very different shapes and obey different statistics. It would be helpful to classify the trajectories quantitatively without looking at their shape. In figure 7 the time series of the distance fluctuation of the oxygen atoms from 3 selected

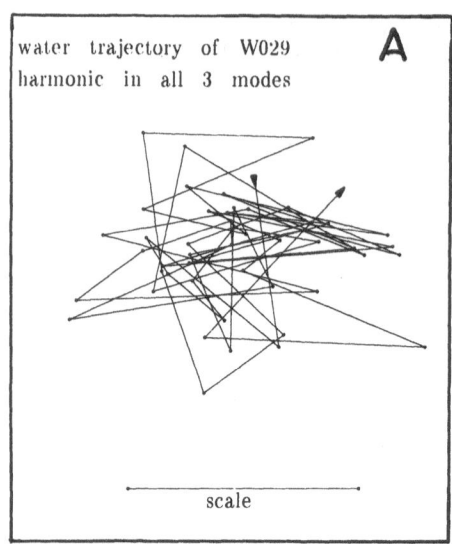

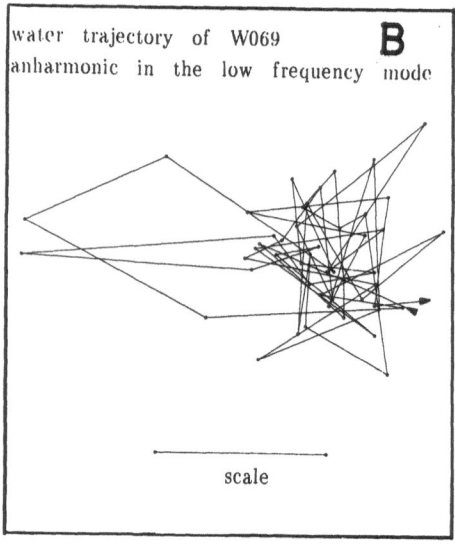

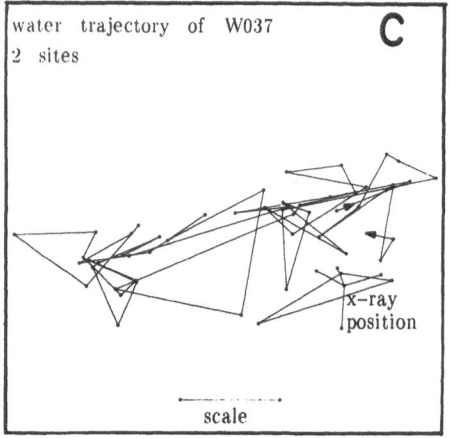

Fig. 8A-C: 2-dimensional projection of the atomic trajectory of the oxygen atom from 3 water molecules referring to figure 7. The trajectories display the last 100ps of the 140ps computer simulation. They are represented by 51 data points which are connected by straight lines corresponding to a time increment of 2.0ps. The bar represents the length of 1Å. The arrows mark the start and end point of the atomic trajectory. The x-ray position of the oxygen atom (4-leg symbol) is displayed if it is close enough at the trajectory. The triangle provides the position of the water molecule after energy minimization of the x-ray structure.

motion. For the 14 atoms of the tyrosine residue L162 the $\langle \tilde{x}^2 \rangle$-values are depicted in figure 4. They show only a mild anisotropy. For the oxygen atoms of the 39 water molecules the $\langle \tilde{x}^2 \rangle$-values are depicted in figure 9. In some cases the anisotropy of the motion of the water molecules can be quite high. The right pannel yields the corresponding frequencies of the harmonic modes according to the relation

$$\omega_k = \left[\frac{k_B T}{m \langle \tilde{x}^2 \rangle} \right]^{1/2}, \quad k = 1,2,3. \tag{10}$$

By using the mass parameter of water (m = 18 a.u.) the frequency values run from less than 10 cm^{-1} to almost 100 cm^{-1}. However, one has to realize that the lowest frequencies refer to water molecules with very extended trajectories which can not well be represented by harmonic motions.

The higher order moments, referring to the 3 main axes, provide further information on the atomic trajectories. The third order moments are a measure for the asymmetry of the potential envelope which provides an effective cage for the atomic trajectory. For the 14 atoms of the tyrosine residue L162 the asymmetry is much smaller (not shown).

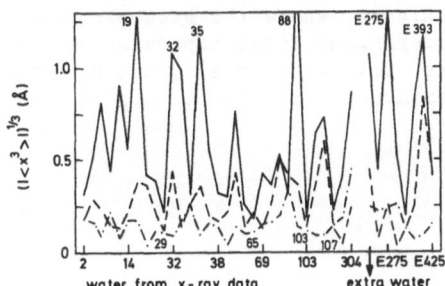

water from x-ray data   extra water

Fig. 10: The third order moments of the 3 main axes are displayed for the trajectories of the oxygen atoms of the 39 water molecules. The numbers label the waters which have very high or very low $|\langle \tilde{x}^3 \rangle|$ -values. Solid, dashed and dashed-dotted line refer to the low, intermediate and high frequency mode as in figure 9.

For harmonic motions and more generally for a Gaussian distribution of atomic positions the ratio of moments

$$Q = \frac{\langle \tilde{x}^4 \rangle}{3 \, (\langle \tilde{x}^2 \rangle)^2} \tag{11}$$

is unity (see appendix). A deviation from unity is a measure of the anharmonicity of the potential envelope of the corresponding atomic trajectory. Values of the ratio of moments (11) smaller or larger than unity refer to different types of anharmonicities of the potential envelope. In figure 11 these deviations are characterized by 1-dimensional potential envelopes which obey a power law

water molecules are depicted. As the analysis will show the motion of
one water molecule is rather harmonic, the second behaves strongly anharmonic and the trajectory of the third water molecule represents a 2 site
system. Only the last situation can be guessed from the time series of
the fluctuations and becomes evident by glancing at the trajectory of
the corresponding water molecule W037 (Fig. 8C). The other two cases are
not so obvious.

A harmonic analysis which employes the second order moments

$$\langle x_k x_l \rangle = \frac{1}{N+1} \sum_{n=0}^{N} [x_k (t_0 + n\Delta t) - \langle x_k \rangle] \ [x_l (t_0 + n\Delta t) - \langle x_l \rangle], \ k,l=1,2,3 \quad (9)$$

of the cartesian components $x_k = (r)_k$ of an atomic trajectory $r(t)$ can
provide such a characterization. Details are given in the appendix. The
essence is that the atomic fluctuations are approximated by a thermalized 3-dimensional harmonic oscillator with arbitrary orientation of
the 3 main axes.

The second order moments are evaluated from the trajectory of an atom
which feels the full potential energy from its neighbours. Hence the resulting $\langle \tilde{x}^2 \rangle$-values of the 3 main axes provide a global approximation of
the potential energy surface with a harmonic potential envelope. The
tilde indicates moments which refer to the main axes. This approximation
is superior to the usual normal mode analysis, where the harmonic potential is obtained from a temperature independent local expansion at the
energy minimum. The present analysis is related to the quasi-harmonic
approximation where the full protein trajectory is used to evaluate the
second order moments of all atoms [44-46].

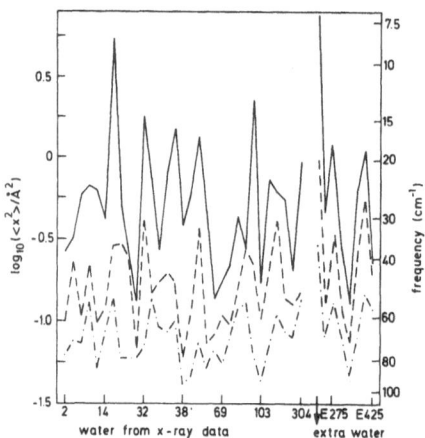

Fig. 9: The $\langle \tilde{x}^2 \rangle$-values of the 3
main axes are displayed. Solid,
dashed and dashed-dotted line
represent the low, intermediate
and high frequency mode of the
oxygen atomic trajectory from
the 39 water molecules.

The difference of the $\langle \tilde{x}^2 \rangle$-values from the 3 main axes provide a measure
of the anisotropy of the potential energy which determines the atomic

$$V_\alpha(x) = a_\alpha\, x^\alpha .$$ (12)

This analytical form of the potential energy has been used by Frauen-felder et al. [47] to interprete the temperature dependence of rms values derived from the x-ray structure of myoglobin. Potential envelopes with α values smaller (larger) than 2 are steep (flat) in the central part and have flat (steep) wings relative to a harmonic potential. The first type of anharmonic potential exhibits phonon softening with increasing temperature, the other type yields increasing frequencies with rising temperature. The limiting case of an infinite α value leads to the box potential with a ratio of moments Q = 3/5 (Fig. 11). A two site system yields an even smaller ratio of moments (11) than the box potential namely Q = 1/3.

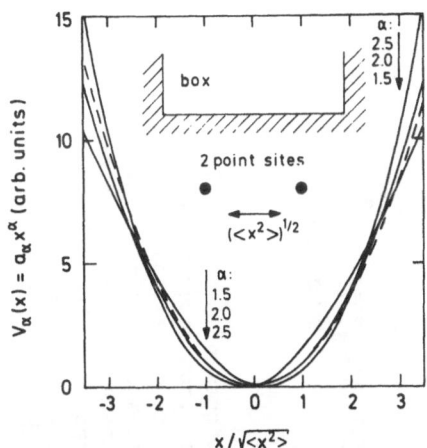

Fig. 11: Several 1-dimensional potential envelopes yielding the same $\langle x^2 \rangle$-value are displayed: power law, eq. (12), with expo-nents α = 1.5, 2.0, 2.5 (solid lines), box potential and two site system. The dashed line exhibits the asymmetric poten-tial: $V_+(x) = 0.92\, x^2$ for x>0 and $V_-(x) = 1.08\, x^2$ for x<0. The value of the third moment is given by $|\langle x^3 \rangle|^{1/3} = 0.58\, \sqrt{\langle x^2 \rangle}$, a value typical for water mole-cules with relative harmonic motions.

The 39 water molecules exhibit moment ratios Q ranging from the value of the 2 site system Q = 1/3 to Q values much larger than 2.0 corresponding to α values smaller than unity (Fig. 12). There are 2 water molecules from the x-ray structure which come close to a 2 site system namely W019 and W037. The latter molecule returns back to its initial site (see fi-gure 8C) and can therefore be considered to be a true 2 site system. The water molecule W019 is leaving its initial site and does not return back within the 140 ps time of computer simulation (see figure 6). If it would return some time later it is a 2 site system involving slow fluc-tuation, else it performs a slow relaxation. Many water molecules have trajectories which exhibit an anharmonic potential envelope with very flat wings corresponding to large values of the ratio of moments, eq. (11). The flat wing of the potential envelope can clearly be recognized by looking at the trajectory of the water molecule W069. This molecule is moving away from its initial position in the x-ray structure only during the first 40 ps which are used for protein structure relaxation. The same is true for the water molecule W029. For the last 100 ps its motion is nearly harmonic according to the higher order moments.

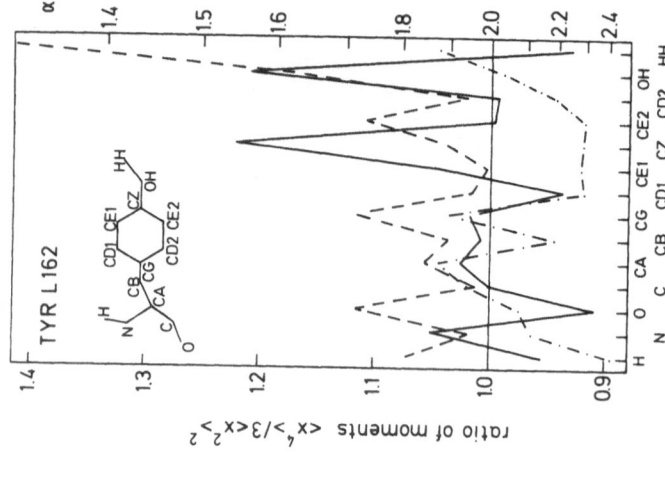

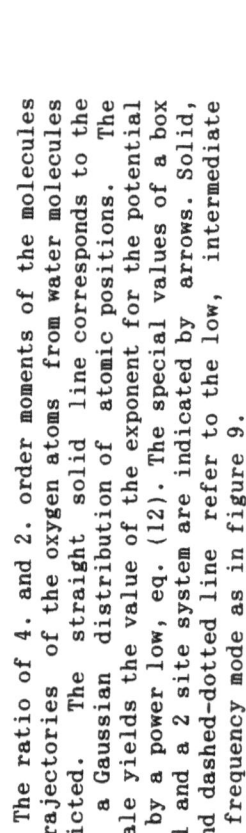

Fig. 13: The ratio of moments, eq. (12), of all atoms of the tyrosine residue L162 are displayed. Solid, dashed and dashed-dotted line refer to the mode with large, intermediate and small $\langle \tilde{x}^2 \rangle$-value as in figure 4.

Fig. 12: The ratio of 4. and 2. order moments of the molecules atomic trajectories of the oxygen atoms from water molecules are depicted. The straight solid line corresponds to the value of a Gaussian distribution of atomic positions. The right scale yields the value of the exponent for the potential governed by a power low, eq. (12). The special values of a box potential and a 2 site system are indicated by arrows. Solid, dashed and dashed-dotted line refer to the low, intermediate and high frequency mode as in figure 9.

In contrast to the motion of the water molecules the motion of the individual atoms of the protein molecule appear to be relatively harmonic. For the tyrosine residue L162 only the hydrogen atom of the phenol oxygen atom performs quite anharmonic motions (Fig. 13). It is not always the low frequency mode with the largest flexibility which shows the largest asymmetries and strongest anharmonicities. An example for the latter case is the hydrogen atom at the phenol oxygen of the tyrosine residue L162. It exhibits a potential envelope with a rather flat wing for the intermediate frequency mode (dashed line in Fig. 13).

6. Fluctuations of electrostatic and Van der Waals energies of the tyrosine residue L162

A quantitiy relevant for the ET process is the fluctuation of the electronic ground state energy of a molecular group which takes part in the transfer process. The relevant energy terms are the electrostatic (estat) and Van der Waals (VdW) interaction of the nonbonded atoms, eq. (4). This information is easily accessible from the data of a computer simulation with CHARMM [34]. We argue that the aromatic tyrosine residue L162 supports the ET process by mixing the electronic states of the donor and acceptor molecules. Hence it is interesting to analyse the energy fluctuations of this residue.

By looking at the time series of the estat and VdW energies of the residue L162 it becomes evident that on the average the VdW energy contributes more than 80% to the total nonbonded interactions. By reordering the energies from the time series according to the energy values a distribution can be derived from the density of energy values. A simple way to obtain a continous distribution from the discrete set is to place Gaussian distributions at the individual energy values (dashed line in figure 14). This method yields a distribution which in the low density regime exhibits a lot of structure. Another method employs the density of the energy values by measuring the length of the energy interval containing a given number (typically 30) of energy values. It provides a rather smooth energy distribution (solid line in figure 14).

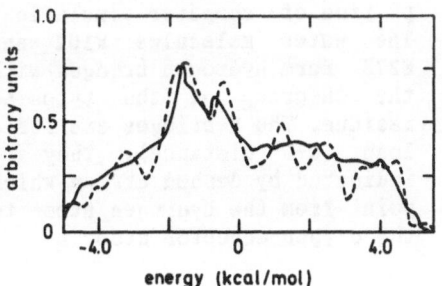

Fig. 14: The distribution of the estat and VdW energies of the tyrosine residue L162 with the rest of the protein (including the water molecules) is displayed. Two different methods are used to evaluate the distribution from a time series of energies with 501 data points. For details see text.

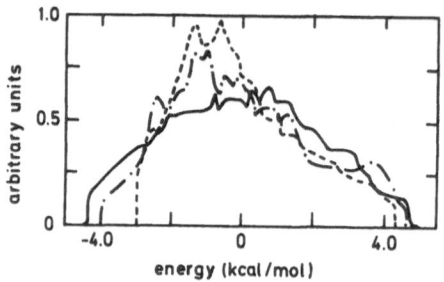

Fig. 15: Three energy distributions: the residue L162 with the protein without the water (solid line), the residue L162 with the water molecules alone (dashed line) and the OH-group of the phenol ring of L162 with the water molecules (dashed-dotted line) are displayed. The density of energy values is used to evaluate the distribution. For details see text.

In figure 14 the spectrum of the estat and VdW energy of the residue L162 with the relevant part of the protein including the 39 water molecules is depicted. The energy spectrum exhibits fluctuations of up to 4 kcal/mol. This energy corresonds to roughly 2000K, a value much larger than $k_B T$. It can be reconciled by considering that the tyrosine residue involves about 40 degrees of freedom. If each degree of freedom is independent and obeys a Gaussian energy fluctuation of width $\Delta E^2 = \langle (E - \langle E \rangle)^2 \rangle = (k_B T)^2$, the total energy fluctuation adds up to $\Delta E = k_B T \sqrt{40} = 1900K$.

In figure 15 it is shown that a considerable portion of the energy fluctuation is due to the water molecules around the tyrosine residue (dashed line). Most of this energy fluctuation is due to the OH-group of the phenol ring of the tyrosine (dashed-dotted line in figure 15). The water molecule W104 and the extra water E275 which are very close to the OH-group of residue L162 contribute most to this energy fluctuation (see figure 16).

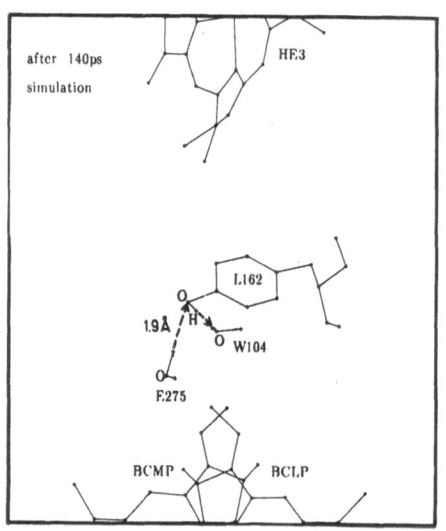

Fig. 16: A close-up of the relevant part of the RC centered around the tyrosine residue L162 is shown. It is a snapshot of the protein structure after 140 ps time of computer simulation. The water molecules W104 and E275 form hydrogen bridges with the OH-group of the tyrosine residue. The H-bridges are 1.9 Å long (O-H distance). They are indicated by dashed arrows which point from the hydrogen atom to the oxygen acceptor atom.

At each instant of time the interaction energy of the residue L162 with the water molecules and with the protein (without water) add up to the total energy of the residue interaction with the protein including the water molecules. However, the corresponding energy spectra do not add up. This can clearly be seen by comparing the partial energy spectra in figure 15 with the total spectrum in figure 14 (solid line). Also the sum of the squares of the widths ΔE from the two partial energy spectra is much larger than the square of the width from the total spectrum. This indicates that the partial energy terms are anticorrelated, in other words at a particular instant of time one group contributes a positive energy value whereas another group in the neighbourhood may contribute a negative energy value to the total energy.

7. Implications for the electron transfer process and outlook.

The parameters of a simple ET model with an electronic state for the donor and acceptor molecule are the electronic coupling V inducing the transfer, the frequency ω of the relevant vibrational mode which couples to the transfer process, the corresponding shift parameter λ, and the difference in the electronic energy between initial and final state $\Delta E = E_i - E_f$ (see figure 17).

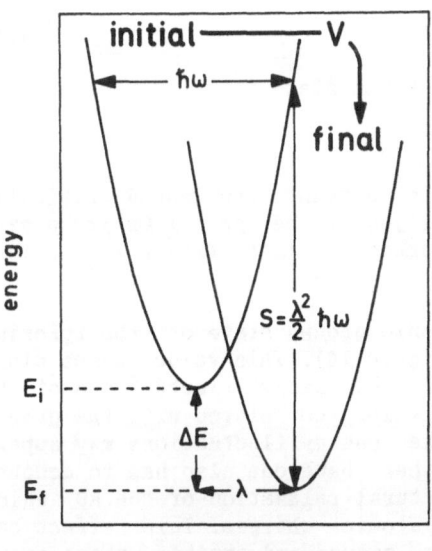

Fig. 17: The potential energy function of a vibrational mode is displayed for the initial and final state of an ET process involving only 2 states. The parameters are explained in the text.

All parameters except the frequency of an intramolecular vibrational mode can depend strongly on the instantaneous configuration in a protein dynamics simulation. The parameters V and λ depend very critically on the initial and final electronic state. Their evaluation requires a full quantum chemical calculation of the donor acceptor system including also specific residues of the RC. Such calculations are now made on the INDO level for specific configurations of the RC [48-50]. Since such calculations are very time consumming they can not easily be performed for the large number of protein structures obtained by the computer simulation.

The energy difference ΔE depends mainly on the partial atomic charges and polarizabilities of the molecular groups which participate in the ET process. In the CHARMM program polarizabilities are accounted for by enlarged values for the partial charges [33,34]. The partial charges are treated time independent. In the frame of this approximation the full time series of the energy difference ΔE of the electronic states of the tyrosine residue which presumably participate in the ET process can be obtained from a single dynamic simulation. The energies are then evaluated with 2 sets of atomic charges at the tyrosine residue. One set refers to the electronic ground state

$$SP(+) + TYR(0) + CYT(2+) \, , \qquad (13)$$

the second refers to the intermediate excited state which can be either

$$SP(0) + TYR(+) + CYT(2+) \, , \qquad (14)$$

where the hole is transferred or

$$SP(+) + TYR(-) + CYT(3+) \, , \qquad (15)$$

where the electron is transferred. The final state is

$$SP(0) + TYR(0) + CYT(3+) \, . \qquad (16)$$

Which one of the intermediate states is relevant can depend critically on the dynamics of the 2 hydrogen bridges of the phenol OH-group with the 2 water molecules in the neighbourhood [51] (see figure 16). A more detailed analysis will follow [38].

The energy fluctuations of the electronic ground state of the tyrosine residue are as large as 4 kcal/mol (figure 14). This value comes close to typical ΔE values for ET systems. The energy fluctuations of the excited state TYR(+) or TYR(-) are certainly correlated with the ground state TYR(0) such that only part of the energy fluctuations may appear in the energy difference ΔE. On the other hand one also has to account for the instantaneous portion of structural relaxation of the RC which goes along with a change of partial atomic charges. This effect can diminish the correlation of electronic ground and excited state energies. The high mobility of the water molecules can play a dominant role for the instantaneous relaxation process. A proper treatment of this effect requires at each instant of time an energy relaxation of the tyrosine residue environment with the new partial atomic charges. The statistical properties of the energy difference ΔE and the other parameters must enter an ET model which includes the influence of the intermediate electronic state at the tyrosine residue explicitly.

At the present state of knowledge we argue that at low temperatures the conformation of the tyrosine residue and its neighbourhood is very rigid and can not support the ET process by fluctuations with low ΔE values. The parameters of the ET process belong to the inverted regime providing a temperature independent rate [52,53] governing the low temperature regime. At higher temperatures the tyrosine residue and the water molecules become mobile. Then energy fluctuations with low ΔE values can lead to enlarged transfer rates. The specificity of the temperature dependence of the energy difference remains to be shown. This view is closely related to earlier work [29] where the increase of the transfer rate was related to an enhanced electronic coupling V mediating the transfer. This mechanism may also be operative. It is difficult to clarify this point because it requires large scale quantum chemical calculations for a large number of protein conformations.

The dynamics of the water molecules in the RC can be so important that a series of carefully set protein dynamics studies with different degrees of extra water will be required. Furthermore it is neccessary to develop criteria which help to decide on the proper amount of extra water and on the suitable places for these water molecules. The tools to characterize the trajectories of the water molecules are provided in terms of the analysis of moments (section 5). In this context it is also useful to investigate the reasons for the discrepancy of the rms values of atomic positions of the protein from computer simulation and x-ray data. It is probably more realistic to have smaller rms values in the computer simulations [43]. The discrepancy may be related to the type of constraints used in the buffer zone and to the degree of extra water.

Acknowledgement

The authors thank Prof. H. Michel and Prof. J. Deisenhofer for kindly providing the x-ray structure of Rps. viridis prior to publication. We also like to express our gratitude to Prof. S. F. Fischer and Prof. R. Rigler for continuous support and valuable discussions. EWK thanks Dr. P. O. J. Scherer for helpful discussions. All computer grafics are made with a PC. Assistance by Dr. P. O. J. Scherer and Mr. G. Thiedemann are gratefully acknowledged. The computer simulations are made on the Cray 1 XMP of the Max-Planck-Institut, Garching and on the Convex C210, donated by the Kjell and Marta Beijer Foundation at the Karolinska Institutet Stockholm. We are very thankful to Prof. Oesterheld for generous allocation of computer time at the Cray.

Appendix: moment analysis of atomic trajectories

In the appendix the analysis of individual atomic trajectories in terms of second order moments is described. The second order moments of all 3 cartesian components x_k, $k = 1,2,3$ of an atomic trajectory can be calculated as time average with respect to their first order moments $\langle x_k \rangle$ according to

$$\langle x_k x_1 \rangle = \frac{1}{N+1} \sum_{n=0}^{N} [x_k(t_0+n\Delta t)-\langle x_k \rangle] [x_1(t_0+n\Delta t)-\langle x_1 \rangle], \quad k,1=1,2,3 , \quad \text{(A1)}$$

where N+1 is the number of equidistant time points. The second order moments are collected in matrix form as follows

$$(\underline{XX})_{k,1} = \langle x_k x_1 \rangle . \tag{A2}$$

A 3-dimensional thermalized classical harmonic oscillator or more specifically a Brownian harmonic oscillator has a spatial distribution
given by a 3-dimensional generalized Gaussian

$$P(x_1,x_2,x_3) = [(2\pi)^3 \text{ Det}(\underline{XX})]^{1/2} \exp[-r_0(\underline{XX})^{-1}_0 r/2] \tag{A3}$$

where the inverse of the matrix of second order moments, eq. (A2), enters. By using solely the second order moments of an atomic trajectory
the motion of the atom is approximated by a 3-dimensional harmonic
oscillator. In the usual normal mode analysis the second order moments
are calculated from the second derivative of the potential energy surface. In contrast to such a local description the present treatment
accounts also for more global features of the potential energy surface.

By transforming to the main axes of the 3-dimensional harmonic oscillator

$$\tilde{r} = \underline{D}_0 r \tag{A4}$$

the 3-dimensional Gaussian distribution factorises in a product of 3
1-dimensional Gaussians as follows

$$P(x_1,x_2,x_3) = \prod_{k=1,2,3} \tilde{p}_k(\tilde{x}_k) , \tag{A5}$$

where

$$\tilde{p}_k(x) = [2\pi\langle \tilde{x}_k^2 \rangle]^{-1/2} \exp\left[-\frac{x^2}{2\langle \tilde{x}_k^2 \rangle} \right] \tag{A6}$$

and $\langle \tilde{x}_k^2 \rangle$ are the second order moments with respect to the 3 main axes.
Higher order moments of a Gaussian $\langle x^n \rangle_G$ can be expressed in terms of
the second order moments as follows

$$\langle x^{2n+1} \rangle_G = 0 , \qquad \langle x^{2n} \rangle_G = \frac{(2n)!}{2^n n!} [\langle x^2 \rangle_G]^n, \quad n=1,2,3... \tag{A7}$$

In particular one has

$$\langle x^3 \rangle_G = 0 \quad \text{and} \quad \langle x^4 \rangle_G = 3 [\langle x^2 \rangle_G]^2 . \tag{A8}$$

In general the moments from the trajectories will not obey the relations
(A7). Deviations from these relations can be used to characterize the
atomic trajectories. Such an analysis is closely related to the quasi-
harmonic approximation [32,44-46] where however the complete protein trajectory is analysed by this method.

References

1. A. Fersht, Enzyme Structure and Mechanism (Freeman, New York, 1984).
2. B. Chance, D. De Vault, H. Frauenfelder, R. A. Marcus, J. R. Schrieffer, and N. Sutin, Tunneling in Biological Systems (Academic, New York, 1979).
3. C. Ho, Electron Transport and Oxygen Utilization (North-Holland, Amsterdam, 1979).
4. Godvindjee, Photosynthesis, Energy Conversion by Plants and Bacteria (Academic, New York, 1982). Vol 1.
5. S. L. Mayo, W. R. Ellis, Jr., R. J. Crutchley, and H. B. Gray, Science 233, 948-952 (1986).
6. H. Michel, J. Mol. Biol. 158, 567-572 (1982).
7. J. Deisenhofer, O. Epp, K. Miki, R. Huber, and H. Michel, J. Mol. Biol. 180, 385-398 (1984).
8. J. Deisenhofer, O. Epp, K. Miki, R. Huber, and H. Michel, Nature 318 618-624 (1985).
9. J. Deisenhofer, O. Epp, I. Sinning, H. Michel, (1989).
10. J. Breton, J. L. Martin, A. Migus, A. Antonetti, and A. Orsay, Proc. Natl. Acad. Sci. 83, 5121-5125 (1986).
11. J. D. Mc Elroy, D. C. Mauzeral, and G. Feher, Biochim. Biophys. Acta 333, 261-277 (1974).
12. F. Parak, E. W. Knapp, and D. Kucheida, J. Mol. Biol. 161, 177 - 194 (1982).
13. F. Parak, J. Heidemeier, and E. W. Knapp, in: Biological and Artificial Intelligence Systems, E. Clementi and S. Chin, editors, (ESCOM, Leiden, 1988) p. 23-48.
14. W. Doster, S. Cusack, and W. Petry, Nature 337, 754 - 756 (1989).
15. F. Fujara and W. Petry, Europhys. Lett. 4, 921-927 (1987).
16. F. Parak, E. N. Frolov, A. A. Kononenko, R. L. Mössbauer, V. I. Goldanskii, and A. B. Rubin, FEBS Letters 117, 368 - 372 (1980).
17. G.R. Fleming, J.L. Martin, and J. Breton, Nature 333,190-192 (1988).
18. M. E. Michel-Beyerle, M. Plato, J. Deisenhofer, H. Michel, M. Bixon, and J. Jortner , Biochim, Biophys. Acta 932, 52-70 (1987).
19. R.A. Marcus, Chem.Phys.Lett. 133, 471-477 (1987); 146, 13-23 (1988).
20. P. O. J. Scherer and S. F. Fischer, Chem. Phys. 131, 115-127 (1989).
21. S. Creighton, J. K. Hwang, A. Warshel, W. W. Parson, and J. Norris, Biochem. 27, 774 - 781 (1988).
22. R. J. Shopes, L. M. A. Levine, D. Holten, and C. A. Wraight, Photosyn. Res. 12, 165-180 (1987).
23. B. Chance, T. Kihara, D. De Vault, W. Hildreth, M. Nishimura, and T. Hiyama, in Photosynthetic Research, ed. by H. Metzner (Laupp, Tübingen, 1969), Vol. III, p. 1321-1346.
24. M. Bixon and J. Jortner, J. Chem. Phys. 89, 3392-3393 (1988).
25. F. Parak, H. Hartmann, K. D. Aumann, H. Reuscher, G. Rennekamp, H. Bartunik, and W. Steigemann, Eur. Biophys. J. 15, 237-249 (1987).
26. D. De Vault and B. Chance, Biophys, J. 6, 825-847 (1966).
27. M. Bixon and J. Jortner, J. Phys. Chem. 90, 3795-3800 (1986).
28. W. Bialek and R. F. Goldstein, Phys. Scr. 34, 273-282 (1986).
29. E. W. Knapp and S. F. Fischer, J. Chem. Phys. 87, 3880-3887 (1987).
30. T. Kihara and P. L. Dutton, Biochim.Biophys.Acta 205,196-204 (1970).

31. T. Kihara and J.A. McCray, Biochim.Biophys.Acta **292**, 297–309 (1972).
32. N. Liang, A. G. Mauk, G. J. Pielak, J. A. Johnson, M. Smith, and B. Hoffman, Science **240**, 311 – 313 (1988).
33. B. R. Brooks, R. E. Bruccoleri, B. O. Olafson, D. J. States, S. Swaminathan, and M. Karplus, J. Comp. Chem. **4**, 187–217 (1983).
34. C. L. Brooks III, M. Karplus, and B. M. Pettitt, Proteins: a Theoretical Perspective of Dynamics, Structure, and Thermodynamics (Wiley, New York, 1988).
35. W. F. van Gunsteren and H. J. C. Berendsen, Molec. Phys. **34**, 1311–1327 (1977).
36. C. L. Brooks III, A. Brünger and M. Karplus, Biopolymers **24**, 843–865 (1985)
37. R. Kubo, Rep. Prog. Theor. Phys. **29**, 235–265 (1966).
38. E. W. Knapp and L. Nilsson, unpublished.
39. W. L. Jorgensen, J. Chandrasekhar, J. D. Madura, R. W. Impley, and M. L. Klein, J. Chem. Phys. **79**, 926 – 935 (1983).
40. H. Treutlein, K. Schulten, J. Deisenhofer, H. Michel, A. Brünger, and M. Karplus, in: The photosynthetic reaction center – structure and dynamics, eds. J. Breton and A. Vermeglio, NATO ASI series A: Life sciences (Plenum Press, New York, 1988).
41. C. B. Post, B. R. Brooks, M. Karplus, C. M. Dobson, P. J. Artymiuk, J. C. Cheetham, and D. C. Phillips, J.Mol.Biol. **190**, 455–479 (1986).
42. M. Levitt and R. Sharon, Proc. Natl.Acad.Sci.US **85**,7557–7561 (1988).
43. F. Parak and E. W. Knapp, Proc.Natl.Acad.Sci.US **81**,7088–7092 (1984).
44. M. Karplus and J. N. Kushick, Macromolecules **14**, 325 – 332 (1981).
45. R. M. Levy, M. Karplus, J. N. Kushick, and D. Perahia, Macromolecules **17**, 1370 – 1374 (1984).
46. R. M. Levy, A. R. Srinivasan, and W. K. Olson, Biopolymers **23**, 1099 – 1112 (1984).
47. H. Frauenfelder, G. A. Petsko, and D. Tsernoglou, Nature **280**, 558 – 563 (1979).
48. M. Plato, K. Möbius, M. E. Michel-Beyerle, M. Bixon, J. Jortner, J. Am. Chem. Soc. **110**, 9279 – 9285 (1988).
49. A. Warshel and W. W. Parson, J. Am. Chem. Soc. **109**, 6143 – 6151; ibid. **109**, 6152 – 6163 (1987).
50. P. O. J. Scherer and S. F. Fischer, Chem. Phys. **131**, 115–127 (1989).
51. S. F. Fischer, private comunication (1989).
52. R. P. Van Duyne and S. F. Fischer, Chem. Phys. **5**, 183 – 197 (1974).
53. S. Efrima and M. Bixon, Chem. Phys. Letters **25**, 34 – 37 (1974).

MÖSSBAUER SPECTROSCOPY ON PHOTOSYNTHETIC BACTERIA: INVESTIGATION   OF
REACTION CENTERS OF RHODOPSEUDOMONAS VIRIDIS

F. PARAK, A.BIRK, E.FROLOV[+], V. GOLDANSKII[+], I.SINNING[*] and H.MICHEL[*]
Institut für Molekulare Biophysik der Universität, Welderweg 26,
6500 Mainz, Fed. Rep. of Germany
[+]
[*] Institute of Chemical Physics of the Academy of Sciences, Moscow,USSR
Max-Planck-Institut für Biophysik, 6000 Frankfurt,Fed.Rep. of Germany

ABSTRACT   Crystals   of ^{57}Fe enriched   reaction   centers   have   been
investigated by Mössbauer spectroscopy. The cytochrome irons are in the
low spin ferric state. The non-heme iron of the electron accepting side
is partly ferrous high spin and partly ferrous low spin (or ferric high
spin). Under the conditions of the experiment sodium ascorbate reduces
only one cytochrome iron into the   ferrous   low   spin   state. Membrane
bound proteins become flexible at higher   temperatures than   proteins
with a hydrophilic surface. They are also less flexible, at least up to
temperatures of about 250 K.

Mössbauer spectroscopy on the iron isotope ^{57}Fe has been proved to
be a powerful tool for the investigation of iron   containing biomolecules. The absorption energies are shifted and split   by the   hyperfine
interactions of the ^{57}Fe   nucleus   with   the   electrons of the atom and
with magnetic fields and charges in the neighbourhood. The   line positions allow the characterization of the charge   and   the   spin state of
the iron. Different iron sites can be distinguished.   In addition,   the
analysis of the areas and the spectral distributions   of the absorption
lines yields valuable information on the dynamic properties of the system. Besides vibrations with respect to its neighbours the iron labels
segmental motions within macromolecules and motions of the whole molecule within a crystal or a membrane. In proteins such motions are often
important for the function.

Photosynthetic organisms use the iron at many steps of photochemical and enzymatic processes. Several photosynthetic systems have,
therefore, already been investigated by Mössbauer spectroscopy [1-9].
In the following we limit ourselves to photosynthetic bacteria. Debrunner et al. [1,3] investigated isolated reaction centers (RC) from Rhodobacter sphaeroides R-26 (formerly called Rhodopseudomonas sphaeroides
[10]). They found one well defined iron species. The Mössbauer   hyperfine parameters were indicative for high spin Fe^{2+} and were insensitive
to a removal of one or both quinones and on change of detergent or addition of o-phenanthroline.

Mössbauer experiments on chromatophores from Rhodospirillum rubrum

413

J. Jortner and B. Pullman (eds.), Perspectives in Photosynthesis, 413–421.

have demonstrated the correlation of the dynamics of the system and its
function [2]. In these chromatophores the kinetics of the dark   reduction after light absorption strongly depends on temperature. At room
temperature the dark reduction is a rather slow process with a characteristic time of several seconds while at low temperatures (150 K and
lower) the reduction occurs several orders of magnitudes faster. Fast
reduction is possible if the excited electron reaches only the acceptor
1. Arriving at the acceptor 2 enough energy is already dissipated to
prevent an easy backdiffusion on the same pathway. The energy dissipation can only occur if the system is flexible. Mössbauer   spectroscopy
demonstrates the correlation between the probability of the electron
transfer to acceptor 2 and the flexibility: Above 150 K the mean square
displacements of the iron components of the chromatophore increase
drastically with temperature.

     In the following discussion we concentrate on some recent investi-
gations on Rhodopseudomonas (Rps.) viridis. Rps. viridis cells have
been grown on cultures enriched in the isotope ^{57}Fe. Sample 1 contains
the whole cells washed several times in water at 4 °C in order to get
rid of Fe ions of the culture adhered to the surface of the cells. For
the preparation of chromatophores (sample 2), cells were suspended in
0.05 M Tris-HCl-buffer, pH 7.6, at 4°C and washed twice. The cells were
broken in an ultrasonic desintegrator and the crude photosynthetic membranes were isolated by differential centrifugation: After removing de-
bris by centrifugation at 15000 g for 15 min, the supernatant was cen-
trifuged at 120000 g for 1 hour. The precipitated pigmented material
was used as absorber. All operations have been done at about 4°C. The
reaction centers were isolated and   crystallized according to [11]
(sample 3; abbreviated RCC$_{ox}$). Some crystals were reduced by soaking
one night in the motherliquid containing 20 mM ascorbate in a N$_2$-atmos-
phere (sample 4; abbreviated RCC$_{red}$).

     Fig. 1 to 4 shows the Mössbauer spectra of the 4 samples   at T =
150 K. The hyperfine pattern of these spectra have been fitted with the
parameters given in Tab. 1 (fit 1). The discussion   of the spectra is
started with sample 3 (RCC$_{ox}$) since the composition is best known from
X-ray crystallography [12]. The crystals contain only the non-heme iron
of the reaction center and the four heme irons of the cytochrome c.
We look on the parameters of fit 1. The Mössbauer data can, however, be
fitted only assuming 3 different iron species. The largest contribution
to the spectrum comes from the iron species 3. The parameters are close
to those of other cytochromes c and are typical for low spin ferric
complexes. The parameters of the iron species 1 clearly indicate the
ferrous high spin state and are close to those obtained by Boso et al.
[3] on the reaction centers from Rhodobacter sphaeroides. On a first
view the isomer shift IS and the quadrupole splitting QS of the iron
component 2 is typical for the ferric high spin state. A comparison
with the investigation on different cytochromes shows, however, that
the ferrous low spin state is also possible. Assuming equal mobility of
the three iron components at 150 K the percentage of the   absorption
areas A give the fractions of the   iron components. The expected ratio
of 20 % non-heme iron and 80 % cytochrome c is not obtained. We will

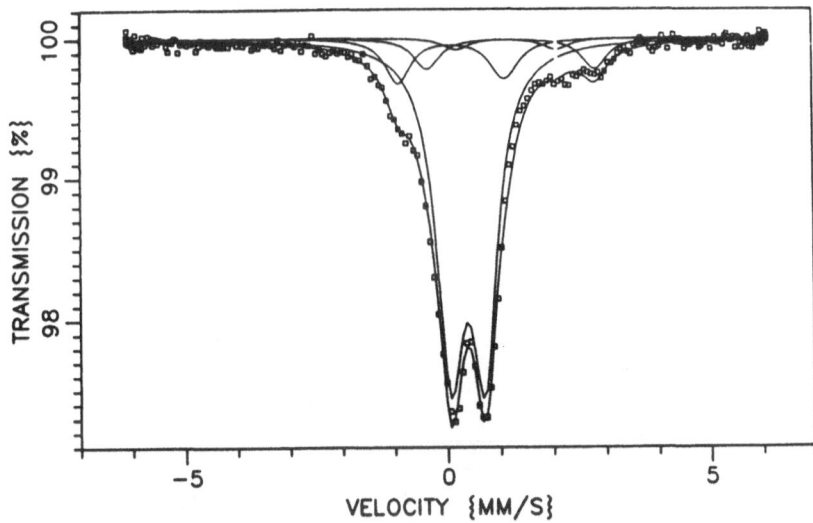

Figure 1. Mössbauer spectrum of the washed cells from Rps. viridis (sample 1), T = 150 K.

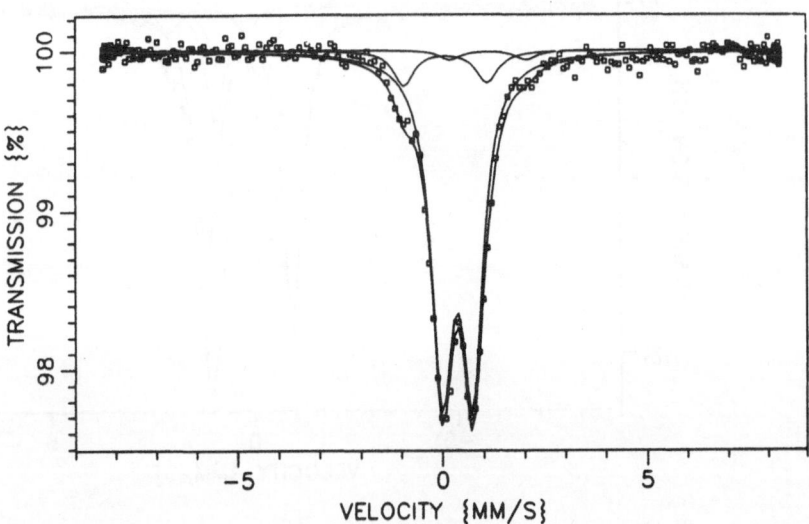

Figure 2. Mössbauer spectrum of the chromatophores from Rps. viridis (sample 2), T= 150 K.

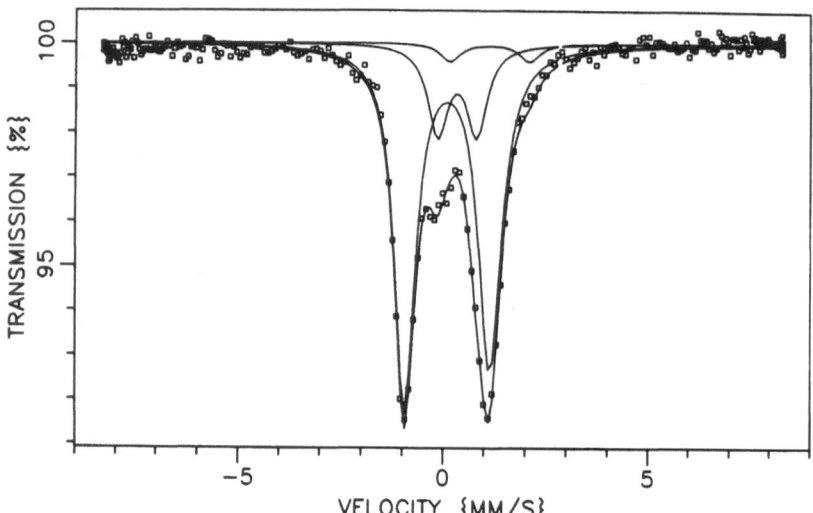

Figure 3. Mössbauer spectrum of the RC-crystals from Rps. viridis (sample 3, RCC_{ox}), T = 150 K.

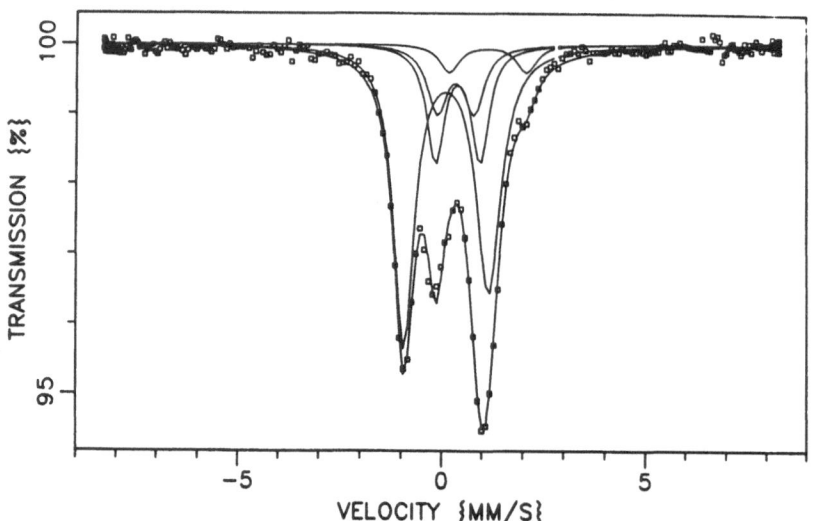

Figure 4. Mössbauer spectrum of the reduced RC-crystals from Rps. viridis (sample 4, RCC_{red}), T = 150 K.

TABLE 1. Isomer shifts, IS, (relative to iron metal), quadrupole splittings, QS, and relative areas, A, at 150 K obtained from the fits 1 and 2 of the Mössbauer spectra (Figs. 1-4).

Sample	1 cells			2 chromatophores			3 crystals (RCC_{ox})			4 crystals (red) (RCC_{red})		
Fe species	IS (mm/s)	QS (mm/s)	A (%)	IS (mm/s)	QS (mm/s)	A (%)	IS (mm/s)	QS (mm/s)	A (%)	IS (mm/s)	QS (mm/s)	A (%)
fit 1												
1	1.25	1.88	3.0	1.26	1.87	2.5	1.21	1.80	3.8	1.21	1.88	5.8
2							0.46	0.93	18.8	0.46	0.93	13.4
3	0.21	2.00	11.2	0.20	2.03	9.7	0.22	2.03	77.4	0.22	2.09	60.4
4				0.47	0.74	87.8				0.51	1.12	20.4
5	0.49	0.64	77.8									
6	1.33	3.15	8.0									
fit 2												
1							1.25	1.87	4.0	1.25	1.87	6.3
2							0.20	2.03	78.0	0.20	2.03	60.3
3							0.51	1.11	18.0	0.51	1.11	33.4
4												
5												
6												

come back to this later.

A comparison with the chromatophores or the cells shows that component 1 and 3 of the sample RCC_{ox} are again present (with slightly changed hyperfine parameters). In addition we find a large amount of one other iron protein (component 5). The similar fraction of this iron species in cells and chromatophores shows that this protein is membrane bound. The Mössbauer parameters are compatible with FeS-proteins like ferredoxin. The differences in the hyperfine parameters of this species in sample 1 and 2 together with the large width of the lines (about 0.7 mm/s) indicate the contribution of unresolved slightly different iron sites. Identifying component 1 and 3 with the non-heme iron and the cytochrome c, respectively, one obtains in cells and chromatophores roughly the expected area ratio of 1 to 4. The remaining component 6 in the cell spectrum represents Fe^{2+} in the high spin state and stems probably from iron ions of the culture not washed from the cell surfaces.

We now compare the samples RCC_{ox} and RCC_{red} and look on fit 2 for RCC_{ox} first. The reduction diminishes the contribution of the component 3 (ferric low spin of cytochrome c) and produces a new iron component 4. The hyperfine parameters are nearly identic to those of the component c of reduced cytochrome cd from Thiobacillus denitrificans [13] and identify a ferrous low spin complex. As it is known from literature [14] sodium ascorbate reduces only the high potential cytochrome. The area ratio of component 3 and 4 shows that only one third of the four cytochrome irons has been reduced. This makes no physical sense although it cannot be excluded that the soaking time has been too short because the diffusion is rather slow in crystals. The fit 2 of the RCC_{red} sample gives a quite satisfying description of the experimental data with three iron species (component 1: non-heme Fe^{2+} high spin; component 2: low potential cytochrome Fe^{3+} low spin and component 4: high potential cytochrome Fe^{2+} low spin). The contribution of the non-heme iron is, however, too small as in the case of the RCC_{ox}-sample.

In the next step the RCC_{ox} spectrum has been fitted with the assumption that part of the cytochrome iron is already reduced and component 2 does not exist (fit 2). The least squares fit gives reasonable results but again the contribution of the non-heme iron is too small. As a cross check we fitted the RCC_{red} sample with the assumption that component 2 is present in addition to the reduced cytochromes (component 4) (RCC_{red}, fit 1). Comparing the two fits for RCC_{ox} and RCC_{red} makes clear that the spectra can be interpreted with a number of different sets of parameters. Also combinations not discussed here give a reasonable agreement with the experimental data. Nevertheless, some features become unambigeously clear:

With the assumption of equal or similar mobility of the different iron species in the crystal it is not possible to obtain the expected contribution of the non-heme Fe2+ high spin. To explain its low contribution one has to assume a drastically increased mobility. This is in clear contradiction to the results of RSMR experiments on single crystals of myoglobin [15]. Below 200 K most of the mobility of the iron in a protein comes from modes involving the whole molecule in the crystal.

The Mössbauer spectra can be understood easily if the iron component 2 belongs to the non-heme iron too. With the parameters of fit 1 the area of component 1 plus 2 is about 20 % for RCC_{ox} and RCC_{red}. The Mössbauer results then show that only one cytochrome iron can be reduced by sodium ascorbate. This is in contradiction to the redox potentials of [14]. It is, however, possible that the detergent needed for crystallization (N,N-dimethyldodecylamine N-oxide) serves as a redox buffer, allowing the reduction of only one heme group with our experimental conditions.

For a discussion of the dynamic properties we use in the following the total areas, A_{tot}, of the Mössbauer spectra of RC_{ox}-crystals. As usual we have taken the average mean square displacements, $\langle x^2 \rangle$, of the irons proportional to $\ln A_{tot}$ where the normalization has been obtained from a linear extrapolation assuming $\langle x^2 \rangle = 0$ at $T = 0$ K. Fig. 5 gives the $\langle x^2 \rangle$-values of RCC_{ox} (open squares), freeze dried myoglobin (open circles) and myoglobin crystals (full circles) [16,17]. Below 180K the three samples behave quite similar. The $\langle x^2 \rangle$-values decrease linear with decreasing temperature. This behaviour can be parameterized by a Debye law. For the reaction center crystals one obtains a Debye temperature of about 220 K. Myoglobin in crystals as well as freeze dried myoglobin shows a pronounced increase of the flexibility above 200 K. In contrast, the RC crystals follow the "Debye" behaviour to higher

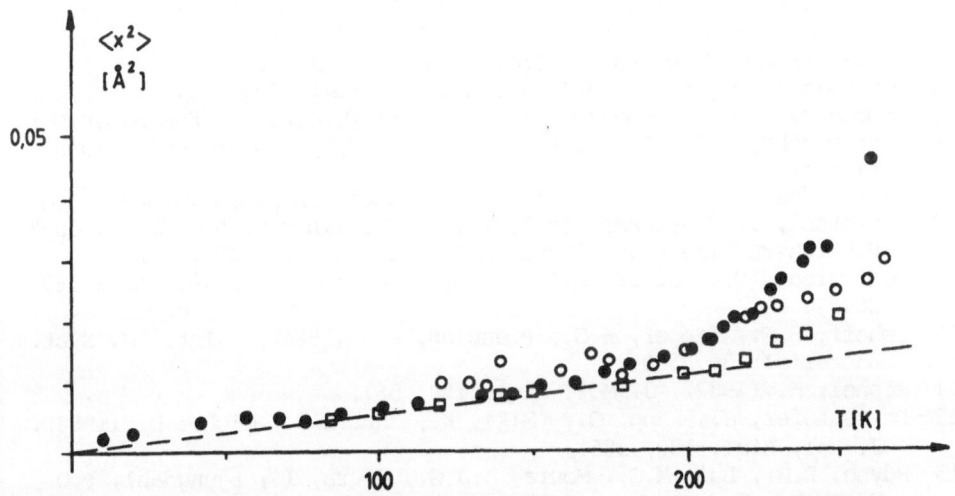

Figure 5. Temperature dependence of the mean square displacement, $\langle x^2 \rangle$. Symbols: open squares: RCC_{ox}, open circles: freeze dried myoglobin, closed circles: myoglobin crystals.

temperatures. Above 210 K the <x²>-values of the   RC-crystals increase
more than linear with temperatures. The additional increase, which is
typical for biomolecules is, however, much smaller than in myoglobin
crystals. The <x²>-values remain even below those   of   freeze   dried
myoglobin which is strongly immobilized because of the missing first
hydration shell. We have   found similar small values   for   bacte-
riorhodopsin [18]. Between 200 K and 250 K membrane bound proteins seem
to be less flexible than water soluble proteins. This is certainly   a
consequence of the different surrounding. The increasing mobility above
a certain temperature shows many features of a glass transition. According to Frauenfelder et al. [19] the glass transition of the protein
is slaved by the glass transitions of the surrounding medium. At present it is not clear if the differences in the protein flexibilities
can be extrapolated to room temperature.

This work was supported by the Deutsche Forschungsgemeinschaft.

References

1 Debrunner, P.G., Schulz, P.G., Feher, C.E., Okamura, M.Y. (1975),
   Biophys. J. 15, 226a.
2 Parak, F., Frolov, E.N., Kononenko, A.A., Mössbauer,R.L., Goldanskii, V.I., Rubin, A.B. (1980), FEBS Lett. 117, 368.
3 Boso, B., Debrunner, P., Okamura, M.Y., Feher, G. (1981), Biochim.
   Biophys. Acta 638, 173.
4 Evans, E.H., Dickson, D.P.E., Johnson, C.E., Rush, J.D., Evans,
   M.C.W. (1981), Eur. J. Biochem. 118, 81.
5 Petrouleas, V., Simopoulos, A., Kostikas, A., Isaakidou, J., Papa-
   georgiou, G., Dismukes, C. (1981) ´Photosynthesis II. Electron
   transport and photophosphorylation´, in G. Akoyunoglou, Balaban
   Int. Scien. Serv. Philadelphia, pp. 677-686.
6 Petrouleas, V., Diner, B.A. (1982), FEBS Lett. 147, 111.
7 Uspenskaya, N.Ya., Novakova,A.A., Aleksandrov,A.Yu., Kuz´min,R.N.,
   Kononenko, A.A., Rubin, A.B. (1983), Biophysics 28, 401 (Transla-
   tion from Biofizika 28, 376).
8 Petrouleas, V., Diner,B.A. (1984) ´Advances in photosynthesis re-
   search´, in C.Sybesma (ed.), Vol. 1, Martinus Nijhoff/Dr.W. Junk
   Publishers, The Hague/Boston/Lancaster, pp. 195-198.
9 Petrouleas, V., Diner, B.A. (1986), Biochim. Biophys. Acta 849,
   264.
10 Imhoff, S.F., Trüper, H.G., Pfenning, N. (1984),   Int. J. Syst.
   Bacteriol. 34, 340.
11 Michel, H. (1982), J. Mol. Biol. 158, 567.
12 Deisenhofer, J., Epp, O., Miki, K., Huber R., Michel H. (1984),
   J. Mol. Biol. 180, 385.
13 Huynh, B.H., Lui, M.C., Moura, J.J.G., Moura, I., Ljungdahl, P.O.,
   Münck,E., Payne, W.J., Peck, H.D., Der Vartanian, D.V., Le Gall,
   J. (1982), Biolog. Chem. 257, 9576.
14 Dracheva,S.M., Drachev,L.A., Zaberezknaya,S.M., Konstantinov,A.A.,
   Semenov, A.Y., Skulachev, V.P. (1986), FEBS Lett. 205, 41.

15 Nienhaus, G.U., Heinzl, J., Huenges, E., Parak, F. (1989), Nature (London) 338, 665.
16 Parak, F., Knapp,E.N., Kucheida,D. (1982), J. Molec. Biol.161,177.
17 Parak, F., Fischer, M., Graffweg,L., Formanek H. (1987) ´Distributions and fluctuations of protein structures investigated by X-ray analysis and Mössbauer spectroscopy´, in E. Clementi and S. Clin (eds.), Plenum Publish. Comp. New York, pp. 139-148.
18 to be published
19 Iben, I.E.T., Braunstein, D., Doster, W., Frauenfelder, H., Hong, M.K., Johnsson, J.B., Cuck,S., Ormos, P., Schulte,A., Steinbach, P.J., Xie, A.H., Young, R.D. (1989), Phys. Rev. Lett. 62, 1916.

MAGNETIC RESONANCE AND MOLECULAR ORBITAL STUDIES OF THE PRIMARY DONOR
STATES IN BACTERIAL REACTION CENTERS

M. PLATO[a], K. MÖBIUS[a], W. LUBITZ[b], J.P. ALLEN[c], and G.
FEHER[c]

a) Institute for Molecular Physics, Free University of
   Berlin, Arnimallee 14, D-1000 Berlin 33, FRG
b) Physical Institute II, University of Stuttgart, Pfaffen-
   waldring 57, D-7000 Stuttgart 80, FRG
c) Department of Physics, University of California, San
   Diego, La Jolla, California CA 92093, USA

ABSTRACT. Spin density calculations were performed on the oxidized
primary electron donor $P_{865}^{+\cdot}$ in reaction centers (RC's) from Rhodobacter sphaeroides R-26. The calculations were based on the all-valence-
electron MO method RHF-INDO/SP and on the three-dimensional structure
determined by X-ray diffraction. Results are compared with experimental isotropic hyperfine couplings (hfc's) obtained from ENDOR/TRIPLE
magnetic resonance spectra and with previous studies on $P_{960}^{+\cdot}$ in RC's
from Rhodopseudomonas viridis. Both bacteriochlorophyll dimer species
$P_{865}^{+\cdot}$ and $P_{960}^{+\cdot}$ show marked deviations of their spin distributions from
C_2-symmetry. However, $P_{865}^{+\cdot}$ is more symmetric than $P_{960}^{+\cdot}$ with respect to
sums of spin densities on the A and B halves of the dimers. The ef-
fects of various environmental (inclusion of surrounding amino acid
residues) and structural refinements (e.g. rotation of acetyl groups)
on hfc's and excited state properties were studied. The latter were
obtained by additional INDO/S calculations. In both species, P_{865} and
P_{960}, the excited electron in the lowest unoccupied molecular orbital
(LUMO) is predicted to be preferentially on the B half of the dimer,
which is close to the accessory bacteriochlorophyll monomer on the
photoactive A branch. Thus the preferential electron transfer along
the A branch originates in the excited state of the dimer.

1.   Introduction

The primary charge separation step in bacterial photosynthetic reaction centers (RC's) is the transfer of an electron from a bacteriochlorophyll dimer to a bacteriopheophytin monomer [1, and references
therein]. The rate of this process is determined by both the three-
dimensional structure and the electronic structure of the pigments.

J. Jortner and B. Pullman (eds.), Perspectives in Photosynthesis, 423–434.
© 1990 Kluwer Academic Publishers.

The <u>three-dimensional structure</u> of the cofactors and protein
subunits have been determined by X-ray diffraction for RC's from two
bacterial species: Rhodopseudomonas (Rps.) viridis [2, 8] and Rhodobacter (Rb.) sphaeroides R-26 [3, 9].

Information on the <u>electronic structure</u> of chlorophyll pigments
in RC's has been obtained from magnetic resonance methods such as
Electron Spin Resonance (ESR), Electron Nuclear Double Resonance
(ENDOR) and Electron Nuclear Nuclear Triple Resonance (Special and
General TRIPLE). These techniques have been applied to illuminated
RC's of Rps. viridis and Rb. sphaeroides R-26 in liquid solution,
yielding isotropic hyperfine couplings (hfc's) - including their
relative signs and multiplicities (i.e. number of contributing nuclei)
- of the oxidized "special pair" electron donors $P_{960}^{+\cdot}$ and $P_{865}^{+\cdot}$, respectively. To relate these hfc's to structural properties, advanced
molecular orbital (MO) theories as well as the X-ray diffraction
results were used. The results of these investigations on $P_{960}^{+\cdot}$ have
been reported previously [4, 5].

This paper presents an analogous investigation on $P_{865}^{+\cdot}$ in Rb.
sphaeroides R-26, including calculations of excited singlet state
properties of P_{865}, with the aim to correlate differences or similarities in the <u>structure</u> with differences or similarities in the <u>function</u>
of these two primary donor species.

2.   Results and Discussion

2.1   REVIEW OF PREVIOUS RESULTS ON $P_{865}^{+\cdot}$ AND $P_{960}^{+\cdot}$

The oxidized donor $P_{865}^{+\cdot}$ in bacteriochlorophyll a (BChl a) containing
RC's of Rb. sphaeroides R-26 exhibits a $1/\sqrt{2}$ narrowing of the ESR line
[18] (not shown) and a reduction of the hfc's from ENDOR by a factor
close to 2, when compared with the monomeric BChl $a^{+\cdot}$ [19, 20] (see
fig. 1). These observations confirmed the proposal that P_{865} is a
dimer or "special pair" built of two BChl a molecules with the electrons equally shared [6]. Generally, an unpaired electron symmetrically delocalized over N identical molecules will result in a linewidth
narrowed by $1/\sqrt{N}$ and hfc's reduced by $1/N$ relative to an isolated
molecule.

The recent X-ray structural analysis of RC's of Rb. sphaeroides
R-26 [3, 9, and references therein] confirms the dimeric nature of the
electron donor P_{865}. The X-ray analysis also shows a highly symmetrical arrangement of the pigments along two branches A and B that are
approximately related by a twofold symmetry axis (see fig. 2).

A very similar arrangement of pigments had been found earlier for
the BChl b containing RC's of Rps. viridis [2]. However, the radical
cation of the primary donor P_{960} in this bacterial species does not
exhibit a $1/\sqrt{2}$ narrowing of the ESR spectrum nor a twofold reduction
of hfc's when compared with the BChl $b^{+\cdot}$ monomer (see fig. 1). A
detailed analysis of ENDOR/TRIPLE spectra of $P_{960}^{+\cdot}$ on native and partially deuterated RC's [4] and MO calculations [5] have been reported.

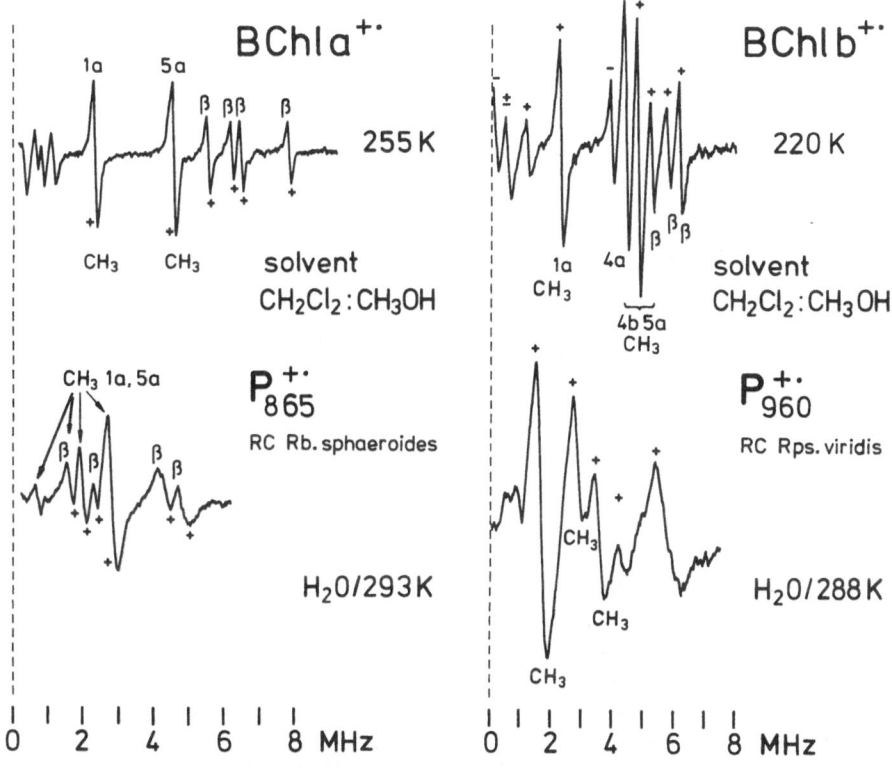

Fig. 1: Electron Nuclear Nuclear Resonance (Special TRIPLE) spectra of
the oxidized donors $P_{865}^{+\cdot}$ and $P_{960}^{+\cdot}$ in RC's from Rb. sphaeroides
R-26 and Rps. viridis in aqueous solutions (below) and the
respective isolated BChl radical cation monomers (above). Line
positions correspond to half the magnitude of corresponding
hfc's (signs were obtained from General TRIPLE).

This work which is based on the semi-empirical all-valence-electrons
MO method RHF-INDO/SP [5, 7] and on the three-dimensional structure of
the RC from Rps. viridis (resolution: 2.9 Å, R = 23 %), concluded that
the spin density distribution, ρ_i, on the various atoms is strongly
asymmetrical with respect to the two BChl's, D_A and D_B, of the dimer.
This asymmetry amounts to $\Sigma\rho_i^A/\Sigma\rho_i^B \cong 2.8$ implying that the unpaired
electron (or associated positive hole) resides predominantly on D_A.
This asymmetry was shown to be largely due to the asymmetric distribution of partial charges of amino acid residues (AAR's) in the immediate surrounding of the donor.

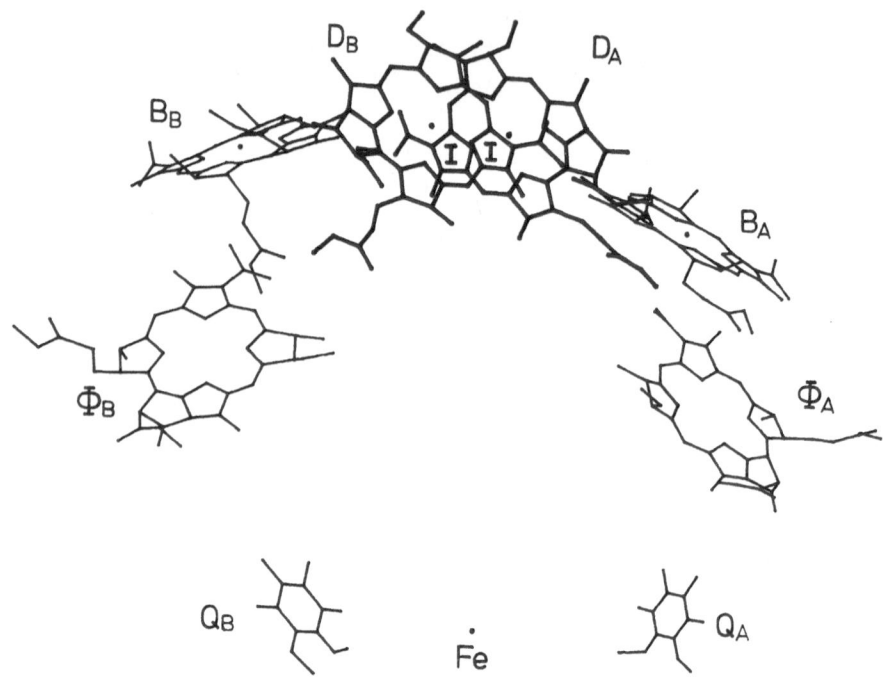

Fig. 2: Cofactor structure of the RC from Rb. sphaeroides R-26. Electron transfer proceeds preferentially along the A branch. B =
bacteriochlorophyll, Φ = bacteriopheophytin, Q = ubiquinone.
Phytyl tails of the cofactors have been omitted for clarity.

    Similar theoretical investigations on $P_{865}^{+\bullet}$ prior to the determi-
nation of the three-dimensional structure had been performed on model
BChl $\underline{a}$ dimers in-vacuo employing the RHF-INDO/SP method in conjunction
with techniques that optimize the geometry by minimizing the total
energy of the structure [7]. Approximate C_2 symmetry was maintained in
most calculations and the monomeric halves were kept planar except for
the two acetyl groups attached to rings I (see fig. 3) which were
allowed to rotate. The most stable dimer geometry was found for a
maximum π-overlap of rings I, with two symmetrical Mg-O bonds between
the acetyl oxygen atoms of $D_A(D_B)$ and the central Mg atoms of $D_B(D_A)$.
The X-ray structures of Rps. viridis and Rb. sphaeroides confirm the
overlap of rings I but show the Mg-O distance to be too large for an
Mg-O bond. This may be due to protein interactions (e.g. hydrogen
bonds with residues that were not included in the in-vacuo calculations) or due to an overall distortion (e.g. enlargement of the plane-
to-plane distance) of the in-vacuo dimer structure. Whereas possible
hydrogen bonds between acetyl oxygen atoms and neighboring AAR's (HIS
L168 and TYR M195) were reported for P_{960} [8], no such bonds were
suggested for P_{865} [9].

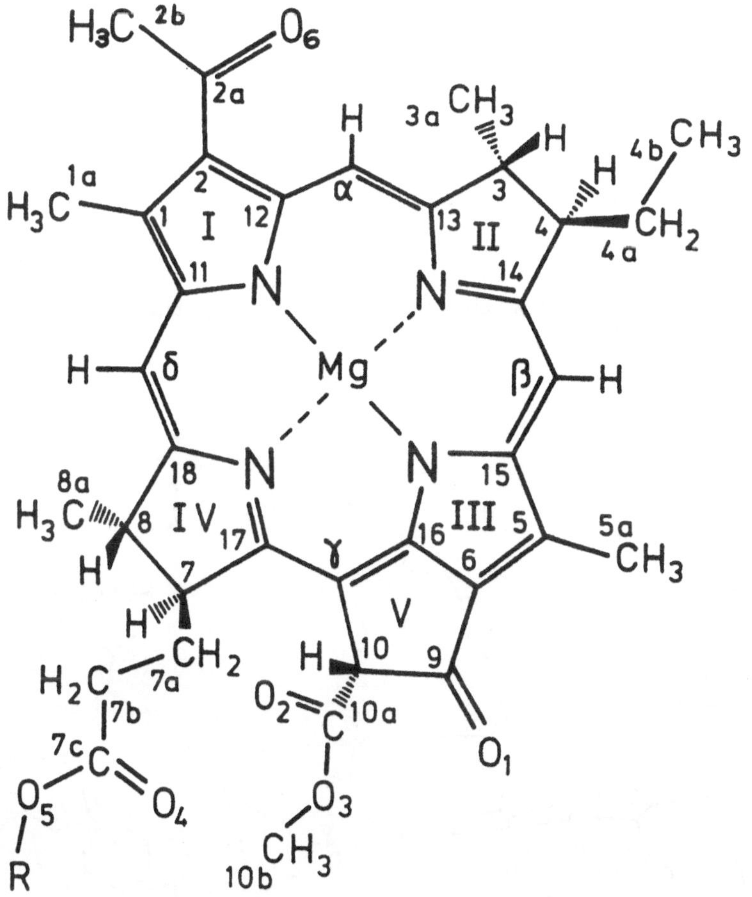

Fig. 3: Molecular structure of bacteriochlorophyll a (BChl a) includ-
ing numbering scheme. The side chain R is phytyl ($-C_{20}H_{39}$) in
Rb. sphaeroides. This chain has been truncated to $-CH_3$ in the
MO calculations.

2.2 MO CALCULATIONS ON $P_{865}^{+\cdot}$ BASED ON THE THREE-DIMENSIONAL STRUCTURE

We present spin density calculations on $P_{865}^{+\cdot}$ analogous to those presented
earlier for $P_{960}^{+\cdot}$ [5]. These RHF-INDO/SP calculations are based on the
X-ray diffraction data of RC's from Rb. sphaeroides R-26. These data
result from a refinement at a resolution of 2.8 Å with an R-value of
26 %. The average rms error in the coordinates was estimated to be
± 0.4 Å [3].
    The calculations proceeded in several steps A, B, ... E with an
increasing level of environmental and structural refinements. As in
$P_{960}^{+\cdot}$, limited structural refinements of the heavy atom skeleton were

performed to minimize the total energy. Due to the high sensitivity of
hfc's on nuclear positions, particularly of "β-protons" on peripheral
methyl groups (e.g. 1a, 5a) and on the saturated pyrrole rings (positions 3, 4, 7, 8; see fig. 3), all structural refinements remained
within the relatively small range of positional changes of ~ ± 0.5 Å.

A) "Bare" dimer: unmodified heavy atom skeleton with positions of
attached H atoms adjusted to minimize the energy.

B) "Dressed" dimer: inclusion of 10 nearest (within 5 Å) AAR's including backbone atoms (see fig. 4) This step calculates electrostatic
effects due to partial atomic charges of AAR's using the CNDO results
by Momany et al. [10].

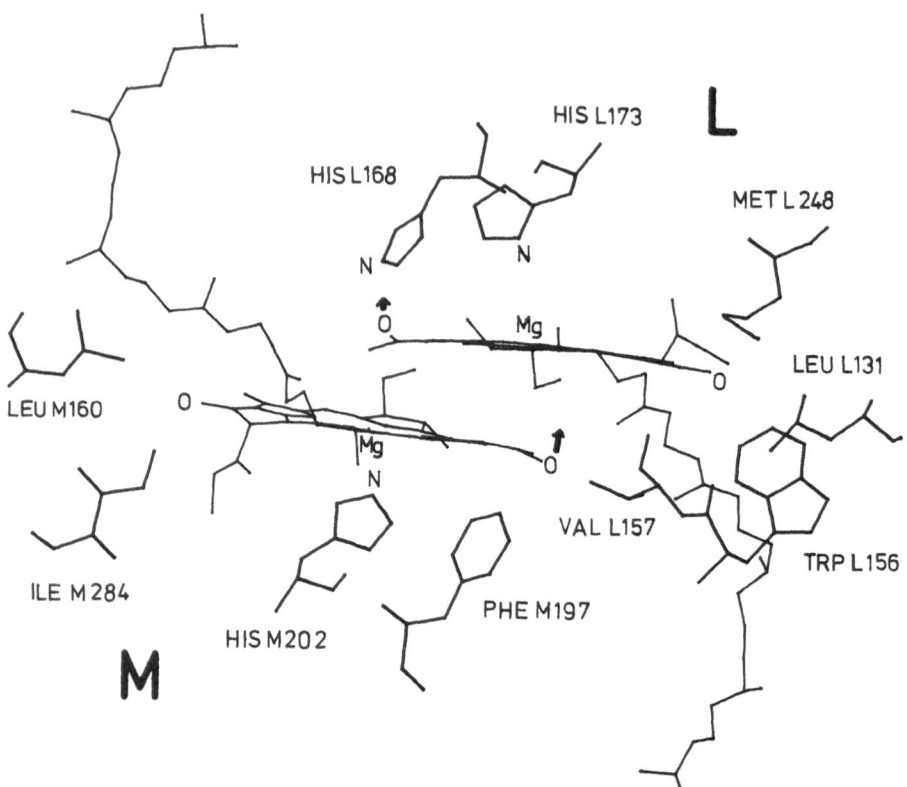

Fig. 4: Amino acid residues of the L and M protein subunits in the
        vicinity of the BChl a dimer in the RC of Rb. sphaeroides.
        Arrows indicate displacements of acetyl O atoms obtained by
        minimizing the total energy (see text).

C) <u>Adjustment of rotational angles</u> of the two acetyl groups on rings I. This step results in a considerable decrease of total energy by ~ 0.5 eV mainly due to the relief of steric hindrance between the methyl groups 1a and 2b (see fig. 3). The rotational out-of-plane angles change from - 12° to - 27° (towards the outside of the dimer) and from + 6° to + 39° (towards the inside of the dimer) on the A- and B-halves, respectively. The corresponding positional changes of the acetyl oxygen atoms are 0.2 and 0.5 Å, respectively. This stabilizes both groups at the same distance of 3.8 Å between the acetyl O atoms and the opposing Mg atoms. This distance is too large for an Mg-O bond, although there is a considerable electrostatic interaction between the oppositely (partially) charged Mg and O atoms. The movement of the acetyl group on D_A towards the outside of the dimer may indicate a hydrogen bond between an AAR and the oxygen atom of this acetyl group. The N_ϵ on HIS L168 and the acetyl O atom are within hydrogen bonding distance in both bacterial species (2.73 Å in P_{960}, 3.45 Å in P_{865}) considering the estimated error of ± 0.4 Å in the coordinates of Rb. sphaeroides.

D) <u>Optimization of C-CH$_3$ bond lengths</u> on rings I and III of both D_A and D_B. This step reduces $r(C-CH_3)$ from roughly 1.54 Å to 1.46 Å.

E) <u>"Ring puckering"</u> through the introduction of a twist angle of ± 4° in all saturated rings II and IV. This step, which has been sucessfully applied to monomer model systems [11], moves β-protons closer (by ~ 0.2 Å) to the plane of the porphyrin skeleton. This reduces the magnitude of calculated β-proton hfc's to values that are close to the observed ones.

The results of the RHF-INDO/SP calculation of the major proton hfc's for the various refinement steps are summarized in table 1. Also given are the ratios of sums of π-spin-densities on the A- and B-halves of the dimer which provide a measure of the asymmetry of the spin density distributions. The last row of table 1 shows the corresponding ratios of orbital electronic charge densities in the LUMO. This quantity measures the charge asymmetry in the first excited singlet state (see section 2.3 below).

The proposed assignments of several of the experimental hfc's (see table 1) should be regarded as tentative. In particular there is a high uncertainty in the <u>individual</u> assignments of the hfc's from the β-protons at positions 3, 4, 7 and 8 on D_A and D_B. This is mainly due to the high sensitivity of the hfc's to proton positions (compare steps D and E) and to the limited resolution of the presently available ENDOR/TRIPLE spectra. Improvements are expected from further structural refinements and improved magnetic resonance experiments which are currently in progress. Notwithstanding these uncertainties the essential features of the ENDOR/TRIPLE spectra are correctly predicted by the calculations. These are:
i) The weighted average of the theoretical dimer hfc's (column E, table 1), defined by $\Sigma N_i a_i / \Sigma N_i$ where N_i is the number of equivalent nuclei contributing to a line (e.g. 3 for methyl), is 5.0 MHz, which is in good agreement with the observed value of 4.9 MHz (column a_i^{exp}

Nucleus	a_i^{theor} [MHz]					a_i^{exp} [MHz]	
(Position)	A	B	C	D	E	Dimer	Monomer
$H_{methyl}(1a_A)$	1.5	2.0	2.3	3.2	2.4	3.2	4.9
$H_{methyl}(1a_B)$	1.6	1.3	0.8	1.1	1.8	1.4	
$H_{methyl}(5a_A)$	2.9	3.6	4.1	5.0	4.0	5.6	9.6
$H_{methyl}(5a_B)$	4.0	3.3	2.8	2.8	3.3	3.8	
$H_\beta(3_A)$	11.5	14.1	16.0	16.4	6.8	5.8	13.5
$H_\beta(3_B)$	20.7	17.4	15.2	14.4	9.7	9.6	
$H_\beta(4_A)$	6.3	7.6	8.6	8.9	5.9	4.6	16.4
$H_\beta(4_B)$	13.0	11.0	9.6	9.2	8.8	6.2	
$H_\beta(7_A)$	14.0	16.8	18.8	19.4	9.2	8.8	13.1
$H_\beta(7_B)$	14.3	12.0	10.3	9.9	8.8	8.4	
$H_\beta(8_A)$	15.1	18.4	20.6	21.2	8.3	6.2	11.8
$H_\beta(8_B)$	10.8	9.0	7.6	7.2	7.6	5.8	
$\Sigma\rho_i^A/\Sigma\rho_i^B$	0.7	1.2	1.7	1.9	0.9		
Q_B^{LUMO}/Q_A^{LUMO}	0.5	0.9	6.8	6.8	5.3		

Table 1: Theoretical (RHF-INDO/SP) hfc's of the oxidized donor $P_{865}^{+\cdot}$ from RC's of Rb. sphaeroides R-26 for the various stages of refinement A, B, ... E (see text) compared with the experimental values (obtained by ENDOR/ TRIPLE) of the dimer as well as of the monomer. The experimental hfc's were obtained by fitting the spectra of fig. 1 with a computer simulation program. The two bottom rows also give the theoretical A/B ratios of sums of spin densities in $P_{865}^{+\cdot}$ and of the electronic orbital charges in the LUMO of $P_{865}^{+\cdot}$.

for dimer, table 1). Furthermore, the weighted average of the monomer hfc's (column a_i^{exp} for monomer, table 1) is 9.8 MHz, which is twice that of the dimer hfc's.
ii) The calculated hfc's of methyl protons which are better defined in their interaction with the π-system than the β-protons due to rotational averaging, correspond well with the observed hfc's in partially deuterated RC's (spectrum not shown) in which only methyl proton hfc's are observed [12]. These spectra reveal a weak line with an hfc of 1.4 MHz (also observed as a well-resolved line in the native samples, see fig. 1). This hfc had been omitted from earlier assignments but can now be clearly assigned to the methyl group $1a_B$. Its weak ENDOR/ TRIPLE intensity can be explained by the reduction of the ENDOR en-

hancement for small hfc's.[1] The two methyl hfc's at 3.2 and 3.8 MHz are not resolved in the deuterated samples whereas the line at 5.6 MHz is well separated.

The above assignment implies a pronounced deviation from C_2-symmetry (earlier assignments were based on C_2-symmetry), at least for the local spin density distribution in the region of the overlapping rings I. As can be seen from table 1, this asymmetry is very sensitive to the position of the acetyl groups (step C).

iii) A striking result of the calculations is that - despite the <u>local</u> asymmetries of the spin densities ρ_i - the <u>sums</u> of ρ_i on the dimer halves are nearly equal, as $\Sigma\rho_i^A/\Sigma\rho_i^B = 0.9$ (see table 1, step E; this value could vary between 0.9 and 1.3 for non-uniform ring puckering). This explains the observed approximate $1/\sqrt{2}$ reduction of the ESR linewidth generally expected for a symmetrical dimer (see section 2.1). This has been explicitly proven by computer simulation of the inhomogeneously broadened ESR line for the assignment given in table 1 with inclusion of a computed average hfc of 2 MHz for the eight ^{14}N nuclei.

2.3 EXITED STATE PROPERTIES OF P_{865}

Additional calculations of excited state properties of P_{865} have been performed using the INDO/S (CI) method developed by Ridley and Zerner [14]. As many as 50 singly excited dimer configurations have been considered in the various stages of refinement A ... E. The calculated position of the long wavelength transition moves from 820 nm in step A to about 850 nm in the final refinement E. These calculations do not include any possible additional structural changes in the excited state. As can be seen from table 1, the B/A ratio of the electronic orbital charges in the LUMO of the dimer, Q_B^{LUMO}/Q_A^{LUMO}, changes dramatically from 0.5 (step A) to 5.3 (step E). The largest increase in this ratio is caused by the rotational movement of the acetyl groups (step C). The final result (E) suggests that the excited electron is preferentially on D_B. The same, though less pronounced result, has been found earlier for the first excited singlet state of P_{960} [15]. This simple orbital picture appears justified since the charge-transfer transition HOMO→LUMO is almost purely (97 %) contained in the multi-configurational lowest excited singlet state.

The similarity of the excited state properties of both bacterial species Rb. sphaeroides R-26 and Rps. viridis points to a common functional property of the two dimers. The X-ray data clearly show that in both species D_B is closer to the "accessory" BChl monomer, B_A, bridging the primary donor with the first acceptor bacteriopheophytin, Φ_A, in the photoactive A branch (see fig. 2). Regardless of the exact nature of the electron-transfer (ET) mechanism, it is generally accepted that the accessory monomer B_A is involved in the primary charge

----

[1] i.e. for $a_i \leq (2\pi T_{2e})^{-1}$, where T_{2e} is the electron spin-spin relaxation time [13].

separation event. Our results support the idea that the universally
observed vectorial ET ("unidirectionality") along only one branch of
the RC [16, 17] is already initiated in the excited state of the donor
itself.

3.   Conclusion

The dimer cation radicals $P_{865}^{+ \cdot}$ and $P_{960}^{+ \cdot}$ in the RC's of Rb. sphaeroides
R-26 and Rps. viridis, respectively, show significant differences in
their average spin density distributions over the dimer halves. In
this respect, $P_{865}^{+ \cdot}$ is clearly more symmetrical than $P_{960}^{+ \cdot}$. However,
like $P_{960}^{+ \cdot}$, $P_{865}^{+ \cdot}$ reveals considerable local asymmetries of corresponding spin densities on the two halves. This is mainly attributed to the
strong π-π interaction between the monomers, which results in the
formation of "super"-MO's extending over the entire dimer ("supermolecule"). In this supermolecule the monomeric contributions to the total
wave function are distorted from their characteristics in the isolated
monomers.   These local distortions are strongly dependent on subtle
structural features such as asymmetric rotational angles of acetyl
groups in the overlap region of the two BChl's (rings I). Although
spin density distributions in the radical ion doublet ground states of
the donor do not generally permit direct conclusions about the electronic structure of the lowest excited singlet donor state, they can
nevertheless serve as sensitive probes for structural details often
not revealed by X-ray analysis.
     Calculations of excited state properties of both donor species
yield a predominant localization of the excited state (characterized
by the electronic charge distribution in the LUMO) on D_B. This common
functional behavior of the two donor species is achieved by different
structural means: In P_{865} the rotation of acetyl groups appears to
play an important role due to the (partial) absence of hydrogen bonding to the protein. This is not the case in P_{960} where the acetyl
groups are assumed to be spatially fixed due to hydrogen bonding [8].
Obviously, P_{960} distorts more in its gross overall structure to produce the required asymmetry. From our results, unidirectionality of
the electron transfer along the A branch of the RC's of both bacterial
species Rb. sphaeroides and Rps. viridis appears to be already established in the excited states of the electron donors.

Acknowledgements:
The authors thank Drs. F. Lendzian and E. Tränkle for helpful discussions, and D.C. Rees and his group for the refinement of the Rb.
sphaeroides structure. Financial help of the Deutsche Forschungsgemeinschaft (SFB 337 and 312) and the National Institutes of Health
(GM41300 and GM13191) is gratefully acknowledged.

References

[1] Kirmaier, C. and Holten, D. (1987) 'Primary Photochemistry of Reaction Centers from the Photosynthetic Purple Bacteria', Photosynth.Res. 13, 225.

[2] Deisenhofer, J, Epp, O., Miki, K., Huber, R. and Michel, H. (1985) 'X-ray Structure Analysis of a Membrane Protein Complex', Nature 318, 618; (1984) J.Mol.Biol. 180, 385.

[3] Allen, J.P., Feher, G., Yeates, T.O., Komiya, H. and Rees, D.C. (1987) 'Structure of the Reaction Center from Rhodobacter sphae-roides R-26: The Cofactors', Proc.Natl.Acad.Sci. USA 84, 5730.

[4] Lendzian, F., Lubitz, W., Scheer, H., Hoff, A.J., Plato, M., Tränkle, E. and Möbius, K. (1988) 'ESR, ENDOR and TRIPLE Resonance Studies of the Primary Donor Radical Cation $P_{960}^{+\cdot}$ in the Photosynthetic Bacterium Rhodopseudomonas viridis', Chem.Phys. Lett. 148, 377.

[5] Plato, M., Lubitz, W., Lendzian, F. and Möbius, K. (1988) 'Magnetic Resonance and Molecular Orbital Studies of the Primary Donor Cation Radical $P_{960}^{+\cdot}$ in the Photosynthetic Bacterium Rhodopseudomonas viridis', Isr.J.Chem. 28, 109.

[6] Norris, J.R., Uphaus. R.A., Crespi, H.L. and Katz, J.J. (1971) 'Electron Spin Resonance of Chlorophyll and the Origin of Signal I in Photosynthesis', Proc.Natl.Acad.Sci. USA 68, 625.

[7] Plato, M., Tränkle, E., Lubitz, W., Lendzian, F. and Möbius, K. (1986) 'Molecular Orbital Investigation of Dimer Formations of Bacteriochlorophyll a. Model Configurations for the Primary Donor of Photosynthesis', Chem.Phys. 107, 185.

[8] Michel, H. Epp, O, Deisenhofer, J. (1986) 'Pigment-Protein Interactions in the Photosynthetic Reaction Center from Rhodopseudomonas viridis', EMBO J. 5, 2445.

[9] Allen, J.P., Feher, G., Yeates, T.O., Komiya, H. and Rees, D.C. (1988) 'Structure of the Reaction Center from Rhodobacter sphae-roides R-26: Protein Cofactor Interactions', Proc.Natl. Acad. Sci. USA 85, 8487.

[10] Momany, F.A., McGuire, R.F., Burgess, A.W. and Scheraga, H.A. (1975) 'Energy Parameters in Polypeptides. VII.' J.Phys.Chem. 79, 2361.

[11] Davis, M.S., Forman, A., Hanson, L.K., Thornber, J.P. and Fajer, J. (1979) 'Anion and Cation Radicals of Bacteriochlorophyll and Bacteriopheophetin b. Their Role in the Primary Charge Separation of Rhodopseudomonas viridis', J.Phys.Chem. 83, 3325.

[12] Lendzian, F. (1982) 'Elektron-Kern-Mehrfachresonanz an Primär-produkten der Photosynthese', Ph.D. Thesis, Department of Physics, Free University of Berlin.

[13] Allendoerfer, R.D. and Maki, A.H. (1970) 'A Phenomenological Description of ENDOR in Solution. Example: The tri-t-butylphen-oxyl Radical', J.Magn.Reson. 3, 396.

[14] Ridley, J. and Zerner, M. (1973) 'An Intermediate Neglect of Differential Overlap Technique for Spectroscopy: Pyrrole and the Azines', Theor.Chim. Acta (Berlin) 32, 111.

[15]   Plato, M., Lendzian, F., Lubitz, W., Tränkle, E. and Möbius, K.
        (1988) 'Molecular Orbital Studies on the Primary Donor P_{960} in
        Reaction Centers of Rps. viridis' in: The Photosynthetic Bacte-
        rial Reaction Center, Breton, J. and Vermeglio, A., Eds., Plenum
        Press, New York, p. 379.
[16]   Michel-Beyerle, M.E., Plato, M., Deisenhofer, J., Michel, H.,
        Bixon, M. and Jortner, J. (1988) 'Unidirectionality of Charge
        Separation in Reaction Centers of Photosynthetic Bacteria',
        Biochim.Biophys. Acta 932, 52.
[17]   Plato, M., Möbius, K., Michel-Beyerle, M.E., Bixon, M. and
        Jortner, J. (1988) 'Intermolecular Electronic Interactions in
        the Primary Charge Separation in Bacterial Photosynthesis',
        J.Am.Chem.Soc. 110, 7279.
[18]   McElroy, J.D., Feher, G. and Mauzerall, D.C. (1969) 'On the
        Nature of the Free Radical Formed During the Primary Process of
        Bacterial Photosynthesis', Biochim.Biophys. Acta 172, 180.
[19]   Feher, G., Hoff, A.J., Isaacson, R.A. and Ackerson, L.C. (1975)
        'ENDOR Experiments on Chlorophyll and Bacteriochlorophyll in
        vitro and in the Photosynthetic Unit', Ann.N.Y.Acad.Sci. 244,
        239.
[20]   Norris, J.R., Scheer, H. and Katz, J.J. (1975) 'Models for
        Antenna and Reaction Center Chlorophylls', Ann.N.Y.Acad.Sci.
        244, 261.

INDEX